# Modeling and Simulation in Science, Engineering and Technology

More information about this series at http://www.springer.com/series/4960

Giovanna Guidoboni • Alon Harris • Riccardo Sacco
Editors

# Ocular Fluid Dynamics

Anatomy, Physiology, Imaging Techniques,
and Mathematical Modeling

 Birkhäuser

*Editors*

Giovanna Guidoboni
Department of Math Sciences
Indiana University – Purdue University I
Indianapolis, IN, USA

Alon Harris
Icahn School of Medicine at Mount Sinai
New York, NY, USA

Riccardo Sacco
Department of Mathematics
Polytechnic University of Milan
Milan, Italy

ISSN 2164-3679          ISSN 2164-3725   (electronic)
Modeling and Simulation in Science, Engineering and Technology
ISBN 978-3-030-25885-6          ISBN 978-3-030-25886-3   (eBook)
https://doi.org/10.1007/978-3-030-25886-3

Mathematics Subject Classification: 97Mxx, 97M50, 97M60

This book is published under the imprint Birkhäuser, www.birkhauser-science.com by the registered company Springer Nature Switzerland AG.
The registered company address is: Gewerbestrasse 11, 6330 Cham, Switzerland

# Preface

The book aims at providing an overview of the current theoretical approaches available to model ocular fluid-dynamics in health and disease, along with the outstanding questions in the related theoretical and clinical fields. The theoretical modeling of ocular biophysics is a fast-growing research field that is attracting more and more scientists from various disciplines. This book will serve as a comprehensive reference for current and future scientists interested in this field, who will find a broad picture of the state-of-the-art with respect to:

- main open questions, controversial issues, and conjectures related to fluid flow in ophthalmology that pose major challenges for future advances in clinical research
- experimental and clinical technologies available to visualize and measure various physical quantities related to ocular fluids
- mathematical and computational models that investigate various aspects of fluid flow in the eye

This book stems from the idea that new answers to outstanding questions may only come from the true interaction among scientists with different expertise, including mathematics, engineering, physics, computer science, biology, chemistry, physiology, ophthalmology, optometry, clinical science, and pharmacology. Since each scientist looks at the same question from a different angle and describes the question using a different language, this book aims at creating a hub where such different perspectives can be explored with open minds and collaborative spirits, thereby providing a genuinely interdisciplinary overview of this diverse research field that will help guide new scientific investigations and spark new ideas.

This book focuses on five fluids, specifically blood, aqueous humor, vitreous humor, tear film, and cerebrospinal fluid. Each fluid is examined individually in a different part of the book, each part sharing the same structure based on four sections summarizing:

- elements of anatomy and physiology pertaining to that fluid
- pathological consequences of alterations in that fluid

- imaging technologies currently available to measure and visualize information pertaining to that fluid
- modeling approaches currently available to study the flow of that fluid in the eye

The book concludes with contributed chapters on future perspectives in the fields of imaging and modeling with application to ophthalmology.

The book integrates contributions by experts who strived to utilize a language accessible to scientists across disciplines, without compromising the accuracy of the presentation. Importantly, the authors of the chapters in this book were the first to read each other's contributions, thereby giving this book the flavor of a monograph rather than a mere collection of independent papers.

We believe that this book will foster an interdisciplinary approach to the study of the eye that combines clinical and experimental methods, data-driven modeling (e.g., based on statistics and machine learning), and physically based modeling (e.g., based on physics and biochemistry). Each of these approaches has advantages and limitations. *Experimental and clinical methods* provide invaluable data and information on living systems, but it is very challenging to isolate and control all the factors that influence the function of the system in vivo. *Data-driven models* allow the identification of patterns and correlations within large datasets, but it is very challenging to elucidate the cause-and-effect mechanisms that give rise to such patterns and correlations. *Physically based models* provide virtual laboratories, where the mechanistic contribution of specific factors on cardiovascular and lymphatic physiology can be investigated theoretically, but it is very challenging to account for all possible factors and their natural variability among individuals. In addition, the relevance of theoretical predictions based on mathematical models is tightly dependent on the quality of the data that was used to calibrate model parameters and run the model simulations.

We hope that this book will serve as a catalyst for the integration of these three approaches in a novel paradigm to address disease prevention, diagnosis, and treatment in a precise and individualized manner, which represents the twenty-first century view of ophthalmology, and, more generally, of medicine.

Columbia, MO, USA — Giovanna Guidoboni
New York, NY, USA — Alon Harris
Milano, Italy — Riccardo Sacco

# Contents

# Contributors

**Julia Arciero** Mathematical Sciences, Indiana University – Purdue University Indianapolis, Indianapolis, IN, USA

**Caroline R. Baumal** Department of Ophthalmology/Vitreoretinal Surgery, New England Eye Center, Tufts University School of Medicine, Boston, MA, USA

**Carolyn G. Begley** School of Optometry, Indiana University, Bloomington, IN, USA

**John Berdahl** Vance Thompson Vision, Sioux Falls, SD, USA

**Peter Bracha** Department of Ophthalmology, Indiana University School of Medicine, Indianapolis, IN, USA

**Richard J. Braun** Department of Mathematical Sciences, University of Delaware, Newark, DE, USA

**Filiz Bunyak** Department of Electrical Engineering and Computer Science, University of Missouri, Columbia, MO, USA

**Lucia Carichino** Mathematical Sciences, Worcester Polytechnic Institute, Worcester, MA, USA

**Simone Cassani** Mathematical Sciences, Worcester Polytechnic Institute, Worcester, MA, USA

**Shyam S. Chaurasia** Department of Electrical Engineering and Computer Science, University of Missouri, Columbia, MO, USA

**G. Chiaravalli** Department of Electrical Engineering and Computer Science Department of Mathematics, University of Missouri, Columbia, MO, USA

**Adam T. Chin** Department of Ophthalmology/Vitreoretinal Surgery, New England Eye Center, Tufts University School of Medicine, Boston, MA, USA

**Thomas A. Ciulla** Midwest Eye Institute, Indianapolis, IN, USA

**Tobin A. Driscoll** Department of Mathematical Sciences, University of Delaware, Newark, DE, USA

**Mariia Dvoriashyna** Department of Applied Mathematics and Theoretical Physics, University of Cambridge, Cambridge, UK

**David Fleischman** Department of Ophthalmology, University of North Carolina at Chapel Hill, Chapel Hill, NC, USA

**Anat Galor** Miami Veterans Administration Medical Center, Miami, FL, USA
Bascom Palmer Eye Institute, University of Miami, Miami, FL, USA

**Gian Paolo Giuliari** Department of Ophthalmology, Indiana University School of Medicine, Indianapolis, IN, USA

**Josh Gross** Eugene and Marilyn Glick Eye Institute, Indiana University School of Medicine, Indianapolis, IN, USA

**Alessandra Guglielmi** Dipartimento di Matematica, Politecnico di Milano, Milano, Italy

**Giovanna Guidoboni** Department of Electrical Engineering and Computer Science; Department of Mathematics, University of Missouri, Columbia, MO, USA

**A. Hajrasouliha** Department of Ophthalmology, Indiana University School of Medicine, Indianapolis, IN, USA

**Alon Harris** Icahn School of Medicine at Mount Sinai, New York, NY, USA

**Jinxin Huang** Department of Physics and Astronomy, University of Rochester, Rochester, NY, USA
Corning Research and Development Corporation, Corning, NY, USA

**Ingrida Januleviciene** Eye Clinic, Lithuanian University of Health Sciences, Kaunas, Lithuania

**Larry Kagemann** Department of Ophthalmology, NYU Langone Medical Center, NYU School of Medicine, New York, NY, USA

**Jing Li** Beijing Institute of Ophthalmology, Beijing Tongren Hospital, Capital Medical University, Beijing, China

**Xiaoxia Li** Beijing Institute of Ophthalmology, Beijing Tongren Hospital, Capital Medical University, Beijing, China

**Sunu Mathew** Department of Ophthalmology, Indiana University School of Medicine, Indianapolis, IN, USA

**Aurelio Giancarlo Mauri** Dipartimento di Matematica, Politecnico di Milano, Milano, Italy

**Frances Meier-Gibbons** Eye Center Rapperswil, Rapperswil, Switzerland

**Kannappan Palaniappan** Department of Electrical Engineering and Computer Science, University of Missouri, Columbia, MO, USA

**Vikram Paranjpe** Miami Veterans Administration Medical Center, Miami, FL, USA
Bascom Palmer Eye Institute, University of Miami, Miami, FL, USA

**Padmanabhan Pattabiraman** Department of Ophthalmology, Indiana University School of Medicine, Indianapolis, IN, USA

**Lam Phung** Miami Veterans Administration Medical Center, Miami, FL, USA
Bascom Palmer Eye Institute, University of Miami, Miami, FL, USA

**Daniele Prada** Istituto di Matematica Applicata e Tecnologie Informatiche "Enrico Magenes" del Consiglio Nazionale delle Ricerche, Pavia, Italy

**Jan O. Pralits** Department of Civil, Chemical and Environment Engineering, University of Genoa, Genova, Italy

**Rodolfo Repetto** Department of Civil, Chemical and Environmental Engineering, University of Genoa, Genova, Italy

**Jannick P. Rolland** The Institute of Optics, University of Rochester, Rochester, NY, USA

**Lucas Rowe** Department of Ophthalmology, Indiana University School of Medicine, Indianapolis, IN, USA

**Riccardo Sacco** Dipartimento di Matematica, Politecnico di Milano, Milano, Italy

**Lorenzo Sala** IRMA, UMR CNRS 7501, Université de Strasbourg, Strasbourg, France

**Fabrizia Salerni** Mathematical, Physical and Computer Sciences, University of Parma, Parma, Italy

**Ilaria Sartori** Dipartimento di Matematica, Politecnico di Milano, Milano, Italy

**Lina Siaudvytyte** Eye Clinic, Lithuanian University of Health Sciences, Kaunas, Lithuania

**A. Bailey Sperry** Tufts University, Medford, MA, USA

**Marcela Szopos** MAP5, UMR CNRS 8145, Université Paris Descartes, Paris, France

**Jenna Tauber** Department of Ophthalmology, NYU Langone Medical Center, NYU School of Medicine, New York, NY, USA

**Carol B. Toris** Department of Ophthalmology and Visual Sciences, University of Nebraska Medical Center, Omaha, NE, USA
Department of Ophthalmology, Case Western Reserve University, Cleveland, OH, USA

**Luca Torriani** Dipartimento di Matematica, Politecnico di Milano, Milano, Italy

**Marc Töteberg-Harms** Department of Ophthalmology, University Hospital Zurich, Zurich, Switzerland

**Jennifer H. Tweedy** Department of Bioengineering, Imperial College London, London, UK

**George Tye** Department of Ophthalmology and Visual Sciences, Case Western Reserve University, Cleveland, OH, USA

**Alice Chandra Verticchio-Vercellin** Department of Ophthalmology, Indiana University School of Medicine, Indianapolis, IN, USA
IRCCS – Fondazione Bietti, Rome, Italy

**Ningli Wang** Beijing Tongren Eye Center, Beijing Tongren Hospital, Capital Medical University, Beijing Ophthalmology and Visual Sciences Key Laboratory, Beijing, China
Beijing Institute of Ophthalmology, Beijing Tongren Hospital, Capital Medical University, Beijing, China

**Zheng Zhang** Beijing Tongren Eye Center, Beijing Tongren Hospital, Capital Medical University, Beijing Ophthalmology and Visual Sciences Key Laboratory, Beijing, China
Beijing Institute of Ophthalmology, Beijing Tongren Hospital, Capital Medical University, Beijing, China

# Part I
# Introduction

# Mathematical and Physical Modeling Principles of Complex Biological Systems

Riccardo Sacco, Giovanna Guidoboni, and Aurelio Giancarlo Mauri

**Abstract** A model of a complex system is a facsimile that can be used to investigate the problem at hand by simulating its behavior under specific conditions. Many modeling approaches are used in the applied sciences, including physical, animal, conceptual, and mathematical models. Specific examples will be provided in the chapter to illustrate the synergistic application of modeling to the simulation of biological fluid flow with relevance to ophthalmology.

## 1 Introduction

Problem complexity in Life Sciences and Engineering is becoming so prohibitive that the classic trial-and-error technique customarily adopted for investigation, design, and parametric optimization of a given system is no longer practicable and effective. Let us make this important concept more concrete by means of two specific examples, namely the pathology of open-angle glaucoma (OAG) and the production of memory devices for data storage.

OAG is a neuropathology of the optic nerve head that, by the end of 2020, will be the cause of blindness for almost 80 million individuals worldwide [28]. OAG is a multifactorial disease in which alterations of intraocular pressure, hemodynamic conditions, and metabolic functions, in conjunction with aging and epigenetic effects, concur in a nontrivial manner to determine the very often asymptomatic occurrence of the pathology and its progression. These multiple pathogenic factors make the search of the causes of OAG a very complex task and prevent the clinical scientist from the possibility of easily disentangling one specific factor from the

R. Sacco (✉) · A. G. Mauri
Dipartimento di Matematica, Politecnico di Milano, Milano, Italy
e-mail: riccardo.sacco@polimi.it; aureliogiancarlo.mauri@polimi.it

G. Guidoboni
Department of Electrical Engineering and Computer Science, Department of Mathematics, University of Missouri, Columbia, MO, USA
e-mail: guidobonig@missouri.edu

© Springer Nature Switzerland AG 2019
G. Guidoboni et al. (eds.), *Ocular Fluid Dynamics*, Modeling and Simulation in Science, Engineering and Technology, https://doi.org/10.1007/978-3-030-25886-3_1

others and quantify its relative impact on the onset of the disease in each individual patient.

Memory devices represent the most advanced level of Information and Communication Technology (ICT) ever since the continuous increase of electronic storage capabilities has initiated the so-called era of cloud computing [1]. The ICT boom is however close to a stop because the shrinking of device dimensions, main responsible of improvement of electronic component performance, has almost reached its ultimate limit predicted by Moore's law [22]. Such a negative perspective is prompting scientists and device designers to search for new technological solutions based on the joint adoption of new architectures and new materials. This challenge is still far from a conclusive answer.

The two above-described examples are striking paradigms of modern problems that must be faced nowadays in Life Sciences and Technology. As it is often the case, difficult questions call for sophisticated answers, thereby requiring the synergistic contribution of different competences and skills spanning from Mathematics to Physics and Engineering. Modeling can play a fundamental role by facilitating the integration of such multidisciplinary contributions to advance the understanding of complex systems.

## 2 A Broad View of Modeling: Meaning, Aims, and Approaches

Modeling means, in a broad sense, to create a sort of facsimile, henceforth referred to as the model, that can be used to study the main features of the system at hand and its behavior under specific conditions. Different approaches are used for investigation in Biology and Applied Sciences, including physical, animal, conceptual, and mathematical models. In the remainder of the chapter we provide a description of each of these modeling methods with specific examples in bioengineering.

### 2.1 Physical Models

Physical models are physical replicas of the system under consideration. Physical models are utilized extensively in engineering, biochemistry, and architecture to visualize important features of the design of an automobile, of the DNA structure, or of a civil structure. With the rapid development of three-dimensional printing, physical models provide medicine with powerful tools that facilitate education and surgical planning, with many potential benefits for training, research, and clinical interventions in ophthalmology [13]. Physical models may also be utilized in biomedical applications to reproduce some of the main features of a living system and study them in a controlled experimental environment. For example, in [7, 8] a mock circulatory flow loop is utilized to investigate the flow conditions in the human abdominal aorta. Another example can be found in [20], where an ultrasound imaging chamber incorporated in a cardiac flow loop allowed two- and three-dimensional Doppler characterizations of both simple and complex models of valvular regurgitation.

## 2.2   Animal Models

Animal models consist of replicating in an animal a disease or an injury similar to a human condition. Animal species utilized for biomedical research include mice, dogs, pigs, and monkeys, whereas conditions of interest may be inbred, induced, or already existing in the animals. Animal models are very often the key to groundbreaking discoveries in Medicine and Life Sciences. For example, the research conducted on dogs by Frederick Banting showed that the isolates of pancreatic secretion were successful in treating diabetes. This led to the discovery of insulin in 1922, jointly with John Macleod and Charles Best [32]. Another example is the study on rhesus monkeys conducted by Jonas Salk in the 1940s, which led to the isolation of the polio virus and the creation of a polio vaccine [4]. In ophthalmology, animal models are utilized for the investigation of many pathological conditions, including glaucoma [5, 30], myopia [9, 24], and age-related macular degeneration [25, 29].

## 2.3   Conceptual Models

Conceptual models provide a scheme of cause-effect relationships that govern the behavior of a complex system. In biology, chemistry, and life sciences, conceptual models are typically used to represent the chains of reactions and mechanisms regulating specific functions. The advantage of conceptual models is that visualization of cause-effect relationships helps interpret the complex interactions existing among them. The limitation of conceptual models is that they are not quantitative and, consequently, they cannot predict how much and to what extent the system behavior will be affected by alterations in specific mechanisms.

## 2.4   Mathematical Models

Mathematical models are constituted by sets of mathematical equations and formulas whose solutions describe the behavior of a complex system or the probability that a specific event occurs. Thus, mathematical models can help translate conceptual models into a solvable problem, whose solution may provide a quantitative tool to study the behavior of a complex system. Mathematical models can be classified into two main categories, namely data-driven models and mechanism-driven models, as described below.

### 2.4.1   Data-Driven Models and Statistical Viewpoint

Data-driven models aim at identifying patterns and trends within a given dataset by employing techniques based on statistical methods. Data-driven models typically consider very large datasets, which include data acquired by means of physical models, animal models, and studies on human subjects. The main outcome of a data-driven model is a set of correlations among relevant factors within the dataset. Despite recent progress to infer causality beyond correlations [16, 31], it is still very difficult to get information on the fundamental cause-effect relationships among physical and biophysical mechanisms that ultimately determine system behavior. An example of data-driven modeling applied to the study of glaucoma progression can be found in chapter "Statistical Methods in Medicine: Application to the Study of Glaucoma Progression".

### 2.4.2   Mechanism-Driven Models and Biophysical Viewpoint

Mechanism-driven models aim at providing quantitative information on the mechanisms that give rise to a given set of data. Mechanism-driven models are the mathematical translation of physical and biophysical principles, such as Newton's laws of dynamics, conservation of mass, electric charge, momentum, and energy. Mechanism-driven models are deterministic in nature, but can also include stochastic effects, and may be constituted by very different mathematical structures, such as algebraic relationships, ordinary differential equations, partial differential equations, and stochastic differential equations. This book provides a review of the mechanism-driven models currently available to investigate the flow of various fluids in the eye, specifically blood (see Chapter "Mathematical Modeling of Blood Flow in the Eye"), aqueous humor (see Chapter "Mathematical Models of Aqueous Production, Flow and Drainage"), vitreous humor (see Chapter "Mathematical Models of Vitreous Humour Dynamics and Retinal Detachment"), tear film (see Chapter "Mathematical Models of the Tear Film") and cerebrospinal fluid (see Chapter "Mathematical Modeling of the Cerebrospinal Fluid Flow and Its Interactions").

### 2.4.3 Synergy Between Statistical and Biophysical Viewpoints

Statistical and biophysical viewpoints are not in competition, rather, they are complementary and synergistic in the quest for a deeper understanding of complex systems. On the one hand, biophysical relationships identified via mechanism-driven models can be used to "inform" the statistical analysis about the existence of linear and/or nonlinear dependencies among covariates, which should be properly taken into account by the statistical analysis to avoid incorrect interpretations of trends in the data. On the other hand, stochastic variations of model parameters can be included in the biophysical models to account for individual variabilities and/or uncertainties in the measurements, whose influence on the biophysical outcomes can be quantified via statistical methods within data-driven models.

Ultimately, the synergy between statistical and biophysical models leads to a theoretical version of the complex system under investigation, which may serve as a *virtual laboratory* to perform theoretical experiments at low cost and in a short time, without the need of expensive equipment and resources. For example, a virtual laboratory could be used to:

- *simulate* several alternative scenarios to produce a quantitative prediction of system behavior, thereby allowing scientists to test and compare conjectures, as well as to formulate new ones;
- *compare* outcomes of existing studies (e.g. measurements on human subjects or physical and animal models) with simulations, thereby allowing scientists to interpret real data based on conjectured physical and biophysical mechanisms;
- *identify* factors that have a major impact on the system behavior, thereby providing a guide to the design of new experimental and clinical studies;
- *explore* levels of detail that current experimental techniques cannot reach due to instrumental limitation, thereby allowing scientists to investigate microscopic variables that may significantly affect the macroscopic function of the system.

## 3 Towards the Development of a Virtual Laboratory

The development of a virtual laboratory in Life Sciences is a highly nontrivial process that calls for a multidisciplinary effort. Conceptually, this process consists of five main tasks:

1. *problem definition*, where the main open questions in the applied field are identified and a strategy to address them is outlined;
2. *multiphysics/multiscale analysis*, where the relevant biophysical processes are identified and their spatial and temporal scales are characterized;
3. *model selection*, where the modeling approach is selected on the basis of its suitability to address the questions of interest and a specific mathematical problem is formalized;

4. *model solution*, where analytical or numerical techniques are utilized to obtain exact or approximated solutions to the specific mathematical problem;
5. *model assessment*, where the solutions of the specific mathematical problem are validated against experimental and clinical data.

In the next sections, each task is described in detail with reference to specific cases of interest in ophthalmology.

## 3.1  Problem Definition

The most stimulating, yet challenging, task of developing a virtual laboratory consists in pinning down important questions of interest in the applied field and devising a realistic strategy to address them. Oftentimes, this task is the most time-consuming, since it requires a dynamic interaction among scientists with different expertise, who often see the same problem from different viewpoints and utilize different languages to express them. The path leading to the definition of specific problems to be addressed through the use of a virtual laboratory is made of many steps, where areas of interest are successively refined until a set of well-formulated questions is shaped.

Let us make this concept more specific by means of an example in glaucoma research. As mentioned in Sect. 1, glaucoma is a multifactorial disease for which the only approved therapies aim at lowering the intraocular pressure (IOP), even though overwhelming evidence shows that IOP is not the only factor contributing to the disease [18, 19]. Thus, one of the main questions troubling ophthalmologists all around the world is.

| Question<br>*Level 0* | Why lowering intraocular pressure (IOP) is not enough to stop glaucoma progression in all patients? |
|---|---|

This question provides the starting point and the overall motivation of the scientific investigation and, therefore, we refer to it as *Level 0*. This Level 0 question, however, is still too broad to be addressed by means of biophysical models. A series of more specific questions or conjectures is then compiled, such as the one below:

Each of the Level 1 questions listed above is further elaborated into specific problems that can be studied by means of a virtual laboratory. For example, starting from question (1a), we can define the following specific problems:

| **Questions** *Level 1* | (1a) Does the same IOP level have the same consequences on different individuals? |
|---|---|
| | (1b) Are there other pathogenic factors, in addition to the level of IOP, that make some patients progress more than others? |
| | (1c) Should glaucoma be understood as a family of diseases sharing similar symptoms despite being characterized by different pathogenic processes? |
| | (1d) … |

| **Questions** *Level 1a* | (1a.i) What is the effect of IOP level on ocular biomechanics, including the distribution of stresses and strains in the tissues? |
|---|---|
| | (1a.ii) What is the effect of IOP level on ocular hemodynamics, including the distribution of pressures and velocities in the blood vessels? |
| | (1a.iii) What is the effect of IOP level on ocular oxygenation, including the distribution of oxygen in the blood vessels and in the tissues? |
| | (1a.iv) What is the effect of IOP level on the functionality of vascular regulation in the eye? |
| | (1a.v) What is the effect of IOP level on the flow of aqueous humor? |
| | (1a.vi) … |

Each of the Level 1a questions listed above can be the starting point for the development of a specific mechanism-driven model that, as discussed in the sections below, may require different mathematical and computational methods. However, when working on a model addressing a specific problem, for example (1.a.ii), it is important to keep in mind that we are working on a piece of a big mosaic representing the *big picture*. In other words, the specific problem, say (1.a.ii), is part of a bigger problem that ultimately aims at understanding why lowering IOP, per se, does not always prevent progression to blindness (see question at Level 0).

## 3.2 Multiphysics/Multiscale Analysis

Once identified a specific problem to address by means of a virtual laboratory, an interdisciplinary effort is required to identify the biophysical factors and processes that are likely to play a role in determining system behavior. Considering again the example of problem (1.a.ii), some of the most relevant factors and processes include (being not limited to):

- systemic factors driving blood flow, such as blood pressure and vascular regulation;

- mechanical properties of blood vessels, such as stiffness and compliance;
- biochemical processes at the cellular levels, such as nitric oxide absorption and myosin phosphorylation in the smooth muscle cells;
- intracellular chemical reactions, such as intracellular calcium uptake-release.

It is important to emphasize that, in the human body, all the factors and processes listed above take place simultaneously, even though at very different scales in space and time. Specifically, we can identify the following hierarchy of spatial scales:

- a *macroscale* (corresponding to the whole body) whose characteristic spatial length is of the order of meters;
- a *mesoscale* (corresponding to a specific organ) whose characteristic spatial length is of the order of centimeters;
- a *microscale* (corresponding to a single cell) whose characteristic spatial length is of the order of micrometers;
- a *nanoscale* (corresponding to the cellular membrane) whose characteristic spatial length is of the order of nanometers.

In addition, we can identify the following hierarchy of temporal scales:

- a *macroscale* (corresponding to a lifetime) whose characteristic temporal length is of the order of years;
- a *mesoscale* (corresponding to a day) whose characteristic temporal length is of the order of hours;
- a *microscale* (corresponding to the heartbeat) whose characteristic temporal length is of the order of seconds;
- a *nanoscale* (corresponding to the cellular reactions) whose characteristic temporal scale is of the order of microseconds (or even less).

Thus, spatial and temporal scales characterizing the biophysical processes within the human body span over 9 orders of magnitude. It is therefore no surprise that accounting for such an enormously wide spectrum of scales within a single mathematical model, despite theoretically possible, may be prohibitive and unaffordable even for the most powerful existing computational facilities. However, the link between the different scales is an essential part of life. For example, the onset and progression of glaucoma occurs over decades as the result of subtle damage to the ocular tissues occurring at every ion exchange within the heartbeat. Thus, the development of a sound mechanism-driven model in the context of life science should account for the multiscale and multiphysics nature of Life, while balancing between biophysical accuracy and model complexity. How to effectively and accurately attain this balance remains one of the biggest challenges in applied mathematics.

## 3.3   Model Selection

Mathematical models can be divided into two main categories: lumped parameter (LP) models and distributed parameter (DP) models. LP models mathematically represent the biophysical problem at hand by the construction of an equivalence with an electrical circuit. In this approach, the biophysical system is described by means of a network of interconnected electrical equivalent elements typically including resistors, capacitors, inductors, and current/voltage sources. The variables determined by a LP model are usually the nodal values of the electric potential and the branch currents in the network, which, in the electric analogy to fluid flow, correspond to the fluid pressure and the flow rate, respectively. DP models mathematically represent the biophysical problem at hand by means of a continuum-based geometrical description of the medium in which phenomena and processes take place. Phenomena are typically governed by fundamental laws of Physics and Mechanics such as mass, charge, and momentum conservation principles. Unlike LP models, the solution variables of a DP model are functions of position and time, rather than nodal values and/or branch currents. LP and DP models have both advantages and disadvantages. LP models provide a systemic view of the problem dynamics at low computational costs, but do not allow a detailed description of local spatially dependent phenomena. Conversely, DP models provide detailed spatial descriptions of the system under investigation, typically at the price of a much higher computational effort and allocation of memory resources. The decision of whether to adopt a DP or LP model is not trivial and depends on numerous factors, including the level of accuracy that is required to the model solution compared to the level of accuracy in the knowledge of model parameters and input data. In the following, we provide two examples for the applications of LP and DP models to the study of different aspects of ocular biophysics, namely the interplaying role of IOP, blood pressure and blood flow in the determination of retinal blood flow (see Sect. 3.3.1), and the interplaying role of electrochemical and fluid dynamical mechanisms in the production of aqueous humor by the ciliary processes (see Sect. 3.3.2).

### 3.3.1   Lumped Parameter Model of Blood Flow in the Retina

In this section we provide an example of the use of a lumped parameter model in ophthalmology, referring to [12] and chapter "Mathematical Modeling of Blood Flow in the Eye" for all the mathematical details, simulation results, and comments. The main goal of the investigation is to utilize a mechanism-driven approach to shed light on the complex interaction among risk factors in glaucoma. To this end, a LP mathematical model is developed to simulate blood flow through the central retinal artery (CRA), central retinal vein (CRV), and retinal microvasculature. In this approach, variable resistances are used to describe active and passive diameter changes due to vascular regulation and intraocular pressure (IOP). In the mechanistic description, blood flow is:

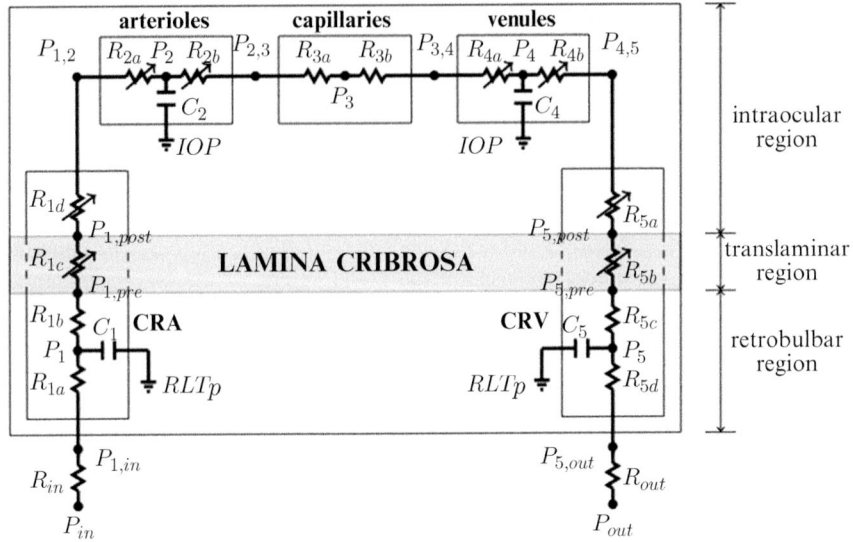

**Fig. 1** Schematic representation of a mechanism-driven, lumped parameter model to study the relationship between IOP, blood pressure, and blood flow in the retina (figure reproduced from [12])

- driven by the difference between input and output pressures (denoted by $P_{in}$ and $P_{out}$, respectively);
- impeded by the combined action of IOP and retrolaminar tissue pressure (RLTp);
- modulated by vascular regulation.

A schematic of the model is reported in Fig. 1. In the LP description of blood flow throughout the retinal microvasculature, the nodal pressures $P_i$ are functions of the sole time variable in such a way that, at each time instant, their value biophysically represents the spatial average of blood pressure in the considered vascular compartment. For example, referring to the scheme of Fig. 1, the nodal pressure $P_1$ is the spatial average of the blood pressure in the CRA, $P_2$ in the arterioles, $P_4$ in the venules, and $P_5$ in the CRV. The mathematical formulation emanating from the electrical equivalent circuit of Fig. 1 consists of the solution of a system of nonlinearly coupled ordinary differential equations for the nodal pressures $P_i$, with $i = 1, 2, 4, 5$.

The proposed lumped parameter model is used to simulate retinal blood flow for three theoretical patients with high, normal, and low blood pressure. The model predicts that patients with high and normal blood pressure can regulate retinal blood flow as IOP varies between 15 and 23 mm Hg and between 23 and 29 mm Hg, respectively, whereas patients with low blood pressure do not adequately regulate blood flow if IOP is 15 mm Hg or higher. Thus, hemodynamic alterations are predicted to impact patients' health conditions only if IOP changes occur out of the regulating range, which, most importantly, depend on blood pressure.

These theoretical predictions have been recently confirmed by the population-based study conducted in [33] over nearly 10,000 individuals (nearly 20,000 eyes), in which it has been found that patients with the highest probability of the occurrence of glaucoma are those exhibiting a combination of low blood pressure and elevated IOP.

### 3.3.2 Distributed Parameter Model of Aqueous Humor Production in the Ciliary Process

In this section we provide an example of the use of a DP model in ophthalmology, referring to [21] for all the mathematical details, simulation results, and comments. The main goal of the investigation is to utilize a mechanism-driven approach to shed light on the role of bicarbonate ion on the active secretion of aqueous humor across the membrane of the nonpigmented epithelial cells of the ciliary process. To this end, a distributed parameter mathematical model is developed to simulate the coupled interaction between ion electrodynamics and aqueous humor flow into the basolateral space adjacent to the nonpigmented epithelial (NPE) cells. In the mechanistic description, ion electrodynamics is driven by the balance between:

- a gradient in ion concentration across the membrane;
- an electric field generated by the transepithelial potential difference across the membrane and by the ions in motion throughout the membrane;
- the translational velocity of the aqueous humor flowing across the membrane.

In the mechanistic description, active secretion of aqueous humor is driven by the balance between:

- a fluid pressure gradient across the membrane;
- a shear stress between fluid elements moving with different velocity;
- an electric pressure due to the ions flowing inside the aqueous humor fluid.

A mathematically simplified cylindrical three-dimensional geometry of the transmembrane channel that has been used in numerical simulations is reported in Fig. 2.

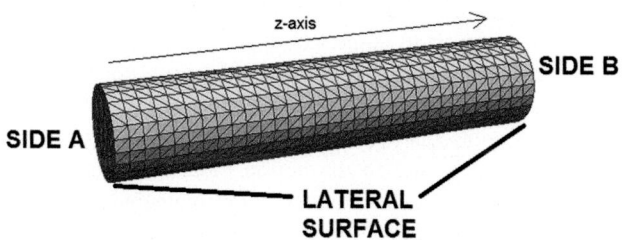

**Fig. 2** Geometry of a NPE transmembrane channel. Side A represents the intracellular NPE region, Side B represents the extracellular region in the basolateral space. The thickness of the channel is 5 nm (figure reproduced from [21])

In the DP description, unlike the case of the LP description, the dependent variables of the problem are also functions of the spatial position, mathematically represented by a three-dimensional vector $\mathbf{x} = (x, y, z)$, where $z$ is the cylinder axial coordinate, whereas $x$ and $y$ are the coordinates in the plane orthogonal to the $z$ axis. Thus, the mathematical formulation consists of the solution of a system of nonlinearly coupled partial differential equations for the electric potential, the ion concentrations of bicarbonate, sodium, potassium, and chloride, the aqueous humor pressure and velocity, which depend both on temporal and spatial coordinates.

The proposed DP model is used to disentangle the contribution of bicarbonate from that of the other ions, which is very difficult to investigate experimentally, in the formation of the transmembrane epithelial potential difference $V_m$ and in the secretion of aqueous humor into the basolateral space. Model predictions indicate that $V_m$ is close to baseline experimental measurements only if bicarbonate is included in the simulation. Model simulations of the sodium-potassium (Na/K) pump indicate an efflux of sodium and an influx of potassium, in accordance with pump physiology. The simulated Na/K ratio is 1.53, which is in very good agreement with the theoretical stoichiometric ratio of 1.5. The above theoretical model predictions suggest that bicarbonate inhibition may prevent physiological baseline values of the nonpigmented transepithelial potential difference and Na/K ATPase function, thus providing useful indication in the design of medications to decrease active secretion of aqueous humor.

## *3.4 Model Solution*

The two examples illustrated in Sect. 3.3 demonstrate that the use of LP and DP models for the simulation of complex biophysical problems leads to sophisticated systems of differential equations. Finding the exact solution of such equations is impossible unless drastic simplifications are introduced in the mathematical formulation, such as, for instance, in the electric equivalent circuit of Fig. 1, transforming the nonlinearly varying resistors into linear resistors. This approach has the advantage of making the analysis treatable at the price, however, of neglecting significant biophysical features of the system under investigation. In order to cope with the mathematical model in its more general integrity, it is therefore mandatory to resort to numerical approximation techniques. We refer to [2, 17] for the numerical treatment of ordinary differential equations and to [15, 27] for the numerical treatment of partial differential equations. In the following, we give a very short introduction to the concept of numerical approximation of a mathematical problem and to the notion of approximation error, which is the difference between the exact solution of the mathematical problem and the solution of its approximation. In addition, we shortly address the issue of the actual implementation of a numerical method into a computing machine environment, with special emphasis on the notion of finite arithmetics and machine precision.

### 3.4.1 The Mathematical Problem

Let us denote by $D$ the space of admissible data and by $d \in D$ a given value of the data. Let also denote by $V$ a vector space. Typically (but not necessarily) $V$ has infinite dimension. The abstract formulation of a mathematical problem is:

*Given $d \in D$, find $x \in V$ such that*

$$F(x, d) = 0 \tag{1}$$

where $F$ is the functional relation between $x$ (the solution of (1)) and $d$ (the input data of (1)). In general, it is not guaranteed that (1) admits a unique solution or that it is even solvable. In what follows we assume that (1) is *well-posed*, meaning that it admits a unique solution and that such solution depends with continuity on the data (see for further details [26, Chap. 2]).

### 3.4.2 The Numerical Problem

As previously anticipated, solving (1) exactly is, in general, very difficult or even impossible. Thus, we associate with problem (1) the following family of numerical problems:

*Given $d_h \in D_h$, find $x_h \in V_h$ such that*

$$F_h(x_h, d_h) = 0 \tag{2}$$

where $D_h$ and $V_h$ are finite dimensional subspaces of $D$ and $V$, respectively, whereas $d_h$ and $x_h$ are the approximation of the input data $d$ and of the exact solution $x$ in $D_h$ and $V_h$, respectively. The quantity $h$ is a positive number usually referred to as discretization parameter. Referring to the examples illustrated in Sect. 3.3, we may think at $h$ as the incremental time step and/or the spatial grid size. As in the case of problem (1), also (2) is assumed to be well-posed. The main, fundamental, difference between (1) and (2) is that $x_h$ is sought for in a finite-dimensional space so that the solution of (2) is computable whereas the solution of (1) is, in general, not.

### 3.4.3 Approximation Error and Convergence

In general, $x_h$ and $x$ obviously do not coincide. Therefore, we define the *approximation error* intrinsically associated with (1) and (2) as

$$e_h := x - x_h . \tag{3}$$

The requirement for (2) to be a good approximation of (1) is the *convergence* of $x_h$ to $x$, which is mathematically stated by the following limit

$$\lim_{h \to 0} e_h = 0. \tag{4}$$

Thus, as intuitively understandable, we expect $x_h$ to get increasingly close to $x$ as the discretization parameter becomes increasingly fine. At the same time, we also expect $d_h$ to converge to $d$ as $h$ becomes small, so that, should $x_h$ converge to $x$, the following property is satisfied

$$\lim_{h \to 0} F_h(x_h, d_h) = F(x, d). \tag{5}$$

This latter relation means that the approximate and the exact problem tend to coincide when we have convergence of *both* data and solution.

### 3.4.4 Computer Implementation

Even if the solution of the numerical problem (2) is computationally affordable, this does not necessarily mean that it is also computationally easy to achieve. In other words, we need in general to *implement the actual computation of $x_h$ within an algorithm* that is then to be run on a computer machine. In the hardware, any real number $y$ is replaced by a *machine representation* called floating-point number and denoted by $fl(y)$. It is important to emphasize that $fl(y)$ is *NOT*, in general, equal to $y$; rather, there is an intrinsic error that adds to the discretization error introduced in (3). This additional source of error is called *round-off error*, and can be estimated as follows

$$fl(y) = y(1 + \delta) \tag{6}$$

where $\delta$ is small quantity of the order of $10^{-16}$. From (6) we see that any input datum in a computer machine may, in general, be affected by a "native" error. This error is a very small quantity and is the result of the *finite precision* of the computer hardware in storing any number in the memory. This is the reason why $\delta$ is also referred to as *machine precision unit*. The fact that the machine precision unit is a very small quantity is, of course, good news. However, it is very important to keep in mind that the effect introduced by machine precision on numerical computations may not always be negligible, as it happens, for example, to the error due to numerical cancellation of significant digits that is introduced by the finite arithmetic of the computer (see [26, Sect. 2.4]).

### 3.4.5 The Issue of Stability

The principal objective in the design of a numerical problem is to ensure the convergence of $x_h$ to $x$, as stated by relation (4). It can be shown (see [26, Sect. 2.2.1]) that a necessary condition for (4) to hold is that the numerical problem (2) is *stable*.

Stability is strictly related to the mathematical notion of *continuous dependence on data*. To better explain this latter concept, let us consider a perturbation $\delta d_h$ in the data $d_h$ such that the modified data is $\tilde{d}_h = d_h + \delta d_h$ and assume that $\tilde{d}_h \in D_h$ as it was for the unperturbed data $d_h$. Correspondingly, consider the *perturbed numerical problem:*

Given $\tilde{d}_h \in D_h$, find $\tilde{x}_h = (x_h + \delta x_h) \in V_h$ such that

$$F_h(\tilde{x}_h, \tilde{d}_h) = 0. \tag{7}$$

The numerical model enjoys the property of continuous dependence on data (equivalently, it is stable), if the perturbation in the solution $\delta x_h$ is small when the perturbation in the data $\delta d_h$ is small too. To quantify this concept, we associate with the numerical problem (2) the *condition number* $K \geq 1$. This number provides an estimate of the amplification that may be introduced to $\delta d_h$ by problem (2). If $K$ is not much larger than 1, then the perturbation $\delta x_h$ will be not much larger than $\delta d_h$, so that we can conclude that problem (2) is *well-conditioned* and the computed solution $x_h$ is reliable. Conversely, if $K \gg 1$, then the perturbation $\delta x_h$ may be much larger than $\delta d_h$, so that we can conclude that problem (2) is *ill-conditioned* and needs to be handled with particular care in order to obtain a reliable numerical approximation. Oftentimes, ill-conditioning is addressed by means of a suitable stabilization of (2). Examples of this latter method are represented by the regularization techniques that allow to transform an ill-conditioned problem into a well-conditioned problem (see [23]) or the stabilized finite element formulations for the numerical approximations of partial differential equations proposed in [6, 10, 14].

### 3.5 Model Assessment

It is extremely important to keep in mind that the numerical solution of a mathematical model does not conclude the process of model development. As a matter of fact, once the problem has been defined and the corresponding mathematical model has been solved, as discussed in Sects. 3.4.1–3.4.4, we obtain as a result some predictions of the behavior of the investigated system. These predicted results must be compared with experimental data that can assess the validity of the assumptions that were made to derive the model in the first place. Depending on whether or not the model results are capable of capturing the essential features of the biophysical system, it may become necessary to revisit the whole definition of the mathematical problem and to modify certain assumptions that proved to be overly simplistic.

An example of the importance of model assessment is provided in Sect. 2.1.2 of chapter "Mathematical Modeling of Blood Flow in the Eye", where the mathematical modeling of venous collapsibility in the retina is discussed. When modeling venules as compliant tubes, thereby adopting the renown Laplace law, the model depicted in Fig. 1 predicts that retinal venules do not collapse even in the case when IOP is higher than their intraluminal pressure. This model prediction is clearly in contrast with the experimental observations on cats reported by Glucksberg and Dunn [11] and Attariwala et al. [3], where open retinal venules, i.e., not collapsed, were seen only for IOP values lower than their intraluminal pressure. This inconsistency between model predictions and experimental observations demanded a reassessment of the model assumptions. In particular, by representing the venules as collapsible tubes, thereby substituting the Laplace law with the law for collapsible tubes (Starling resistors), the model predictions demonstrated to be consistent with the experimental findings. A comparison between the intraluminal pressures computed when adopting the Laplace law and the law for collapsible tubes is reported in Fig. 13 of chapter "Mathematical Modeling of Blood Flow in the Eye".

We would like to emphasize that this example also illustrates the importance of gathering reliable experimental data, which, in addition, should be correctly interpreted and utilized when developing the mathematical model. Thus, the construction of a mathematical model for the study of a biophysical system is a truly interdisciplinary endeavor that calls for team work across disciplines and competencies.

## 4   Conclusions and Perspectives

Problem complexity requires new tools for achieving a satisfactory solution and producing significant advances in knowledge and technology. In this perspective, the adoption of mathematical modeling can be a valuable approach, especially when it is based on physical principles and calibrated against a set of real data. In this chapter, we provided a short introduction to the paradigm of modeling and to the rationale that leads from the phase of *problem definition* to the phase of *model assessment*. Specific examples have been included to illustrate the use of mathematical modeling in the study of different aspects of ocular biophysics, namely the interplaying role of IOP, blood pressure and blood flow in the determination of retinal blood flow, and the interplaying role of electrochemical and fluid dynamical mechanisms in the production of aqueous humor by the ciliary processes.

The fascinating conclusion that can be drawn from the content of this chapter is that mathematical modeling is a very general and powerful technique to address the solution of a complex problem in Engineering and Biology. A particular feature, which makes it unique and versatile, is the ability to develop an *abstract picture* (the model), which relates the specific problem (for example, the blood flow in the retina) to a general framework by means of connections and analogies (for example, the electric analogy to fluid flow depicted in the circuit of Fig. 1). The benefits of drawing this abstract picture are twofold. On the one hand, the model user can take advantage of existing algorithms, possibly developed for other types of applications and yet sharing the same structure. On the other hand, the model developers may discover new theoretical and computational challenges that call for new methodologies to be devised. Thus, the development and utilization of mathematical models as virtual laboratories is a genuine interdisciplinary endeavor that has significant impacts across disciplines, from engineering to medicine.

# References

1. M. Armbrust, A. Fox, R. Griffith, A. D. Joseph, R. Katz, A. Konwinski, G. Lee, D. Patterson, A. Rabkin, I. Stoica, and M. Zaharia. A view of cloud computing. *Commun. ACM*, 53(4): 50–58, 2010.
2. U.M. Ascher and L.R. Petzold. *Computer Methods for Ordinary Differential Equations and Differential-Algebraic Equations*. SIAM, 1998.
3. Rajpaul Attariwala, Claudine P Giebs, and Matthew R Glucksberg. The influence of elevated intraocular pressure on vascular pressures in the cat retina. *Investigative ophthalmology & visual science*, 35(3):1019–1025, 1994.
4. Anda Baicus. History of polio vaccination. *World journal of virology*, 1(4):108, 2012.
5. R.A. Bouhenni, J. Dunmire, A. Sewell, and D.P. Edward. Animal models of glaucoma. *Journal of Biomedicine and Biotechnology*, 2012:Article ID 692609, 11 pages, 2012.
6. F. Brezzi, L.P. Franca, T.J.R. Hughes, and A. Russo. b = integral of g. *Computer Methods in Applied Mechanics and Engineering*, 145:329–339, 1997.
7. S. Canic, C.J. Hartley, D. Rosenstrauch, Tambača J, Guidoboni G., and A. Mikelic. Blood flow in compliant arteries: an effective viscoelastic reduced model, numerics, and experimental validation. *Ann Biomed Eng*, 34(4):575–592, 2006.
8. S. Canic, J. Tambaca, G. Guidoboni, A. Mikelic, C. J. Hartley, and D. Rosenstrauch. Modeling viscoelastic behavior of arterial walls and their interaction with pulsatile blood flow. *SIAM J Appl Math*, 67(1):164–193, 2006.
9. Schaeffel F and Feldkaemper M. Animal models in myopia research. *Clinical and Experimental Optometry*, 98(6):507–517, 2015.
10. Leopoldo P. Franca and Alessandro Russo. Mass lumping emanating from residual-free bubbles. *Computer Methods in Applied Mechanics and Engineering*, 142(3):353–360, 1997.
11. Matthew R Glucksberg and Robert Dunn. Direct measurement of retinal microvascular pressures in the live, anesthetized cat. *Microvascular Research*, 45(2):158–165, 1993.
12. Giovanna Guidoboni, Alon Harris, Simone Cassani, Julia Arciero, Brent Siesky, Annahita Amireskandari, Leslie Tobe, Patrick Egan, Ingrida Januleviciene, and Joshua Park. Intraocular pressure, blood pressure, and retinal blood flow autoregulation: a mathematical model to clarify their relationship and clinical relevance. *Investigative Ophthalmology & Visual Science*, 55(7):4105–4118, 2014.
13. Wenbin Huang and Xiulan Zhang. 3D Printing: Print the Future of Ophthalmology3D Printing. *Investigative Ophthalmology & Visual Science*, 55(8):5380–5381, 2014.
14. Thomas J.R. Hughes, Gonzalo R. Feijóo, Luca Mazzei, and Jean-Baptiste Quincy. The variational multiscale method—a paradigm for computational mechanics. *Computer Methods in Applied Mechanics and Engineering*, 166(1):3–24, 1998. Advances in Stabilized Methods in Computational Mechanics.
15. T.J.R. Hughes. *The Finite Element Method: Linear Static and Dynamic Finite Element Analysis*. Dover Civil and Mechanical Engineering. Dover Publications, 2000.
16. Guido W Imbens and Donald B Rubin. *Causal inference in statistics, social, and biomedical sciences*. Cambridge University Press, 2015.
17. J. D. Lambert. *Numerical Methods for Ordinary Differential Systems: The Initial Value Problem*. John Wiley & Sons, Inc., New York, NY, USA, 1991.

18. M. Cristina Leske. Open-angle glaucoma—an epidemiologic overview. *Ophthalmic Epidemiology*, 14(4):166–172, 2007.
19. M. Cristina Leske, Anders Heijl, Leslie Hyman, Boel Bengtsson, LiMing Dong, and Zhongming Yang. Predictors of long-term progression in the early manifest glaucoma trial. *Ophthalmology*, 114(11):1965–1972, 2007.
20. Stephen H. Little, Stephen R. Igo, Marti McCulloch, Craig J. Hartley, Yukihiko Nosé, and William A. Zoghbi. Three-dimensional ultrasound imaging model of mitral valve regurgitation: Design and evaluation. *Ultrasound in Medicine & Biology*, 34(4):647–654, 2008.
21. A. G. Mauri, L. Sala, P. Airoldi, G. Novielli, R. Sacco, S. Cassani, G. Guidoboni, B. A. Siesky, and A. Harris. Electro-fluid dynamics of aqueous humor production: simulations and new directions. *Journal for Modeling in Ophthalmology*, 2:48–58, 2016.
22. G. E. Moore. Cramming more components onto integrated circuits. *Electronics*, 38(8):4pp, 1965.
23. V. Morozov. *Methods for Solving Incorrectly Posed Problems, PUBLISHER =*.
24. T. T. Norton. Animal models of myopia: learning how vision controls the size of the eye. *Ilar Journal*, 40(2):59–77, 1999.
25. Mark E. Pennesi, Martha Neuringer, and Robert J. Courtney. Animal models of age related macular degeneration. *Molecular Aspects of Medicine*, 33(4):487–509, 2012. New insights into the etiology and treatments for macular degeneration.
26. A. Quarteroni, R. Sacco, and F. Saleri. *Numerical Mathematics*, volume 37 of *Texts in Applied Mathematics*. Springer, 2007.
27. A. Quarteroni and A. Valli. *Numerical Approximation of Partial Differential Equations*. Springer-Verlag, New York, Berlin, 1994.
28. Harry A Quigley and Aimee T Broman. The number of people with glaucoma worldwide in 2010 and 2020. *British journal of ophthalmology*, 90(3):262–267, 2006.
29. P. Elizabeth Rakoczy, Meaghan J.T. Yu, Steven Nusinowitz, Bo Chang, and John R. Heckenlively. Mouse models of age-related macular degeneration. *Experimental Eye Research*, 82(5):741–752, 2006.
30. Weinreb RN and Lindsey JD. The importance of models in glaucoma research. *Journal of glaucoma*, 14(4):302–304, 2005.
31. Donald B. Rubin. Teaching statistical inference for causal effects in experiments and observational studies. *Journal of Educational and Behavioral Statistics*, 29(3):343–367, 2004.
32. Robert D Simoni, Robert L Hill, and Martha Vaughan. The discovery of insulin: the work of Frederick Banting and Charles Best. *Journal of Biological Chemistry*, 277(26):e15–e15, 2002.
33. Yih-Chung Tham, Sing-Hui Lim, Preeti Gupta, Tin Aung, Tien Y Wong, and Ching-Yu Cheng. Inter-relationship between ocular perfusion pressure, blood pressure, intraocular pressure profiles and primary open-angle glaucoma: the Singapore epidemiology of eye diseases study. *British Journal of Ophthalmology*, 102(10):1402-1406, 2018.

# Part II
# Blood

# Vascular Anatomy and Physiology of the Eye

Daniele Prada, Alon Harris, Giovanna Guidoboni, Lucas Rowe,
Alice Chandra Verticchio-Vercellin, and Sunu Mathew

**Abstract** This chapter provides an overview of the main structural and functional properties of the ocular vasculature. Four major circulatory systems within the eye are considered, namely those nourishing the retina, the optic nerve head, the choroid, and the anterior segment. Some aspects related to vascular regulation and innervation are also discussed, along with outstanding questions that remain a matter of debate.

## 1 Introduction

Blood circulation in the eye is structured in a very complex way in order to nourish the tissues without interfering with visual function. Interestingly, some components are extremely rich in blood, such as the choroid, whereas others are completely avascular, such as the vitreous humor, lens, and central regions of the cornea and fovea. Some of the most relevant ocular components are schematized in Fig. 1.

---

D. Prada (✉)
Istituto di Matematica Applicata e Tecnologie Informatiche "Enrico Magenes" del Consiglio Nazionale delle Ricerche, Pavia, Italy
e-mail: daniele.prada@imati.cnr.it

A. Harris
Icahn School of Medicine at Mount Sinai, New York, NY, USA

L. Rowe · S. Mathew
Department of Ophthalmology, Indiana University School of Medicine, Indianapolis, IN, USA

G. Guidoboni
Department of Electrical Engineering and Computer Science, University of Missouri, Columbia, MO, USA

Department of Mathematics, University of Missouri, Columbia, MO, USA

A. C. Verticchio-Vercellin
Department of Ophthalmology, Indiana University School of Medicine, Indianapolis, IN, USA

IRCCS – Fondazione Bietti, Rome, Italy

© Springer Nature Switzerland AG 2019
G. Guidoboni et al. (eds.), *Ocular Fluid Dynamics*, Modeling and Simulation in Science, Engineering and Technology, https://doi.org/10.1007/978-3-030-25886-3_2

23

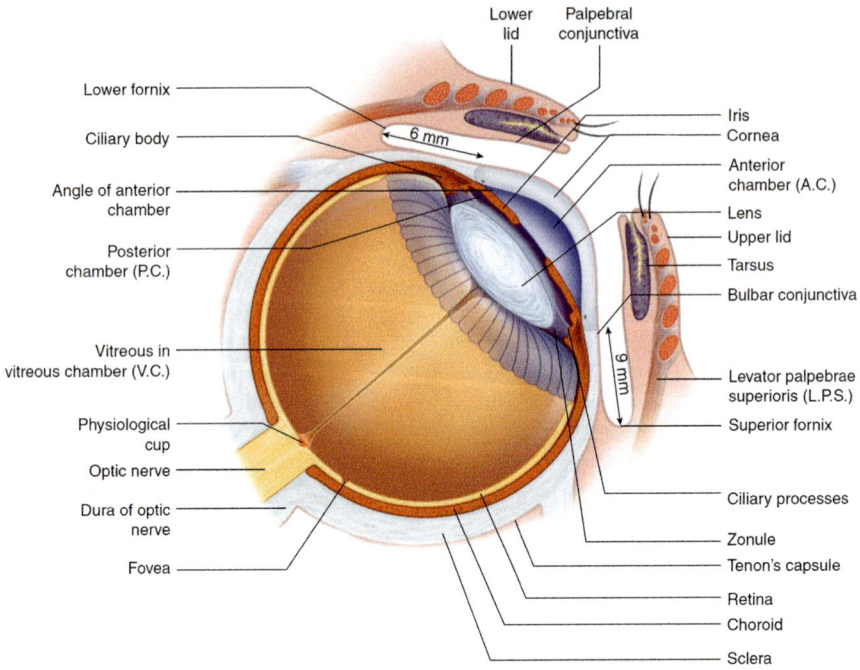

**Fig. 1** Schematic representation of the eye and its components (courtesy of Ansari M.W., Nadeem A. (2016) The Eyeball: Some Basic Concepts. In: Atlas of Ocular Anatomy. Springer [5])

The primary goal of this chapter is to provide an overview of the main structural and functional properties of the ocular vasculature. In Sect. 2, we outline the main vascular pathways that supply and drain the eye. Then, we describe four major circulatory systems within the eye, namely those nourishing the retina (Sect. 3), the optic nerve head (Sect. 4), the choroid (Sect. 5), and the anterior segment (Sect. 6). Finally, we review some aspects related to vascular regulation and innervation (Sect. 7), and we discuss some outstanding questions that remain a matter of debate (Sect. 8).

## 2   From the Heart to the Eye and Back

Oxygenated blood leaves the heart through the aorta, which is the largest artery in the human body. The first two arteries branching from the aorta are the brachiocephalic artery and the left common carotid artery. The brachiocephalic divides into the right subclavian (which supplies the right arm) and right common carotid artery. The right and left common carotid arteries supply the head and neck following symmetrical courses. Each common carotid artery branches into internal and external carotid arteries at the level of the neck.

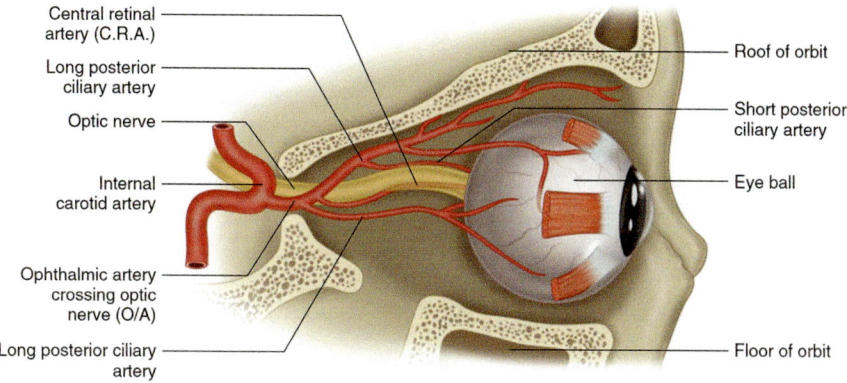

Central retinal artery (C.R.A.)
Long posterior ciliary artery
Optic nerve
Internal carotid artery
Ophthalmic artery crossing optic nerve (O/A)
Long posterior ciliary artery
Roof of orbit
Short posterior ciliary artery
Eye ball
Floor of orbit

**Fig. 2** Blood supply to the eye by the ophthalmic artery, a branch of the internal carotid artery (courtesy of Ansari M.W., Nadeem A. (2016) The Blood Supply to the Eyeball. In: Atlas of Ocular Anatomy. Springer, Cham)

The main vascular supply to the eye is the ophthalmic artery (OA), which is the first branch of the internal carotid artery and the only branch of the internal carotid artery outside the cranium (see Fig. 2). Smaller vascular contributions come from the external carotid artery via the internal maxillary artery and the facial artery. The OA enters the optic canal and runs along the optic nerve. Its major branches include branches to the extraocular muscles, the central retinal artery, and the posterior ciliary arteries. The OA runs through different anatomical spaces where it experiences various external pressures.

After its origin from the internal carotid artery, the OA follows a short intracranial course before it pierces the dura mater of the optic nerve and enters the optic canal. In this intracranial course, the OA is affected by intracranial pressure (ICp) from the surrounding cerebrospinal fluid (CSF). After entrance into the optic canal, the OA follows its extracranial, or intraorbital, course where it travels in close relation and parallel to the optic nerve before separating in a complex pattern into its many branches to supply the eyeball and periophthalmic tissues. In its extracranial course, the OA is affected by an external pressure applied to the surrounding tissues of the orbit [49]. This external pressure can also be utilized to measure ICp non-invasively [66], thereby offering an attractive alternative to invasive measurements of CSF pressure via lumbar puncture or pressure sensor implantation into the brain's ventricles.

The central retinal artery (CRA) branches off the OA and penetrates the optic nerve approximately 10–20 mm behind the globe [34] (see Fig. 3). The CRA runs adjacent to the central retinal vein (CRV) within the central portion of the optic nerve. At the level of the sclera, the CRA and the CRV pierce through the lamina cribrosa, a structure consisting of fenestrated connective tissue beams, which also allow the retinal ganglion cell (RGC) axons to pass on their path from the retina to the optic nerve. The CRA and the CRV then emerge from the optic nerve within the

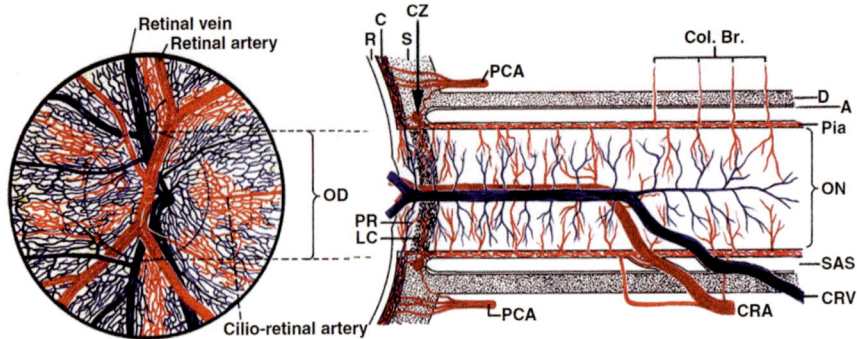

**Fig. 3** Schematic representation of blood supply of the optic nerve and retina. Left half shows retinal appearance. Abbreviations used: *A* arachnoid, *C* choroid, *CRA* central retinal artery, *Col. Br.* collateral branches, *CRV* central retinal vein, *CZ* circle of Zinn-Haller, *D* dura, *LC* lamina cribrosa, *OD* optic disc, *ON* optic nerve, *P* pia mater, *PCA* posterior ciliary artery, *PR* prelaminar region, *R* retina, *S* sclera, *SAS* subarachnoid space (courtesy of Hayreh S.S. (2015) Blood Supply of the Optic Nerve Head. In: Ocular Vascular Occlusive Disorders. Springer, Cham)

eye; here, the CRA branches into four major vessels, which supply the inner two-thirds of the retina and the most anterior portion of the optic nerve head (ONH), the superficial nerve fiber layer (SNFL). A medial and a lateral posterior ciliary artery (PCA) branch from the OA. Another PCA may be present occasionally. Each medial and lateral PCA divides into one long posterior ciliary artery (LPCA) and seven to ten short posterior ciliary arteries (SPCAs) before penetrating the sclera (see Fig. 4).

The vascular segments are exposed to various external pressures depending on their position in the network. The intraocular segments are exposed to the intraocular pressure (IOP), the retrobulbar segments are exposed to the retrolaminar tissue pressure (RLTp), and the translaminar segments are exposed to an external pressure that depends on the internal state of stress within the lamina cribrosa. The RLTp is related to the pressure of the cerebrospinal fluid (CSFp) in the subarachnoid space immediately posterior to the lamina cribrosa. Morgan et al. showed that RLTp and CSFp are related in dogs [50], but it is still unclear to which extent a similar relation also applies to humans. The difference between the IOP and the RLTp is called translaminar pressure difference and it is borne by the lamina cribrosa.

Venous drainage from the eye occurs primarily via the vortex veins and the central retinal vein. They merge with the superior and inferior ophthalmic veins that drain into the cavernous sinus, the pterygoid venous plexus, and the facial vein. The existence of valves in these veins is under investigation. The lack of valves was frequently stated as a reason for spread of infection from the mid-face to the cavernous sinus. Recently, evidence has been found of the existence of valves in the superior ophthalmic vein and its two main tributaries [76]. Valves were also seen in the facial vein. The spread of infection from the face could then be due to the

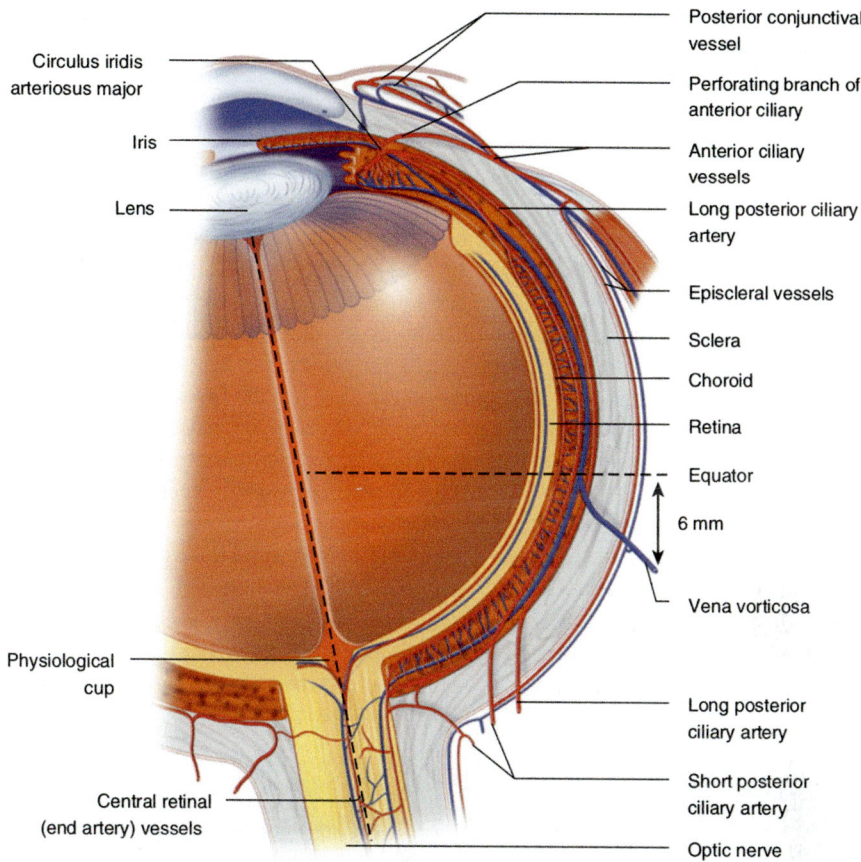

Circulus iridis
arteriosus major

Iris

Lens

Physiological
cup

Central retinal
(end artery) vessels

Posterior conjunctival
vessel

Perforating branch of
anterior ciliary

Anterior ciliary
vessels

Long posterior ciliary
artery

Episcleral vessels

Sclera

Choroid

Retina

Equator

6 mm

Vena vorticosa

Long posterior
ciliary artery

Short posterior
ciliary artery

Optic nerve

**Fig. 4** The blood supply to the eye: branches of the ophthalmic artery. The peripapillary and posterior choroid up to the equator is primarily supplied by the SPCAs, whereas the more anterior regions of the choroid are supplied by the LPCAs and the anterior ciliary arteries. The anterior ciliary arteries arise from the muscular branches of the OA and run to the front of the eyeball along with the tendons of the recti muscles. They supply the recti muscles, conjunctiva, sclera, and iris (courtesy of Ansari M.W., Nadeem A. (2016) The Blood Supply to the Eyeball. In: Atlas of Ocular Anatomy. Springer, Cham)

existence of communications between the facial vein and cavernous sinus and the direction of blood flow, rather than the absence of venous valves.

The cavernous sinus drains via the superior and inferior petrosal sinuses, which then join the sigmoid sinus to form the internal jugular vein. The pterygoid venous plexus becomes the maxillary vein, which joins the superficial temporal vein to become the retromandibular vein. The posterior branch of the retromandibular vein unites with the posterior auricular vein to form the external jugular vein, while the anterior branch of the retromandibular vein unites with the facial vein to form the common facial vein that enters the internal jugular vein. The internal and external jugular veins drain into the subclavian veins, which empty into the superior vena cava that delivers the deoxygenated blood to the right atrium of the heart.

## 3   Retina

The retina is a multilayered sheet of neural tissue lining the choroid (see Figs. 4, 5, and 6). Its anterior edge is called the ora serrata. The retina is thinnest at the ora serrata anteriorly and at the fovea centralis posteriorly, thereby explaining why, in retinal detachment, traumatic retinal holes develop at these sites frequently. Most externally, in contact with the pigment epithelium is a neural epithelium with photoreceptors, i.e., rods and cones. Moving inward are the external nuclear layer (with nuclei of photoreceptors), the external plexiform layer (having synapses), the internal nuclear layer (nuclei of bipolar cells), the internal plexiform layer (having synapses), the ganglion cell layer, and, finally, the nerve fiber layer made of axons of ganglion cells, which ultimately leave the eye through the lamina cribrosa and form the optic nerve. All these are bound together by special vertical cells called the fiber of Müller. At the posterior pole of the eye, a specially differentiated spot is called the fovea centralis, which is avascular and contains only cones. There are no rods here. The fovea is surrounded by a small area called the macula lutea. The spot where the axons of the ganglion cells leave the eye through the lamina cribrosa is

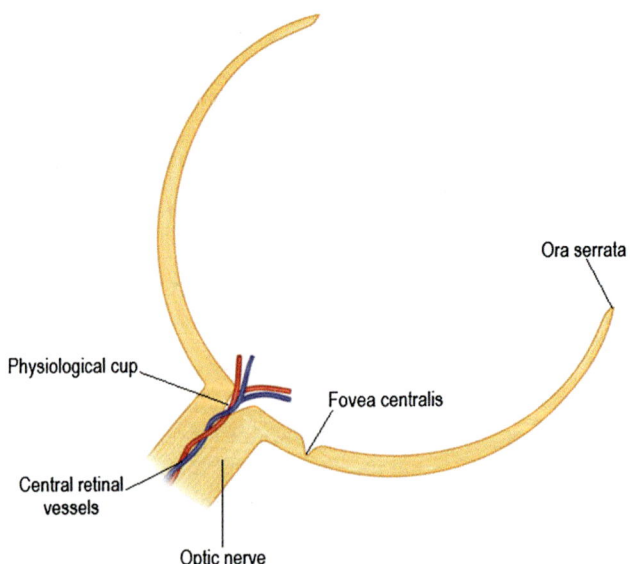

**Fig. 5** Schematic view of the retina (courtesy of Ansari M.W., Nadeem A. (2016) The Blood Supply to the Eyeball. In: Atlas of Ocular Anatomy. Springer, Cham)

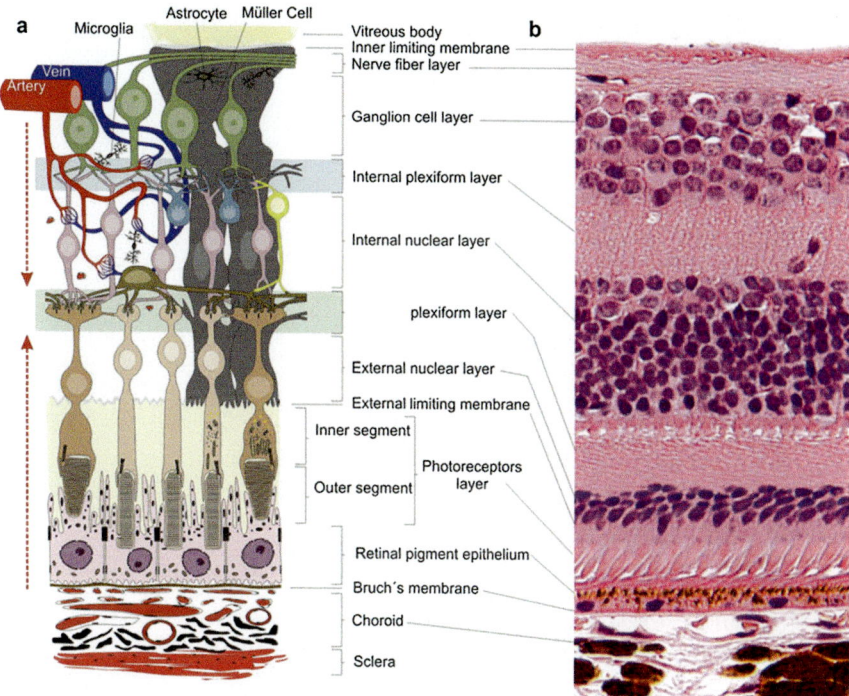

**Fig. 6** Organization of the retina. (Left) Modified schematic drawing of the different layers and cell types of the retina. (Right) Corresponding plastic section of the macular retina (hematoxylin–eosin stain, medium magnification) (courtesy of Leonardi A., Vasconcelos-Santos D.V., Nogueira J.C., McMenamin P.G. (2016) Anatomy. In: Zierhut M., Pavesio C., Ohno S., Orefice F., Rao N. (eds) Intraocular Inflammation. Springer, Berlin, Heidelberg)

called the optic disc, usually characterized by a small depression in its center called the physiological cup (Figs. 5 and 6).

The CRA supplies the inner two-thirds of the retina. The deeper outer layers, including photoreceptors and bipolar cells, are nourished by the choriocapillaris, which are part of the uveal system. The retinal pigment epithelial layer, which separates the retina from the choroid, actively transports metabolites and waste to and from the deep layers of the retina to the choroid (see Fig. 6).

Before emerging from the optic nerve within the eye, the CRA pierce through the lamina cribrosa. Guidoboni et al. showed that the compression exerted on the CRA by the lamina cribrosa may be responsible for the clinically observed velocity reduction in the CRA after acute IOP elevation [29]. Their model also predicts that the same IOP elevation would induce different CRA hemodynamic responses in eyes with different anatomical features. In another theoretical study, the authors found that the ability of IOP to induce noticeable changes in retinal hemodynamics depends on the levels of blood pressure and autoregulation (see Sect. 7) of the individual [28]. These different individual responses may help explaining why

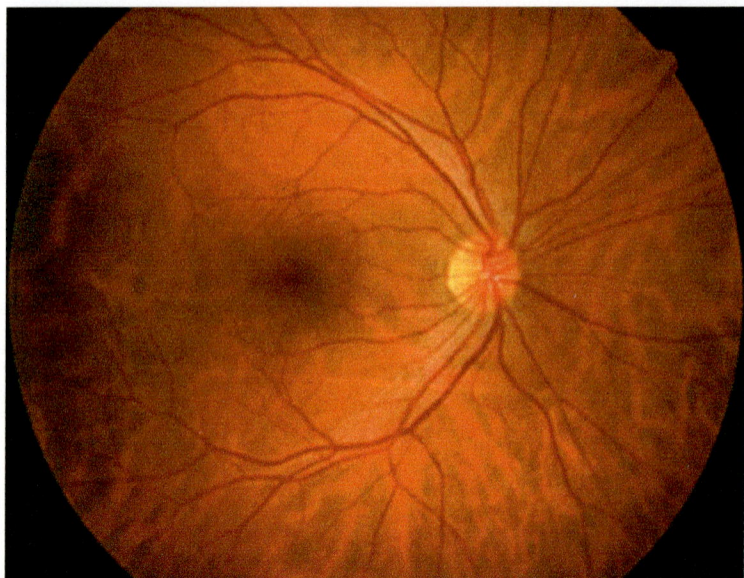

**Fig. 7** Fundus photograph of a 35-year-old Caucasian male. The major arteries and veins emanate from the optic nerve head. The veins appear of darker red color and thicker than the arteries. Each quadrant of the retina is supplied by a major artery and vein. The nasal vessels (right) have a straighter and shorter course. The temporal blood vessels run an arcuate path around the macula. The major vessels divide by dichotomous and side arm branches (courtesy of Lutty G.A., Bhutto I., McLeod D.S. (2012) Anatomy of the Ocular Vasculatures. In: Schmetterer L., Kiel J. (eds) Ocular Blood Flow. Springer, Berlin, Heidelberg)

susceptibilities to a given level of IOP differ among individuals and may help revealing how retinal ischemic damage may occur during IOP elevations in certain individuals.

The CRA typically branches into four major arterioles, each of which nourishes a quadrant of the retina (see Fig. 7). Retinal arterioles and venules lie within the superficial nerve fiber layer of the retina. Retinal capillaries form several layers parallel to the retina surface; they are more concentrated in the proximity of the optic nerve than in the peripheral retina.

Sometimes, in the peripapillary region, the retina also exhibits a superficial network of fine capillaries. This network, known as the radial peripapillary capillary plexus (RPCP), has a unique organization as it runs parallel to the densely packed nerve fiber layer axon bundles and is present only in this retinal region [8]. Three other retinal vascular networks are present through the retinal thickness and differ from the RPCP in that they display lobular configurations rather than running parallel to nerve fiber layer axons. The most superficial of the three networks, the superficial vascular plexus (SVP), is supplied by the central retinal artery and composed of larger arteries, arterioles, capillaries, venules, and veins primarily in the ganglion cell layer (see Fig. 8). The two deeper networks, the intermediate

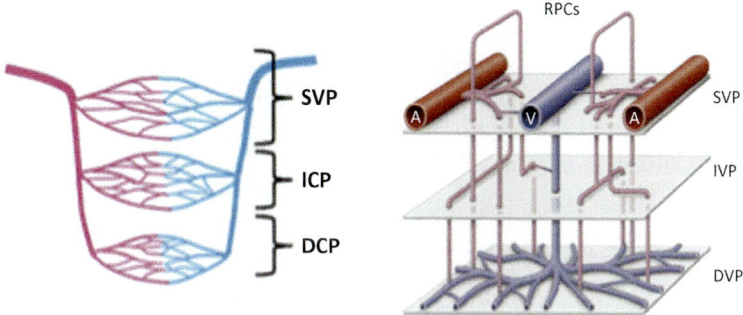

**Fig. 8** (Left) Pictorial representation of the parallel connection between retinal capillary beds as hypothesized by Campbell and coworkers (extracted from Fig. 6 in [8], licensed under CC BY 4.0 license: https://creativecommons.org/licenses/by/4.0/) (Right) 3D model of the retinal microcirculation as proposed by Fouquet and coworkers (extracted from Fig. 8 in [20], licensed under CC BY 4.0 license: https://creativecommons.org/licenses/by/4.0/)

capillary plexus (ICP) and deep capillary plexus (DCP), are supplied by vertical anastomoses from the SVP. The retinal vascular networks are spread throughout the retina; however, they are segmented differently depending on the region of the retina [8]. In the peripheral retina, the SVP and the deep vascular plexus (DVP) are the only two distinguishable layers. In the macula, the DVP can be further separated into the ICP and DCP. In the peripapillary region, the SVP can be further separated into the RPCP and SVP. Around the foveal avascular zone (FAZ), the retinal layers converge to form a single capillary loop that defines the borders of the FAZ. Including the fovea, there are three specific areas of the retina that are devoid of vascular networks. The fovea, which contains only photoreceptors, is the center of a 400-μm-wide capillary-free region. Additionally, the very posterior retinal edge, about 1.5 mm around the edge, receives no retinal circulation. Lastly, the areas of the retina adjacent to the major retinal arteries and veins lack a capillary bed [8].

Retinal venous drainage occurs via the CRV, which exits the eye passing through the optic nerve and runs parallel to the CRA. Usually, the CRV drains into the superior ophthalmic vein or, sometimes, directly into the cavernous sinus (see Fig. 9). The exact course through which retinal capillaries converge into the venous side of the circulation is currently matter of debate. Recent literature has led to a conversation about whether the three capillary beds of the SVP, ICP, and DCP run in parallel or in series. A study that used projection-resolved optical coherence tomography angiography (PR-OCTA) found the deeper ICP and DCP to be supplied by vertical anastomoses from the SVP, which supports a parallel connection between the vascular networks [8]. Conversely, a study that used high density confocal microscopy found that the bulk of arterial flow runs through the SVP where it divides this flow to RPCP on one side, and to the serially arranged ICP and DCP on the other side [20]. Additionally, a study that utilized optical coherence tomographic angiography (OCTA) found that all collateral vessels associated with retinal vein occlusion coursed through the DCP and the absence of collateral vessels isolated

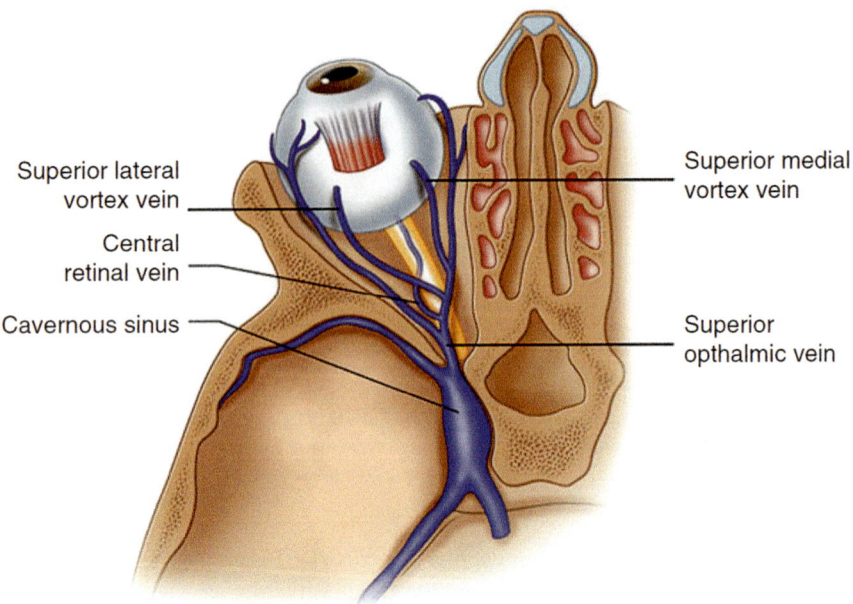

**Fig. 9** Diagram of the venous drainage from the eye (courtesy of Browning D.J. (2012) Anatomy and Pathologic Anatomy of Retinal Vein Occlusions. In: Retinal Vein Occlusions. Springer, New York, NY)

to the SVP supports a serial arrangement of the three capillary beds [23]. The connection of the three capillary beds is important to decipher as it may help explain why the three vascular networks respond differently to physiological changes. Functional hyperemia has shown to invoke relatively large dilation of intermediate layer capillaries in comparison to smaller dilation in the superficial and deep layer capillaries [41]. All three layers have been shown to display a decrease in density due to large increases in IOP [77]. Mathematical modeling could be utilized as a virtual laboratory to simulate the implications of different network architectures on hemodynamic responses in health and disease [11, 12].

## 4 Optic Nerve Head

The anatomy and vascular supply of the ONH is best divided into four regions, from anterior to posterior segments (see Figs. 3 and 10). The most anterior part of the ONH is the *SNFL*. Immediately behind the SNFL is the *prelaminar region*, which lies adjacent to the peripapillary choroid. Posterior to the prelaminar region, the *laminar region* is composed of the lamina cribrosa, a structure consisting of fenestrated connective tissue beams through which the RGC axons pass on their

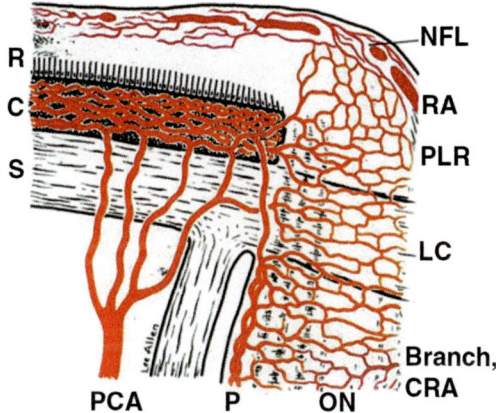

**Fig. 10** Schematic representation of blood supply of the optic nerve head. Abbreviations used: *C* choroid, *CRA* central retinal artery, *LC* lamina cribrosa, *NFL* surface nerve fiber layer of the disc, *ON* optic nerve, *P* pia, *PCA* posterior ciliary artery, *PLR* prelaminar region, *R* retina, *RA* retinal arteriole, *S* sclera (courtesy of Hayreh S.S. (2015) Blood Supply of the Optic Nerve Head. In: Ocular Vascular Occlusive Disorders. Springer, Cham)

path from the retina to the optic nerve. Finally, the *retrolaminar region* lies posterior to the lamina cribrosa. It is marked by the beginning of axonal myelination and is surrounded by meninges.

The vascular system nourishing the ONH is quite complex [34, 35, 68] and shows high interindividual and intraindividual variability [36]. An important anatomic distinction between the different portions of the ONH is that blood flow to the ONH is primarily supplied by the PCAs, whereas the SNFL receives oxygenated blood primarily from retinal arterioles [51]. In approximately 30% of all people, a cilioretinal artery may be present and supply the temporal SNFL [34].

The prelaminar region is mainly supplied by direct branches of the short PCAs and by branches from the circle of Zinn-Haller (see Figs. 10 and 11). The circle of Zinn-Haller, if present, is a complete or incomplete ring of arterioles within the perineural sclera formed by the confluence of branches of the short PCAs. The arterial circle branches into the prelaminar region, lamina cribrosa, and retrolaminar pial system and supplies the peripapillary choroid. These vessels exhibit an anastomotic blood exchange [6], but it is unclear whether this exchange can counterbalance an insufficiency of a single PCA. There is also evidence of direct arterial supply to the prelaminar layer arising from the choroidal vasculature [35], even though the extent to which it contributes to the perfusion of the region is still a matter of debate.

Blood flow to the laminar region is provided by centripetal branches of the short PCAs (see Fig. 10). The centripetal branches arise either directly from the short PCAs or from the circle of Zinn-Haller. These precapillary branches perforate the outer boundary of the lamina and then branch into an intraseptal capillary network, which runs inside the laminar beams. It is still unclear whether there are

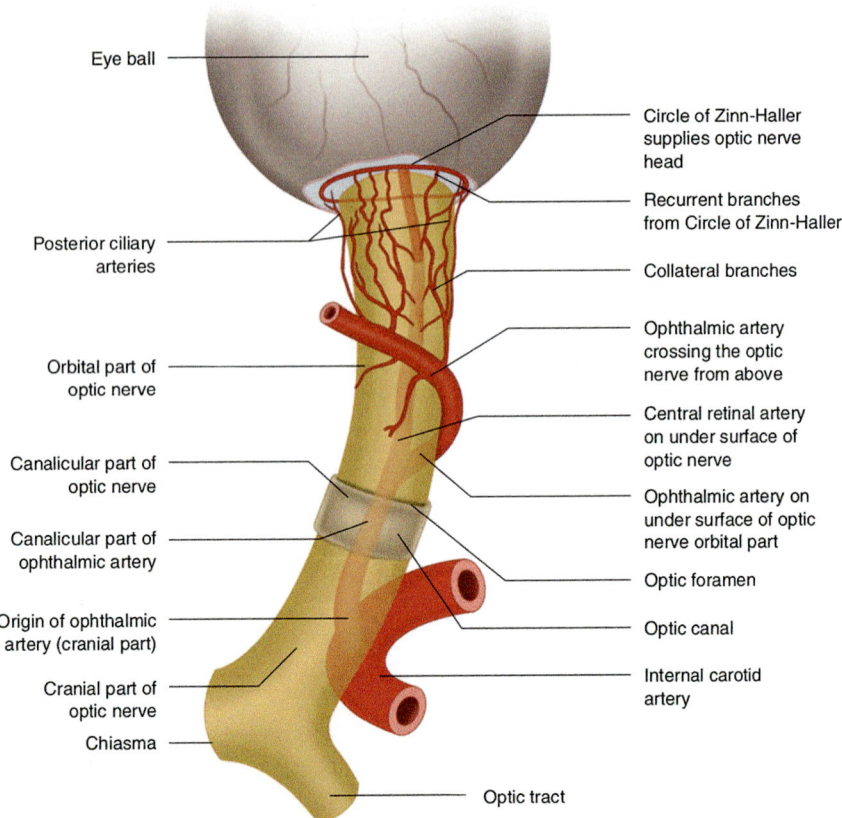

**Fig. 11** Blood supply to various parts of the optic nerve: the optic disc receives blood from the circle of Zinn-Haller (courtesy of Ansari M.W., Nadeem A. (2016) The Blood Supply to the Eyeball. In: Atlas of Ocular Anatomy. Springer, Cham)

anastomoses between the capillary or precapillary bed of the laminar region, the prelaminar region, and the SNFL region. If these anastomoses exist, they might play a role when a sudden (or slowly progressive) vascular occlusion on the precapillary or intracapillary level happens, but this hypothesis requires further investigation.

The retrolaminar region is supplied by the CRA and the pial system (see Fig. 3). The pial system is an anastomosing network of capillaries located immediately within the pia mater. The pial system originates from the circle of Zinn-Haller and may also be fed directly by the short PCAs. The branches of the pial system extend centripetally to nourish the axons of the optic nerve. The CRA may supply several small intraneural branches in the retrolaminar region. Some of these branches may also anastomose with the pial system.

In the ONH, the capillaries form a continuous network throughout its entire length, being continuous posteriorly with those in the rest of the optic nerve and anteriorly with the adjacent retinal capillaries [4]. It is unclear whether this implies that blood flow regulation is similar [43] or not [35] in both vascular regions, independent of the arterial source.

Venous drainage of the ONH occurs primarily through the CRV. In the SNFL, blood is drained directly into the retinal veins, which then join to form the CRV. In the prelaminar, laminar, and retrolaminar regions, venous drainage occurs via the CRV or axial tributaries to the CRV.

Several critical questions concerning the ONH blood supply remain unanswered. The CRA within the intraorbital optic nerve is innervated, but innervation stops (at least) anterior to the lamina cribrosa, and it does not follow the branches of the CRA inside the eye [75]. Neurotransmitter receptors, however, are present on the surface of retinal vessels [38]. In addition, normal retinal vessels lack fenestrations [52]. Hence, vasoactive hormones cannot leak from capillaries and reach the muscular coat of nearby arterioles where they can influence blood flow. The branches of the PCA that feed the intrascleral portion of the optic nerve may or may not be innervated and/or fenestrated. Such knowledge is crucial to understand how blood flow is regulated in the ONH.

## 5   Choroid

The outer choroid is composed of large non-fenestrated vessels, whereas the inner choroid is characterized by a much smaller vessel caliber. The innermost layer of the choroid, the choriocapillaris, is composed of anastomotic fenestrated capillaries beginning at the border of the optic disc (see Fig. 12). These capillaries are separate and distinct from the capillary beds of the anterior optic nerve. The choriocapillaris posterior to the equator of the eye is supplied by the SPCAs, whereas the anterior

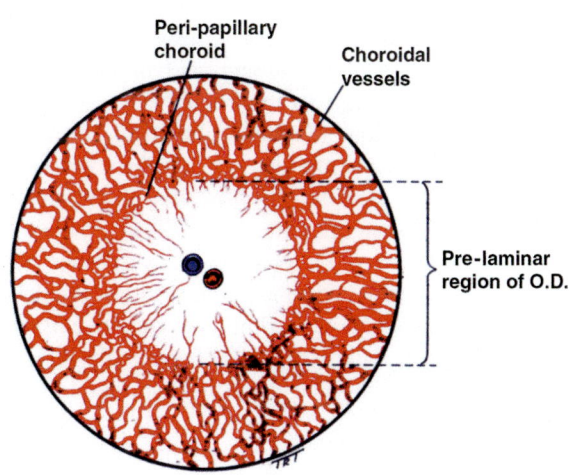

**Fig. 12** Schematic representation of the choroidal and peripapillary arteries and their centripetal contribution to the prelaminar region of the optic nerve head. OD optic disc (courtesy of Hayreh S.S. (2015) Blood Supply of the Optic Nerve Head. In: Ocular Vascular Occlusive Disorders. Springer, Cham)

Peri-papillary choroid

Choroidal vessels

Pre-laminar region of O.D.

choriocapillaris is fed by the LPCAs and the anterior ciliary arteries (ACAs). The LPCAs enter the sclera and run to the front of the eye through the suprachoroidal space to branch near the ora serrata. Each LPCA sends some branches to supply the anterior choriocapillaris. The ACAs accompany the rectus muscles anteriorly to supply the major circles of the iris (see Fig. 4). Before reaching the iris, some branches pass posteriorly to feed the anterior choriocapillaris together with the LPCAs. There is no evidence of anastomoses between the choriocapillaris anterior and posterior to the equator.

The choriocapillaris drains into the vortex veins mainly, and, to a lesser extent, into the anterior ciliary veins, passing through the ciliary body. The vortex veins drain into the inferior ophthalmic vein (IOV) and the superior ophthalmic vein (SOV). The SOV exits the eye through the superior orbital fissure and drains into the cavernous sinus. The IOV sends a branch to the SOV and then exits the orbit through the inferior orbital fissure into the pterygoid plexus.

# 6  Anterior Segment

The posterior segment of the eye comprises the retina, ONH, and choroid, discussed above. The anterior segment includes the cornea, iris, ciliary body, lens, and the anterior and posterior chambers filled with aqueous humor (see Figs. 1 and 4). The cornea is a transparent avascular tissue that acts as the primary barrier and anterior refractive surface of the eye. The cornea is avascular to provide for proper visual function; however, it still relies on oxygen and nutrients to remain healthy. These are provided via diffusion and/or convection directly by the air (oxygen), tear film, aqueous humor, and by tiny vessels on the outermost edge of the cornea [5, 14].

The iris is the pigmented muscular curtain between the cornea and lens which controls the size of the pupil to regulate the amount of light that reaches the posterior segment of the eye. The ciliary body is a wide ring of tissue extending from the sclera to the iris that is responsible for aqueous humor production and outflow, secretion of hyaluronic acid into the vitreous and also the process of accommodation. The iris and ciliary body are both supplied by the ACAs, the LPCAs, and anastomotic connections from the anterior choroid [39]. The ACAs follow the extraocular muscles to pierce the sclera near the limbus and join the major arterial circle of the iris. Similarly, the LPCAs pierce the sclera near the posterior pole, travel between the sclera and choroid, and join the major arterial circle of the iris. The major arterial circle of the iris branches off to supply the iris and ciliary body. The majority of venous drainage from the iris and ciliary body is directed posteriorly to the choroid and into the vortex veins (see Figs. 4 and 13).

The lens is the largest avascular organ in the body [47]. The lens is transparent to properly refract light onto the retina and therefore lacks blood vessels that would interfere with this process. Internal circulating currents of ions coupled to fluid movement are believed to substitute blood circulation in the role of providing the organ with nutrients to remain healthy [47].

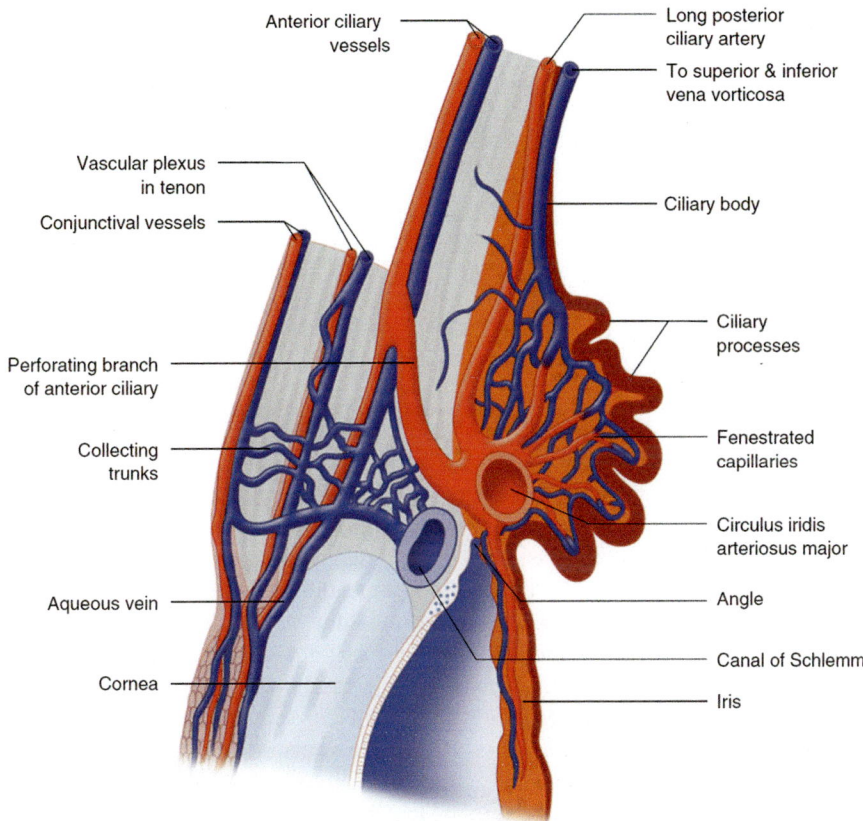

**Fig. 13** Blood vessels to the anterior segment of the eye (courtesy of Ansari M.W., Nadeem A. (2016) The Blood Supply to the Eyeball. In: Atlas of Ocular Anatomy. Springer, Cham)

## 7 Vascular Regulation and Innervation

As in other tissues—including the brain [53], heart [65], kidney [60], skeletal muscle [19], and gut [44]—blood flow to the ocular vasculature is regulated. Ocular vascular regulation causes changes in vascular tone in order to maintain relatively constant blood flow over a substantial range of IOP and systemic blood pressures, with the aim of providing adequate nutrients, oxygen, and tissue turgor at all times. In particular, mechanisms regulating blood flow to a specific tissue that act within the local tissue are named "autoregulatory" mechanisms. Moreover, in experiments assessing hemodynamic responses to light stimulation [25, 62, 63], blood flow in the retina and ONH seems to be highly correlated with increased feedforward neural activity. This "long range" phenomenon is called neurovascular coupling [46].

Blood flow regulation is evaluated most often on a flow versus pressure graph, where pressure may be expressed as mean arterial pressure (MAP), IOP, or ocular perfusion pressure (OPP). OPP refers to the arteriovenous pressure difference driving blood flow through the intraocular vasculature. The intraocular venous pressure is very close to IOP [13] and thus OPP is usually estimated as the difference between the mean arterial BP and IOP in the upright position. OPP may also be defined as mean systolic (sOPP) or diastolic (dOPP) [13]. Mean OPP is typically estimated as

$$Mean\ OPP = \frac{2}{3}MAP - IOP,$$

where

$$MAP = diastolic\ BP + \frac{1}{3}\left(systolic\ BP - diastolic\ BP\right).$$

The brachial arterial pressure has been often considered as representative of systemic BP. The factor 2/3 accounts for the drop in BP between the brachial and ophthalmic artery when the subject is seated [13] and the fact that the orbital arteries are further downstream. It is not clear how accurately the above formula approximates OPP. Nevertheless, even if more reliable formulas for computing the OPP have been proposed [13], the previously mentioned relations are consistently used in clinical studies. A reliable, direct measure of OPP would of course be desirable, but without this, care is needed when interpreting blood flow regulation studies. Autoregulatory capacity is conventionally assessed in two steps: blood flow or other hemodynamic parameters are measured before and after OPP is artificially modified by a step challenge in either IOP or the systemic arterial blood pressure. If blood flow changes significantly from normal after the pressure challenge, then autoregulation is said to be impaired; if blood flow remains nearly constant over the pressure change (this feature is known as "autoregulation plateau"), autoregulation is said to be intact. When a pressure step challenge is applied rapidly, we can distinguish two phases in the hemodynamic response: (1) an initial transient, or dynamic, phase lasting a few seconds during which the vasculature tries to return blood flow to its original level by adjusting vascular resistance (dynamic autoregulatory phase) and (2) a steady-state phase when transient blood flow changes have equilibrated to a steady-state level (static autoregulatory phase) (Fig. 14). While various features, such as metabolic, myogenic, and neurogenic factors, are involved in static regulation over longer durations, contractive factors are involved in the instantaneous reaction of dynamic regulation. These pathways are important as defective autoregulation has been linked to exert an important role in the pathophysiology of ophthalmic vascular diseases including open-angle glaucoma and diabetic retinopathy [45, 69].

In the eye, the circulation in the retina [15, 18, 32, 41, 55, 62, 72], ONH [7, 25, 26, 42, 56, 63, 70, 74], and retrobulbar vessels [64, 67] is known to exhibit regulatory capability, though to different extents. The extent of autoregulation in the choroidal vascular bed is still unknown [52]. Moreover, only few published studies address autoregulation in the ciliary body [2, 40] and the iris [2, 9, 73]. In these studies, several perturbations have been induced to assess ocular blood flow regulation

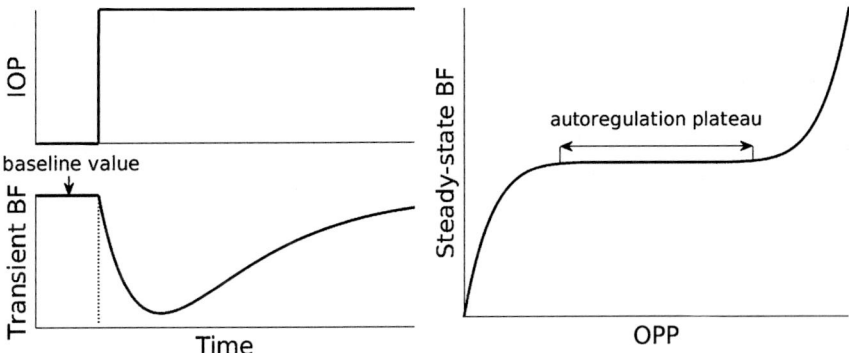

**Fig. 14** (Left) Schematic representation of a dynamic autoregulation response following a step increase in IOP during which the vasculature tries to return ocular BF to its baseline value. (Right) Illustration of a characteristic static autoregulation curve describing the relationship between steady-state BF responses in the eye and OPP

including light flicker, $CO_2$ breathing, IOP alterations, exercise, positional changes, and changes in blood pressure. Consensus around a clinical definition of ocular vascular regulation, and how to best assess it, is still missing.

The physiology of ocular blood flow regulation is an active area of research. Blood flow to a particular tissue is usually controlled by resistance arterioles (arterial branches smaller than 40 mm), which change diameter actively to achieve autoregulation [57]. Capillaries have smaller diameters than arterioles. Because resistance to flow is inversely proportional to the fourth power of the vessel radius (Poiseuille's law) [24], capillaries exhibit high resistance individually; however, they are numerous and in parallel array, and this outweighs the reduction in vessel size. Their large cross-sectional area does not contribute much to the net resistance between arteries and veins. Flow to a local capillary network is also controlled by a precapillary sphincter (where it exists), a band of smooth muscle located where capillaries originate from small arterioles [3]. Finally, blood flow may also be controlled by changes in the resistance of the capillaries themselves due to the contractile properties of pericytes [3]. The retinal and optic nerve capillaries lack precapillary sphincters but have abundant pericytes [21, 22]. It has been shown that pericytes respond to carbon dioxide concentrations (mediated by pH), oxygen levels (affected by changes in NO concentrations), and adenosine levels [10, 30, 48]. Vascular smooth muscle cells respond in a very similar manner. Pericytes on isolated retinal capillaries have been found to constrict or dilate in response to neurotransmitters, following $Ca^{2+}$ alterations [58]. In situ in brain slices, pericytes constrict in response to noradrenaline and dilate in response to glutamate [54]. In the isolated retina, blocking ionotropic receptors of the neurotransmitter gamma-aminobutyric acid has been shown to constrict capillaries, suggesting that endogenous transmitter release could regulate capillary diameter [31, 54]. Thus, capillaries are equipped to participate in the control of ocular blood flow through metabolic feedback and neuronal feedforward coupling mechanisms, but this still needs to be further investigated.

In addition to the contractile features of the vasculature, it is also important to consider the nature of the endothelium. With this perspective, capillaries may be classified as fenestrated or non-fenestrated. In the retinal tissue, in analogy with other parts of the central nervous system, the endothelial cells lack fenestration, and are joined by tight junctions that prevent substances within the plasma from diffusing into the surrounding tissue, thereby forming a blood–tissue barrier [61], also known as blood–retinal barrier. Circulating vasoconstrictive hormones, such as angiotensin and epinephrine, raise intracellular $Ca^{2+}$ when they occupy receptors [72]. In vascular smooth muscle, as well as in pericytes, an increase of intracellular $Ca^{2+}$ induces contraction of the cell. Consequently, endothelial cells produce more NO, which relax vascular smooth muscle and pericytes. In this way, the circulating hormones leak through fenestrated vessels, reach the muscular coat of vessels in the region, and cause vasoconstriction in the skin and many visceral organs. In contrast, vessels in the central nervous system, where hormones can only affect endothelial cells and are prevented from reaching the contractile coat of the vessels to the blood–tissue barrier, exhibit vasodilation [72]. In contrast to retinal capillaries, choroidal capillaries are fenestrated and have sparse pericytes [52].

The blood flow to the choroid is believed to be primarily regulated by hormonal and autonomic neuronal mechanisms. The autonomic nervous system ends at the lamina cribrosa so that retinal and ONH vessels do not exhibit innervation. Yet, retinal vessels do retain receptors for various neurotransmitters on their surface. Consequently, retinal blood flow is believed to be locally regulated in response to metabolic needs, with capillary recruitment and derecruitment presumably providing constant oxygen delivery. The mediators of these local regulation mechanisms include oxygen, carbon dioxide, angiotensin-II, adenosine, nitric oxide (NO), and endothelin-1 [45]. The roles of angiotensin-II and endothelin-1 are disputable due to the blood–retinal barrier that is supposed to exclude circulating hormones from transport to smooth muscle cells. Despite this, these two hormones could diffuse from the choroid to retinal tissue through the incomplete blood–retinal barrier.

The extent of choroidal autoregulation is not fully understood. In response to changes in OPP, the choroid has shown to regulate blood flow through sympathetic vasoconstriction while parasympathetic vasodilation must occur through anterior tissues of the eye [71]. A study using laser speckle flowgraphy and optical coherence tomography determined that the choroid is able to autoregulate in response to decreases in OPP induced by increased IOP; however, at a threshold of higher IOPs, there is no autoregulation. The intrinsic choroidal neurons receive both parasympathetic and sympathetic innervations and are assumed to play a role in blood flow regulation, although these mechanisms were not determined [1]. The extent of choroidal response to physiological changes continues to be studied. Furthermore, another topic that continues to be discussed is the role of diurnal variations in ocular autoregulation of blood flow. Autoregulation of ocular blood flow is difficult to assess, and there exists little specific criteria on classifying autoregulation status as impaired or functional [45]. Further investigation of these areas of ocular vascular autoregulation could help delineate the role of autoregulation in the pathophysiology of ophthalmic vascular diseases.

# 8 Discussion

In this chapter, we presented a brief overview of the vascular anatomy and physiology of the eye. Not only the eye's blood supply is quite complex, but it also shows high interindividual and intraindividual variability [27, 34, 37, 71]. Marked variation is observed in the vascular patterns of the anterior optic nerve, peripapillary retina, and the posterior choroid across individuals. The predominant difference is in the arterial supply. Varying numbers of branches have been found in the PCAs, the SPCAs, the LPCAs, and the ACAs. Due to the different anatomy, depth, and physiology of these vessels, different imaging devices are needed to measure blood flow in them. Recent advances in ocular imaging are extending our grasp of the role of ocular hemodynamics in health and disease.

Vascular changes are also associated with increasing age [16]. With age, there is a decrease in both capillary diameter (due to the thickening of the basement membrane) and density [59], thereby causing an increase in blood velocity and a decrease in the total volume available to flow. Moreover, the increasing incidence of atherosclerosis with age reduces arterial distensibility in all vascular beds, including the ocular vasculature [17]. Evidence has been found of a decreased compliance of the retrobulbar arteries [33], which, in turn, results in a decrease in retinal and neuroretinal rim microcirculation. Decreases in endothelial nitric oxide synthase (ENOS) activity with age can lead to reduced nitric oxide (NO) availability, possibly compromising the endothelium-dependent balance between vasodilation and vasoconstriction. All of these changes can impair oxygen supply, reduce nutrient exchange, diminish metabolic waste removal, and weaken the response to local changes in metabolic demand, ultimately exposing the eye to risk of ischemic episodes. Stressors, such as age-associated systemic diseases, environmental factors, or genetic predispositions, can further disrupt ocular blood flow and leave the optic nerve vulnerable to additional damage and disease formation.

Further investigation is needed to better understand the different aspects of ocular blood supply. As imaging techniques are advancing, new information is becoming available and concepts that were deemed established are now being re-examined. A synergistic alliance across all studies—animal studies, experimental studies, clinical studies, and mathematical modeling studies—will play a crucial role in this direction.

# References

1. Akahori T, Iwase T, Yamamoto K, Ra E, Terasaki H. Changes in Choroidal Blood Flow and Morphology in Response to Increase in Intraocular Pressure. Invest Ophthalmol Vis Sci. 2017 Oct 1;58(12):5076-5085.
2. Alm A, Bill A. Ocular and optic nerve blood flow at normal and increased intraocular pressures in monkeys (Macaca irus): a study with radioactively labelled microspheres including flow determinations in brain and some other tissues. Exp Eye Res. 1973 Jan 1;15(1):15-29.

3. Anderson DR. Glaucoma, capillaries and pericytes. 1. Blood flow regulation. Ophthalmologica. 1996;210(5):257-62
4. Anderson DR. Ultrastructure of the optic nerve head. Arch Ophthalmol. 1970;83(1):63-73.
5. Ansari MW, Nadeem A. Atlas of Ocular Anatomy. Springer Switzerland, 2016.
6. Awai T. Angioarchitecture of intraorbital part of human optic nerve. Jpn J Ophthalmol. 1985;29(1):79-98.
7. Boltz A, Schmidl D, Werkmeister R, et al. Regulation of optic nerve head blood flow during combined changes in intraocular pressure and arterial blood pressure. J Cereb Blood Flow Metab. 2013;33(12):1850-6
8. Campbell JP, Zhang M, Hwang TS, Bailey ST, Wilson DJ, Jia Y, Huang D. Detailed Vascular Anatomy of the Human Retina by Projection-Resolved Optical Coherence Tomography Angiography. Sci Rep. 2017; 7: 42201. doi: https://doi.org/10.1038/srep42201.
9. Chamot SR, Movaffaghy A, Petrig BL, Riva CE. Iris blood flow response to acute decreases in ocular perfusion pressure: a laser Doppler flowmetry study in humans. Exp Eye Res. 2000 Jan;70(1):107-12.
10. Chen Q, Anderson DR. Effect of $CO_2$ on intracellular pH and contraction of retinal capillary pericytes. Invest Ophthalmol Vis Sci. 1997;38(3):643-51
11. Chiaravalli G. A virtual laboratory for retinal physiology: a theoretical study of retinal oxygenation in healthy and disease. MS Thesis, Politecnico di Milano (Italy). December 2018
12. Chiaravalli G, Guidoboni G, Sacco R, Ciulla T, Harris A. A theoretical study of the vascular configuration of the capillary plexi inside the retinal tissue based on OCTA data. Presented at 2019 Annual Meeting of the Association for Research in Vision and Ophthalmology
13. Costa V, Harris A, Anderson D, et al. Ocular perfusion pressure in glaucoma. Acta Ophthalmol. 2014;92(4):e252-66
14. DelMonte DW, Kim T. Anatomy and physiology of the cornea. J Cataract Refract Surg. 2011 Mar;37(3):588-98.
15. Dumskyj MJ, Eriksen JE, Doré CJ, Kohner EM. Autoregulation in the human retinal circulation: assessment using isometric exercise, laser Doppler velocimetry, and computer-assisted image analysis. Microvasc Res. 1996 May;51(3):378-92.
16. Ehrlich R, Kheradiya NS, Winston DM, Moore DB, Wirostko B, Harris A. Age-related ocular vascular changes. Graefes Arch Clin Exp Ophthalmol. 2009;247(5):583–91.
17. Embleton SJ, Hosking SL, Roff Hilton EJ, Cunliffe IA (2002) Effect of senescence on ocular blood flow in the retina, neuroretinal rim and lamina cribrosa, using scanning laser Doppler flowmetry. Eye 16:156–162
18. Foke GT, Pasquale LR. Retinal blood flow response to posture change in glaucoma patients compared with healthy subjects. Ophthalmology. 2008 Feb;115(2):246-52. Epub 2007 Aug 8.
19. Folkow B, Sonnenschein RR, Wright DL. Loci of neurogenic and metabolic effects on precapillary vessels of skeletal muscle. Acta Physiol Scand. 1971;81(4):459-71
20. Fouquet S, Vacca O, Sennlaub F, Paques M. The 3D Retinal Capillary Circulation in Pigs Reveals a Predominant Serial Organization. Invest Ophthalmol Vis Sci. 2017 Nov; 58(13): 5754-63. doi: https://doi.org/10.1167/iovs.17-22097.
21. Frank RN, Dutta S, Mancini MA. Pericyte coverage is greater in the retinal than in the cerebral capillaries of the rat. Invest Ophthalmol Vis Sci. 1987;28(7):1086-91
22. Frank RN, Turczyn TJ, Das A. Pericyte coverage of retinal and cerebral capillaries. Invest Ophthalmol Vis Sci. 1990;31(6):999-1007
23. Freund KB, Sarraf D, Leong BCS, Garrity ST, Vupparaboina KK, Dansingani KK. Association of Optical Coherence Tomography Angiography of Collaterals in Retinal Vein Occlusion With Major Venous Outflow Through the Deep Vascular Complex. JAMA Ophthalmol. 2018 Nov 1;136(11):1262-1270.
24. Fung YC. Biomechanics: circulation. NewYork, Springer; 1997
25. Garhöfer G, Resch H, Weigert G, et al. Short-term increase of intraocular pressure does not alter the response of retinal and optic nerve head blood flow to flicker stimulation. Invest Ophthalmol Vis Sci. 2005;46(5):1721-5

26. Geijer C, Bill A. Effects of raised intraocular pressure on retinal, prelaminar, laminar, and retrolaminar optic nerve blood flow in monkeys. Invest Ophthalmol Vis Sci. 1979 Oct;18(10):1030-42.
27. Guidoboni G, Harris A, Arciero JC, Siesky BA, Amireskandari A, Gerber AL, Huck AH, Kim NJ, Cassani S, Carichino L. Mathematical modeling approaches in the study of glaucoma disparities among people of African and European descents. Journal of Coupled Systems and Multiscale Dynamics. 2013 Apr 1;1(1):1-21.
28. Guidoboni G, Harris A, Cassani S, et al. Intraocular pressure, blood pressure, and retinal blood flow autoregulation: a mathematical model to clarify their relationship and clinical relevance. Invest Ophthalmol Vis Sci. 2014 May 29;55(7):4105-18.
29. Guidoboni G, Harris A, Carichino L, et al. Effect of intraocular pressure on the hemodynamics of the central retinal artery: A mathematical model. Mathematical Biosciences and Engineering, 2014, 11(3): 523-546.
30. Haefliger IO, Chen Q, Anderson DR. Effect of oxygen on relaxation of retinal pericytes by sodium nitroprusside. Graefes Arch Clin Exp Ophthalmol. 1997;235(6):388-92
31. Hall CN, Reynell C, Gesslein B, et al. Capillary pericytes regulate cerebral blood flow in health and disease. Nature. 2014;508(7494):55-60
32. Harris A, Arend O, Bohnke K, Kroepfl E, Danis R, Martin B. Retinal blood flow during dynamic exercise. Graefes Arch Clin Exp Ophthalmol. 1996 Jul;234(7):440-4.
33. Harris A, Harris M, Biller J, Garzozi H, Zarfty D, Ciulla TA, Martin B (2000) Aging affects the retrobulbar circulation differently in women and men. Arch Ophthalmol 118:1076–1080
34. Harris A, Jonescu-Cuypers CP, Kagemann L, et al. Atlas of ocular blood flow. Oxford, UK, Butterworth-Heinemann; 2ed, 2010.
35. Hayreh SS. The blood supply of the optic nerve head and the evaluation of it - myth and reality. Prog Retin Eye Res. 2001;20(5):563-93.
36. Hayreh SS. The 1994 Von Sallman Lecture. The optic nerve head circulation in health and disease. Exp Eye Res. 1995;61(3):259-72.
37. Hayreh SS. Inter-individual variation in blood supply of the optic nerve head. Its importance in various ischemic disorders of the optic nerve head, and glaucoma, low-tension glaucoma and allied disorders. Doc Ophthalmol. 1985;59(3):217-46
38. Hoste AM, Boels PJ, Brutsaert DL, De Laey JJ. Effect of alpha-1 and beta agonists on contraction of bovine retinal resistance arteries in vitro. Invest Ophthalmol Vis Sci. 1989;30(1):44-50.
39. Kiel JW. The Ocular Circulation. San Rafael (CA), Morgan & Claypool Life Sciences; 2010.
40. Kiel JW, Hollingsworth M, Rao R, Chen M, Reitsamer HA. Ciliary blood flow and aqueous humor production. Prog Retin Eye Res. 2011 Jan;30(1):1-17.
41. Kornfield TE, Newman EA. Regulation of Blood Flow in the Retinal Trilaminar Vascular Network. J Neurosci. 2014 Aug 20; 34(34): 11504–11513.
42. Liang Y, Downs JC, Fortune B, Cull G, Cioffi GA, Wang L. Impact of systemic blood pressure on the relationship between intraocular pressure and blood flow in the optic nerve head of nonhuman primates. Invest Ophthalmol Vis Sci. 2009 May;50(5):2154-60. doi: https://doi.org/10.1167/iovs.08-2882. Epub 2008 Dec 13.
43. Lieberman MF, Maumenee AE, Green WR. Histologic studies of the vasculature of the anterior optic nerve. Am J Ophthalmol. 1976;82(3):405-23.
44. Lundgren O. Autoregulation of intestinal blood flow: physiology and pathophysiology. J Hypertens Suppl. 1989;7(4):S79-84
45. Luo X, Shen YM, Jiang MN, Lou XF, Shen Y. Ocular Blood Flow Autoregulation Mechanisms and Methods. J Ophthalmol. 2015; 2015: 864871.
46. Mackenzie PJ, Cioffi GA. Vascular anatomy of the optic nerve head. Can J Ophthalmol. 2008;43(3):308-12
47. Mathias RT, Kistler J, Donaldson P. The Lens Circulation. J. Membr. Biol. 2007 Mar;216(1):1-16.
48. Matsugi T, Chen Q, Anderson DR. Adenosine-induced relaxation of cultured bovine retinal pericytes. Invest Ophthalmol Vis Sci. 1997;38(13):2695-701

49. Michalinos A, Zogana S, Kotsiomitis E, Mazarakis A, Troupis T. Anatomy of the Ophthalmic Artery: A Review concerning Its Modern Surgical and Clinical Applications. Anat Res Int. 2015; 2015: 591961.
50. Morgan WH, Yu DY, Alder VA, et al. The correlation between cerebrospinal fluid pressure and retrolaminar tissue pressure. Invest. Ophthalmol. Vis. Sci. 1998, 39, 1419-1428
51. Onda E, Cioffi GA, Bacon DR, Van Buskirk EM. Microvasculature of the human optic nerve. Am J Ophthalmol. 1995;120(1):92-102.
52. Pasquale L, Jonas JB, Anderson DR. Anatomy and physiology, in Weinreb RN, Harris A (eds) Ocular Blood Flow in Glaucoma. Amsterdam, Kugler Publications; 2009, pp 3-13.
53. Paulson OB, Strandgaard S, Edvinsson L. Cerebral autoregulation. Cerebrovasc Brain Metab Rev. 1990;2(2):161-92
54. Peppiatt CM, Howarth C, Mobbs P, Attwell D. Bidirectional control of CNS capillary diameter by pericytes. Nature. 2006;443(7112):700-4
55. Pillunat LE, Stodtmeister R, Wilmanns I, Christ T. Autoregulation of ocular blood flow during changes in intraocular pressure. Preliminary results. Graefes Arch Clin Exp Ophthalmol. 1985;223(4):219-23.
56. Prada D, Harris A, Guidoboni G, et al. Autoregulation and neurovascular coupling in the optic nerve head. Surv Ophthalmol. 2016 Mar-Apr;61(2):164-86.
57. Pries AR, Secomb TW, Gaehtgens P. Biophysical aspects of blood flow in the microvasculature. Cardiovasc Res. 1996;32(4):654-67
58. Puro DG. Physiology and pathobiology of the pericyte containing retinal microvasculature: new developments. Microcirculation. 2007;14(1):1-10
59. Ramrattan RS, van der Schaft TL, Mooy CM, de Bruijn WC, Mulder PG, de Jong PT (1994) Morphometric analysis of Bruch's membrane, the choriocapillaris, and the choroid in aging. Invest Ophthalmol Vis Sci 35(6):2857–2864
60. Rein H. Vasomotorische regulationen. Ergebnisse der Physiologie. 1931;32(1):28-72
61. Resch H, Garhofer G, Fuchsjager-Mayrl G, et al. Endothelial dysfunction in glaucoma. Acta Ophthalmol. 2009;87(1):4-12
62. Riva CE, Grunwald JE, Petrig BL. Autoregulation of human retinal blood flow. An investigation with laser Doppler velocimetry. Invest Ophthalmol Vis Sci. 1986;27(12):1706-12
63. Riva C, Hero M, Titze P, Petrig B. Autoregulation of human optic nerve head blood flow in response to acute changes in ocular perfusion pressure. Graefe's Arch Clin Exp Ophthalmol. 1997;235(10):618-26
64. Roff EJ, Harris A, Chung HS, Hosking SL, Morrison AM, Halter PJ, Kagemann L. Comprehensive assessment of retinal, choroidal and retrobulbar haemodynamics during blood gas perturbation. Graefe's archive for clinical and experimental ophthalmology. 1999 Dec 1;237(12):984-90.
65. Rubio R, Berne RM. Regulation of coronary blood flow. Progress in cardiovascular diseases. 1975 Sep 1;18(2):105-22.
66. Siaudvytyte L, Januleviciene I, Daveckaite A, Ragauskas A, Bartusis L, Kucinoviene J, Siesky B, Harris A. Literature review and meta-analysis of translaminar pressure difference in open-angle glaucoma. Eye (Lond). 2015 Oct;29(10):1242-50.
67. Sines D, Harris A, Siesky B, Januleviciene I, Haine CL, Yung CW, Catoira Y, Garzozi HJ. The response of retrobulbar vasculature to hypercapnia in primary open-angle glaucoma and ocular hypertension. Ophthalmic research. 2007;39(2):76-80.
68. Singh S, Dass R. The central artery of the retina. II. A study of its distribution and anastomoses. Br J Ophthalmol. 1960;44:280-99
69. Schmidl D, Garhofer G, Schmetterer L. The complex interaction between ocular perfusion pressure and ocular blood flow-relevance for glaucoma. Exp Eye Res. 2011;93(2):141-55
70. Sossi N, Anderson DR. Effect of elevated intraocular pressure on blood flow. Occurrence in cat optic nerve head studied with iodoantipyrine I 125. Arch Ophthalmol. 1983 Jan;101(1):98-101.
71. Steinle JJ, Krizsan-Agbas D, Smith PG. Am J Physiol Regul Integr Comp Physiol. 2000 Jul; 279(1):R202-9.

72. Tachibana H, Gotoh F, Ishikawa Y. Retinal vascular autoregulation in normal subjects.Stroke. 1982 Mar-Apr;13(2):149-55.
73. Tomidokoro A., Araie M., Tamaki Y., Tomita K. In vivo measurement of iridial circulation using laser speckle phenomenon Invest Ophthalmol Vis Sci. 1998; 39: 364–71.
74. Weinstein JM, Duckrow RB, Beard D, Brennan RW. Regional optic nerve blood flow and its autoregulation. Invest Ophthalmol Vis Sci. 1983 Dec;24(12):1559-65.
75. Ye XD, Laties AM, Stone RA. Peptidergic innervation of the retinal vasculature and optic nerve head. Invest Ophthalmol Vis Sci. 1990;31(9):1731-7.
76. Zhang J, Stringer MD. Ophthalmic and facial veins are not valveless. Clin Exp Ophthalmol. 2010 Jul;38(5):502-10.
77. Zhi Z, Cepuma W, Johnson E, Jayaram H, Morrison J, Wang RK. Evaluation of the effect of elevated intraocular pressure and reduced ocular perfusion pressure on retinal capillary bed filling and total retinal blood flow in rats by OMAG/OCT. Microvasc Res. 2015 Sep;101:86-95.

# Pathological Consequences of Vascular Alterations in the Eye

Daniele Prada, L. Rowe, A. Hajrasouliha, T. Ciulla, I. Januleviciene,
G. Chiaravalli, G. Guidoboni, and A. Harris

**Abstract** This chapter reviews the abundant evidence of correlations between vascular alterations and ocular diseases. In particular, we discuss retinal diseases, including age-related macular degeneration, diabetic retinopathy and retinal vessel occlusions, glaucoma, and non-arteritic ischemic optic neuropathy. Current inconsistencies among studies and outstanding controversial questions are emphasized to bring the reader up to date with respect to the main challenges in the field.

## 1 Introduction

Vascular alterations have been associated with several pathological conditions affecting ocular tissues and visual function. In this chapter, we will review abnormal vascular conditions associated with retinal diseases, including age-related macular

D. Prada (✉)
Istituto di Matematica Applicata e Tecnologie Informatiche "Enrico Magenes" del Consiglio
Nazionale delle Ricerche, Pavia, Italy
e-mail: daniele.prada@imati.cnr.it

L. Rowe · A. Hajrasouliha
Department of Ophthalmology, Indiana University School of Medicine, Indianapolis, IN, USA

T. Ciulla
Midwest Eye Institute, Indianapolis, IN, USA

I. Januleviciene
Eye Clinic, Lithuanian University of Health Sciences, Kaunas, Lithuania

G. Chiaravalli · G. Guidoboni
Department of Electrical Engineering and Computer Science, University of Missouri, Columbia,
MO, USA

Department of Mathematics, University of Missouri, Columbia, MO, USA

A. Harris
Icahn School of Medicine at Mount Sinai, New York, NY, USA

© Springer Nature Switzerland AG 2019      47
G. Guidoboni et al. (eds.), *Ocular Fluid Dynamics*, Modeling and Simulation in
Science, Engineering and Technology, https://doi.org/10.1007/978-3-030-25886-3_3

degeneration and diabetic retinopathy, glaucoma, non-arteritic ischemic optic neuropathy, and retinal vessel occlusions.

Pathological vascular changes may be due to changes in geometry (e.g., vessel diameter and vessel tortuosity), mechanical properties (e.g., stiffness and compliance of vessel walls and blood rheology), functionality (e.g., vascular regulation), and pressure levels (e.g., blood pressure, intraocular pressure, and cerebrospinal fluid pressure). Understanding and characterizing such alterations is important for diagnostic purposes, monitoring disease progression, and identifying pathogenic causes of diseases. However, often the conclusions from clinical and experimental studies are not unanimous, thereby giving rise to controversies and debates. In the following sections, we aim to provide an overview of the current knowledge regarding potential associations between vascular alterations and ocular diseases, while emphasizing outstanding questions and crucial challenges in the field.

## 2   Vascular Alterations in Retinal Diseases

The retinal tissue is organized in a specific layered structure, where each layer is characterized by different levels of oxygen consumption. Three major capillary plexi perfuse the retinal tissues, namely the superficial, intermediate, and deep capillary plexi, denoted by SCP, ICP, and DCP, respectively, in Fig. 1. The three capillary layers are primarily differentiated at the macula and reduced in the peripheral retina. The layer conformation in the foveal region is remarkably different from the rest of the retina, as shown in Fig. 1. The central region, also called foveal avascular zone (FAZ), is characterized by the absence of blood vessels, thereby requiring oxygen to diffuse in from the choroid, the vitreous, and the outer macular rim.

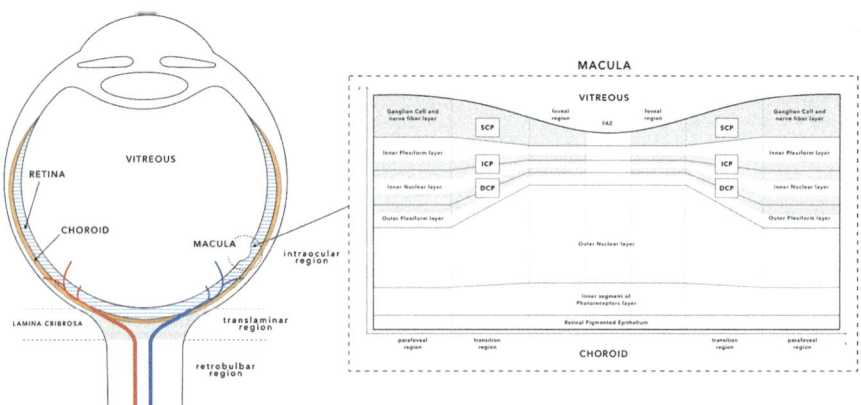

**Fig. 1** Schematic representation of the layered structure of the retinal tissue. The superficial, intermediate, and deep capillary beds are denoted by SCP, ICP, and DCP, respectively. The foveal avascular zone (FAZ) is also reported in the center of the macula [17]

Alterations in the vasculature of the retina have shown to play a prominent role in the development and progression of retinal disease, most notably in relation to age-related macular degeneration (AMD) and diabetic retinopathy (DR). AMD is the leading cause of adult blindness in developed countries and presents as a major health problem for individuals over 65 years of age [61, 92]. DR represents a leading cause of impaired vision in individuals between the ages of 25 and 75, and it is crucial for diabetic patients to be regularly screened for development of retinal disease due to the disease's rapid progression [38].

## 2.1  Vascular Alterations in Age-Related Macular Degeneration

Alterations to choroidal hemodynamics in AMD have been investigated in several studies [21, 48, 93]. There are two types of AMD: non-neovascular (traditionally known as "dry") and neovascular (traditionally known as "wet") AMD [62, 77]. Neovascular AMD is characterized by the development of choroidal neovascularization (CNV), which results in exudation with or without hemorrhage that ultimately leads to fibrotic scar formation and disruption of the architecture of the retinal pigment epithelium (RPE) and photoreceptors (see Fig. 2). Both types of AMD cause progressive loss of central vision [62, 77, 78]. Data concerning the estimated incidence and prevalence of AMD in different countries are available [39, 91]. In some cases, patients with early AMD show drusen underneath the RPE, with areas of mottled RPE hyperpigmentation and hypopigmentation. Drusen are yellow deposits made up of fatty proteins.

A reduced blood supply to the choroid may decrease the amount of oxygen available to nourish the FAZ, thereby making it more susceptible to ischemic damage. For example, the combination of increased pulsatility and decreased velocity of the short posterior ciliary arteries is evidence of increased vascular resistance observed in eyes with AMD. The resulting impaired choroidal perfusion is likely related to the degradation of metabolic transport function of the RPE [41]. It has been demonstrated that systematic decreases in choroidal circulatory parameters correlate with increased severity of AMD [49]. A decrease of blood flow in the choroid may be secondary to arteriosclerotic effects due to aging, which may lead to the partial obstruction of vortex veins, therefore to an increase in the resistance of the choroidal circulation [40, 42]. This may also be responsible for the delayed and heterogenous filling of the choroid that has been displayed in eyes with non-neovascular AMD [20]. Impaired autoregulation has also been proposed as a major determinant of reduced choroidal perfusion [93]. The nervous sympathetic activity controlling autoregulatory mechanisms is affected by aging, and it may compromise the ability of the vasculature to maintain the necessary levels of blood supply to the tissue. Additionally, the release of vasoconstricting factors by the vascular endothelium may be affected by aging and disrupt the smooth muscle of the arterioles or the contractile state of pericytes surrounding the capillaries [50, 51].

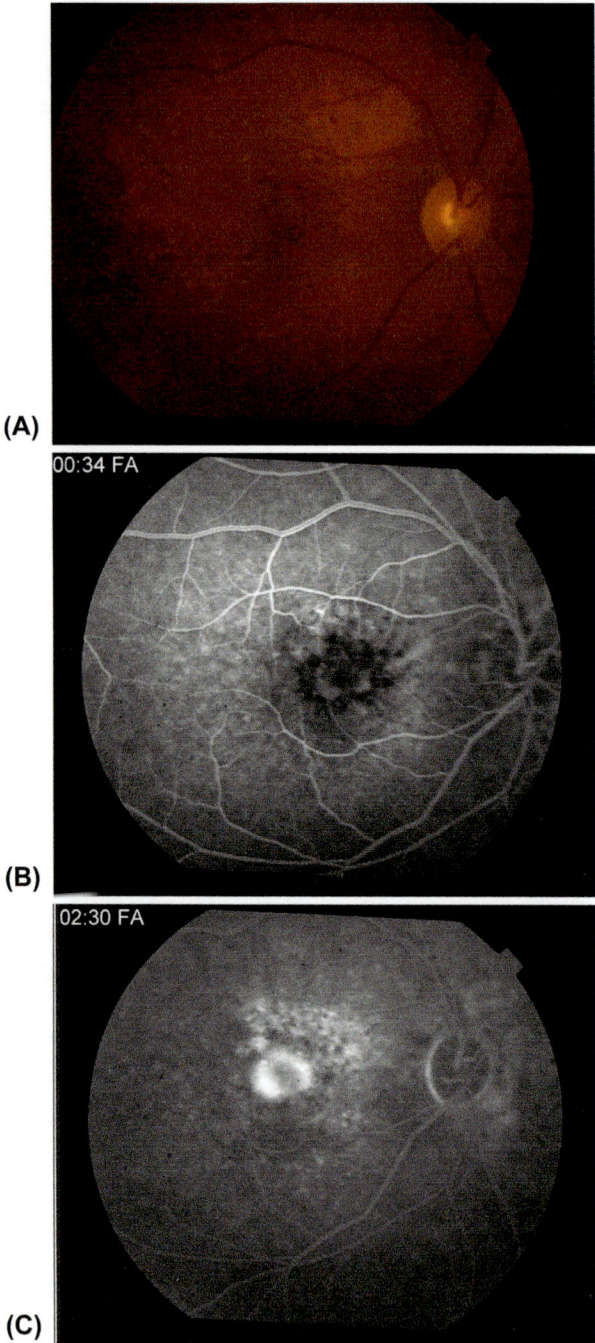

**Fig. 2** Neovascular age-related macular degeneration. This 65-year-old man complained of painless progressive blurring of central vision, right eye. (**a**) Color Fundus Photography, Right Eye showed scattered drusen and central pigment clumping. (**b**) Fluorescein Angiography (FA)

   The relationship between vascular alterations and non-neovascular and neo-vascular forms of AMD has attracted particular attention, since the pathogenic mechanisms that underlie the two conditions are still poorly understood. It has been proposed that one explanation for the pathogenesis of both forms of the disease is photoreceptor and RPE ischemia secondary to poor choroidal or Bruch's membrane (BM) perfusion [42]. It has been suggested that the different forms of AMD may be associated with different ratios between choroidal and cerebral vascular resistance [42]. Specifically, choroidal vascular resistance appears to be lower than cerebral resistance in non-neovascular AMD and larger in the neovascular case. A second hypothesis, presented in [49], relates AMD to the decrease of choroidal blood flow and the eventual presence of Presumed Macular Watershed Filling (PMWF) [56, 104], which eventually lead to the development of localized CNV [109].

   The traditional classification of AMD into a "dry" and "wet" form is evolving with findings based on optical coherence tomography angiography (OCTA). Fluorescein angiography (FA) and indocyanine green angiography (ICGA) are used to image neovascularization in the retina and choroid [2, 105]. FA was once considered to be the gold standard for visualizing CNV in AMD [26, 27, 75], but with optical coherence tomography (OCT), the retinal layers can be noninvasively and accurately segmented in a three-dimensional fashion, unlike FA [46]. The advent of OCTA (Fig. 3) has proved to be a valuable tool in the diagnosis and management of neovascular AMD and other retinal disorders, as it detects flow through blood vessels and has the ability to detect and localize CNV [84, 87]. Using FA as ground truth, OCTA was found to have 91% sensitivity and 50% specificity [117], but it has the advantage of being noninvasive, rapid, and less expensive.

   OCTA reveals cases of neovascular AMD which are non-exudative or non-leaky (cases in which CNV does not leak) on FA, and cases of neovascular AMD which are exudative or leaky (cases in which CNV does leak) on FA. Cases of non-leaky neovascular AMD do not show manifestations of exudation on OCT such as subretinal fluid or intraretinal cystoid spaces, and are likely much more common than previously imagined. Most importantly, clinicians do not know how to follow or treat neovascular AMD patients who are non-leaky with no exudates. Treating these patients with anti-vascular endothelial growth factor (VEGF) agents may cause regression of the CNV, but may also lead to ischemia and subsequent macular atrophy [16, 85]. Presence of macular atrophy at baseline and increased number of injections especially monthly versus *pro re nata* treatment have been reported to be associated with progression of macular atrophy. Thus, some clinicians believe that treating these non-leaky neovascular AMD patients may actually be harmful as

---

**Fig. 2** (continued) revealed a predominantly subfoveal choroidal neovascular membrane (CNV) in the early frames. Classic CNV, also known as Type 2 CNV, is located subretinally in the layer between RPE and photoreceptors. (**c**) FA revealed leakage from the CNV in the later frames, as well as staining of drusen. There is intense stippled staining superior to the classic CNV suggestive of an occult component, which are also known as type 1 CNV, located under the RPE, in the layer between Bruch's membrane (BM) and RPE

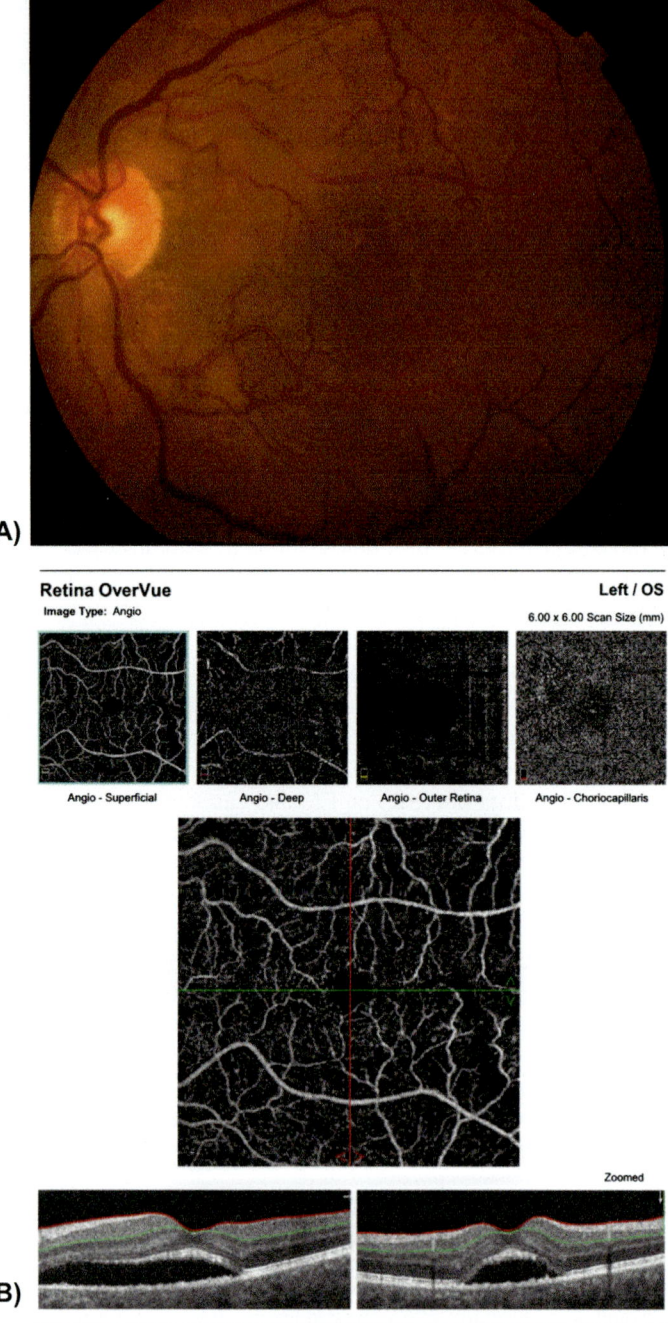

**Fig. 3** Central serous retinopathy. This 55-year-old man complained of blurred central vision, metamorphopsia, and micropsia. (**a**) Color Fundus Photography, Left Eye, revealed mild drusen superior and temporal to his fovea. (**b**) Optical Coherence Tomography (OCT) and OCT

the CNV may be supplying ischemic areas of retina without causing symptomatic exudation.

The pathophysiology behind the development of CNV has been much debated. While some believe that immune response following stress or damage to the RPE leads to the production of pro-angiogenic factors, which induce CNV formation [1], another theory suggests that degeneration within the choroidal vasculature promotes the development of CNV [19, 60, 112, 113]. Shirinifard et al. [109] used computer simulations to study the formation of CNV in great detail and showed that after penetrating BM, the choriocapillaries can expand in three patterns: Type 1, sub RPE (the layer between BM and RPE), Type 2, subretinal (the layer between RPE and photoreceptors), and Type 3, combined (both loci simultaneously). Thus, they hypothesize that the formation of CNV is due to a defect in adhesion of any of the following three layers: (a) RPE–RPE junction, (b) RPE basement membrane to BM, and (c) RPE to photoreceptor outer segment layer [109]. Type 1 CNV is not well demarcated from the surrounding vasculature. On OCTA, type 1 CNV is visible at the mid choroid, choriocapillaries, and RPE slab [35]. In contrast, Type 2 CNV is small and sharply demarcated from the surrounding vasculature and is seen in the outer retina [35]. It appears as a medusa-shaped complex, often with distinctly visible vessel-like structures. It is surrounded by a dark ring and may have one or more feeder vessels extending into the choroidal layer [111]. Type 3 CNV appears as a clear tuft-shaped network with high flow in the outer retinal layer [96].

OCTA is primarily considered a qualitative tool. Thus, while OCTA cannot measure the size of the CNV directly, it can measure active CNV with blood flow and determine response to therapy [117]. The CNV lesions may be associated with areas of alteration in the choriocapillaries, which could be atrophy or flow impairment [64, 84]. While studies have tried to use OCTA as a quantitative tool [64, 128], this approach is limited by irregularity in the shape of red blood cells, the dependence of backscattering on the orientation of the red blood cells, and high anisotropy of the red blood cell scattering phase [128]. Nevertheless, through OCTA, much more will be learned about the development of CNV and progression from non-exudative to exudative forms, which may someday lead to more effective neovascular AMD treatments.

---

**Fig. 3** (continued) Angiography (OCTA). (**b**—Bottom Row): OCT revealed subfoveal fluid on the horizontal and vertical cuts. (**b**—Top Row): OCTA was ordered to assess for choroidal neovascularization (CNV). OCTA uses motion of blood cells as "contrast" to image flow through retinal vessels. Note that the superficial and deep vasculature of the retina can be imaged with high resolution. The outer retina and choriocapillaris are also imaged, in order to detect CNV, which is absent in this image

## 2.2   Vascular Alterations in Diabetic Retinopathy

Vascular alterations are also critical to the development and progression of DR, the leading cause of blindness in middle-aged individuals in the USA [15]. The alterations in vessel structure are classified as non-proliferative and proliferative based on the presence or absence of neovascularization [25, 29]. Even at the stage of non-proliferative diabetic retinopathy, the vascular changes (i.e., presence of microaneurysm, capillary drop outs, beading of the veins and intraretinal macrovascular abnormality) are evident at the clinical level and are graded from mild to severe based on Early Treatment Diabetic Retinopathy Study retinopathy severity scheme.

Although the pathogenesis of DR is multifactorial, it is primarily attributed to the metabolic effects of chronic hyperglycemia, leading to selective loss of pericytes around the capillary network and subsequent retinal ischemia and injury [73]. Peri-cytes function as regulators of blood flow and their absence increases vascular per-meability and capillary loss [82]. Furthermore, the vascular alterations contributing to the pathogenesis of DR include retinal microthrombosis. It has been observed that the retinal vascular network of diabetic patients contains more and larger platelet–fibrin thrombi as compared to matched controls [8]. These microthrombi lead to retinal capillary occlusion and damage. Capillary loss triggers release of growth factors (i.e., IGF-1, platelet-derived growth factor, fibroblast growth factor, and vascular endothelial growth factor), thereby promoting neovascularization. [11, 69]

With the advent of OCTA, vascular changes in the retina are visible at an earlier stage with ease [103, 127]. In addition, wide-field imaging and fluorescein angiography can delineate the changes present in almost the entire retina (Fig. 4). Innovations in imaging techniques have truly changed the field and impacted our basic knowledge of retinal vasculature function and anatomy. Non-invasive, real time imaging modalities, such as Retinal Function Imager, can detect early functional abnormalities and may further impact our clinical knowledge with a potential to improve outcomes in future [114].

## 2.3   Vascular Alterations in Retinal Vessel Occlusions

Thrombosis is the formation of a blood clot, known as a thrombus, within a blood vessel. It prevents blood from flowing normally through the circulatory system. Retinal vein occlusion is a disease that involves the thrombosis, that is, the formation of a blood clot, known as thrombus, within a retinal vein that inhibits blood flow to corresponding retinal ganglion cells (Fig. 5). Retinal ganglion cell death results in subsequent visual field loss. Retinal vein occlusion is the second most common cause of visual loss from retinal vascular disease, trailing only diabetic retinopathy [24]. There are three distinct classifications of retinal vein occlusion separated by the location of the occlusion in the retina. Branch retinal vein occlusion occurs when

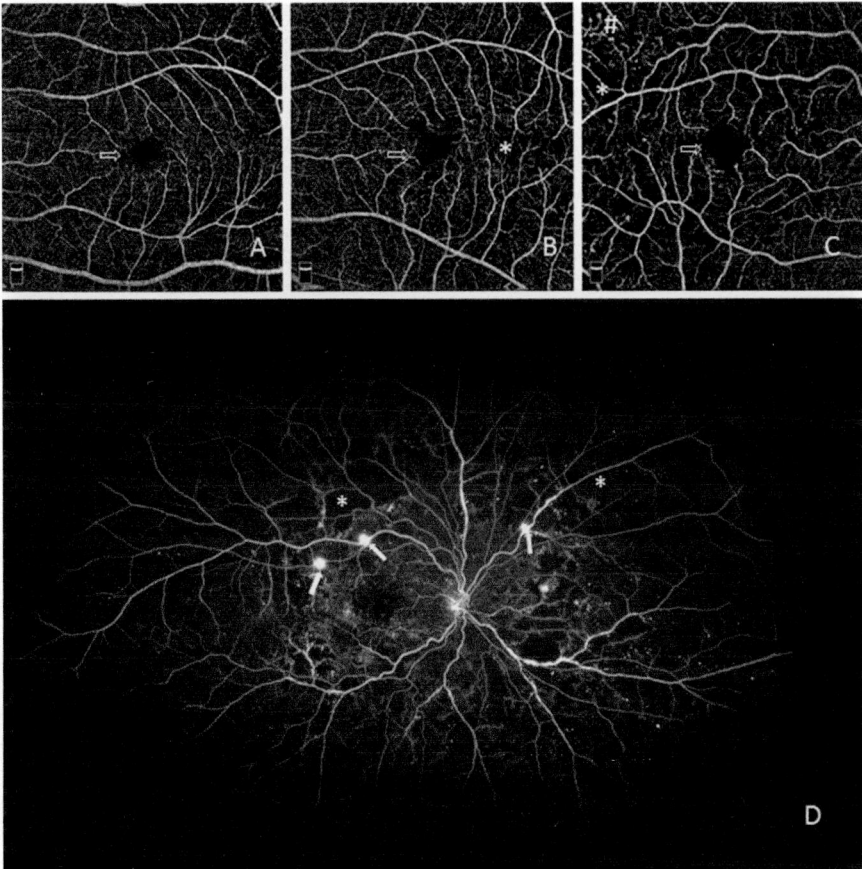

**Fig. 4** New imaging modalities have improved the assessment of retinopathy in diabetic patients. Optical coherence tomography angiography (OCTA) (**A–C**) of normal retina (**a**). Note the uniform capillary structure and normal size of foveal avascular zone (FAZ) (empty arrow). Mild (**b**) and severe (**c**) non-proliferative diabetic retinopathy with evident loss of capillary loss ($_*$), tortuosity of vessels, and microaneurysm formation (#). (**d**) Wide-field fluorescein angiography revealed early neovascularizations (solid arrow) both in the superior arcade and superior nasally close to the large area of nonperfusion and capillary loss ($_*$)

a vein in the distal retinal venous system is occluded. Central retinal vein occlusion occurs when the central retinal vein experiences a thrombosis at the level of the lamina cribrosa of the optic nerve [67]. Lastly, hemiretinal vein occlusion occurs when the occlusion is located in a vein that drains the superior or inferior hemiretina.

The subtypes of retinal vein occlusion all occur due to vascular alterations that involve thrombosis of veins that inhibit blood flow to retinal ganglion cells. However, the specifics of the occlusion differ between the subtypes. Branch retinal vein occlusion occurs through the compression of a branch vein by retinal arterioles at arteriovenous crossing points. Autopsies of eyes of patients with branch retinal

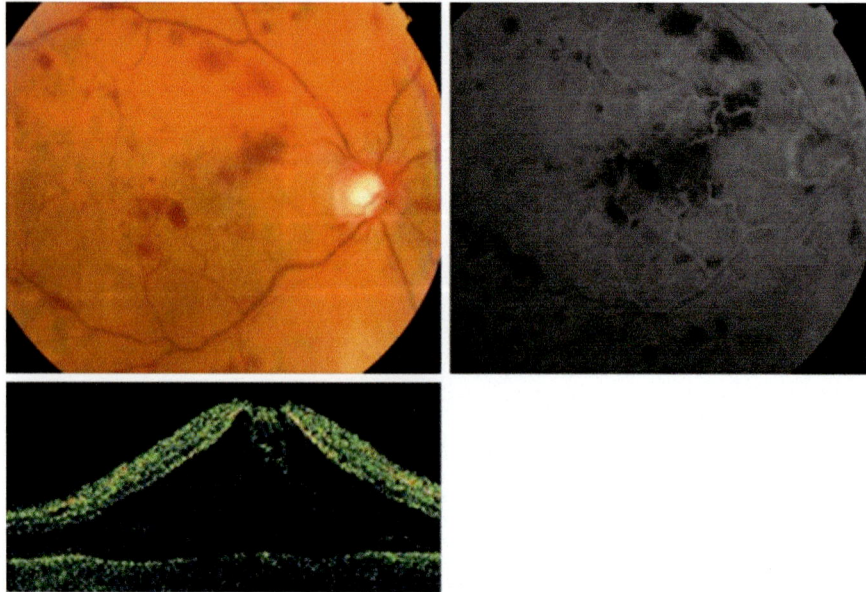

**Fig. 5** Fundus photograph (top left), fluorescein angiogram, and ocular coherence tomography image of a 59-year-old man with recent central retinal vein occlusion for 2 days [80]

vein occlusion revealed that the occlusion occurred at arteriovenous crossings and were associated with sclerotic retinal arterioles [12]. Branch retinal vein occlusions have also been shown to be caused by degenerative changes within venous walls and hypercoagulability [63].

The vascular alterations that occur in central retinal vein occlusions differ from those seen in branch retinal vein occlusions. The central retinal vein and central retinal artery run in parallel to one another in a common tissue sheath prior to crossing through the lamina cribrosa. Prior to piercing this opening, the central retinal vein branches off many collateral vessels that are subject to compression from mechanical stretching of the lamina cribrosa, increases in intraocular pressures, and posterior bowing of the lamina. Impingement of the central retinal vein is a common cause of the vascular alterations seen as a result of occlusion. These hemodynamic alterations are responsible for the stagnant flow and resulting thrombus in the central retinal vein that lead to decreased blood flow, increased blood viscosity, and an altered lumen wall. There are also two categories of perfusion status among central retinal vein occlusions that often determine the extent of the visual field loss that patients experience. A perfused central retinal vein occlusion displays <10 disc areas of retinal capillary nonperfusion as seen from fluorescein angiography. Perfused status generally experiences a lesser degree of intraretinal hemorrhage. On the other hand, nonperfused central retinal vein occlusion demonstrates ≥10 disc areas of retinal capillary nonperfusion on angiography and often experiences a greater degree of intraretinal hemorrhage. Arterial hypoperfusion secondary to outflow obstruction from a central retinal vein occlusion is likely the most common cause of paracentral acute middle maculopathy found in eyes with retinal vascular obstruction [45]. It is hypothesized that a less hemorrhagic, more perfused central retinal vein occlusion is a result of occlusion of the central retinal vein at a location more posterior. A more posterior occlusion allows for normal collateral channels to provide alternative pathways for venous drainage [89]. Recent literature supports this by demonstrating a significant correlation between automatically quantified macular vascular density on OCTA and peripheral nonperfusion on fluorescein angiography, thereby identifying OCTA as a useful tool for identifying high-risk retinal vein occlusion patients who may benefit from further evaluation with fluorescein angiography [107].

The least common subtype of retinal vein occlusion is hemiretinal vein occlusion. This subtype is often regarded as a variant of central retinal vein occlusion due to the fact that it involves hemorrhage of either the superior or inferior branch of the central retinal vein. A hemispheric occlusion blocks one of these branches near the optic disc and results in sudden altitudinal visual field defects. Prognosis depends on the severity of retinal ischemia, and severe ischemia presents a risk for neovascular glaucoma [13].

In addition to retinal vein occlusion, the rare and much less common retinal artery occlusion is another vascular ocular disease. Similar to retinal vein occlusion, there are both artery occlusions of the central and branched types. Central retinal artery occlusion involves infarction of the central retinal artery after branching from the ophthalmic artery and entering the orbit. The central retinal artery divides into multiple branches to supply the inner layers of the retina; therefore, occlusion of these branches is diagnosed as the branched type. Central retinal artery occlusion is considered an ocular emergency as the duration of the occlusion dramatically influences the risk of recovering vision [126].

# 3   Vascular Alterations in Glaucoma

Glaucoma is an optic neuropathy characterized by progressive death of retinal ganglion cells (RGCs) within the optic nerve head (ONH) and irreversible visual loss. Its etiology and treatment are still unclear. The main known modifiable risk factor in glaucoma patients is elevated intraocular pressure (IOP) [118]. Lowering IOP with topical ocular hypotensive medications might be effective in delaying or preventing the onset of glaucoma [47, 68]; however, a high percentage of individuals with elevated IOP do not exhibit reproducible optic disc deterioration attributable to glaucoma [59, 68], and many glaucoma patients continue to experience disease progression despite lowering IOP to target levels. Some glaucoma patients have no history of elevated IOP—a condition called normal tension glaucoma (NTG) [108].

Several studies suggest correlations between impaired ocular blood flow (OBF) and glaucoma [36, 54]. The arteriovenous pressure difference driving blood flow through the intraocular vasculature is known as ocular perfusion pressure (OPP).

The intraocular venous pressure is very close to IOP [23] and thus OPP is usually estimated as the difference between the arterial blood pressure (BP) and IOP. BP is positively correlated with IOP, mainly in elderly populations [119]. It is unclear whether the BP level is a risk factor for the development or progression of open-angle glaucoma (OAG) in an individual patient. It has been hypothesized that low BP and sustained BP drops during sleep are risk factors for patients with abnormal OBF regulation [83]. Some studies found that lower OPP, especially diastolic OPP, is a risk factor for primary OAG [9, 76, 97, 119]. On the other hand, in the Beijing Eye Study, OPP was not significantly associated with OAG [124]. Other studies found that circadian OPP fluctuations were found to be related to glaucoma severity in NTG [18].

OCTA has shown to provide a quick and reproducible way to qualitatively and quantitatively display areas of decreased or interrupted perfusion in the eye [65, 79] (Fig. 6). Recently, OCTA has been used to find that decreased ocular blood vessel density was associated with the severity of glaucomatous visual field loss, independent of structural loss [123, 125]. It has also been shown that glaucoma affects perfusion and vessel density in the superficial vascular complex of the macula more than the deeper plexuses [116].

Evidence has been found of an increased resistance to BF (Fig. 7) and decreased BF velocities in glaucoma patients when comparing with healthy controls or subjects with ocular hypertension [14, 66]. Results in literature are controversial on whether disturbed OBF affects only NTG patients or if all glaucoma patients, including those affected by primary OAG, exhibit it.

Inconsistent findings may be due to the complex relationship between OPP, IOP, BP, and optic disc structure. The translamina cribrosa pressure difference TLpD, defined as the difference between IOP and cerebrospinal fluid pressure (CSFp), may affect this relationship as well. As measured by OCTA, ocular blood vessel density was significantly lower in glaucoma eyes with focal lamina cribrosa defects than in glaucoma eyes without them [115]. Since CSFp and BP are correlated, lower BP leads to lower CSFp to allow cerebral perfusion, and higher BP indicts higher CSFp to prevent cerebral hemorrhage. If BP is medically reduced, then CSFp is also reduced. However, if IOP is relatively constant, this leads to an elevated TLpD that may cause optic disc damage. Ren et al. [100] found that abnormally low CSFp is associated with low arterial BP in many NTG patients. Decreased neuroretinal rim area and ophthalmic artery blood flow parameters were found in NTG patients having lower CSFp compared with NTG patients with higher CSFp [110].

Thus, evidence suggests that OBF instabilities are likely to be caused by both ocular factors, like IOP, and systemic factors, like BP. This could also be explained by theoretical considerations. OBF is proportional to the ratio between OPP and the vascular resistance R. In view of Poiseuille's law, $R$ is inversely proportional to the fourth power of the vessel radius $r$, and directly proportional to its length $L$ and to the viscosity of the fluid $\mu$

$$R = 8\,\mu\,L/\left(\pi\,r^4\right).$$

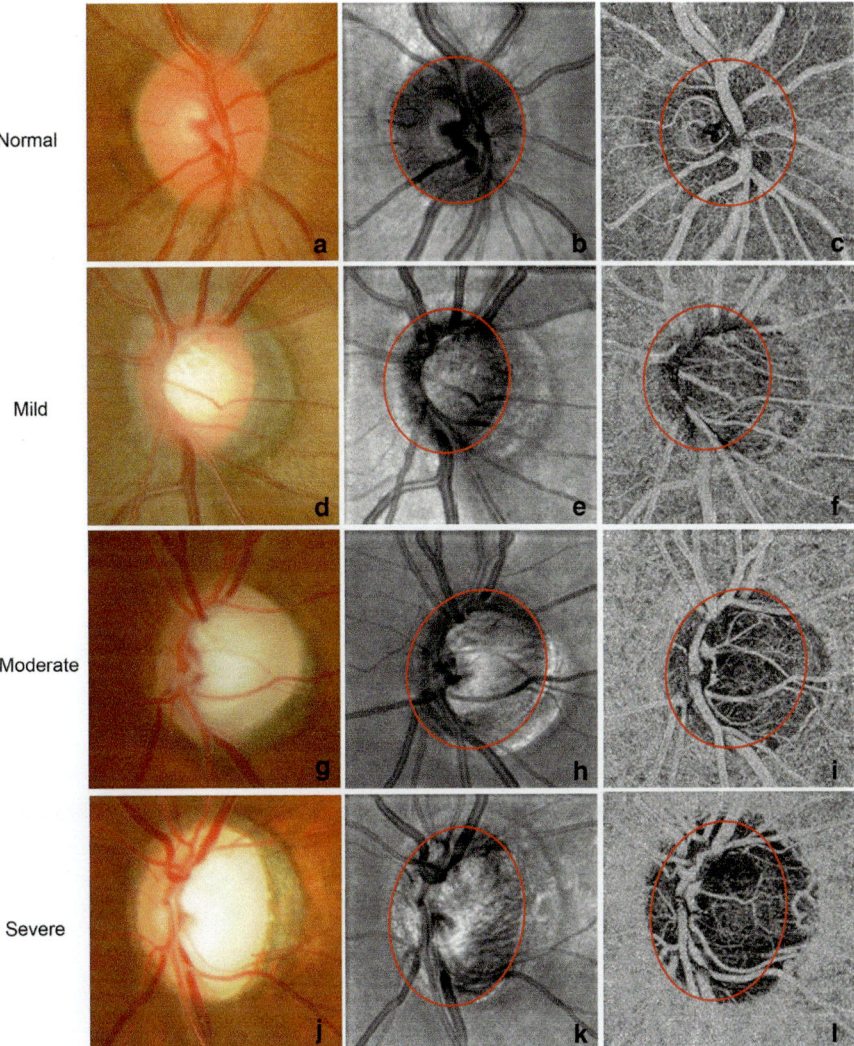

**Fig. 6** Disc photographs (**a, d, g, j**), optic coherence tomography (OCT) reflectance (**b, e, h, k**), and OCT angiograms (**c, f, i, l**) in the eyes of normal subjects (**a–c**) and OAG patients (**d–l**). Disc margins are marked by the red elliptical outlines (second and third columns). A dense microvascular network was visible on the OCTA of the normal disc (**c**). This network was greatly attenuated from mild to severe in the glaucomatous disc (**f, i, l**) [124]

In particular, vessel radius, and vascular tone in general, is controlled by a complex combination of myogenic, metabolic, hormonal, and neurogenic mechanisms [94]. Any ocular or systemic condition influencing these mechanisms, like vascular dysregulation, arteriosclerosis, systemic hypertension, systemic hypotension, rheological factors, alterations in the autonomic nervous system, the immune and

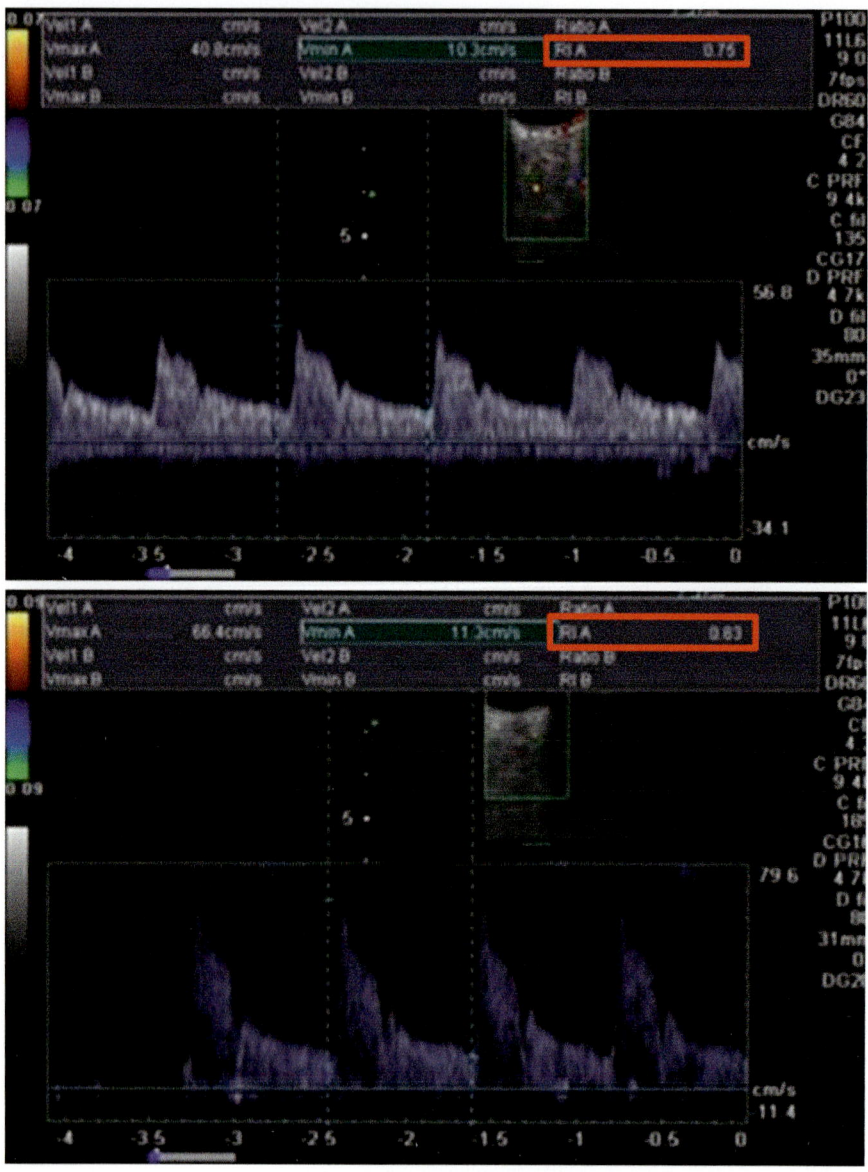

**Fig. 7** Color Doppler imaging of the ophthalmic artery (OA) taken with a 7.5-MHz linear probe (Toshiba Aplio 80) from two patients with primary open-angle glaucoma. The resistivity index (0.75) of the stable patient (top image, red box) is much lower than that observed (0.83) in the patient that showed visual field progression (bottom image, red box) [81]

endocrine systems may impair OBF, thereby rendering the ONH more susceptible to damage by compromising oxygen delivery (leading to hypoxia) in potentially dangerous situations of reduced BP, increased IOP, and/or increased local metabolic demands [83]. However, the mechanisms through which the damage occurs are still not completely understood [98, 122]. Evidence suggests that ischemia and related hypoxia might influence the astrocytes in the ONH [58] and/or the mitochondria of RGC axons [90], and that the ultimate death of RGCs can occur via apoptosis [90, 98] and/or autophagy [78].

As outlined above, among the ocular factors, IOP and IOP fluctuations may play a huge role in OBF instabilities. However, what is unclear from IOP elevation is whether ganglion cell dysfunction is impaired due to direct mechanical compression or whether the defect is secondary to the vascular insufficiency caused by the reduced OPP. At least in a subset of patients, more likely in those with lower IOP, OBF instabilities are closely related to vascular dysregulation [32, 34, 44]. Vascular dysregulation may be induced, for example, by the primary vasospastic syndrome [37] or by endothelial dysfunction due to oxidative stress leading to a decrease in the synthesis of NO and an excess of endothelin production [101].

There appears to be some evidence for a pathogenic role of an abnormal hemorheology, especially in NTG [52, 72]. Blood or plasma viscosity, established parameters for chronic vascular disease, are elevated and red blood cell function and deformability seem to be decreased.

Certain IOP-lowering medications, formulated as eye drops or as systemic drugs, may impact OBF and its regulation. However, it is not easy to discriminate between the effects of the lowered IOP and those of the improved circulation. In 2006, the interpretation of initial data from the Thessaloniki Eye Study suggested that both antihypertensive treatment and low DBP were independently associated with optic disc structural changes [120]. However, the subgroup analyses led to more accurate findings, namely, that only patients on antihypertensive medications with a low DBP (<90 mmHg) had a significantly associated change in optic disc structure [121]. The authors hypothesized that OPP status (meaning OPP with or without antihypertensive treatment) may be more relevant to glaucoma pathogenesis than OPP alone.

Many studies have investigated the impact of medications altering OBF on the development and progression of glaucoma. Some utilized sample sizes with fewer than 50 subjects, or focused on short-term effects on OBF [10, 33, 71, 74, 86, 95, 106]. Moreover, experiments were designed using a large variety of technologies, rather than a unique standardized protocol. All these limitations demand further investigation before implementing blood flow modulation in glaucoma management.

Age is another independent risk factor for glaucoma [28], and increasing age is also associated with vascular changes [30]. With age, there is a decrease in both capillary diameter (due to the thickening of the basement membrane) and density [99], thereby causing an increase in blood velocity and a decrease in the total volume available to flow. Moreover, the increasing incidence of atherosclerosis with age reduces arterial distensibility in all vascular beds, including the ocular vasculature [31]. Evidence has been found of a decreased compliance of the retrobulbar arteries [53], which, in turn, results in decrease in retinal and neuroretinal rim microcirculation. Decreases in endothelial nitric oxide synthase (ENOS) activity with age can lead to reduced nitric oxide (NO) availability, possibly compromising the endothelium-dependent balance between vasodilation and vasoconstriction. All these changes can impair oxygen supply, reduce exchange of nutrients, diminish removal of metabolic waste, and weaken response to local changes in the metabolic demand, possibly rendering the ONH more susceptible to ischemic damage. Other systemic factors might be possibly involved in ocular vascular dysregulation, like diabetes and cardiovascular diseases. Their association with OAG is still under investigation.

In conclusion, it is now widely accepted by the scientific community that ocular vascular insults in glaucoma pathophysiology are important to consider in certain individuals. However, whether improved perfusion has any beneficial long-term effect on the visual field, and what treatments may best accomplish this physiological effect, remains unclear. More specifically, the questions and future research directions that need to be answered include the following:

- Which technology is the most useful to detect changes in blood flow?
- What affects the nutrient diffusion within the optic nerve head?
- What is the relationship between biomechanics, tissue remodeling, and blood flow?
- Which blood vessels should be targeted for treatment?
- What is the best medication to modify blood flow?
- Do different ocular vessels respond differently to a given medication? And if so, what is the physiological mechanism?

Longitudinal, longer term studies with larger patient cohorts utilizing standardized measurement methods are needed to confirm the relationship between OBF and structural and functional changes in glaucoma patients as well as the causative relationship between them.

## 4   Vascular Alterations in Non-Arteritic Ischemic Optic Neuropathy

Ischemic optic neuropathy is the most common form of persistent monocular vision loss in individuals over 50 years of age. Furthermore, non-arteritic ischemic optic neuropathy (NAION) is the most common clinical presentation of acute ischemic damage to the optic nerve (Fig. 8). Arteritic anterior ischemic optic neuropathy (AAION) differs from NAION in that it primarily occurs in patients over 70 years old and is due to giant cell arteritis [5]. Hypertension, diabetes, hyperlipidemia, and end-stage renal disease have been associated with NAION. Additionally, small optic disc, obstructive sleep apnea, nocturnal hypotension, and phosphodiesterase inhibitors have been identified as risk factors [3]. Currently, there is no generally accepted treatment or secondary prevention of NAION.

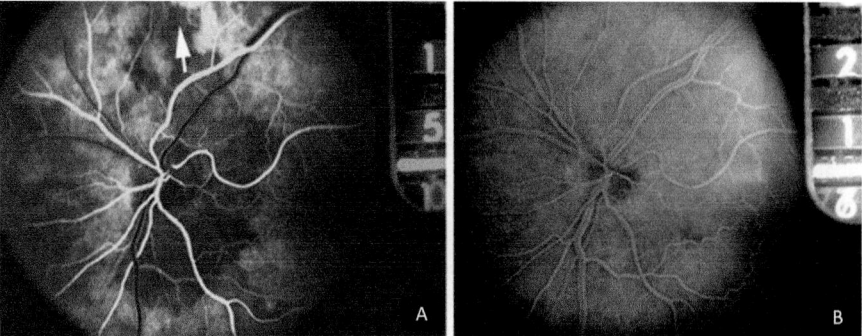

**Fig. 8** Fluorescein angiograms of an eye 12 days after development of NAION. (**a**) Angiogram 15 s after the injection of fluorescein shows no filling of the temporal, superior, and inferior peripapillary choroid (main source of blood supply to the optic nerve head), superior choroidal watershed zone (arrow), and the optic disc, with normal filling of both the medial and lateral posterior ciliary arteries. (**b**) Angiogram 6 s after (**a**) shows complete filling of the peripapillary choroid, the watershed zone, and upper part of the disc [55]

The pathogenesis of NAION is believed to be attributed to flow impairment in the prelaminar area of optic nerve. The exact mechanism of this process, however, is not fully understood. The occlusion is believed to more specifically occur to the branches of the peripapillary choroidal arterial system [102]. Several studies have investigated the vascular alterations of the eye in patients with NAION. The blood flow velocities of the nasal posterior ciliary artery and the central retinal artery have shown to be considerably reduced in NAION patients in comparison to age-matched controls [70]. Bertram et al. revealed increased arteriovenous passage time in the retina with fluorescein angiography [7]. Arnold and Hepler discovered significantly delayed filling of optic nerve head capillaries [4]. Additionally, patients with NAION displayed reduced blood flow velocities in the capillaries of the optic nerve head as measured by laser Doppler velocimetry [22]. The vascular alterations reported in these studies are responsible for the pathogenesis and visual defects seen in patients with NAION.

The lack of sufficient blood flow to the optic nerve results in the abrupt and painless onset of visual loss. The field defect is generally altitudinal and involves the area of central fixation, which accounts for the severity of visual acuity loss. The visual field defect is almost always unilateral, but presents with a 15% risk of future development in the other eye.

Recent literature utilizing OCTA has revealed shifts in the superficial capillary network of the optic nerve head that correlate strongly with the functional vision loss in both the acute and non-acute phases of NAION [43]. Furthermore, OCTA has been demonstrated that rarefaction of the peripapillary and macular retinal capillary layers is correlated with visual field and visual acuity loss [6]. Understanding the vascular alterations that arise and are responsible for the loss of vision seen in

NAION patients may help to prevent the onset and progression of the disease, and could potentially lead to the development of future treatments.

Although the vast majority of NAION cases are diagnosed as anterior, or affecting the optic disc, there are some cases that are posterior. Posterior ischemic optic neuropathy (PION) is distinguished clinically from the anterior form by a normal-appearing optic nerve head, which makes it difficult to diagnose. The posterior segment of the optic disc is supplied by a pial capillary plexus that is derived from branches of the ophthalmic artery. Only a small number of capillaries penetrate the nerve and extend to the central portion of the optic nerve, which makes the posterior segment much less vascularized than the anterior segment. Clinical cases of PION reveal that the vascular alterations likely involve intraorbital infarction of these capillaries [57].

## 5   Conclusions

The research discussed in this chapter shows extensive evidence of correlations between ocular diseases and vascular and hemodynamic alterations. The cause-to-effect relationships giving rise to these correlations, however, remain elusive, thereby posing a formidable challenge to our current capabilities to the early diagnosis of ocular diseases and their effective treatments.

Significant advances in ocular imaging techniques have made available new, valuable information regarding structural and functional properties of the eye. However, the clinical interpretation of such information is arduous. Findings are not always consistent across studies, with not all patients suffering from the same disease actually exhibiting similar vascular and hemodynamic alterations. A major question that remains to be addressed is whether the hemodynamic alterations observed in a given patient are primary or secondary to the disease. Should they be primary, they could be regarded as pathogenic factors causing the disease and could therefore be utilized as targets of therapeutic strategies. Conversely, should hemodynamic alterations be secondary to the disease, they could be regarded as mere consequences of the disease and would not serve as therapeutic targets. Identifying patient subgroups for which hemodynamic alterations are primary to the disease process would provide new hope for the millions of people currently progressing irreversibly to blindness.

## References

1. Aiello LP, Northrup JM, Keyt BA, et al. Hypoxic regulation of vascular endothelial growth factor in retinal cells. Arch Ophthalmol. 1995;113:1538–154.
2. Albert DM, Miller JW, Azar DT, Blodi BA. Albert & Jakobiec's Principles & Practice of Ophthalmology. 3rd ed. Philadelphia, PA: Saunders Elsevier; 2008.

3. Aminoff MJ, Greenberg DA, Simon RP. Neuro-Ophthalmic Disorders. In: Clinical Neurology, 9e New York, NY: McGraw-Hill; 2015.
4. Arnold AC, Hepler RS. Fluorescein angiography in acute nonarteritic anterior ischemic optic neuropathy. Am J Ophthalmol. 1994 Feb 15; 117(2):222-30.
5. Atkins et al. Treatment of Nonarteritic Anterior Ischemic Optic Neuropathy. Surv Ophthalmol. 2010 Jan-Feb; 55(1): 47–63.
6. Augstburger E, Zéboulon P, Keilani C, et al. Quantitative analysis of optical coherence tomographic angiography (OCT-A) in patients with non-arteritic anterior ischemic optic neuropathy (NAION) corresponds to visual function. PLoS One. 2018 Jun 28;13(6):e0199793.
7. Bertram B, Hoberg A, Wolf S. et al Videofluoresceinangiographic findings in acute anterior ischemic optic neuropathy. Klin Mbl Augenheilk 1991199419–423.
8. Boeri D, Maiello M, Lorenzi M. Increased prevalence of microthromboses in retinal capillaries of diabetic individuals. Diabetes. 2001;50(6):1432.
9. Bonomi L, Marchini G, Marraffa M, et al. Vascular risk factors for primary open angle glaucoma: the Egna-Neumarkt Study. Ophthalmology. 2000 Jul;107(7):1287-93.
10. Bose S, Piltz JR, Breton ME. Nimodipine, a centrally active calcium antagonist, exerts a beneficial effect on contrast sensitivity in patients with normal-tension glaucoma and in control subjects. Ophthalmology. 1995 Aug;102(8):1236-41.
11. Boulton M, Foreman D, Williams G, McLeod D. VEGF localisation in diabetic retinopathy. Br J Ophthalmol. 1998;82(5):561.
12. Bowers DK, Finkelstein D, Wolff SM, Green WR. Branch retinal vein occlusion. A clinicopathologic case report. Retina. 1987;7(4):252.
13. Bowling B. Kanski's Clinical Ophthalmology. Eighth Edition. Edinburgh: Elsevier; 2016. 13: Retinal vascular disease; 519-577.
14. Butt Z, O'Brien C, McKillop G, et al. Color Doppler imaging in untreated high- and normal-pressure open-angle glaucoma. Invest. Ophthalmol. Vis. Sci. 1997;38(3):690-696.
15. Centers for Disease Control and Prevention. 2003 National Diabetes Fact Sheet. http://www.cdc.gov/diabetes/pubs/estimates.htm#complications (Accessed on March 28, 2008).
16. Channa R, Sophie R, Bagheri S, et al. Regression of choroidal neovascularization results in macular atrophy in anti-vascular endothelial growth factor-treated eyes. Am J Ophthalmol. 2015;159:9-19.
17. Chiaravalli G. A virtual laboratory for retinal physiology: a theoretical study of retinal oxygenation in healthy and disease. Master's thesis, Politecnico di Milano (Italy). Master in Engineering Physics, final examination held on 12/20/2018. Main Advisor: R. Sacco (Mathematics, Politecnico di Milano). Co-advisor: G. Guidoboni.
18. Choi j, Kim KH, Jeong J et al. Circadian Fluctuation of Mean Ocular Perfusion Pressure Is a Consistent Risk Factor for Normal-Tension Glaucoma. Invest. Ophthalmol. Vis. Sci. 2007;48(1):104-111.
19. Chong NH, Keonin J, Luthert PJ, et al. Decreased thickness and integrity of the macular elastic layer of Bruch's membrane correspond to the distribution of lesions associated with age-related macular degeneration. Am J Pathol. 2005;166:241–251.
20. Ciulla TA, Harris A, Kagemann L, et al. Choroidal perfusion perturbations in non-neovascular age related macular degeneration. Br J Ophthalmol. 2002 Feb;86(2):209-13.
21. Ciulla TA, Harris A, Martin BJ. Ocular Perfusion and age-related macular degeneration. Acta Ophthalmol Scand. 2001 Apr;79(2):108-15.
22. Collignon-Robe NJ, Feke GT, Rizzo JF 3rd. Optic nerve head circulation in nonarteritic anterior ischemic optic neuropathy and optic neuritis. Ophthalmology. 2004 Sep; 111(9):1663-72.
23. Costa V, Harris A, Anderson D, et al. Ocular perfusion pressure in glaucoma. Acta Ophthalmol. 2014;92(4):252-66.
24. Cugati S, Wang JJ, Rochtchina E, Mitchell P. Ten-year incidence of retinal vein occlusion in an older population: the Blue Mountains Eye Study. Arch Ophthalmol. 2006;124(5):726.
25. Diabetic Retinopathy Study Research Group. Preliminary report on effects of photocoagulation therapy Am J Ophthalmol, 81 (1976), pp. 383-396.

26. Do DV, Gower EW, Cassard SD, et al. Detection of new-onset choroidal neovascularization using optical coherence tomography: the AMD DOC study. Ophthalmology 2012;119:771-778.
27. Do DV. Detection of new-onset choroidal neovascularization. Curr Opin Ophthalmol. 2013;24:224-227.
28. Doucette LP, Rasnitsyn A, Seifi M, Walter MA. The interactions of genes, age, and environment in glaucoma pathogenesis. Surv Ophthalmol. 2015;60(4):310–26.
29. Early Treatment Diabetic Retinopathy Study Research Group. Early treatment diabetic retinopathy study design and baseline patient characteristics. ETDRS report number 7. Ophthalmology, 98 (1991), pp. 741-756.
30. Ehrlich R, Kheradiya NS, Winston DM, et al. Age-related ocular vascular changes. Graefes Arch Clin Exp Ophthalmol. 2009;247(5):583–91.
31. Embleton SJ, Hosking SL, Roff Hilton EJ, Cunliffe IA. Effect of senescence on ocular blood flow in the retina, neuroretinal rim and lamina cribrosa, using scanning laser Doppler flowmetry. Eye (Lond). 2002 Mar;16(2):156-62.
32. Emre M, Orgül S, Gugleta K, Flammer J. Ocular blood flow alteration in glaucoma is related to systemic vascular dysregulation. Br J Ophthalmol. 2004 May; 88(5): 662–666.
33. Engin KN, Engin G, Kucuksahin H, et al. Clinical evaluation of the neuroprotective effect of alpha-tocopherol against glaucomatous damage. Eur J Ophthalmol. 2007 Jul-Aug;17(4):528-33.
34. Evans DW, Harris A, Garrett M, et al. Glaucoma patients demonstrate faulty autoregulation of ocular blood flow during posture change. Br J Ophthalmol. 1999;83(7):809-13.
35. Farecki ML, Gutfleisch M, Faatz H, et al. Characteristics of type 1 and 2 CNV in exudative AMD in OCT-Angiography. Graefes Arch Clin Exp Ophthalmol. 2017;255:913-921.
36. Flammer J, Orgul S, Costa V, et al. The impact of ocular blood flow in glaucoma. Prog Retin Eye Res. 2002;21(4):359-93.
37. Flammer J, Pache M, Resink T. Vasospasm, its role in the pathogenesis of diseases with particular reference to the eye. Prog Retin Eye Res. 2001 May;20(3):319-49.
38. Fraser CE, D'Amico DJ. Diabetic retinopathy: Classification and clinical features. Mulder JE, ed. UpToDate. Waltham, MA: UpToDate Inc.
39. Friedman DS, O'Colmain BJ, Munoz B, et al. Prevalence of age-related macular degeneration in the United States. Arch Ophthalmol. 2004;122: 564-572.
40. Friedman E, Ivry M, Ebert E, et al. Increased Scleral Rigidity and Age-related Macular Degeneration. Ophthalmology. 1989 Jan;96(1):104-8.
41. Friedman E, Krupsky S, Lane AM, et al. Ocular blood flow velocity in age-related macular degeneration. Ophthalmology. 1995 Apr;102(4):640-6.
42. Friedman E. A Hemodynamic Model of the Pathogenesis of Age-related Macular Degeneration. Am J Ophthalmol. 1997 Nov;124(5):677-82.
43. Gaier ED, Wang M, Gilbert AL, et al. Quantitative analysis of optical coherence tomographic angiography (OCT-A) in patients with non-arteritic anterior ischemic optic neuropathy (NAION) corresponds to visual function. PLoS One. 2018 Jun 28;13(6):e0199793.
44. Gasser P, Flammer J. Blood-cell velocity in the nailfold capillaries of patients with normal-tension and high-tension glaucoma. Am J Ophthalmol. 1991 May 15;111(5):585-8.
45. Ghasemi Falavarjani K, Phasukkijwatana N, Freund KB, et al. En Face Optical Coherence Tomography Analysis to Assess the Spectrum of Perivenular Ischemia and Paracentral Acute Middle Maculopathy in Retinal Vein Occlusion. Am J Ophthalmol. 2017 May;177:131-138.
46. González-López A, Ortega M, Penedo MG, Charlón P. Automatic Vessel Shade-Robust Segmentation of Retinal Layers in OCT Images. Stud Health Technol Inform. 2014;207:47-54.
47. Gordon MO, Beiser JA, Brandt JD, et al. The Ocular Hypertension Treatment Study: baseline factors that predict the onset of primary open-angle glaucoma. Arch Ophthalmol. 2002 Jun;120(6):714-20; discussion 829-30.
48. Grunwald JE, Hariprasad SM, DuPont J, et al. Foveolar choroidal blood flow in age-related macular degeneration. Invest Ophthalmol Vis Sci. 1998 Feb;39(2):385-90.

49. Grunwald JE, Metelitsina TI, Dupont JC, et al. Reduced foveolar choroidal blood flow in eyes with increasing AMD severity. Invest Ophthalmol Vis Sci. 2005 Mar;46(3):1033-8.
50. Haefliger IO, Meyer P, Flammer J, Lüscher TF. The vascular endothelium as a regulator of the ocular circulation: a new concept in ophthalmology?. Surv Ophthalmol. 1994;39:123–132.
51. Haefliger IO, Zschauer A, Anderson DR. Relaxation of retinal pericyte contractile tone through the nitric oxide-cyclic guanosine monophosphate pathway. Invest Ophthalmol Vis Sci. 1995;35:991–997.
52. Hamard P, Hamard H, Dufaux J, Quesnot S. Optic nerve head blood flow using a laser Doppler velocimeter and haemorheology in primary open angle glaucoma and normal pressure glaucoma. Br J Ophthalmol. 1994 Jun;78(6):449-53.
53. Harris A, Harris M, Biller J, et al. Aging affects the retrobulbar circulation differently in women and men. Arch Ophthalmol. 2000 Aug;118(8):1076-80.
54. Harris A, Rechtman E, Siesky B, et al. The role of optic nerve blood flow in the pathogenesis of glaucoma. Ophthalmol Clin North Am. 2005;18(3):345-53, v.
55. Hayreh SS. Non-arteritic anterior ischemic optic neuropathy versus cerebral ischemic stroke. Graefes Arch Clin Exp Ophthalmol (2012) 250: 1255-60.
56. Hayreh SS. Posterior ciliary artery circulation in health and disease: the Weisenfeld lecture. Invest Ophthalmol Vis Sci. 2004 Mar;45(3):749-57; 748.
57. Hayreh SS. Posterior ischaemic optic neuropathy: clinical features, pathogenesis, and management. Eye (Lond). 2004;18(11):1188.
58. Hernandez MR, Miao H, Lukas T. Astrocytes in glaucomatous optic neuropathy. Prog Brain Res. 2008;173:353-73.
59. Hollows FC, Graham PA. Intra-ocular pressure, glaucoma, and glaucoma suspects in a defined population. Br J Ophthalmol. 1966;50(10):570-86.
60. Hussain AA, Starita C, Marshall J. Transport characteristics of aging human Bruch's membrane: implications for age-related macular degeneration. In: Ioseliani O, editor. Focus on Macular Degeneration Research (AMD) Nova Biomedical Books; 2004. pp. 59–113.
61. Hyman L. Epidemiology of eye disease in the elderly. Eye (Lond). 1987;1 ( Pt 2):330.
62. Iroku-Malize T, Kirsch S. Eye Conditions in Older Adults: Age-Related Macular Degeneration. FP Essent. 2016;445:24-8.
63. Jaulim A, Ahmed B, Khanam T, Chatziralli IP. Branch retinal vein occlusion: epidemiology, pathogenesis, risk factors, clinical features, diagnosis, and complications. An update of the literature. Retina. 2013 May;33(5):901-10.
64. Jia Y, Bailey ST, Wilson DJ, et al. Quantitative optical coherence tomography angiography of choroidal neovascularization in age-related macular degeneration. Ophthalmology. 2014;121:1435-44.
65. Jia Y, Wei E, Wang X, et al. Optical coherence tomography angiography of optic disc perfusion in glaucoma. Ophthalmology. 2014;121(7):1322–32.
66. Kaiser HJ, Schoetzau A, Stumpfig D, Flammer J. Blood-flow velocities of the extraocular vessels in patients with high-tension and normal-tension primary open-angle glaucoma. Am J Ophthalmol 1997; 123: 320-327.
67. Karia N. Retinal vein occlusion: pathophysiology and treatment options. Clin Ophthalmol. 2010; 4: 809–816.
68. Kass MA, Heuer DK, Higginbotham EJ, et al. The Ocular Hypertension Treatment Study: a randomized trial determines that topical ocular hypotensive medication delays or prevents the onset of primary open-angle glaucoma. Arch Ophthalmol. 2002 Jun;120(6):701-13; discussion 829-30.
69. Katsura Y, Okano T, Noritake M, et al. Hepatocyte growth factor in vitreous fluid of patients with proliferative diabetic retinopathy and other retinal disorders. Diabetes Care. 1998;21(10):1759.
70. Kaup M, Plange N, Arend KO, Remky A. Retrobulbar haemodynamics in non-arteritic anterior ischaemic optic neuropathy. Br J Ophthalmol. 2006 Nov; 90(11): 1350–1353.
71. Kitazawa Y, Shirai H, Go FJ. The effect of Ca2+-antagonist on visual field in low-tension glaucoma. Graefes Arch Clin Exp Ophthalmol 1989; 227: 408-412.

72. Klaver JH, Greve EL, Goslinga H, et al. Blood and plasma viscosity measurements in patients with glaucoma. Br J Ophthalmol. 1985 Oct;69(10):765-70.
73. Kohner EM, Patel V, Rassam SM. Role of blood flow and impaired autoregulation in the pathogenesis of diabetic retinopathy. Diabetes. 1995;44(6):603.
74. Koseki N, Araie M, Yamagami J, et al. Effects of oral brovincamine on visual field damage in patients with normal-tension glaucoma with low-normal intraocular pressure. J Glaucoma. 1999 Apr;8(2):117-23.
75. Kotsolis AI, Killian FA, Ladas ID, Yannuzzi LA. Fluorescein angiography and optical coherence tomography concordance for choroidal neovascularization in multifocal choroiditis. Br J Ophthalmol. 2010;94:1506-1508.
76. Leske MC, Wu SY, Nemesure B, Hennis A. Incident open-angle glaucoma and blood pressure. Arch Ophthalmol. 2002 Jul;120(7):954-9.
77. Lim LS, Mitchell P, Seddon JM, et al. Age-related macular degeneration. Lancet 2012;379:1728-1738.
78. Lin WJ, Kuang HY. Oxidative stress induces autophagy in response to multiple noxious stimuli in retinal ganglion cells. Autophagy. 2014;10(10):1692-701.
79. Liu L, Jia Y, Takusagawa HL, et al. Optical Coherence Tomography Angiography of the Peripapillary Retina in Glaucoma. JAMA ophthalmology. 2015;133(9):1045-52.
80. Martinet V, Guigui B, Glacet-Bernard A, et al. Macular edema in central retinal vein occlusion: correlation between optical coherence tomography, angiography and visual acuity. Int Ophthalmol. 2012 Aug;32(4):369-77.
81. Martinez A. Retrobulbar Ocular Blood Flow Evaluation in Open-Angle Glaucoma. In: Ferreras A. (eds) Glaucoma Imaging. Springer, Cham, 2016.
82. Mizutani M, Kern TS, Lorenzi M. Accelerated death of retinal microvascular cells in human and experimental diabetic retinopathy. J Clin Invest 1996;97:2883-2890.
83. Moore D, Harris A, WuDunn D, et al. Dysfunctional regulation of ocular blood flow: a risk factor for glaucoma? Clin Ophthalmol. 2008;2(4):849–61.
84. Moult E, Choi W, Waheed NK, et al. Ultrahigh-speed swept-source OCT angiography in exudative AMD. Ophthalmic Surg Lasers Imaging Retina. 2014;45:496-505.
85. Munk MR, Ceklic L, Ebneter A, et al. Macular atrophy in patients with long-term anti-VEGF treatment for neovascular age-related macular degeneration. Acta Ophthalmol. 2016;94:e757-e764.
86. Netland PA, Chaturvedi N, Dreyer EB. Calcium channel blockers in the management of low-tension and open-angle glaucoma. Am J Ophthalmol. 1993 May 15;115(5):608-13.
87. Novais EA, Adhi M, Moult EM, et al. Choroidal Neovascularization Analyzed on Ultrahigh-Speed Swept-Source Optical Coherence Tomography Angiography Compared to Spectral-Domain Optical Coherence Tomography Angiography. Am J Ophthalmol. 2016;164:80-8.
88. Nowak JZ. AMD–the retinal disease with an unprecised etiopathogenesis: in search of effective therapeutics. Acta Pol Pharm. 2014;71:900-16.
89. Oellers P, Hahn P, Fekrat S. Ryan's Retina. Sixth Edition. Edinburgh: Elsevier; 2018. 57: Central Retinal Vein Occlusion; 1166-1179.
90. Osborne NN. Mitochondria: their role in ganglion cell death and survival in primary open angle glaucoma. Exp Eye Res. 2010;90(6):750-7.
91. Owen CG, Jarrar Z, Wormald R, et al. The estimated prevalence and incidence of late stage age related macular degeneration in the UK. Br J Ophthalmol. 2012;96:752-6.
92. Pascolini D, Mariotti SP, Pokharel GP, et al. 2002 global update of available data on visual impairment: a compilation of population-based prevalence studies. Ophthalmic Epidemiol. 2004;11:67-115.
93. Pournaras CJ, Logean E, Riva CE, et al. Regulation of subfoveal choroidal blood flow in age-related macular degeneration. Invest Ophthalmol Vis Sci. 2006 Apr;47(4):1581-6.
94. Prada D, Harris A, Guidoboni G, et al. Autoregulation and neurovascular coupling in the optic nerve head. Surv Ophthalmol. 2016 Mar-Apr;61(2):164-86.
95. Quaranta L, Bettelli S, Uva MG, et al. Effect of Ginkgo biloba extract on preexisting visual field damage in normal tension glaucoma. Ophthalmology. 2003 Feb;110(2):359-62; discussion 362-4.

96. Querques G, Miere A, Souied EH. Optical Coherence Tomography Angiography Features of Type 3 Neovascularization in Age-Related Macular Degeneration. Dev Ophthalmol. 2016;56:57-61.

97. Quigley HA, West SK, Rodriguez J, et al. The prevalence of glaucoma in a population-based study of Hispanic subjects: Proyecto VER. Arch Ophthalmol. 2001 Dec;119(12):1819-26.

98. Quigley HA. Neuronal death in glaucoma. Prog Retin Eye Res. 1999;18(1):39-57

99. Ramrattan RS, van der Schaft TL, Mooy CM, et al. Morphometric analysis of Bruch's membrane, the choriocapillaris, and the choroid in aging. Invest Ophthalmol Vis Sci. 1994 May;35(6):2857-64.

100. Ren R, Jonas JB, Tian G, et al. Cerebrospinal fluid pressure in glaucoma: a prospective study. Ophthalmology. 2010;117(2):259–66.

101. Resch H, Garhofer G, Fuchsjäger-Mayrl G, et al. Endothelial dysfunction in glaucoma. Acta Ophthalmol. 2009 Feb;87(1):4-12.

102. Ropper AH, Samuels MA, Klein JP. Chapter 13. Disturbances of Vision. In: Adams & Victor's Principles of Neurology, 10e New York, NY: McGraw-Hill; 2014.

103. Rosen RB, Andrade Romo JS, Krawitz BD, et al. Earliest Evidence of Preclinical Diabetic Retinopathy Revealed using OCT Angiography (OCTA) Perfused Capillary Density. Am J Ophthalmol. 2019 Jan 25. pii: S0002-9394(19)30025-X. doi: 10.1016/j.ajo.2019.01.012. [Epub ahead of print].

104. Ross RD, Barofsky JM, Cohen G, et al. Presumed macular choroidal watershed vascular filling, choroidal neovascularization, and systemic vascular disease in patients with age-related macular degeneration. Am J Ophthalmol. 1998 Jan;125(1):71-80.

105. Ryan S, Schachat A, Wilkinson C, et al. Retina. 5th ed. Philadelphia, PA: Saunders Elsevier; 2013.

106. Sawada A, Kitazawa Y, Yamamoto T, et al. Prevention of visual field defect progression with brovincamine in eyes with normal-tension glaucoma. Ophthalmology. 1996 Feb;103(2):283-8.

107. Seknazi D, Coscas F, Sellam A, et al. OPTICAL COHERENCE TOMOGRAPHY ANGIOG-RAPHY IN RETINAL VEIN OCCLUSION: Correlations Between Macular Vascular Density, Visual Acuity, and Peripheral Nonperfusion Area on Fluorescein Angiography. Retina. 2018 Aug;38(8):1562-1570.

108. Shah R, Wormald RP. Glaucoma. BMJ Clin Evid. 2011.

109. Shirinifard A, Glazier JA, Swat M, et al. Adhesion Failures determine the pattern of choroidal neovascularization in the eye: A computer simulation study. PLoS Comput Biol. 2012;8(5):e1002440. doi: https://doi.org/10.1371/journal.pcbi.1002440. Epub 2012 May 3.

110. Siaudvytyte L, Januleviciene I, Daveckaite A, et al. Neuroretinal rim area and ocular haemo-dynamic parameters in patients with normal-tension glaucoma with differing intracranial pressures. Br J Ophthalmol. 2016 Aug;100(8):1134-8.

111. Souied EH, El Ameen A, Semoun O, et al. Optical Coherence Tomography Angiography of Type 2 Neovascularization in Age-Related Macular Degeneration. Dev Ophthalmol. 2016;56:52-6.

112. Spraul CW, Grossniklaus HE. Characteristics of drusen and Bruch's membrane in postmortem eyes with age-related macular degeneration. Arch Ophthalmol. 1997;115:267–273.

113. Spraul CW, Lang GE, Grossniklaus HE, Lang GK. Histologic and morphometric analysis of the choroid, Bruch's membrane, and retinal pigment epithelium in postmortem eyes with age-related macular degeneration and histologic examination of surgically excised choroidal neovascular membranes. Surv Ophthalmol. 1999;44:10–32.

114. Su D, Garg S. The retinal function imager and clinical applications. Eye Vis (Lond). 2018 Aug 12;5:20. doi: 10.1186/s40662-018-0114-1. eCollection 2018.

115. Suh MH, Zangwill LM, Manalastas PI, et al. Optical Coherence Tomography Angiography Vessel Density in Glaucomatous Eyes with Focal Lamina Cribrosa Defects. Ophthalmology. 2016;123(11):2309–17.

116. Takusagawa HL, Liu L, Ma KN, et al. Projection-Resolved Optical Coherence Tomography Angiography of Macular Retinal Circulation in Glaucoma. Ophthalmology. 2017;124(11):1589-1599.
117. Talisa E, de Carlo BA, Marco A, et al. Spectral-Domain Optical Coherence Tomography Angiography of Choroidal Neovascularization. Ophthalmology 2015;122:1228-1238.
118. The Advanced Glaucoma Intervention Study (AGIS): 7. The relationship between control of intraocular pressure and visual field deterioration. The AGIS Investigators. Am J Ophthalmol. 2000;130(4):429-40.
119. Tielsch JM, Katz J, Sommer A, et al. Hypertension, perfusion pressure, and primary open-angle glaucoma. A population-based assessment. Arch Ophthalmol. 1995 Feb;113(2):216-21.
120. Topouzis F, Coleman AL, Harris A, et al. Association of blood pressure status with the optic disk structure in non-glaucoma subjects: the Thessaloniki Eye Study. Am J Ophthalmol. 2006;142(11):60–7.
121. Topouzis F, Wilson MR, Harris A, et al. Association of open-angle glaucoma with perfusion pressure status in the Thessaloniki Eye Study. Am J Ophthalmol. 2013;155(5):843–51.
122. Vasudevan SK, Gupta V, Crowston JG. Neuroprotection in glaucoma. Indian J Ophthalmol. 2011;59(Suppl):S102-13.
123. Wang X, Jiang C, Ko T et al. Correlation between optic disc perfusion and glaucomatous severity in patients with open-angle glaucoma: an optical coherence tomography angiography study Graefes Arch Clin Exp Ophthalmol (2015) 253: 1557-1564.
124. Xu L, Wang YX, Jonas JB. Ocular perfusion pressure and glaucoma: the Beijing Eye Study. Eye (Lond). 2009 Mar;23(3):734-6.
125. Yarmohammadi A, Zangwill LM, Diniz-Filho A, et al. Relationship between Optical Coherence Tomography Angiography Vessel Density and Severity of Visual Field Loss in Glaucoma. Ophthalmology. 2016;123(12):2498–508.
126. Yuzurihara D, Iijima H. Visual outcome in central retinal and branch retinal artery occlusion. Jpn J Ophthalmol. 2004;48(5):490.
127. Zeng Y, Cao D, Yu H, et al. Early retinal neurovascular impairment in patients with diabetes without clinically detectable retinopathy. Br J Ophthalmol. 2019 Jan 23. pii: bjophthalmol-2018-313582. doi: 10.1136/bjophthalmol-2018-313582. [Epub ahead of print].
128. Zhu J, Merkle CW, Bernucci MT, et al. Can OCT Angiography Be Made a Quantitative Blood Measurement Tool? Appl Sci (Basel). 2017;7:687.

# Measurement of Geometrical and Functional Parameters Related to Ocular Blood Flow

Josh Gross and Daniele Prada

**Abstract** This chapter examines the assessment of ocular hemodynamics in health and disease. Beginning with a discussion on ocular perfusion pressure and the physical principles, we systematically present the conceptual basis and details of blood flow measurement technology, paying particular attention to the scientific and clinical strengths and weaknesses of each technique.

## 1  Introduction

Advancements in imaging modalities of ocular blood flow have improved the understanding of ocular vascular physiology, and have helped to describe the role of alterations in blood flow in ocular pathophysiology. It is important to acknowledge that no single technology is capable of assessing all significant vascular beds. Therefore, having an understanding of the physical basis of different devices, the limitations of their measurements, and the resulting limitations in the interpretation of the data they produce is crucial. The purpose of this chapter is to provide the reader that understanding.

In any clinical, experimental, or theoretical study, it is important to bear in mind that the behavior of an ocular vascular bed does not predict the behavior of the other [29]. This implies that the status of ocular hemodynamics in a patient should be determined by evaluating the contribution of each vascular bed, in principle. Moreover, in order to allow accurate inter- and intra-individual comparisons, blood flow measurements should be made in absolute units, i.e., in millimeters per minute per gram of tissue. However, only few techniques are capable of

J. Gross
Eugene and Marilyn Glick Eye Institute, Indiana University School of Medicine, Indianapolis, IN, USA

D. Prada (✉)
Istituto di Matematica Applicata e Tecnologie Informatiche "Enrico Magenes" del Consiglio Nazionale delle Ricerche, Pavia, Italy
e-mail: daniele.prada@imati.cnr.it

© Springer Nature Switzerland AG 2019
G. Guidoboni et al. (eds.), *Ocular Fluid Dynamics*, Modeling and Simulation in Science, Engineering and Technology, https://doi.org/10.1007/978-3-030-25886-3_4

providing volumetric blood flow measurements in absolute units, whereas a vast percentage of them provides surrogates describing various aspects of ocular blood flow. Such surrogates are usually expressed in "arbitrary units." An arbitrary unit is a dimensionless number showing the ratio of amount of substance, intensity, or other quantities to a predetermined reference measurement. The reference measurement is defined by the specific experimental setting. Clear comparisons of measurements in arbitrary units obtained by different investigators can be made only if the respective reference conditions can be reproduced exactly. Nevertheless, arbitrary units can be used to compare multiple measurements performed in similar settings.

## 2   Ocular Perfusion Pressure

Blood flow through the intraocular vasculature is driven by the pressure difference between its arterial and venous supply, known as ocular perfusion pressure (OPP). The arterial supply is represented by the ophthalmic artery (OA) and its branches, whereas venous drainage occurs via the ophthalmic vein or, rarely, directly into the cavernous sinus [29].

In clinical studies, the pressure in the OA is often estimated via the brachial arterial pressure, whereas the intraocular venous pressure is approximated by the intraocular pressure (IOP) [6, 14, 24, 27]. Since the extent to which supine position affects intraocular venous pressure via the hydrostatic water column effect is not clearly understood [14], OPP is usually estimated as the difference between the arterial blood pressure (BP) and IOP in the upright position. OPP may be defined as mean, systolic, or diastolic OPP. Mean OPP is typically calculated as

$$\text{Mean OPP} = \frac{2}{3} \text{ mean arterial pressure (MAP)} - \text{IOP}$$

where $\text{MAP} = \text{diastolic BP} + \frac{1}{3} (\text{systolic BP} - \text{diastolic BP})$. The factor $\frac{2}{3}$ is introduced to account for the drop in BP between the brachial and ophthalmic artery when the subject is seated and the fact that the orbital arteries are further downstream [61].

This estimate of OPP involves potential errors. For example, it is not clear how accurately the above formula approximates the difference between the ocular arterial BP and the brachial arterial pressure because of the hydrostatic column effect when an individual is sitting [10]. We also cannot assume that the difference between the ocular and brachial arterial pressures is the same in normal and diseased vascular beds. Moreover, blood flow is determined not only by OPP but also by vascular tone. Regulation of blood flow may occur through changes in vascular resistance (vasoconstriction and vasodilation) independently of changes in OPP [57]. Equating venous pressure to IOP is also sometimes misleading. For example, if the cerebrospinal fluid pressure is higher than IOP, the venous pressure must exceed cerebrospinal fluid pressure in the subarachnoid space in order to avoid obstruction of the central retinal vein (CRV), and thus venous pressure could not be approximated by IOP.

In this way, the estimate for OPP involves systematic errors; however, even if more reliable formulas for computing the OPP have been proposed [14], the previously mentioned relations are consistently used in clinical studies. A reliable, direct measure of OPP would of course be desirable, but without this, care is needed when interpreting blood flow studies.

## 3    Principles of Operation of Ocular Blood Flow Measurement Techniques

In this section, we present the physical processes and principles of operations of various imaging techniques used in clinical or academic settings.

### 3.1    Optical and Color Doppler Imaging Techniques

In this section, we will be discussing techniques based on the reflection and scattering of electromagnetic or sound waves.

#### 3.1.1    Basic Principles: Scattering, Doppler Effect, Interference

Scattering is a physical phenomenon where a moving particle or a wave, such as light or sound, gets deviated from a straight trajectory due to non-uniformities in the medium which it passes through. More specifically, the term "scattering" is used when the deviated trajectory cannot be described by the law of reflection. There are many kinds of non-uniformities that can cause reflection and/or scattering, such as particles, density fluctuations in fluids, surface roughness, crystallites in solids, tissue structure and composition, and cells in organisms.

If a wave is reflected or scattered by a moving source, it undergoes a change in frequency or wavelength for an observer in relative motion with respect to the source. This is called "Doppler effect." The Doppler effect manifests itself, for example, in the increased pitch of the siren of an approaching ambulance. The same principle applies when a red blood cell (RBC), moving at a velocity vector $\overline{V}$, is hit by a laser beam of single frequency $f_i$ at an angle $\alpha_i$ with $\overline{V}$, and scatters it in various directions. The light scattered in the direction of the incident beam differs from $f_i$ by an amount $\Delta f$ which is simplified to

$$\Delta f = 2\,n\,V cos\,(\alpha_i)\,/\lambda_0 \qquad (1)$$

where n is the refractive index of the medium, $V = \left| \overline{V} \right|$ is the magnitude of $\overline{V}$ and $\lambda_0$ is the central wavelength of the laser beam [43].

Another typical phenomenon of waves which imaging techniques are based on is "interference." When interfering, two waves can add together to create a wave of greater (constructive interference), lower (destructive interference), or same amplitude. This can happen for all types of waves, such as light, sound, or matter waves, provided that the interacting waves are coherent, that is, have constant or nearly constant phase difference. Constructive interference occurs when the phase difference between the waves is an even multiple of $\pi$, whereas destructive interference occurs when the difference is an odd multiple of $\pi$. If the phase difference is intermediate between these two cases, then the amplitude of the resulting wave lies between the minimum and maximum values. Thus, by carefully analyzing the interference signal, it is possible to extract information about the original state of the waves.

### 3.1.2 Description

Laser Doppler flowmetry (LDF) is a noninvasive method of assessing blood flow and perfusion in the optic nerve head (ONH), iris, and subfoveal choroid [62]. LDF is based on the Doppler effect. It measures the shift in frequency $\Delta f$ that occurs when light is scattered by the RBCs moving through blood vessels. LDF uses a fundus camera and a computer system to detect these changes in frequency. This information is used to calculate three hemodynamic parameters: "velocity," "volume," and "flow." According to the theory of Bonner and Nossal [9, 50, 62], "velocity" is proportional to the mean change in Doppler frequency and represents the average speed of RBCs traveling through a vessel; "volume" is the fraction of photons that are Doppler-shifted, and is indicative of the number of RBCs in the given sample; "flow" is a quantity proportional to the "effective Doppler shift" and is indicative of the distance traveled by all moving RBCs inside the sample volume per unit time. "Velocity" is measured in Hertz, whereas "volume" and "flow" are expressed in arbitrary units.

Laser Doppler velocimetry (LDV) is a method related to LDF that has been used to quantify blood velocity in large retinal vessels [35]. By combining this technique with measurement of retinal vessel diameters from fundus photographs, blood flow in absolute units has been estimated using Poiseuille law under the assumption of laminar flow regime. However, the validity of such an assumption depends heavily on the measurement location. Areas with a high degree of tortuosity or near branch points will tend to be more turbulent, whereas vascular areas that are relatively linear and free of branches will tend to be closer to laminar flow.

The Heidelberg retinal flowmeter (HRF, Heidelberg Engineering GmbH, Dossenheim, Germany) is a confocal scanning version of the LDF (see Fig. 1). The system consists of a camera head and a computer system [29]. Unlike the stationary laser point of the LDF, the HRF laser quickly scans the fundus. Each line is divided

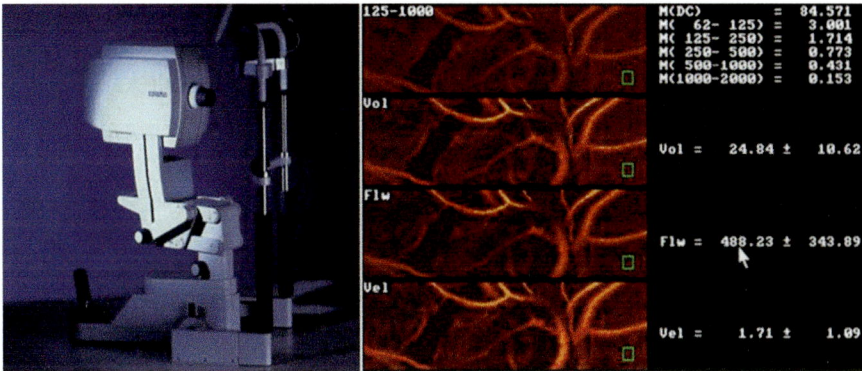

**Fig. 1** Confocal scanning laser Doppler flowmetry (Heidelberg retinal flowmeter) of optic nerve head and peripapillary retina. The left picture: Heidelberg retinal flowmeter. The patient places his chin on the chinrest and his forehead against the bar. The technician aligns the laser with the pupil, which does not need to be dilated. The right picture: A conventional $10 \times 10$ pixel measurement window is positioned in an area without large vessels to collect the flow values in arbitrary units from the retina (courtesy of Harris A., Siesky B. (2016) Other Tests in Glaucoma: Optic Nerve Blood Flow II. In: Giaconi J., Law S., Nouri-Mahdavi K., Coleman A., Caprioli J. (eds) Pearls of Glaucoma Management. Springer, Berlin, Heidelberg)

into 256 individual points. Scattered light from each point is quantified as with LDF, however, only scattered light from the point of illumination is analyzed by the HRF. HRF measurements tend to be concentrated on retinal surface vasculature. After a scan is complete, "velocity," "volume," and "flow" are computed for each pixel as for LDF and displayed as colored maps. Each pixel represents a $10 \times 10 \times 400\,\mu m$ volume of retinal tissue. Several methods of analyzing HRF data are available. One method uses a variable size pixel box to select a part of the scanned area and compute mean values of velocity, volume, and flow [8]. Another method utilizes manual pixel-by-pixel analysis to collect individual pixel measurement points of sufficient quality and display them by histogram and cumulative percentages. The distribution of pixels can be described by identifying 0, 10, 25, 50, 75, and 90th percentile values. In an HRF flow map, pixels with a flow measurement of "0" may represent areas of tissue located between capillaries, and, thus, they are used to represent vascular density [29].

Laser Speckle Flowgraphy (LSFG) is a noninvasive method of estimating blood flow and velocity in the ONH, choroid, retina, and iris [74] by using the laser speckle phenomenon, which is an interference event that occurs when monochromatic light scatters of a diffusing surface, such as paper, rough surfaces, or in media with a large number of scattering particles, like blood. When many waves having approximately the same frequency, but different phases and amplitudes, interfere with each other, the amplitude of the resulting wave varies randomly. This is observed as a speckled pattern of light, which, in the case of LSFG, varies proportionally to the velocity of RBCs and thus provides information on ocular blood flow. The faster the velocity of RBCs, the greater the rate of pattern variation. Although the velocity

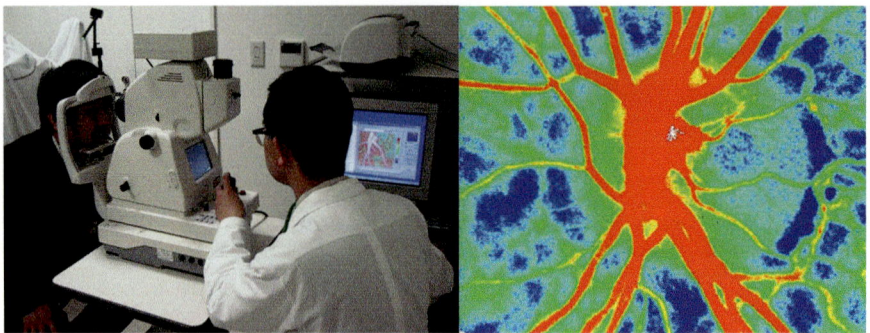

**Fig. 2** Left: Photograph of the laser speckle system. Right: A color-coded map showing the SBR values of ocular fundus around ONH in a healthy subject on the display of the laser speckle system (courtesy of Garhöfer G., Schmetterer L. (2012) Other Approaches. In: Schmetterer L., Kiel J. (eds) Ocular Blood Flow. Springer, Berlin, Heidelberg)

cannot be measured directly, it can be estimated by two indices, the normalized blur (NB) and square blur ratio (SBR) [74]. These parameters are in arbitrary units, so they should not be used to make comparison between different patients or different sites in the same eye. NB values are well correlated with blood flow measurements simultaneously taken with the hydrogen gas clearance method, colored microspheres technique, and other methods in the ONH, iris, choroid, and retina [74]. The distribution of blood flow can be displayed in a 2-dimensional color-coded map, which reflects the time variation of the speckles at each pixel point (see Fig. 2). This allows for visualization of blood flow in real time. Recently, another index for estimating blood flow in the retina has been developed, the relative flow volume [66].

Optical coherence tomography (OCT) is a technique that takes cross-sectional and 3-dimensional images of a biological tissue using a low-coherence interferometer (see Fig. 3). It has become one of the most rapidly accepted new technologies in ophthalmology [29]. In OCT, a low-coherence beam is directed at the target tissue. The light signals reflect off of the tissue back to the interferometer, which stacks a series of cross-sectional scans to give rise to a 3-dimensional image. Early versions of OCT were based on time-domain techniques, which produce an image of internal tissue microstructure in a way that is analogous to ultrasonic pulse-echo imaging [33]: light sent from a single point and reflected by the tissue (an "A" scan) provides information on the time-of-flight delay from scattering sites in the tissue. The delay information is used to identify the depths of various reflective layers. If a number of lateral A-scan probes are arranged in a line, the resulting depth measurements can be assembled to form a cross-sectional image (a "B" scan).

More recent OCT technologies based on the Fourier transform have shown superior levels of resolution and imaging speed with respect to time-domain OCT systems [42]. In Fourier-domain OCT, the interference pattern is recorded in the spectral domain for each lateral position, and cross-sectional images are

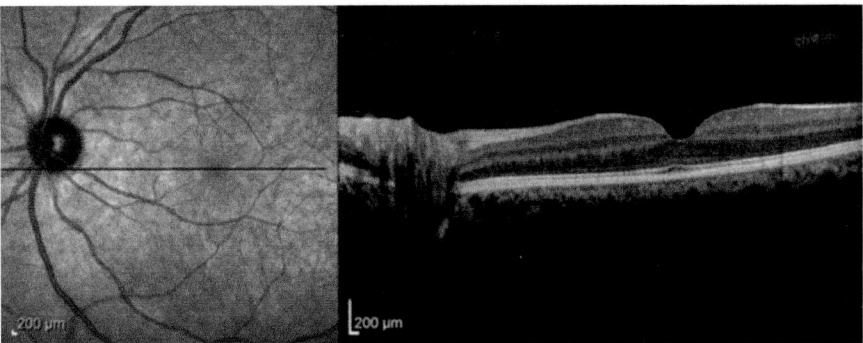

**Fig. 3** Wide field OCT scans of a healthy eye by the Spectralis OCT with OCT-1 module (courtesy of Ahmet Akman, Atilla Bayer, Kouros Nouri-Mahdavi. Optical Coherence Tomography in Glaucoma. A Practical Guide, 2018, Springer International Publishing)

reconstructed via Fourier transform. There are two variants of Fourier-domain OCT: spectral-domain OCT (SD-OCT) and swept-source OCT, each with its own advantages and limitations [30]. Fourier-domain OCT has pushed toward the development of functional extensions of OCT.

Doppler OCT, a commonly used functional extension of OCT, can detect the Doppler frequency shift $\Delta f$ of the reflected light to provide additional information on blood flow. This frequency shift induces a phase shift $\Delta \varphi$ in the interference pattern that could be obtained by analyzing the spectrum within an axial scan (A − scan), between sequential A − scans, or between sequential B − scans (cross − sectional images). If the time T between two measurements is known, we have $\Delta \varphi = 2\,\pi\,\Delta f\,T$. From Eq. (1), the cross − sectional velocity component $V_{//} = V cos\,(\alpha)$ can be computed as $V_{//} = \lambda_0 \Delta \varphi / (4n\,\pi T)$. Thus, the phase signal allows us to obtain cross-sectional velocity profiles of blood in a vessel. Vessel diameter could be directly measured from the velocity profile or from the amplitude of the interference signal. Volumetric blood flow rate in absolute units can be calculated by integrating the velocity over the vessel cross section. Adjustments are required to take into account possible error sources [32]. For example, one critical issue is the accurate estimation of the vessel diameter [51]. Using the OCT signal to measure diameter is delicate since blood is highly scattering and absorptive, resulting in shadowing effects and obscuring the rear vessel boundary. Moreover, depending on the system, it could be hard to distinguish between the phase noise and the small phase differences associated with low velocity values at the border of a vessel, possibly causing vessel diameters to be underestimated erroneously. Hence, some investigators use fundus photographs to extract vessel diameters [43].

Other critical issues are the upper limit of velocity measurement and the problem of automatically determining the relative angle $\alpha$ between the probe beam and the velocity vector of moving blood. As mentioned before, in Doppler OCT, blood motion is detected by calculating the phase difference $\Delta \varphi$ between sequential scans. In order to be uniquely characterized, $\Delta\varphi$ is conventionally restricted to the interval $[-\pi, \pi]$. According to the Nyquist theorem, digital samples must be obtained at least twice per wave in order to reconstruct the wave adequately from the samples. The maximum quantifiable phase difference is then $\pi$, and, consequently, the maximum longitudinal speed is $V_{//} = \pm \lambda_0/(4nT)$, yielding a velocity range $\Delta V_{//} = \lambda_0/(2nT)$. Care has to be taken for speeds close to the maximum detectable velocity, because, in this case, averaging filters usually applied in OCT could produce a lower mean value of the velocity. This phenomenon is known as "phase wrapping" [43]. Regarding automatic determination of the angle between the probe beam and the flow direction, observe that, for $\alpha = \pi/2$, the velocity cannot be measured because the phase shift becomes zero. Indeed, there is a certain angular range, called angle bandwidth, for which structures moving at a given speed are visible. The angle bandwidth can be estimated by solving Eq. (1) for $cos(\alpha)$ and considering a first order approximation about $\alpha = \pi/2$:

$$\Delta\left(\cos(\alpha)\right) \alpha - \Delta V / V^2.$$

Several approaches have been developed to calculate vessel orientation [43]. Doppler OCT is also used for noninvasive angiography. However, in accordance with the discussion about the angle bandwidth, care is needed when using Doppler OCT to study vessels that are nearly perpendicular to the OCT beam, such as the retinal and choroidal vasculatures [22] (Fig. 4).

OCT angiography (OCTA) is a Fourier-domain OCT method for visualizing blood vessels down to the capillary level (see Fig. 5). Unlike Doppler OCT, OCTA is more concerned about separating moving scatters from static background tissue to create angiograms. Nevertheless, it can also be used to estimate ocular blood flow. OCTA approaches can be classified based on the use of Doppler effect or variation of the speckle pattern produced by the scattered signal and whether they use full-spectrum or, rather, split the OCT signal into different spectral bands [22]. Different "speckle variance" approaches use either the phase information of the OCT signal or its amplitude, or both. Recently, Jia and coworkers [38] developed an efficient signal processing algorithm called split-spectrum amplitude-decorrelation angiography (SSADA), with which they were able to produce high-quality angiograms of the human macula and ONH. As with all measuring techniques, OCTA has limitations. Motion error and improper software correction can lead to vessel duplication, residual motion lines, vessel discontinuities. Also, flow from superficial layers can be projected to deeper layers, thereby incorrectly indicating that the imaged blood flow is a few layers deeper than its location in vivo. Several metrics for estimating blood flow have been computed with OCTA: flow index and vessel density [37], capillary dropout or non-perfusion area [36], fractal dimension of vessel lines [65], and perfusion density mapping [2]. All these indices are surrogates of blood flow.

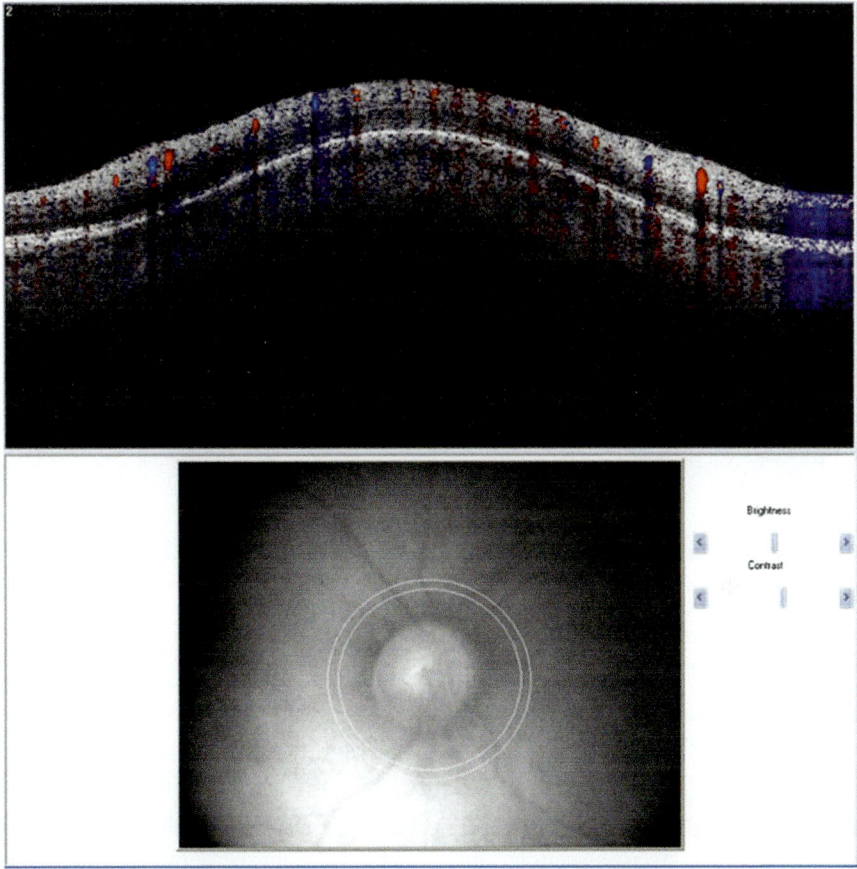

**Fig. 4** FD-OCT image showing the unfolded cross section from a circular scan. Arteries and veins could be distinguished by the direction of flow as determined by the signs (blue or red color) of the Doppler shift and the angle θ (courtesy of Harris A., Siesky B. (2016) Other Tests in Glaucoma: Optic Nerve Blood Flow II. In: Giaconi J., Law S., Nouri-Mahdavi K., Coleman A., Caprioli J. (eds) Pearls of Glaucoma Management. Springer, Berlin, Heidelberg)

Current SD-OCT machines are equipped with spectrometers and wide bandwidth light sources, introducing the potential for spectrographic analysis of structural images. The application of this technique to measure hemoglobin oxygen saturation will be described in Sect. 3.3.

Color Doppler Imaging (CDI), or ultrasound, is an imaging technology commonly used in radiology, cardiology, and obstetrics (see Fig. 6). It is based on the reflection and scatter of sound, rather than light. The CDI probe emits sound waves and uses the time taken by them to return to quantify the exact location of the sources of reflection within a tissue and produce a structural image, similarly to what is done in time-domain OCT. Similarly to OCT imaging, ultrasound performed on a

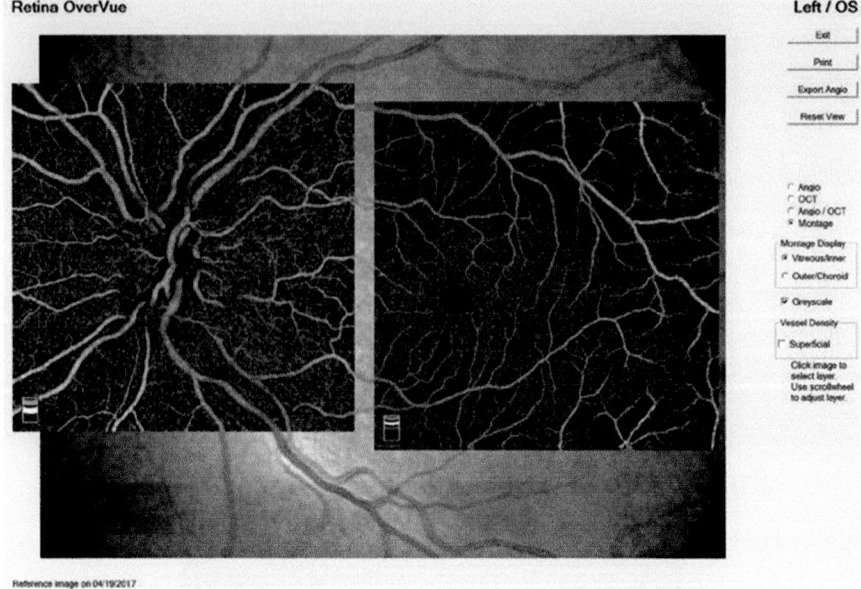

**Fig. 5** The combined report of peripapillary and macular optical coherence tomography angiography (OCTA) scans, showing the superficial capillary plexus in a normal eye. The scan was performed with the AngioVue HD OCTA system by Optovue (courtesy of Ahmet Akman, Atilla Bayer, Kouros Nouri-Mahdavi. Optical Coherence Tomography in Glaucoma. A Practical Guide, 2018, Springer International Publishing)

line from a single point (an "A" scan) is able to quantify the depths of the various reflective layers. A cross-sectional image obtained from the depth measurements coming from a number of A-scans arranged on a line is called "B" scan. Moreover, the same Doppler effect described for the light wave-based technologies also applies to sound waves reflected from moving sources. The frequency of sound waves reflected by moving sources is Doppler-shifted according to Eq. (1). CDI measures Doppler shifts to quantify velocities in absolute units. Various transducers are used to convert measurements into color pixels. The color red represents blood flowing toward the ultrasound probe, whereas blue represents blood moving away from the probe. The velocity waveform is different from vessel to vessel. Its maximum and minimum amplitudes represent the peak systolic (PSV) and end diastolic velocities (EDV), respectively. Once PSV and EDV have been determined, they can be used to calculate Pourcelot's resistive index RI

$$RI = (PSV{-}EDV)\,/PSV,$$

which is an indicator of resistance in the vascular beds being perfused by the artery being measured [29].

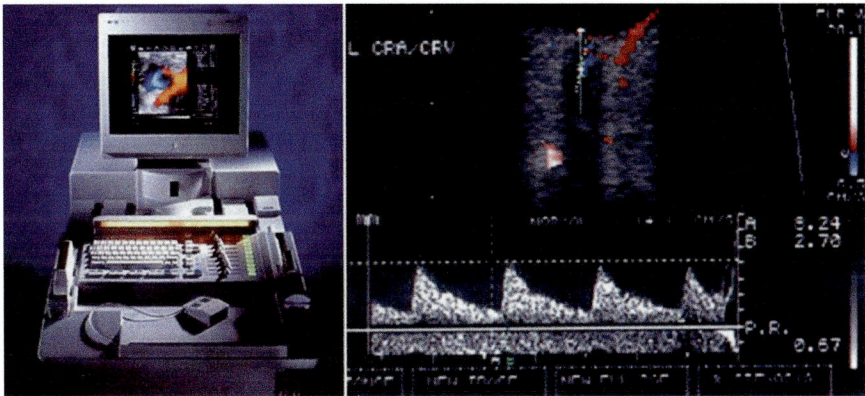

**Fig. 6** Color Doppler imaging. The left picture: A color Doppler machine. The patient is seated comfortably in a half supine position. An ultrasound probe is placed on the closed eyelid and the optic nerve shadow is identified. The vessels sampled include the ophthalmic artery, central retinal artery, and the nasal and temporal short posterior ciliary arteries. The right picture: A color Doppler image of the central retinal artery and vein taken with a 7.5-MHz linear probe. The Doppler-shifted spectrum (time velocity curve) is displayed at the bottom of the image. The red and blue pixels represent blood flow movement toward and away from the transducer, respectively (courtesy of Harris A., Siesky B. (2016) Other Tests in Glaucoma: Optic Nerve Blood Flow II. In: Giaconi J., Law S., Nouri-Mahdavi K., Coleman A., Caprioli J. (eds) Pearls of Glaucoma Management. Springer, Berlin, Heidelberg)

## 3.2   Microsphere and Dye Tracing Techniques

In this section, we will be discussing techniques based on the use of labeled microspheres or dye as markers for regional blood flow measurement.

### 3.2.1   Basic Principles: Gamma Radiation, Fluorescence

Under certain circumstances, labeled microspheres or dyes emit some sort of radiation that allows them to be detected. Four types of labeled microspheres are commonly used for in vitro measurements [55]:

- Radioactive microspheres (RM): they are made of radioactive nuclides that emit electromagnetic radiation in the form of gamma rays. The use of RM has declined over the years for several reasons: personnel is exposed to radiation, RM decay during storage, disposal of radioactive waste is difficult and expensive, environmental hazards cannot be excluded.
- Colored (CM) and fluorescent (FM) microspheres: they are made of colored or fluorescent compounds. After being excited by laser light, fluorescent substances emit light of longer wavelengths. Their waste disposal is simpler than for RM, but there are other health hazards for the personnel, such as the use of sodium or

potassium hydroxide for tissue analysis. Usually, colored labels are less sensitive than fluorescent ones.

- Neutron-activated microspheres (NAMs): they contain stable nuclides that emit gamma radiation when activated by neutron irradiation. NAMs require little tissue processing, and reduce health and environmental hazards with respect to RM.

Similarly to FM, in digital scanning laser ophthalmoscope angiography (SLOA), fluorescent dyes are used to monitor blood flow in vivo.

## 3.2.2   Description

The microsphere method has the ability of measuring blood flow directly, even in small pieces of ocular tissue [55]. It is based on the fact that microspheres injected into the systemic circulation get distributed and entrapped in tissues proportionally to the blood flow through the tissues. By collecting a reference blood sample and counting how many microspheres it contains, It is possible to calculate tissue blood flow values in absolute units from the following equation

$$Qt = Qr * Nt/Nr$$

where $Nr$ and $Nt$ are the number of microspheres in the reference sample and the dissected tissue, respectively, $Qr$ is flow in a reference sample, and $Qt$ is flow through the tissue. In some studies, the dimensionless ratio $Qt/Qr = Nt/Nr$ is used in place of $Qt$.

The number and size of microspheres should be optimized for the tissue being investigated in order to maximize the level of precision in the measurements and, at the same time, minimizing the impact on central and local hemodynamics. For example, increasing the number of injected microspheres would improve the level of precision, but using too many of them would affect blood flow. Similarly, a too small microsphere would not be captured by the tissue under investigation, whereas, if it is too big, it could cause vessel occlusions. Variations of reported values on ocular blood flow measured with the microspheres method can be due to several factors, such as diseases, age, hormonal variations, differences in arterial blood pressure, anesthesia, and arterial blood gases. Finally, biological variations among species highly contribute to the error in the measurements.

In SLOA, a fluorescent dye is injected into an antecubital vein and observed as it fills the retinal and choroidal vasculature. A scanning laser illuminates the fundus in a raster scan pattern. The fluorescent dye becomes excited by the laser light and produce light of longer wavelengths than those of the stimulation light. SLOA systems are equipped with filters that, for each point of the image, block reflected laser light and light scattered from surrounding tissue. Only light emitted by the stimulated dye from the point of interest is allowed to pass and reach a photodetector. Fluorescein dye is used to study retinal hemodynamics, whereas indocyanine green (ICG) is used for the choroid. In the first case, the fundus is illuminated with blue light at an approximate wavelength of 490 nm. Being in the range of visible spectrum, this light is absorbed by the retina. Fluorescein is then chosen because it has an absorption maximum very close to the wavelength of blue light (494 nm). On the other hand, ICG has a peak spectral absorption at about 800 nm. Radiations at these frequencies penetrate retinal layers, allowing ICG angiography to visualize also the choroidal circulation. Moreover, since the ratio of choroidal to retinal blood is approximately 6:1, it is usually assumed that ICG angiograms primarily represent choroidal hemodynamics.

In SLOA, a number of parameters have been developed to quantify retinal and choroidal hemodynamics. All these parameters are based on graphs of fluorescence level within vessels as a function of time. In fact, the fluorescence level is related to the concentration of dye, which, in turn, is affected by blood flow. These graphs are called "dye dilution curves." In fluorescein angiography, one simple parameter is "mean dye velocity," which represents the speed of blood moving through the retinal vasculature. It is determined by graphing fluorescence level at two different positions on a retinal vessel. This is accomplished by manually placing a sample window on the center of the vessel, and then calculating the average value of the pixels contained within the vessel. The separation in time between the two curves represents the amount of time required for the dye to travel from the first to the second position on the vessel. If this is combined with a distance measurement, it is possible to calculate a mean dye velocity. If two dilution curves are obtained, one from an artery and the other from the corresponding retinal vein, the difference in time between the curves represents how long it takes to blood to traverse the retinal vasculature in that quadrant. This is the "arteriovenous passage time" (AVP). By converting distance measurements on SLO images from pixels to micrometers, large vessel diameter and blood velocity can be computed. Unlike fluorescein angiography, where it makes sense to take measurements on individual vessels, in ICG angiography fluorescence quantification is performed on group of vessels because of the complexity and the overlap of arterial and venous networks in the choroid. The most useful parameter that can be obtained from ICG angiography is the delay between the first appearance of dye in the peripapillary and the macular regions of the choroid.

## 3.3   Retinal Oxygenation Evaluation Techniques

In this section, we will be discussing techniques for evaluating retinal oxygenation in vivo.

### 3.3.1   Basic Principles

To better characterize the metabolic state of a tissue, it is crucial to consider tissue oxygenation in addition to blood flow through it. At present, few technologies are available to measure retinal oxygenation: oxygen-sensitive microelectrodes, digital spectral retinal oximetry, magnetic resonance oximetry.

An oxygen-sensitive electrode, also known as *Clark electrode*, estimates oxygen partial pressure ($PO_2$) in blood vessels. The electrode has several components: a platinum terminal, a silver terminal, electrolyte (a solution, typically potassium chloride, KCl), a membrane permeable to oxygen, and a voltage source. The two terminals are submersed in the electrolyte solution. When a voltage is applied across the two electrodes, the platinum becomes negative (i.e., becomes the cathode), and the silver becomes positive (the anode). As oxygen diffuses through the membrane, the silver starts interacting with the KCl and gets oxidized. The platinum cathode utilizes the electrons produced by this reaction to reduce the oxygen. This flow of electrons gives rise to a current that can be measured. The more oxygen is available to carry out the reaction, the greater the flow of electrons, i.e., the higher the current.

Digital spectral retinal oximetry is based on optical methods. Measurement involves passing two wavelengths of light through (or bouncing them off) the tissue of interest. The penetration of light depends, among other factors, on whether the blood is oxygenated or deoxygenated (see Fig. 7). In fact, oxygenated ($HbO_2$) and deoxygenated (Hb) hemoglobin possess different light absorption properties. This is the same principle exploited in spectroscopy to analyze, for example, the chemical composition of a material.

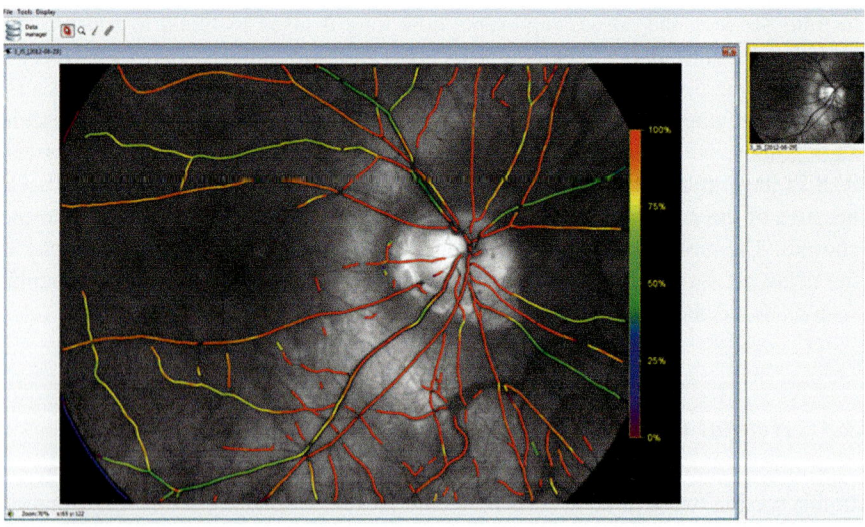

**Fig. 7** Color overlay of oxygen saturation in the retinal vessels of a glaucoma subject showing decreased arterial venous difference in a fundus photograph taken from a spectral retinal oximeter (courtesy of Harris A., Siesky B. (2016) Other Tests in Glaucoma: Optic Nerve Blood Flow II. In: Giaconi J., Law S., Nouri-Mahdavi K., Coleman A., Caprioli J. (eds) Pearls of Glaucoma Management. Springer, Berlin, Heidelberg)

A third technique for oxygenation measurements is based on magnetic resonance (MR). MR signals stem from the interaction of radio waves with atomic nuclei in a tissue [60]. A resonance phenomenon will occur when a radio wave of appropriate frequency reaches the nuclei; nuclei located in an equilibrium state of energy will then be transferred to an unstable state of high energy. At the molecular level, the return to equilibrium depends on the local magnetic and electric conditions at the excited nuclei. One of the processes characterizing the return to the equilibrium state is called the "spin-lattice relaxation process" or "longitudinal relaxation process." Such process is characterized by the relaxation time $T_1$, which is the time required for the system to recover to 63% of its equilibrium value after it has been exposed to certain types of radio waves. MR images of a given tissue depend, among many other factors, on the different relaxation times $T_1$ of its components. Oxygen is a paramagnetic compound that influences $T_1$, allowing MR imaging to be used for oximetry measurements.

### 3.3.2 Description

In order to measure $PO_2$, oxygen-sensitive electrodes need to be calibrated in saline whose $PO_2$ is known a priori. The value of oxygen current in this saline is equated to its $PO_2$. For retinal oximetry, an electrode is surgically inserted into the vitreous chamber and then advanced to just touch a retinal artery, vein, or an intermediate location. Absolute $PO_2$ values can be computed by comparing the current generated by oxygen reduction at the polarized electrode tip with the corresponding saline reference value. Alternatively, oxygen measurements can be retained as currents [3].

In spectral retinal oximetry, images of vessels are recorded at oxygen-sensitive and -insensitive wavelengths. Wavelengths of interest are located in the spectrum from 450 nm to 650 nm. In order to estimate oxygen saturation $SO_2$, two wavelengths of light are passed through (or bounced off) the retinal tissue. One color, say X, is at a wavelength where there is a large difference in the absorption of light between $HbO_2$ and Hb, whereas the other color, say Y, is at a wavelength on which $HbO_2$ and Hb have identical effects. For each wavelength, the optical density (OD) is computed as the log of the ratio between the brightness levels just outside ($I_0$) and inside (I) a vessel

$$OD = \log\left(I_0/I\right)$$

It has been shown that the ratio $ODR = OD_X/OD_Y$ has an approximately linear relationship to oxygen saturation $SO_2$ [28].

$$SO_2 = a + k \cdot ODR,$$

where $a$ and $k$ are constants. Volumetric oxygen delivery to retinal tissue can be estimated by combining measurements of oxygen saturation within arterial and venous blood with retinal blood flow measurements [73].

In MR oximetry, measurements are limited to the posterior vitreous near the retina to eliminate other factors that can affect the observed $T_1$. Oxygenation of the posterior vitreous near the retina is used as a measure of inner retinal oxygenation. An increase in partial oxygen pressure in the vitreous over the room air value ($\Delta PO2$), while breathing either carbogen (95 % O2 – 5 % CO2 mixture) or 100 % oxygen, is detected as an increase in the signal intensity on a T1 – weighted image. To quantify the change in oxygen tension, the baseline and carbogen images are subtracted and then converted to a $\Delta PO2$. The MR – measured $\Delta PO_2$ is similar to that determined using an oxygen electrode in healthy rat retina [7].

# 4 Vascular Beds

In this section, we critically review advantages, limitations, and clinical data of the measurement techniques previously described, according to the ocular vascular bed under investigation.

## 4.1 Anterior Segment Vasculature

The anterior segment of the eye consists of ocular structures such as the sclera, cornea, iris, and ciliary body [19]. The blood supply to these structures originates from the anterior ciliary and long posterior ciliary arteries, which are branches of the OA. For the purposes of ocular blood flow assessment in the anterior segment, there has been limited investigation. However, some data has been generated using multiple imaging modalities in recent years. In enucleated bovine, pig, and human eyes, aqueous angiography using fluorescein and ICG dye with concurrent anterior segment OCT showed segmental angiography outflow patterns [31]. Li et al. used anterior segment structural OCT and OCTA to noninvasively visualize the three-dimensional microstructural and microvascular properties of the limbal region and aqueous outflow pathway in an in vivo human eye [44]. They found they were able to map out and in part 3D-reconstruct the aqueous outflow pathway and limbal vasculature, specifically the episcleral and aqueous vein [44]. Ang et al. evaluated the corneal and limbal vasculature in 20 normal subjects and five patients with abnormal corneal vasculature using the AngioVue OCTA system (Optovue, Inc., Fremont, CA). They found good repeatability for image quality and good interobserver agreement for vascular measurements. Additionally, vascular abnormalities were visualized in patients with various corneal pathologic features such as graft-associated neovascularization, post-herpetic keratitis scarring, lipid keratopathy, and limbal stem cell deficiency [4]. MacKenzie et al. used a retinal fundus camera fitted with a custom image-replicating imaging spectrometer to visualize the bulbar conjunctival and episcleral microvasculature in ten healthy human subjects at normoxia (21% $FiO_2$) and acute mild hypoxic (15% $FiO_2$) conditions

[46]. In conditions of acute mild hypoxia, blood oxygen saturation ($SO_2$) decreased compared to normoxia in both vascular beds ($p < 0.05$). When changing from normal to hypoxic conditions, episcleral vessel diameter increased from $78.9 \pm 8.7$ μm to $97.6 \pm 14.3$ μm ($p = 0.03$). However, vessel diameters in the bulbar conjunctival vessels did not change when changing the oxygen environment. Additionally, when exposed to ambient air, hypoxic bulbar conjunctival vessels exponentially reoxygenated due to oxygen diffusion from the ambient air. Conversely, episcleral vessels did not undergo any significant oxygen diffusion, instead behaving similarly to pulse oximetry measurements.

## 4.2  Retrobulbar Vasculature

Imaging of the retrobulbar vascular beds is accomplished with CDI. As discussed in Sect. 3.1.2, CDI is a noninvasive ultrasound technique that measures the PSV and EDV of the OA, temporal and nasal posterior ciliary arteries (PCAs), and the central retinal artery (CRA) in centimeters per second [18, 21]. Additionally, Pourcelot's resistive index (RI) is calculated (RI = PSV–EDV/PSV) and provides a metric of vascular resistance distal to the point of measurement [29]. CDI has been used extensively, and has demonstrated impactful results; however, it does have its limitations. It does not measure the diameter of vessels; therefore it is unable to measure absolute blood flow volume. Additionally, it requires a skilled examiner to operate and obtain results of high quality [29].

Reproducibility of velocity measurements varies among vascular beds, the CRA, and the PCAs having the worst and best reproducibility, respectively [18, 21]. Many ophthalmic diseases with a known or hypothesized vascular component have been investigated using CDI, including primary open-angle (POAG) and normal tension glaucoma (NTG), diabetic retinopathy (DR), nonarteritic ischemic optic neuropathy (NAION), retinopathy or prematurity (ROP), retinal detachment (RD), and central artery (CRAO) and vein (CRVO) occlusion. In glaucoma patients, CDI has been used to demonstrate differences in blood flow velocity between healthy subjects, and even among different demographics. For instance, the Leuven Eye Study, which enrolled 614 subjects (POAG:214, NTG:192, OHT:27, healthy controls:140), found that glaucoma patients had lower retrobulbar blood flow velocities compared to ocular hypertensive and healthy subjects [1]. Further, Moore et al. studied 112 open-angle glaucoma patients over 4 years and found that patients who had glaucomatous structural and functional progression had lower baseline mean OA PSV and EDV compared to those who did not progress [53]. In a meta-analysis by Xu et al., NTG patients demonstrated significantly decreased velocities in all retrobulbar vessels, as well as significant increases in RI in 50% of vessels [82]. When investigating the racial influences among POAG patients, Siesky et al. found that open-angle glaucoma patients of African descent who had structural changes in the ONH after 4 years demonstrated stronger associations with EDV and RI in the short PCAs [69]. In patients with DR, Sood et al. studied 50 type II diabetic patients with DR and found baseline RI was higher than the normal population, and there was significant increases in RI of the PCA and CRA after 6 months [70]. However, there was no association between DR progression and CDI findings. A meta-analysis in diabetics with and without DR agreed with these findings. Meng et al. showed that diabetics without retinopathy had lower blood flow velocities and higher RI compared with healthy controls. Moreover, patients with DR showed lower velocities and higher RI compared to diabetics without DR [49]. Ozcan et al. studied infants with and without ROP, and found the mean OA PSV of ROP patients to be significantly lower than subjects without ROP ($p < 0.05$) [56]. Further, infants with stage 2 ROP demonstrated significant differences in mean OA PSV compared to stage 1 ROP and eyes without ROP ($p = 0.03$). In patients with acute unilateral NAION, blood flow velocities in the CRA of the affected eye were significantly decreased compared to controls, and CRA velocities tended to be reduced in both eyes of patients with unilateral NAION [85]. In patients with CRAO, CRVO, branch vessel occlusions, and ocular ischemic syndrome, decreases or even absence of flow in all retrobulbar vessels have been demonstrated [17].

## 4.3   Retinal Vasculature

Retinal blood flow is a major target for ocular vascular imaging, and thus many modalities exist in an effort to capture more information about these vascular beds and their role in retinal pathophysiology. One modality is fluorescein angiography (FA), introduced in Sect. 3.2.1. This technique utilizes a fundus camera and fluorescein dye that is injected into a peripheral vein to directly visualize the superficial capillaries as fluorescein dyes transit through. Two of the most common measurements are arteriovenous passage time and the mean transit time. Reproducibility of FA has been characterized by Wolf et al. [81]. FA also allows determination of capillary integrity as evidenced by fluorescein leakage, as well as regional information comparing one area to another. Specifically, areas of nonperfusion can be determined. This technique, however, is two-dimensional and provides little depth information. Moreover, boundaries of capillary non-perfusion or neovascularization can be blurred by dye leakage. Lastly, this technique is time consuming, and is based upon intravenous dye injection, which has a risk of generating a severe reaction in some patients.

A second retinal vasculature imaging technique is the HRF. As described in Sect. 3.1.2, the HRF utilizes laser Doppler flowmetry principles to estimate blood volume, velocity, and flow in arbitrary units. It has subcapillary resolution, is sensitive to small changes in blood flow, and is highly reproducible [29, 48, 86]. Disadvantages of HRF include increased analysis time with respect to other imaging techniques (it takes hours to fully analyze HRF data), photodetector sensitivity to light and background brightness, measurements in arbitrary units, and, finally, the device is no longer manufactured.

OCTA is a novel technique capable of visualizing all the capillary tissue networks throughout the retina in 3D with good reliability and repeatability, and with high resolution (18 μm in X & Y planes, 2–3 mm axial depth, at present) [40]. Current commercially available technology employs software that enables layer segmentation of the retinal vasculature into the superficial plexus (inner limiting membrane to inner plexiform layer), deep plexus (inner nuclear layer to outer plexiform layer), and outer retinal zone (outer nuclear layer to basement membrane) [11]. As mentioned in Sect. 3.1.2, several metrics for estimating blood flow have been computed with OCTA, all in arbitrary units: flow index, vessel density, capillary dropout, fractal dimension of vessel lines, and perfusion density mapping. OCTA is characterized by relatively fast acquisition times [22], and is noninvasive due to the lack of intravenous dye. This modality does find limitation due to the inability to assess for vascular leakage, a limited field of view in current systems, and inability to distinguish between tissue loss and ischemia when vessel density is low [22]. Other limitations have been described in Sect. 3.

LSFG is a technique that is capable of computing two parameters, the normalized blur and the square blur ratio, which are representative of blood velocity in the retina, though in arbitrary units [66, 74]. This modality finds advantage in measuring changes over time, as flow maps are created and can be followed. However, its use is limited by a lack of understanding of its measurements [5] and the fact that the technology is not commercially available [29].

The Canon Laser Blood Flowmeter (CLBF) incorporates LDV and simultaneous vessel densitometry, as well as utilizes an image stabilization system to minimize the impact of eye movement. CLBF has been shown to be a reproducible and repeatable technique to measure retinal volumetric blood flow velocity (mm/s) and vessel diameter (mm) in order to calculate total blood flow in absolute units (μL/min) in the retina [52]. Although the ability to measure flow in absolute units is a clear advantage, this modality finds limitation in that it can only measure flow in vessels of 60 μm diameter or greater, preventing it from measuring capillary flow [35]. Further, clear optical media and pupil dilation are required, and it cannot be used to measure ONH circulation. Finally, the CLBF instrument is no longer commercially available.

As discussed in Sect. 3.1.2, Doppler OCT allows for the ability to reliably measure velocity and total retinal blood flow in absolute units (μL/min) in retinal branch vessels, as well as direct measurement of vessel dimensions from a cross-sectional velocity profile [32, 76, 80]. The peak and average velocity can be analyzed as a function of time along the cardiac cycle from the cross-sectional velocity profile. However, Doppler OCT is limited by only measuring the major retinal branch vessels, and finds difficulty in measuring capillary flow in the retina. Additional limitations have been discussed in Sect. 3.1.2.

Retinal oximetry provides an assessment of ocular tissue metabolic status by measuring light absorption of hemoglobin in the retinal vessels and calculating oxygen saturation ($SO_2$). It is a highly reproducible technique [25], and is a step toward understanding the true impact of ischemia. The biggest disadvantages of retinal oximetry include the need for clear optical media, and the lack of longitudinal studies which limits our understanding of oxygen utilization in patients with eye disease.

A plethora of clinical studies of the retina in patients with retinopathy have been performed. Huber et al. demonstrated using fluorescein angiography that POAG and NTG patients have higher AVP times compared with healthy controls [34]. In diabetic patients, reduced blood flow in the perifoveal and nasal areas was associated with the severity of DR using HRF [15]. In patients with neovascular age-related macular degeneration (N-AMD), HRF showed higher mean blood flow compared to control eyes at baseline, and 1 week after photodynamic treatment (PDT) N-AMD patients had a decrease in retinal blood flow [79]. Interestingly, 1 month after PDT treatment, N-AMD patients had an increase in retinal blood flow toward baseline. LSFG was used in diabetic patients who underwent intravitreal injection of bevacizumab, and showed decreases of approximately 30% in the area of the retina that underwent neovascularization [41]. OCTA has been utilized in many investigations of retinopathy. In patients with DM, OCTA has shown decreasing vascularity of the superficial and deep capillary plexus as the disease progresses [16, 26, 77]. In patients with non-neovascular AMD, OCTA showed significant choroidal depletion and fibrotic replacement, suggesting pathophysiological mechanisms in disease progression [12]. In patients with N-AMD, measurement of new vessel complexes using OCTA has been shown to have interreader agreement comparable to FA [45]. Farecki et al. demonstrated utilizing OCTA that different types of choroidal neovascularization in exudative AMD can be visualized, and lesion can be detected in the choroid into the outer retina [20]. Similarly, Doppler OCT has demonstrated reduced retinal blood flow in patients with NAION, non-proliferative and proliferative DR, and branch retinal vein occlusion (BRVO) compared to normal subjects [71, 80]. Reduced retinal blood flow using LDV has also been described in type II diabetic patients with and without DR compared to healthy subjects [54]. Studies of oxygen saturation in the retinal vessels have shown an association between DR and CRVO and increased retinal venous $SO_2$, suggesting disturbed oxygen utilization in these patients [59]. Geirsdottir et al. showed that exudative AMD patients had higher retinal venous $SO_2$, and therefore a smaller arteriovenous oxygenation difference [23]. POAG and NTG patients have also demonstrated higher retinal venous $SaO_2$ compared to healthy subjects [1].

## 4.4 Choroidal Vasculature

Imaging of the choriocapillaris [58] uses similar techniques described to visualize the retina, and often times they are the same. The current gold standard of choroidal imaging is ICG angiography due to greater penetration through the retinal pigment epithelium and hemorrhages compared to fluorescein angiography, as explained in Sect. 3. ICG angiography also binds more completely with blood proteins, and therefore does not leak out of the fenestrated choriocapillaris vessels like fluorescein. ICG angiography was first introduced in the 1970s, and there is more expertise and comfort with interpretation among ophthalmologist compared to newer techniques. Compared to the more recent OCTA technology, ICG angiography is less prone to artifact [72]. However, it does require intravenous injection of dye, which inhibits its use in some patient populations, such as those with kidney disease and women who are pregnant. ICG also is limited by poor structural analysis, and is relatively expensive.

OCTA has been utilized to visualize the choroidal vasculature, and provides distinct, depth-resolved, 3D imaging. Additionally, this technique is quick, noninvasive, and less expensive than ICG angiography. As already mentioned, OCTA is limited by a relatively small field of view, inability to show vascular leakage, and a tendency for image artifact due to patient movement. As the information provided by OCTA continues to be evaluated, ICG angiography remains the gold standard for detecting choroidal neovascularization. Additionally, Doppler OCT has been used to visualize the choroidal vasculature; however, due to limitations of poor resolution in this area little research is available [43].

Clinical studies of the choriocapillaris using ICG angiography have shown its utility in visualizing multiple vascular lesions including identification of polypoidal choroidal vasculopathy, occult choroidal neovascularization, neovascularization associated with pigment epithelial detachments, and recurrent choroidal neovascular membranes. ICG angiography has also been used to predict recurrent disease in AMD by monitoring choroidal neovascularization size change [63]. Additionally, AMD patients showed delayed macular choroidal filling times compared to age-matched controls. OCTA has been used in patients with AMD to visualize the choroidal vasculature, and has demonstrated identification of perfusion abnormalities and retinal or choroidal neovascular lesions [13].

## 4.5   Optic Nerve Head Vasculature

Imaging of the ONH vasculature has been performed with similar techniques used to capture the retinal vasculature; however, these studies have mainly been performed in patients with glaucoma. Recently, the emergence of OCTA has allowed high resolution imaging of the microvasculature within and around the ONH. Specifically, visualization of the radial peripapillary, optic nerve head, prelaminar, and lamina cribrosa vascular networks is possible, along with layer segmentation of the optic disc vasculature when analyzed in conjunction with structural OCT [11].

However, imaging of the ONH and peripapillary vascular region has proven to be challenging due to the anatomical structures found there. For instance, the emergence of the CRA and vein (CRV) cause interference signals when attempting to visualize the deeper vascular networks in the ONH. This is true for OCTA. Specifically, a shadow is cast from the CRA and CRV to the deeper vascular networks in and around the ONH, causing decreased resolution of images that are directly below these large vessels. Further, longitudinal studies of the ONH using OCTA have yet to be conducted, limiting its prognostic ability. Additionally, ONH imaging using OCTA finds difficulty in differentiating between retinal and ONH microcirculation, and again has a limited field of view in current systems. Imaging of the ONH by HRF has been widely used, and investigations by Jonescu-Cuypers et al. found that due to variations in the Doppler signal between the peripapillary retina and neuroretinal rim tissue, measurements in the neuroretinal rim area may provide inaccurate data because of improper sensitivity settings for the weaker rim signal. Therefore, accurate images of the neuroretinal rim should be taken separately with an appropriately higher sensitivity setting [39].

Clinical studies of the ONH using OCTA have shown that measurements of vessel density were able to distinguish between healthy subject, glaucoma suspects, and glaucomatous eyes, and decreased vessel density was significantly associated with the severity of visual field damage independent of structural loss [83, 84]. Mansoori et al. found that radial peripapillary capillary density was lower in temporal sectors with corresponding RNFL defects in early POAG compared to controls using OCTA [47]. Similarly, Suh et al. utilized OCTA and found that parapapillary deep retinal layer microvascular dropout in 37 of 71 POAG patients was associated with structural and functional defects, as well as low blood pressure [75]. Using HRF, POAG patients were found to have increased areas of retinal non-perfusion that was associated with increasing cup-to-disc ratio over 18 months [78]. Siesky et al. studied POAG patients of African descent for 4 years and found reductions in retinal capillary density in the inferior retina correlated with a reduction in macular thickness [69]. Additionally, NTG patients demonstrated reduced blood flow in the neuroretinal rim that corresponded to regional visual field defect [56, 64]. Similarly, POAG patients studied with fluorescein angiography were found to have delayed disc-filling times compared to healthy controls [68]. LSFG has also demonstrated reduced ONH blood flow in eyes with preperimetric glaucoma [67].

# 5   Conclusions

In this chapter we presented the conceptual basis and details of blood flow measurement technology. Table 1 summarizes the main features, advantages, and limitations of each technique.

Over the past century, medical science has uncovered many factors that influence the development and progression of eye disease, and many are not identified as being multifactorial. However, current statistical approaches and clinical awareness have not fully capitalized on the growth in identified risk biomarkers to prevent irreversible vision loss. Whether improved perfusion has any beneficial long-term effect on the visual field, and what treatments may best accomplish this physio-logical effect, remains unclear. What is established, however, is the awareness that ocular vascular insults in eye disease are important to consider in certain individuals. Advancements in imaging modalities and efforts to establish longitudinal data,

**Table 1** Techniques for in vivo studies of ocular hemodynamics

| Technique | Measurement | Location | Advantages | Disadvantage |
|---|---|---|---|---|
| Anterior segment OCTA | – Vessel density<br>– 3D structure | – Limbus<br>– Aqueous outflow pathway | – Fast acquisition<br>– 3D structure of vascular beds | Cannot identify vascular leakage |
| Retinal fundus camera with imaging spectrometer | – Oximetry<br>– Vessel diameter | – Bulbar conjunctiva<br>– Episcleral vessels | – Measures $O_2$ concentration in varying conditions<br>– Measures vessel diameter response to changing $O_2$ conditions | Custom imaging technique not widely available |
| Fluorescein angiography | – Arterio-venous passage time<br>– Mean transit time | – Anterior segment vasculature<br>– Retinal vasculature | – Gold standard for retinal imaging<br>– Demonstrates vessel compromise via leakage, dropout, decreased perfusion | – Invasive<br>– Risk of anaphylaxis<br>– Only visualize superficial retinal capillaries<br>– Images may be blurred by media opacities |
| Laser Doppler flowmetry (Heidelberg retinal flowmeter) | Retinal blood volume, velocity, and flow (arbitrary units) | Retinal vasculature down to subcapillary level | – Noninvasive<br>– Reproducible<br>– Sensitive to small changes in blood flow | – Increased analysis time<br>– Arbitrary units<br>– Photodetector sensitive to ambient light<br>– No longer manufactured |
| Retinal optical coherence angiography | – Vessel density<br>– Flow index<br>– Fractional dimension of vessels<br>– Perfusion density mapping | – All levels of retinal and ONH capillary networks<br>– Choriocapillaris | – Fast acquisition<br>– Repeatable & reproducible<br>– Noninvasive<br>– High resolution | – Arbitrary units<br>– Unable to demonstrate vessel leakage<br>– Unable to distinguish between tissue loss vs. ischemia<br>– Poor resolution of deep ONH capillaries |

(continued)

by the LA, SA, and C compartments according to a Krogh-type cylinder model (see Appendix 2) assuming constant oxygen demand [92]. $S_{CR}$ is a conducted response signal that depends on the release of ATP by red blood cells at a rate that depends on oxyhemoglobin saturation (for complete details, see [5]). The metabolic rate of $CO_2$ is assumed to be proportional to oxygen by a factor of $-0.81$ [140]. This factor is used to calculate the blood content of carbon dioxide, and the tissue carbon dioxide content is calculated and converted into $P_{CO_2}$ using carbon dioxide dissociation curves. The resulting value of tissue carbon dioxide is used as the value of the signal, $S_{CO_2}$, for the carbon dioxide response. $C_{myo}$, $C_{shear}$, $C_{meta}$, and $C_{CO_2}$ are weights that define the relative contributions of the mechanisms to vascular tone; the positive or negative sign preceding each term corresponds to the increase or decrease in tone generated by each mechanism. $C_{tone}''$ represents a combination of other factors, such as the retinal relaxation factor, that influences vascular tone but not modeled explicitly [33]. $C_{myo}$ is fit to data from Jeppesen et al. [71], $C_{shear}$ is taken directly from [6], and a range of values for $C_{meta}$ and $C_{CO_2}$ are used since there is currently not enough experimental data to quantify these parameters. All parameter values are given in [6].

The model predicts that metabolic responses are most important for maintaining constant blood flow despite changes in ocular perfusion pressure. Figure 5 shows the model predictions of normalized blood flow (normalized with respect to flow at $P = 40\,\mathrm{mm\,Hg}$) as a function of incoming arterial pressure ($P_a$) with various autoregulation mechanisms assumed to be active or inactive. As seen in the figure, the autoregulation plateau is lost if the metabolic and/or $CO_2$ responses are missing.

This model [6] has served as a foundational model upon which several other models [21, 25, 26] have been developed. For example, the model has also been

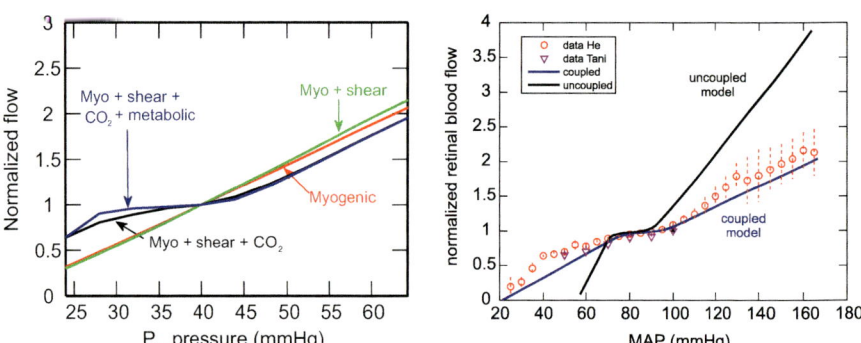

**Fig. 5** *Left*: change in blood flow as pressure at the downstream end of the CRA, $P_a$, is increased. The effects of the myogenic, shear, conducted metabolic, and carbon dioxide mechanisms are evaluated given a healthy level of IOP = 15 mm Hg [6]. (Left image copyright holder: Association for Research in Vision and Ophthalmology) *Right*: change in blood flow as the MAP is increased. Results are obtained using the model proposed in [6] (uncoupled model) and the model proposed in [25] (coupled model). The numerical results are compared with experimental data from [66, 130]

applied to a set of oximetry data obtained from healthy individuals and glaucoma patients to propose possible explanations for the clinically observed increases in venous blood oxygen saturation in advanced glaucoma patients [21]. The model is used to show that the mechanisms that lead to increased venous saturation in primary open-angle glaucoma patients likely differ from the mechanisms leading to increased venous saturation observed in normal tension glaucoma patients. Such a hypothesis could lead to a major breakthrough in understanding whether the vascular changes observed among glaucoma patients are the cause or effect of glaucomatous damage.

Cassani et al. [25] couple the effects of the pressure in the central retina artery and central retinal vein to the mechanistic description of blood flow regulation in the retinal microcirculation described above and in [6]. The model improves upon [6] by incorporating more specific IOP effects by including the deformation of the lamina cribrosa and vessel tone of venous compartments due to IOP. As demonstrated by comparisons between model predictions and experimental data collected by He et al. [66] and Tani et al. [130], this coupled model provides a more accurate representation of overall hemodynamic behavior of the vascular supply to the retina (e.g., for a wide range of ocular perfusion pressures obtained by varying IOP and/or mean arterial pressure (MAP)), as seen in Fig. 5 (right panel).

Cassani et al. [26] also use this theoretical model to investigate the response of retinal blood flow to changes in oxygen demand using the coupled model described in [25]. Oxygen delivery to retinal tissue is modeled using a Krogh cylinder model. The Krogh model is used as a first approximation and could be improved upon based on the known geometry of the retinal tissue layers, as discussed in Sect. 2.2.2. Nevertheless, the model is used to show that the increase in blood flow predicted for increased oxygen demand is not proportionally the same when there is a decrease in oxygen demand, suggesting that the vascular regulatory mechanisms respond differently to different levels of oxygen demand. Extending this model to account for the different hemodynamic demands and roles of the inner and outer retina and improving the vascular geometry will allow for improved model predictions and more accurate representations of the metabolic component to flow regulation.

The models presented by Guidoboni et al. [55], Arciero et al. [6], and Cassani et al. [25] yield predictions of blood flow that are consistent with experimental measures, but these models do not capture the spatial variation of autoregulatory response mechanisms due to varying oxygen levels in retinal tissue. Other models, such as Causin et al. [28], have modeled retinal flow and oxygenation in a realistic geometry (although they do not account for vascular regulation mechanisms). Causin et al. [28] present a model that incorporates a simplified three-dimensional representation of the retina, consisting of tissue layers supplied by the arteriolar network on the surface proximal to the vitreous, capillary plexi embedded in two distinct tissue layers, and arteriolar and venular networks represented by fractal trees. Their work couples a wall mechanics model with a model for oxygen transport in the retina and quantifies the effects of blood pressure, blood rheology, arterial permeability to oxygen, and tissue oxygen demand on the distribution of oxygen in retinal blood vessels and tissue.

*Two-Dimensional Models*

Several non-invasive techniques provide two-dimensional (2D) images of the retinal vasculature. Indeed, the retina can be thought of as a thin 2D surface, thereby motivating the utilization of 2D models to study blood flow. Liu et al. [85] develop a 2D model of the retinal circulation, where the vascular network is reconstructed directly from fundus camera images. The Navier–Stokes equations for blood flow and the convection–diffusion equation for oxygen transport are solved numerically to obtain detailed blood flow and oxygen distribution patterns in a patient-specific retinal arterial tree. A structured tree model for the distal peripheral vessels is combined with the retinal arterial network. The non-Newtonian rheological properties of blood are incorporated by using an empirical viscosity model to account for the Fahraeus–Lindqvist effect. The model successfully captures some of the main features of retinal circulation, including pressure drops in the range of 11–14.6 mm Hg between the inlet and outlets of the reconstructed network and non-uniform oxygen tension, which varied with the vessel diameter and distance from the optic disc. Specifically, the model predicts a mean oxygen saturation in retinal arteries of 93.1% for vessels larger than 50 μm in diameter and 82.2% for smaller arterioles. Even though blood flow regulation is not the focus of the work by Liu et al., it could be included by utilizing some of the methods described above.

*Three-Dimensional Models*

Aletti et al. [1] propose a model of retinal autoregulation within a three-dimensional (3D) network of retinal arteries that is patient-specific. Fundus images are used to reconstruct a patient-specific vasculature using modern segmentation tools (e.g., see Fig. 6). The fluid (blood) is modeled with time-dependent Stokes equations. The structure (blood vessel) is modeled as an elastic structure that interacts with the fluid and is time dependent. The equations for the structure dynamics are embedded as a boundary condition of the fluid problem. The model includes a 1D idealization of the smooth muscle cell fiber layer. There is a parameter defined in the model that describes the activation of smooth muscle cells. This parameter is assumed to depend directly on the blood pressure feeding the system; increasing this parameter simulates an activation of smooth muscle cells. The model is used to predict velocities in a real network of 25 segments of retinal arteries, and the model predictions are in good agreement with published data. Overall, the model is the first successful implementation of a 3D simulation of retinal blood flow in a realistic network of retinal arteries that includes an autoregulation mechanism by describing the effect of smooth muscle fibers on blood flow. Extensions of this work will include specific physiological mechanisms that trigger vessel constriction or relaxation in order to provide an even more complete 3D model of the hemodynamics in the eye. Such 3D models are a major contribution to the field because they eliminate the geometric simplifications inherent in lower dimensional models of the ocular circulation.

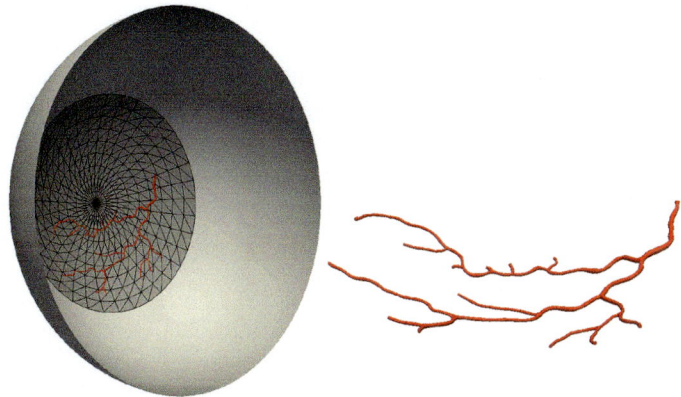

**Fig. 6** Snapshot of vascular geometry obtained for computations using retinal fundus image [1]

## 2.2 Oxygen Transport

### 2.2.1 Overview

The purpose of the circulatory system is to transport nutrients and remove wastes from every tissue in the body. Since blood must flow within a very small distance of every tissue point in the body to allow for passive diffusion, the circulatory system provides a means of transporting blood throughout the body via convection along a complex network of blood vessels before reaching the site of diffusive exchange (e.g., capillaries). Several studies have shown that retinal oxygenation plays a fundamental role in ocular physiology. Pathological changes in retinal oxygenation have been associated with several disease conditions, including glaucoma, diabetic retinopathy, and AMD [134]. However, the difficulty of measuring oxygen levels within the retina of humans or animals has prevented a more complete understanding of the role of oxygenation in ocular diseases. Retinal oximetry measures the level of oxygen bound to hemoglobin within retinal blood vessels, but it does not provide any information on how oxygen is distributed within the tissue. Profiles of the partial pressure of oxygen across the retinal thickness have been acquired in rabbits, guinea pigs, and rats by means of microelectrode-based technologies [80, 141]. However, these measurements do not capture the spatial heterogeneity of the partial pressure of oxygen within tissue and are fairly invasive, thereby precluding their application to humans. In this context, mathematical modeling can help bridge the gaps between measurements of blood flow and blood oxygenation and provide quantitative methods to relate changes in oxygen levels with specific disease conditions.

### 2.2.2 Mathematical Modeling Approaches

Theoretical modeling provides a very useful tool to describe the processes of convection, diffusion, and permeation of oxygen that occur over multiple time and space scales in different tissues. The transport of oxygen by single or multiple vessels has been reviewed previously [48, 110]. This section describes fundamental models of oxygen transport in tissue and blood and provides details on oxygen transport studies in the retina.

*Krogh Cylinder Model and Its Extensions*

The Krogh cylinder model [77] defines an array of parallel, evenly spaced oxygen-delivering vessels (e.g., capillaries) that supply oxygen exclusively to tissue cylinders surrounding each vessel (see Fig. 7). Oxygen is delivered to the nearest tissue via diffusion according to:

$$K \left[ \frac{1}{r} \frac{d}{dr} \left( r \frac{dP_{O_2}}{dr} \right) \right] = M(P_{O_2}) \tag{5}$$

where $K$ is the constant tissue diffusivity, $P_{O_2}$ is the tissue partial pressure of oxygen at a radial distance $r$ within the tissue cylinder, and $M(P_{O_2})$ defines the oxygen consumption per volume and is often assumed to be constant or to depend on $P_{O_2}$ via a Michaelis–Menten relationship. In the Krogh model, the $P_{O_2}$ at the capillary wall equals the average capillary $P_{O_2}$, axial diffusion of oxygen is neglected, and tissue oxygen solubility and diffusivity are uniform. This radial diffusion equation can be solved explicitly for the partial pressure of oxygen in the tissue (see Appendix 2 for details):

$$P_{O_2}(r) = P_{O_2}(r_v) + \frac{M_0}{K} \left[ \frac{r^2 - r_v^2}{4} + \frac{r_t^2}{2} \ln \left( \frac{r_v}{r} \right) \right] \tag{6}$$

where $M_0$ is tissue oxygen demand, $K$ is the diffusion coefficient, $r_v$ is the vessel radius, and $r_t$ is the radius of the tissue region surrounding each vessel, as schematized in Fig. 7.

Although the Krogh model is based on several simplifications, it provides reasonable predictions of oxygen exchange with tissue and is a model that has been used and improved upon by several subsequent studies. For example, the Krogh model has been extended to include other important circulatory effects such as time-dependent transport [124], myoglobin-facilitated tissue transport [92, 93, 137], and intravascular resistance to radial oxygen diffusion [40, 67]. The motion of the myoglobin molecule (which can bind and release oxygen) can lead to an increase in oxygen diffusion, which is a process known as myoglobin-facilitated tissue transport. Several models have been altered to account for this effect [92, 93, 137], which in many instances causes increased oxygen diffusion to low $P_{O_2}$ regions. Capillary intravascular resistance occurs due to the drop in $P_{O_2}$ between the center

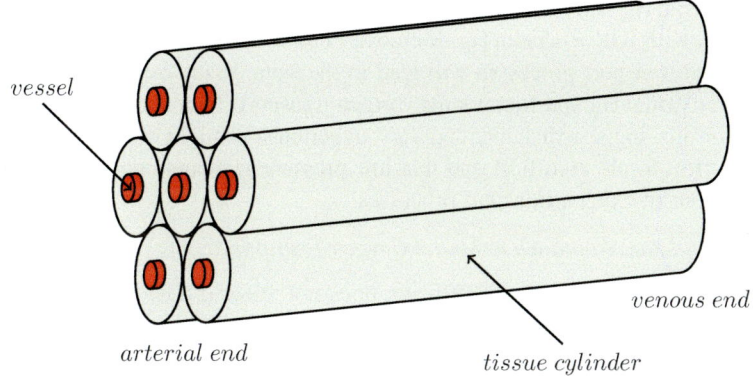

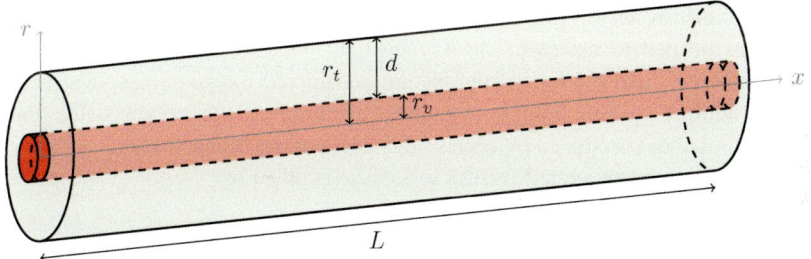

**Fig. 7** Schematic representation of Krogh cylinder model. *Top:* in a Krogh cylinder model, vessels (red) run along the center axis of a tissue cylinder (gray). *Bottom:* schematic of a single vessel (red) supplying a cylindrical region of tissue (gray) with oxygen, where $r$ is the radial coordinate, $x$ is the distance along the network, $r_t$ is the radius of the tissue region, $r_v$ is the radius of the vessel, $d$ is the tissue width, and $L$ is the vessel length [21]

of a red blood cell and the capillary wall [40, 67] and is included in models by using a flux boundary condition on tissue $P_{O_2}$ at the outer edge of the capillary wall. Additional reviews of the Krogh model and its adaptations are given in [48, 106, 110, 127].

In [43], Friedland developed a one-dimensional Krogh cylinder model to study the transmural transport of oxygen to metabolize the human retina. The model accounts for tissue metabolism, time-varying concentrations, and hydrostatic trans-mural pressure gradients, while assuming constant oxygen concentration at the arteriolar end of retinal capillaries and neglecting intravascular and extravascular diffusion. The model is solved analytically via a dimensional analysis based on the small ratio of capillary radius to the Krogh cylinder tissue radius. The steady state model results are used to investigate the effect of IOP and intercapillary hypertension on oxygen concentration. Intercapillary hypertension is modeled by increasing the arteriolar capillary pressure, i.e., the blood pressure at the arterial end of the capillaries. The results suggest that a hypertensive individual with

IOP $= 30$ mm Hg will have the same tissue oxygen concentration as a normotensive individual with IOP $= 15$ mm Hg. Moreover, IOP $= 15$ mm Hg and dysfunctions in the oxygen transport processes will lead to the same tissue oxygen concentration as IOP $= 30$ mm Hg and functioning oxygen transport. In conclusion, the reported model results suggest that intercapillary hypertension might create a counterbalancing effect to elevated IOP, and that low pressure glaucoma might be related to dysfunctions in oxygen transport processes.

*Green's Function Approach to Model Oxygen Transport*

The Krogh model for oxygen diffusion does not allow diffusion of oxygen from multiple vessels to a region of tissue. Since tissue regions are supplied in reality by heterogeneous networks, the Krogh model may underestimate the calculation of tissue hypoxia. Secomb et al. [68, 123, 125] introduced a model of oxygen transport based on a Green's function method which explicitly represents the interactions among vessels and tissue in networks with non-uniform geometry. This modeling approach utilizes techniques from potential theory and reduces the number of unknowns needed to represent the oxygen field. Vessels are modeled as discrete oxygen sources, and the tissue regions are considered oxygen sinks. The oxygen concentration at a tissue point is calculated by summing the oxygen fields (called Green's functions) produced by each of the surrounding blood vessels.

The equation for oxygen diffusion in a tissue is given by:

$$D\alpha\nabla^2 P_{O_2} = M(P_{O_2}) \tag{7}$$

$$D\alpha\nabla^2 G(x, x_i) = \delta(x - x_i) \tag{8}$$

$$P_{O_2} = \sum_i G(x, x_i)q_i \tag{9}$$

where $D$ is the diffusion coefficient, $\alpha$ is the solubility of oxygen in tissue, and $M(P_{O_2})$ is the oxygen consumption rate in the tissue that depends on the local tissue $P_{O_2}$. The Green's function $G(x, x_i)$ is the solution of Eq. (8) and represents the $P_{O_2}$ at a point $x$ resulting from a unit point source at $x_i$. The solution for the tissue $P_{O_2}$ is given in Eq. (9), where $q_i$ represents the distribution of source strengths. Figure 8 shows an example solution of the Green's function model for oxygen saturation in blood and tissue for a complex arterial network in mouse retina [45].

*Models of Oxygen Transport in the Retinal Vasculature and the Retinal Tissue Layers*

The retinal tissue is characterized by distinct layers of different cells whose activity leads to the conversion of a light stimulus into a neural signal that is sent to the brain for decoding. Each cell type is characterized by a specific level of oxygen demand, thereby calling for a detailed mathematical representation of oxygen dynamics across the retinal layers.

In [64], Haugh et al. developed a simplified mathematical model of oxygen diffusion and consumption in the outer layer of the cat retina. Since the $P_{O_2}$ in

**Fig. 8** Sample output of blood vessel and tissue oxygenation using the Green's function method on the arteriolar network of the retina [45]

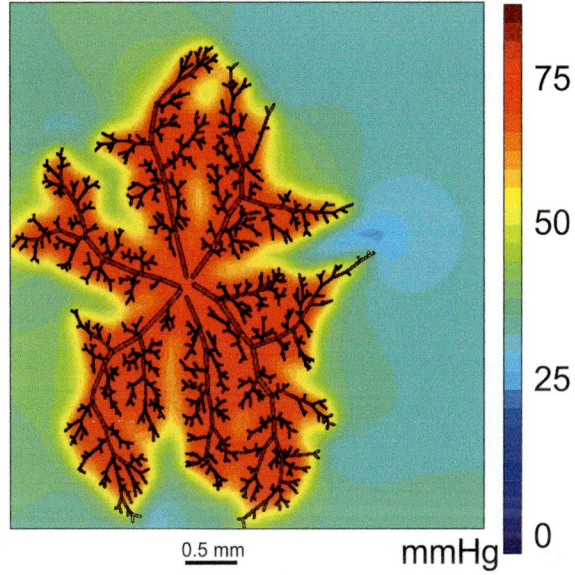

this region depends only on diffusion from the circulation of the choroid and retina, a simple three-layer steady state diffusion model is used in which oxygen diffusion is equated with constant tissue consumption. A cubic spline routine is used to fit experimental $P_{O_2}$ profiles. Interface conditions at the boundary of each layer are imposed to assure continuous oxygen tension and flux. The model was adapted to describe a *general* three-layer diffusion model and a *special* three-layer diffusion model. In the general model, oxygen consumption was assumed constant in each of the three layers. The model was fitted to data from dark-adapted and light-adapted retinas, and model parameters were estimated using nonlinear regression. In the special three-layer model, oxygen consumption was assumed constant in the middle layer and was assumed to be zero in the other layers near the choroid and inner retina since experimental measures showed oxygen consumption rates in those layers to be either zero or negative. This reduced the number of parameters in the model, and nonlinear regression was used to estimate the best fits of the parameters. The results of the two different models were then compared via an analysis of variance (F-test). Model predictions showed that oxygen consumption in light was about 60% of that in the dark. The special three-layer diffusion model (consumption only in the middle layer) showed a good fit with both dark and light adapted profiles.

A coupled description of the oxygen dynamics in the retinal microvasculature and in the retinal tissue layers has been proposed in [24, 28]. This model provides an improved description of the retinal anatomy compared to the Krogh model by considering: (1) six different retinal layers with different metabolism and oxygen sources, (2) two distinct layers of retinal capillaries that exchange oxygen at different retinal depths, and (3) an oxygen exchange mechanism driven by the

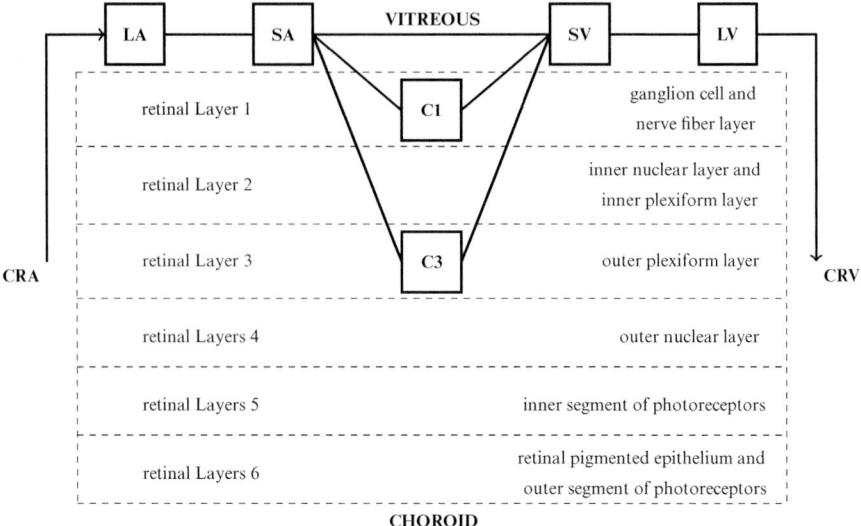

**Fig. 9** Model schematic of retinal tissue and retinal microvasculature. The retinal microvasculature is represented by the following compartments: the large arterioles (LA), small arterioles (SA), intermediate capillaries (C1), deep capillaries (C3), small venules (SV), and large venules (LV). The central retinal artery (CRA) is upstream of the model, and central retinal vein (CRV) is downstream of the model

difference in $P_{O_2}$ levels in the vasculature and in the tissue. A schematic of the model is presented in Fig. 9.

The transport of oxygen in the retinal tissue is described in each layer by a diffusion–reaction equation in the direction of the retinal depth $z$:

$$-\alpha_t D_t \frac{d^2 \overline{p}_t}{dz^2} = \overline{\mathscr{D}}(p_t, p_c) \qquad (10)$$

where $\alpha_t$ represents the oxygen solubility coefficient in the tissue, $p_t$ represents the partial pressure of oxygen in the retinal tissue layer, and therefore $\alpha_t p_t$ represents the oxygen concentration inside the tissue. $D_t$ represents the oxygen diffusion coefficient in the tissue, $\overline{\mathscr{D}}(p_t, p_c)$ represents oxygen supply and consumption, and $p_c$ is the partial pressure of oxygen in the capillaries within the retinal layer. The term $\overline{\mathscr{D}}(p_t, p_c) := \overline{\mathscr{S}}(p_t, p_c) - \overline{M}(p_t)$ is the sum of an oxygen source term $\overline{\mathscr{S}}$, which is different from zero only in those layers where capillaries are present (layers 1 and 3), and an oxygen consumption term $\overline{M}$, which is modeled by Michaelis–Menten kinetics [118]. The full system consists of a set of six differential equations, one for each retinal layer, with layer-specific parameters. The bar over the quantities in Eq. (10) indicates that they are obtained by averaging over a surface of the retina at a fixed depth $z$. The system is closed by imposing boundary conditions to guarantee the continuity of the partial pressure distribution and its first derivative at each interface between layers. The partial pressure of oxygen at the lower end of layer 6 is set equal to the partial pressure of oxygen at the level of the choroid.

The transport of oxygen in the microvasculature is described by a diffusion–advection equation. The equation describes the radial diffusion of oxygen dissolved in plasma and red blood cells, and the advection of oxygen in the axial direction $s$ of the vessel. In this formulation, the radial diffusion of oxygen bound to hemoglobin and the axial diffusion of oxygen are neglected, since they have a minor effect on oxygen dynamics compared to the other terms. The system is closed by adding a boundary condition that regulates the exchange of oxygen between the microvasculature and the retinal tissue surrounding the vessel. The exchange of oxygen is proportional to the weighted difference between the partial pressure of oxygen inside and outside the vessel, and the effect of the wall is incorporated in the permeability constant $\mathscr{L}_p$ using a technique called a *wall-free* model [142]. The equation is then reduced (by computing the average on the cross-section orthogonal to the direction $s$) to obtain an ordinary differential equation in the direction $s$. The reduced equation is coupled with the retinal tissue to obtain (11). Note that the contribution of the radial diffusion of oxygen is included in the term $q(s)$ of Eq. (11) as a result of the reduction procedure:

$$Q_c \frac{\mathrm{d}}{\mathrm{d}s} \left( \alpha_c \overline{p}_c(s)\omega_{u*} + H_D c_0 \overline{S}_{O_2}(s) \right) = -q(s), \quad s \in (0, L] \tag{11}$$

where $Q_c$ is equal to the blood flow in the vasculature, $\alpha_c p_c + H_D c_0 S_{O_2} = c$ is the total oxygen concentration in the blood, which is the sum of the oxygen dissolved in plasma and red blood cells, denoted by $\alpha_c p_c$, and the oxygen bound to hemoglobin in red blood cells, denoted by $H_D c_0 S_{O_2}$. The quantities $p_c$ and $S_{O_2}$ represent the partial pressure of free oxygen and the blood oxygen saturation which is related to $p_c$ through the Hill equation, $H_D$ is the blood hematocrit, and $c_0$ is the oxygen carrying capacity of red blood cells. The quantity $q$

$$q(s) = 2\pi r_j \alpha_c \mathscr{L}_p \left( \overline{p}_c(s) - \beta p_{t,w} \right) f^*(r), \tag{12}$$

is the oxygen volumetric rate per unit of length, where $r$ is the vessel radius, $p_{t,w}$ is the partial pressure of oxygen in the tissue surrounding the vessel and $\beta = \alpha_t / \alpha_c$. The parameter $\mathscr{L}_p$ regulates the exchange of oxygen between the blood inside the vessel and the tissue around it. The bar over the quantities in Eqs. (11) and (12) indicates that they are obtained by averaging over the vessel cross-section at a fixed axial coordinate $s$. The quantities $\omega_{u*}$ and $f^*(r)$ are related to the shape functions assumed for the oxygen partial pressure profile inside the vessel. Equation (11) is closed with an inflow boundary condition for $p_c(s = 0)$. Equations (11) and (12) are derived for all the vascular compartments of the retina (depicted in Fig. 9) and describe the variation of $c$ along the vasculature. Equations (10) and (11) are coupled through Eq. (12) that regulates the oxygen exchange between blood and retinal tissue depending on the values of $p_c$ and $p_t$. The solution of the coupled system of Eqs. (10) and (11) will result in the oxygen partial pressure distribution in the retinal vasculature and tissue. Additional details on the derivation of the model are available in [24, 28].

The modeling techniques described in this section are used in the work by Causin et al. [28] to develop a model that couples retinal blood flow to oxygen transport in the blood and in the retinal tissue. The model consists of two parts: (1) a model for the retinal microcirculation, composed of a vascular network including arterioles, venules, and capillaries, and (2) a model for the retinal tissue, composed of six different layers characterized by specific oxygen consumption rates. The model is used to perform a sensitivity analysis to quantify the influence of changes in some physiological parameter on the oxygen distribution in the vessels and in the retinal tissue. The considered parameters are blood pressure, blood rheology, oxygen arterial permeability, and tissue oxygen demand. The model predicts that a sudden increase in arterial blood pressure would cause an increase in the level of blood flow, and consequently an increase in the oxygen levels in both the vasculature and retinal tissue. These results are in agreement with the experimental observations in [101], and are supported by clinical studies that identify arterial hypertension as a risk factor for retinal diseases. An increase in the value of plasma viscosity resulted in a strong decrease in the level of oxygen tension in the retinal layer containing the retinal ganglion cells. These results support the findings of clinical studies that identify an elevated value of blood viscosity as a risk factor for the occlusion of retinal vessels and diseases such as glaucoma and diabetic retinopathy. Changes to the level of oxygen arterial permeability, $\mathscr{L}_p$ in Eq. (12), resulted in significant changes in the level of oxygen saturation in the arterial vessels of the network and minor changes in the oxygen concentration in the vessel walls and in the retinal tissue. The effect of variations of the tissue oxygen demand (metabolic activity) on the oxygen distribution in the retinal tissue is reported in Fig. 10.

The model by Causin et al. leverages a previous work by Cringle and Yu, who proposed a multilayer mathematical model of oxygen supply and consumption in the rat retina [32]. Cringle and Yu assume that the retina is divided into eight layers, each with a distinct oxygen consumption or supply rate. When applied to the available data from intraretinal oxygen measurements in the rat under normal physiological conditions, a close fit between the model and the data is achieved. The model is then used to investigate recent evidence of oxygen regulating mechanisms in the rat retina during systemic hyperoxia. Fitting the model to the experimental data allowed the relative oxygen delivery or consumption of the key retinal layers to be determined. Two factors combine to produce the relative stability of inner retinal oxygen levels in hyperoxia. The retinal layer containing the outer plexiform layer/deep retinal capillaries switches from being a new supplier to being a net consumer of oxygen, and the oxygen consumption of the outer region of the inner plexiform layer increases significantly. The model provides a useful tool for examining oxygen consumption and supply in all retinal layers, including, for the first time, those layers within the normally perfused inner retina.

In [105], Olson et al. developed a one-dimensional model of oxygen transport along the thickness of the retinal tissue to investigate the effect of progressive and chronic ischemia, associated with prolonged hyperglycemia, on retinal oxygen

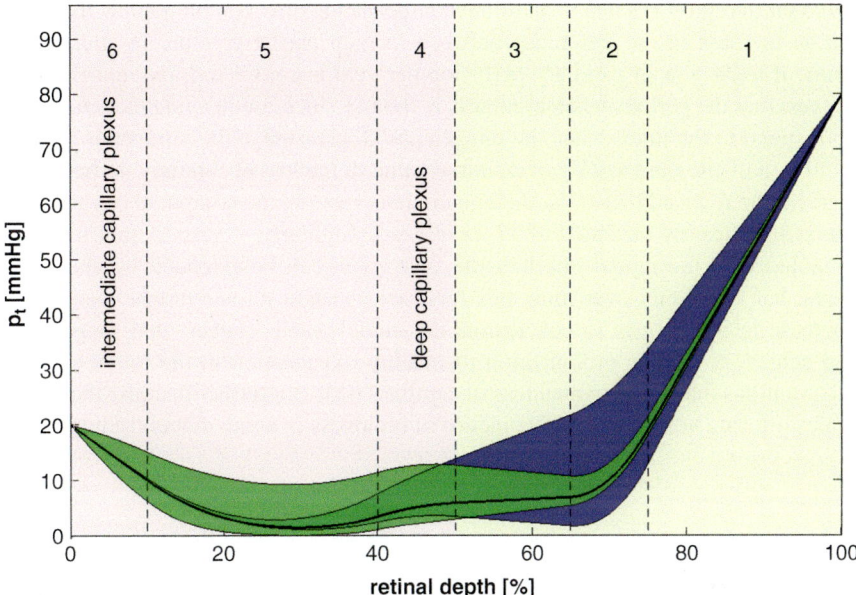

**Fig. 10** Model predicted oxygen profiles in retinal tissue for variations of tissue oxygen demand in inner retina (green area) and outer retina (blue area) [28]. The baseline profile is represented by the black thick curve. Note that the numbering of the layers follows the opposite order of Fig. 9

saturation. In this one-dimensional model, the retinal structure extending from the choroidal to the vitreous interfaces is approximated as four distinct regions with three different oxygen sources and sinks: (1) the majority of the outer retina oxygen consumption is assumed in region 2, which corresponds to the inner segmental layer of the retinal photoreceptors; (2) the inner retina corresponds to region 4 and it is modeled as uniformly distributed sources and sinks of oxygen, accounting for the oxygen consumption of the plexiform layers and for the oxygen source coming from the capillaries; (3) in the remaining retinal regions, purely diffusion and no consumption of oxygen is assumed. In particular, sinks are modeled via a Michaelis–Menten equation with different maximum consumption rate in regions 2 and 4, and sources are modeled based on Fick's principle in region 4. Continuity of oxygen partial pressure and its derivative are assumed at the choroidal border, while a zero flux condition is assumed at the vitreous border. The model also accounts for retinal arteriolar vasodilation feedback mechanisms, in response to low levels of oxygen partial pressure, via a two parameter model of the sources in region 4. These feedback mechanisms also take into account capillary degradation due to hyperglycemia.

When the vasodilation feedback is absent, the steady state model results suggest that oxygen consumption in region 2 is independent of retinal blood flow, hence its main oxygen source is from the choroid via diffusion, while in region 4 the oxygen consumption decreases nonlinearly as blood flow decreases. The nonlinear behavior in region 4 is due to an initial compensating effect on blood flow reduction via an increase in the difference between oxygen partial pressure in arteries and veins. If a 2% loss of capillary perfusion per year is considered, the model results suggest that the periods when blindness is found to increase in aging diabetic adults correspond to the times when the oxygen partial pressure at the vitreous is reduced to 50% (half-life) and to 25% of its initial value. If the loss of capillary perfusion per year can be reduced to 1% via therapeutic intervention, the computed oxygen half-life is significantly extended to 37 more years. Similarly, if vasodilation feedback is included in the model, the half-life of oxygen can be extended by about 3–6 years, but eventually over time, this feedback mechanism is overwhelmed by the decrease in blood flow. In conclusion, the model results suggest that vasodilation can achieve a notable prolongation of healthy oxygenation in the retina and that interventions that lower annual loss of capillary beds can further increase these time scales offering hope for lower incidence of blindness in aging diabetic adults.

## 2.3  Angiogenesis

### 2.3.1  Overview

Angiogenesis is the formation of new blood vessels from pre-existing vessels. This type of vascular growth has been observed to play a particularly important role in developmental processes (embryogenesis), organ growth, wound healing, and pathological processes. The sprouting of new vessels is typically initiated by growth factors, such as vascular endothelial growth factor (VEGF), platelet-derived growth factor (PDGF), or basic fibroblast growth factor (bFGF). Angiogenesis is a two-step process: first, a dense and immature network of new vessels developed by the recursive sprouting and fusion of sprouts; and second, the dense network is remodeled into a mature and hierarchical network by adaptive pruning methods. The entire process of angiogenesis integrates phenomena over numerous biological scales and involves multiple cell types and chemical signals. Thus, mathematical modeling has emerged as a useful tool for understanding and quantifying the complex process of angiogenesis in numerous contexts. Most mathematical models are based on partial differential equation (PDE) models that provide a spatial and temporal description of cell migration and vessel formation [7, 91, 135]. Blood perfusion is considered in some models [135], and continuum approaches and agent-based modeling approaches have also been used [10, 69].

### 2.3.2  Mathematical Modeling Approaches

The retinal vasculature is formed predominantly by angiogenesis. Four to five days before the first angiogenic sprouts, directed movement of endothelial cells is coupled to the migration of astrocytes from the optic nerve region to the periphery of the retina. Poorly oxygenated astrocytes produce VEGF, which leads to the expansion of a new and immature vascular network over the inner surface of the retina. On the day of birth, endothelial cells sprout from the ophthalmic vein, and a dense vasculature feeds the periphery of the retina 8 days following birth. Once blood arrives, the growth factors are down-regulated, and extensive vascular pruning takes place to mature the retinal network. Mathematical models that simulate this process of retinal vascular development must represent a number of angiogenic mechanisms mathematically, such as the proliferation and migration of endothelial cells in response to growth factors (chemotaxis) and insoluble extracellular membrane molecules (haptotaxis) [7, 91, 135].

Many early models of angiogenesis assume that there is only one type of endothelial cell or that endothelial cells are indistinguishable [3, 90]. Thus, a sprout tip cell is modeled using a diffusion–advection equation:

$$\frac{\partial e}{\partial t} = D_e \nabla^2 e - \nabla \cdot (\chi(c) e \nabla c) \tag{13}$$

where $e$ denotes epithelial cell density, $c$ represents the concentration of a growth factor such as VEGF, $D_e$ is the diffusion coefficient, and $\chi(c)$ represents the chemotactic sensitivity of endothelial cells to VEGF. Later models divide the vessels into tips and sprouts, in which the tips of the vessels proliferate, migrate, fuse, and branch while the capillary sprouts follow the tips [8, 15, 16, 108, 122]. Typical models for tips ($n$) and sprouts ($b$) have been given by:

$$\frac{\partial n}{\partial t} = -\nabla \cdot J_n + (\lambda_1 cb + \lambda_2 cn) - (\lambda_3 b + \lambda_4 n)n \tag{14}$$

$$\frac{\partial b}{\partial t} = \|J_n\| + D_b \nabla^2 b + \lambda_5 b(b_0 - b) \tag{15}$$

where flux $J_n = -D_n \nabla n + \chi(c) n \nabla c$, as described in Eq. (13). The first term of Eq. (15) indicates that the sprouts increase at the same rate as the Euclidean norm of the flux of the tips ($\|J_n\| = \sqrt{J_n \cdot J_n}$). These models provide some of the first attempts at explaining angiogenesis and have been adapted in several contexts.

For example, Aubert et al. [7] combine mathematical and experimental techniques to study physiological angiogenesis in the retina during development in the mouse eye. The study focuses on the formation of the superficial plexus and the evolution of VEGF and its effect on tip migration (depicted in Fig. 11). In the mathematical model of the retinal vascular network, three partial differential equations are used to track the average capillary tip density, blood capillary density, and VEGF concentration. The model is used to describe the migration of endothelial tip cells and blood vessels and includes the movement of astrocytes in response to a gradient of PDGF and the secretion of VEGF by hypoxic astrocytes. Capillary tips are assumed to migrate via a biased random walk in which migration is biased to occur in the direction of decreased blood capillary density and in the direction of increasing VEGF concentration. Model parameters are estimated from experimental

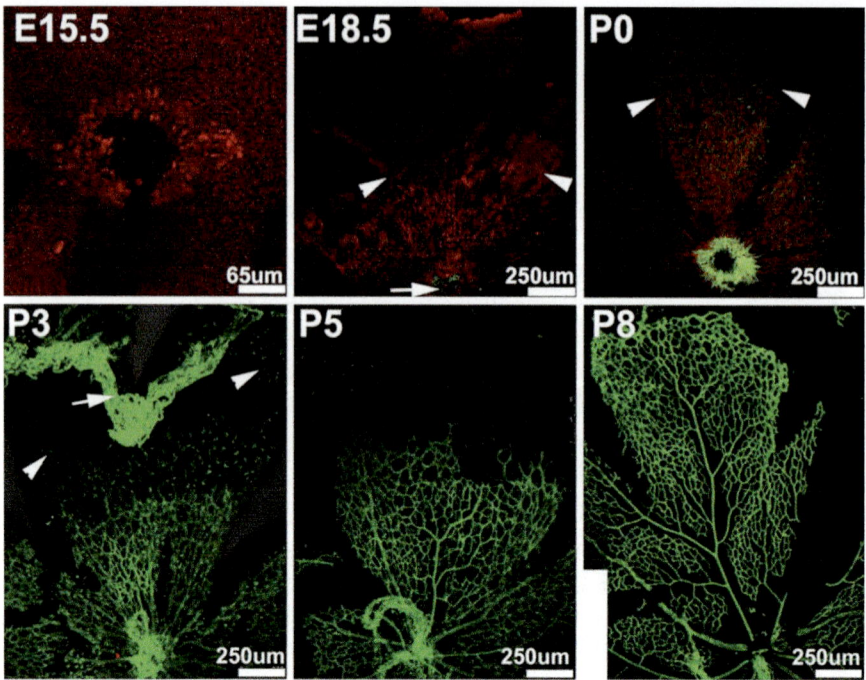

**Fig. 11** Depiction of murine retinal astrocyte migration and retinal vascular plexus formation at pre- and post-natal stages: from embryonic day (E)15.5 to post-natal day (P)8. Endothelial cell sprouting from the optic nerve head begins at the time of birth (P0). By P5, defined arterioles and venules have formed, and by P8, remodeling has led to a sparse network [7]

data collected in the study, and a sensitivity analysis is performed to obtain the most accurate model.

McDougall et al. [91] introduce a two-dimensional hybrid PDE-discrete model to track the migration of astrocytes and endothelial tips toward the outer boundary of the retina. The model includes the effects of VEGF and PDGF on cell migration, and blood perfusion is included throughout the plexus development. Equations for blood flow take into account the ability of the vasculature to adapt to changes in wall shear stress, intravascular pressure, and metabolic responses. Model simulations allow for the resulting retinal trees to adapt and remodel in response to various biological factors. The model predictions of the retinal vascular plexus are shown to agree very well with experimental data collected in the retina (whole-mounted retinal vasculature).

Although angiogenesis in the mammalian retina is well-ordered and highly reproducible, there are only a few modeling attempts of angiogenesis specific to the retina. As outlined by Watson et al. [135], PDE models by Maggelakis and Savakis [87, 88] have modeled the interplay between VEGF, oxygen, and capillary density in pathological cases; other hemodynamic models have predicted flow, shear stress, or pressure values throughout the retinal circulation without emphasizing angiogenesis. In Watson et al. [135], an experimentally based two-dimensional hybrid PDE-discrete model is presented to describe angiogenesis during murine retinal development. The outward migration of astrocytes and endothelial tips in response to various growth factors are modeled, and blood perfusion, vessel adaptation, and remodeling are included as well. Comparisons between their model with experimental data taken from various stages of development show excellent agreement between predicted and observed vascular morphologies. This model provides great potential for understanding how cellular, molecular, and metabolic cues regulate growth and formation of the retinal circulation and offers a useful platform for investigating vascular-related diseases of the eye.

Connor et al. [30] develop a mathematical model of the roles of VEGF and bFGF on angiogenesis in the corneal vasculature. The study is conducted with an ultimate goal of understanding how tumor secretion of angiogenic growth factors results in pathological vessel formation. Since the cornea is highly visible and normally avascular, it provides an ideal environment for research to determine the potency of angiogenic factors. The mathematical modeling approach adopted in their work extends from theoretical models by Balding and McElwain [8] and Byrne and Chaplain [16]. In short, the mathematical model of angiogenesis in [30] neglects detailed cell–cell and chemical interactions and instead consists of PDEs describing the spatio-temporal evolution of sprout tips, vessels, and a generic angiogenic factor (such as VEGF or bFGF). Balance laws yield a system of equations in which the rate of accumulation of each of these three model populations in a control volume depends on the net flux into the control volume and the net production within the control volume. Sprout tips are assumed to move via chemotaxis, haptotaxis, and random motion; vessels are assumed static; and angiogenic factors move via diffusion. The production of sprout tips is assumed to be stimulated by the binding of antigenic factors (e.g., VEGF and bFGF) to immature blood vessels; motion of the tips dictates the length of the vessel (snail-trail approach). In addition to diffusion, the distribution of antigenic factors is assumed to depend on endothelial uptake, drainage throughout the vasculature, and natural decay. Model simulations are run for varied initial levels of VEGF and bFGF to determine the impact of each factor in isolation and in combination on angiogenesis. The model predicted a more sustained angiogenic response in bFGF experiments.

Jackson and Zheng [69] also develop a mathematical model to study mechanisms associated with corneal angiogenesis. A cell-based model that integrates a mechanical model of elongation with a biochemical model of cell phenotype variation regulation by angiopoietins within a developing sprout is used to understand the roles of endothelial cell migration, proliferation, and maturation in angiogenesis. Although based on observations made in the cornea, this study provides important insight into the process of angiogenesis and is not tissue specific. For example, understanding the importance of proliferation in angiogenesis has been identified as a great challenge to those studying tumor angiogenesis. Discrete points are used to represent each cell in a developing sprout. A viscoelastic equation (spring-dashpot system) describes the elongation of tip cells. Continuous differential equations

**Fig. 12** Comparison between
the values of peak systolic
velocity (PSV) and end
diastolic velocity (EDV) in
the central retinal artery
(CRA) for IOP between
15 mm Hg (healthy value) and
45 mm Hg (elevated value) as
measured by Harris et al. [63]
and predicted by the
mathematical model proposed
in [55]

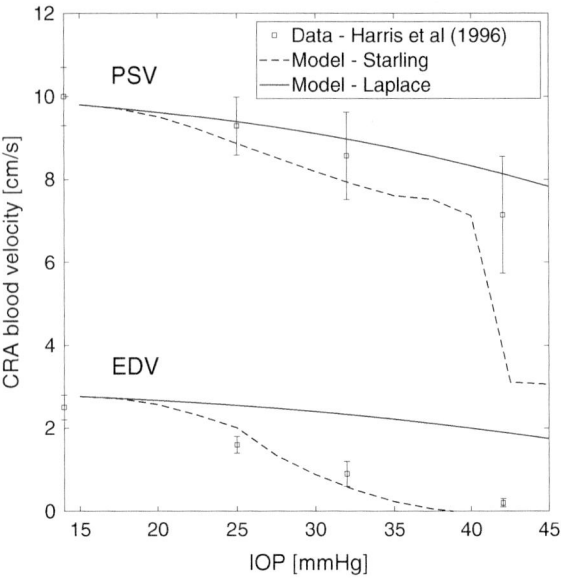

while the EDV is measured at the end of the diastolic phase of the cardiac cycle, when the pressure inside the vessels is at its lowest level. When the pressure inside the vessels is high (and the transmural pressure difference is positive) the behavior of collapsible and compressible tubes is the same; consequently, the model predicts a similar behavior in the PSV response to IOP elevation regardless of whether the Starling or Laplace models are adopted. When the pressure inside the vessels is low, the transmural pressure difference will become negative as IOP increases and the vessels will tend to collapse. The Starling model is able to capture this natural progressive collapse of the vessels, but the Laplace model is not, thereby explaining why the Starling model provides a better agreement with the EDV data than the Laplace model.

Figure 13 shows the model predictions for the time profiles of blood pressure in different vascular compartments for IOP = 15, 25, and 35 mm Hg when the Starling and Laplace models are adopted for venules and CRV. The differences between Starling and Laplace models are amplified with higher IOP levels. As the transmural pressure difference becomes negative, the venous collapse associated with the Starling resistor generates an increase in pressure upstream of the collapse, which makes it easier for the vessels to counterbalance the external pressure, as shown in Fig. 13. This mechanism, consistent with experimental observations, acts as a passive feedback that helps the retinal microvasculature withstand increased IOP levels and sustain the perfusion of the tissue. This passive feedback mechanism is negligible in the Laplace model.

The results discussed in this section emphasize the importance of adopting a Starling resistor modeling technique for retinal venules and veins, especially in conditions where the collapse is highly likely to happen.

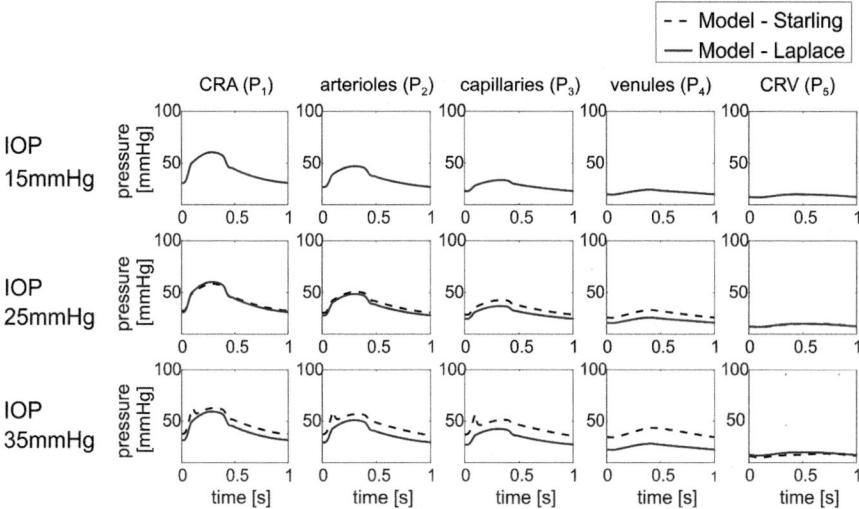

**Fig. 13** Model predicted values of blood pressure in CRA ($P_1$), arterioles ($P_2$), capillaries ($P_3$), venules ($P_4$), and CRV ($P_5$) for IOP = 15, 25, and 35 mm Hg for the Starling and Laplace cases

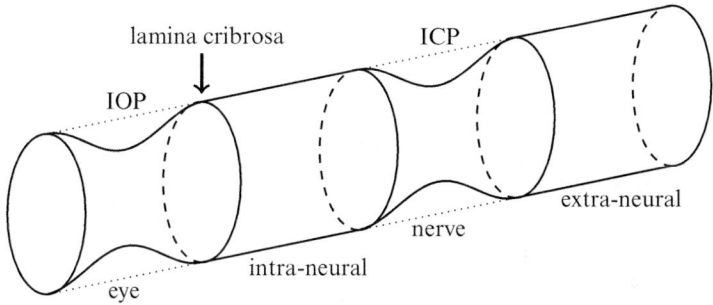

**Fig. 14** Schematic of the different portions of the CRV corresponding to the four two-dimensional compartments adopted in the model developed in [128]

## Two-Dimensional Modeling of Venous Collapse

Stewart et al. considered the hemodynamics of the CRV from the inner region of the eye to the ophthalmic vein using a two-dimensional time-dependent model [128]. The model describes the CRV as a series of four planar two-dimensional compartments representing different portions of the CRV, namely the *eye* portion, the *intra-neural* portion, the *nerve* portion, and the *extra-neural portion*, depicted in Fig. 14. The *intra-neural* and *extra-neural* portions of the vessel are modeled as rigid compartments. The *eye* and *nerve* portions are modeled as collapsible compartments and are exposed to IOP and intracranial pressure (ICP), respectively. In the collapsible compartments, the external wall is assumed to be flexible to mimic the deformability of the corresponding portion of the vessel. All remaining walls are considered rigid.

The model describes the blood flow in the CRV via the two-dimensional incompressible Navier–Stokes equations. The blood flow is driven by a physiological value of flow rate at the inlet of the system, and a pressure value is assigned at the outlet of the system where the CRV drains into the ophthalmic vein. The collapsibility of the *eye* and *nerve* regions is modeled by applying the following boundary condition at the flexible walls:

$$p_{ext,j} - p_j = T_j \frac{\partial^2 d_j}{\partial x^2}, \qquad j = eye, nerve \qquad (19)$$

where $p_{ext,j}$ represents the external pressure to the compartment $j$, namely the IOP and ICP value for the *eye* and *nerve* portions, respectively, $p_j$ is the pressure inside the vessel, $T_j$ is the vessel wall tension, $d_j$ is the vessel diameter, and $x$ is the axial coordinate. Equation (19) can be seen as a tube law relating the transmural pressure difference $p_{ext,j} - p_j$ to the value of wall tension and to the changes in the diameter of the vessel. The system is closed by a von Karman–Pohlhausen approximation for the liquid flow in each vessel portion, where the streamline velocity $u$ is expressed as a function of the radial coordinate $y$, the diameter $d_j$, and the two-dimensional flow rate $Q$ as

$$u = \frac{6\,Q\,y\,(d_j - y)}{d_j^3}, \qquad j = eye, nerve. \qquad (20)$$

The proposed CRV model is used to investigate the possibility of estimating ICP non-invasively by varying IOP and measuring the onset of CRV pulsation. For more details, refer to Sect. 3.2.

## 3   Retrobulbar Vasculature

In this section we will focus on the mathematical aspects related to blood flow in the vessels upstream and downstream of the retinal vasculature, namely the ophthalmic artery (OA), central retinal artery (CRA), and central retinal vein (CRV). These vessels nourish and drain the retina and other ocular vascular beds and are characterized by a larger diameter compared to the retinal vessels. Moreover, vascular changes in the OA, CRA, and CRV have been associated with several ocular diseases, such as glaucoma, diabetic retinopathy, CRA and CRV occlusion, OA stenosis, and uveitis [63, 102, 109]. Blood flow velocity in retrobulbar vessels can be measured non-invasively in vivo via Doppler ultrasound [53].

Despite the importance of the OA, CRA, and CRV hemodynamics in relation to ocular disease, the mathematical modeling of these vessels has started only recently. The main challenge in modeling these vessels is represented by their anatomy, since they run through different cranial/ocular compartments and are hence exposed to different pressurized ambients along their length. In particular, the OA begins inside the cranium and then passes through the optic nerve canal going into the eye socket; hence it can be divided into three segments [65]:

- the *intracranial segment* that precedes the dura mater;
- the *intracanalicular segment* that runs in the optic nerve canal (partly in the subdural space and partly within the substance of the dural sheath); and
- the *extracranial segment* that enters the orbit.

The CRA and CRV start inside the eye socket and then pass through the optic nerve going into the eye; hence they can be divided into three segments [62]:

- the *extra-neural segment* that lies inside the eye socket;
- the *intra-neural segment* that runs inside the optic nerve; and
- the *intraocular segment* that lies inside the eye.

Overcoming these mathematical challenges and developing a mathematical model of OA, CRA, and CRV hemodynamics is crucial for investigating the effects of the different pressures, such as intraocular pressure (IOP), cerebrospinal fluid (CSF) pressure, and intracranial pressure (ICP), on retinal hemodynamics.


## 3.1   Ophthalmic Artery

The OA has the unique feature that it is composed of an intracranial segment (affected by the intracranial pressure) and an extracranial segment (affected by the pressure in the eye socket), see chapter "Vascular Anatomy and Physiology of the Eye". This OA feature has been essential in the development of a non-invasive method to measure intracranial pressure (ICP), for more details see chapter "Instruments to Measure and Visualize Geometrical and Functional Parameters Related to the Fluid Dynamics of Cerebrospinal Fluid in the Eye". The main idea of this method is to compare blood velocity profiles, measured via Doppler ultrasound, in the intracranial and extracranial segments of the OA varying the pressure applied externally to the eye socket [116]. When the velocity profiles in the two segments are similar, then the external pressure ($Pe$) balances the ICP. This novel non-invasive ICP measuring technique has been tested using a combination of mathematical models and clinical results.

Ragauskas et al. [115] developed a one-dimensional model of the OA fluid–structure interaction system, modeling the blood as a Newtonian viscous fluid and the artery wall as a viscoelastic material. The wall of the OA deforms due to ICP, acting as an external pressure in the intracranial segment, and a pressure $Pe$ compressing the extracranial segment. The OA wall is assumed to be rigid in the intracanalicular segment. Numerical simulations, together with clinical measurements, suggest that the proposed non-invasive technique has a negligible systematic error (within limits of $[-3, +1]$ mm Hg), and that the error does not depend on the value of ICP, namely this ICP measuring technique does not require calibration.

The model proposed by Ragauskas et al. in [115] has been extended by Misiulis et al. in [95] to account for the interaction between the OA and the internal carotid artery, and for three-dimensional geometries of both arteries. The blood is modeled as a Newtonian viscous fluid, and an arbitrary Lagrangian–Eulerian formulation is used to couple the fluid flow to the deformation of the arterial wall, assumed to be a neo-Hookean material. This model provides better insight into the effect of ICP and $Pe$ on the hemodynamics in the OA and the internal carotid artery. In particular, the model suggests that there is a difference in the blood flow rate (averaged over a heartbeat) measured at the levels of the internal carotid artery and the extracranial OA segment, and that the hydrodynamic pressure drop between the two measuring points, namely the intracranial and extracranial segments, must be taken into account when evaluating the balance between ICP and $Pe$. Thus, the condition for ICP and $Pe$ balance is described by the following equation:

$$ICP = Pe + \Delta Pe_k,$$  (21)

where $\Delta Pe_k$ is the extra balance pressure term that incorporates the hydrodynamic pressure drop between the two measuring points (intracranial and extracranial segments). With this corrected balance condition, the numerical simulations suggest an absolute systematic error in the ICP measurements of 0.65 mm Hg.

## 3.2   Central Retinal Vessels

There is clinical evidence that CRA and CRV hemodynamics are strongly affected by the level of IOP inside the eye globe and by the level of cerebrospinal fluid (CSF) pressure in the subarachnoid space within the optic nerve, but the mechanisms through which this occurs are still elusive. A collagen structure called the lamina cribrosa within the optic nerve head separates the intraocular region, subjected to the IOP, from the optic nerve, subjected to the retrolaminar tissue pressure (RLTp), which is related to the CSF pressure [97]. Thus, changes in IOP and/or CSF pressure induce a deformation of the lamina cribrosa and affect the central retinal vessels running through it.

Guidoboni et al. proposed the first mathematical model that combines the blood flow in the CRA with the lamina cribrosa deformation [54]. The lamina model incorporates material and geometrical nonlinearities. More precisely, the lamina is modeled as a nonlinear, homogeneous, isotropic, elastic circular plate of constant finite thickness $h_{lc}$, satisfying the equilibrium equations

$$\nabla \cdot \mathbf{S} = \mathbf{0} \quad \text{in} \quad \Omega \subset \mathbb{R}^3,$$  (22)

where

$$\mathbf{S} = \lambda_{lc}(\sigma_e)\, \text{tr}(\mathbf{E})\mathbf{I} + 2\mu_{lc}(\sigma_e)\mathbf{E}$$  (23)

is the stress tensor,

$$\mathbf{E} = \frac{1}{2}[\nabla\mathbf{u} + (\nabla\mathbf{u})^T + (\nabla\mathbf{u})^T\nabla\mathbf{u}] \tag{24}$$

is the Green–Saint–Venant strain tensor, $\mathbf{u}$ is the displacement vector, $\mathbf{I}$ is the identity tensor, and $\lambda_{lc}$ and $\mu_{lc}$ are the Lamé parameters that vary with the effective stress $\sigma_e$ [103, 138]. The lamina is assumed to deform under the combined action of IOP, RLTp, and scleral tension. RLTp is assumed to be a function of CSF as described in [97]. The CRA is modeled as fluid–structure interaction system: a stationary flow of an incompressible Newtonian viscous fluid filling a linearly elastic compliant tube. The CRA walls deform under the action of an external pressure that varies along the vessel length to include RLTp, IOP, and the effect of lamina cribrosa deformation, as described in Fig. 15. More precisely, the external pressure $Pe(z)$ varies along the axial coordinate $z$ as follows:

$$Pe(z) = \begin{cases} \text{RLTp} & \text{for } 0 \leq z < z_{lc,c}, \\ -S_{ss}(0, z) & \text{for } z_{lc,c} \leq z \leq z_{lc}, \\ \text{IOP} & \text{for } z_{lc} < z < L, \end{cases} \tag{25}$$

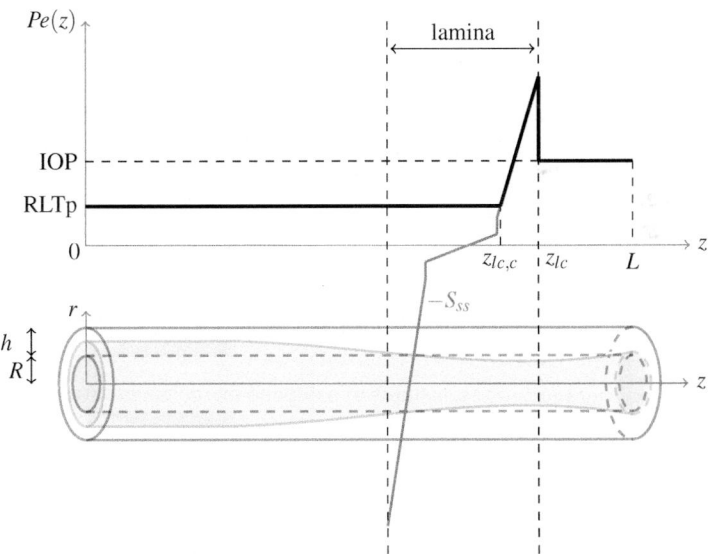

**Fig. 15** Representation of the external pressure $Pe$ acting on the wall of the CRA overlapping a sketch of the domain occupied by blood flowing inside the CRA (shaded area) and the domain of the CRA wall (hollow circular cylinder). $Pe$ varies along the vessel accounting for the intraocular pressure (IOP), retrolaminar tissue pressure (RLTp), and the compression from the lamina cribrosa $S_{ss}$. In the figure, $R$, $L$, and $h$ represent radius, length, and wall thickness of the CRA, respectively

where $S_{ss}$ is the radial component of the stress of the lamina cribrosa evaluated at the lamina centerline, the coordinate $z = z_{lc}$ indicates the relative position of the anterior surface of the lamina cribrosa with respect to the CRA axis, and the coordinate $z = z_{lc,c}$ indicates the lower end of the compressive region in the lamina.

In [54], the CRA model is solved analytically via a dimensional analysis based on the smallness of the aspect ratio $\epsilon = R/L$, where $R$ and $L$ represent radius and length of the CRA, respectively. The 0th order solution of the fluid structure interaction problem in the CRA can be expressed as

$$U_r(r, z) = \left(a_1 \gamma(z) p(z) - a_2 Pe(z)\right) r + \left(\frac{\gamma(z) p(z)}{R} - Pe(z)\right) \frac{a_3}{r}, \tag{26}$$

$$U_z(r, z) = 0, \tag{27}$$

$$p(z) = \frac{1}{b_2} \left[ 1 - \left( N - M \int_0^z (1 - b_1 Pe(t))^{-4} dt \right)^{-1/3} \right], \tag{28}$$

$$\gamma(z) = R\left(1 - b_1 Pe(z)\right) \left( N - M \int_0^z (1 - b_1 Pe(t))^{-4} dt \right)^{1/3}, \tag{29}$$

$$v_r(r, z) = 0, \tag{30}$$

$$v_z(r, z) = \frac{1}{4\mu_b} \left[ r^2 - (\gamma(z))^2 \right] \frac{dp}{dz}, \tag{31}$$

where $r$ and $z$ are the radial and axial coordinates, $U_r$ and $U_z$ are the radial and axial components of the displacement of the vessel wall, $p$ is the blood pressure along the vessel, $\gamma$ is the blood-wall interface, $v_r$ and $v_z$ are the radial and axial components of the blood velocity, and $\mu_b$ is the blood viscosity. $M$ and $N$ are constants that depend on the values $P_0$ and $P_L$ of the blood pressure at the inlet and outlet sections of the vessel, and on the vessel external pressure $Pe$ as follows:

$$N = (1 - b_2 P_0)^{-3}, \qquad M = \frac{N - (1 - b_2 P_L)^{-3}}{\int_0^L (1 - b_1 Pe(t))^{-4} dt}. \tag{32}$$

Here, $a_1, a_2, a_3, b_1,$ and $b_2$ are constants that depend on the geometrical and material properties of the vessel as follows:

$$a_1 = \frac{1}{2(\lambda + \mu)} \frac{R}{h(h + 2R)}, \qquad b_1 = a_2 + \frac{a_3}{R^2}, \tag{33}$$

$$a_2 = a_1 \frac{(h + R)^2}{R}, \qquad b_2 = a_1 R + \frac{a_3}{R^2}, \tag{34}$$

$$a_3 = a_2 \left(1 + \frac{\lambda}{\mu}\right) R(h + R)^2, \tag{35}$$

where $\lambda$ and $\mu$ are the Lamé parameters of the CRA walls.

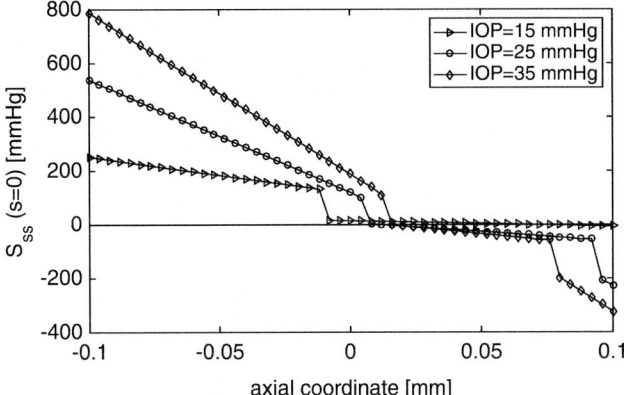

**Fig. 16** Effect of IOP elevation on the lamina cribrosa radial stress $S_{ss}$ along the central axis of the lamina ($s = 0$) as a function of the axial coordinate across the thickness of the lamina, for different IOP values

This model is used to test the conjecture that the lamina might exert compression on the central retinal vessels, since stresses in the lamina are impossible to measure directly in vivo. The model predicts a region of negative radial compressive stress in the center of the lamina cribrosa that increases with IOP both in terms of magnitude and depth penetration in the lamina's thickness, see Fig. 16. This compressive region causes a constriction of the CRA segment passing through the lamina, and therefore an increase in its resistance to blood flow.

In order to further investigate the mechanical factors that contribute to the influence of IOP and CSF pressure on retinal hemodynamics, the model proposed in [54] has been extended to account for the CRV hemodynamics and for the retinal microcirculation [19, 20], as shown in Fig. 17. The CRV is modeled as a compliant tube using the same model utilized for the CRA in [54] with appropriate geometric and material parameters. The vasculature upstream of the CRA and downstream of the CRV, as well as that in between, is modeled as a network of resistances, in the same fashion as in [55, 129]. The resistances are assumed to be constant, with the exception of the resistance representing the effect of retinal venules, denoted by $\mathscr{R}_3$ in Fig. 17, which varies as a function of IOP via Poiseuille's law (Appendix 1) and Laplace's law (Table 1) as follows:

$$\mathscr{R}_3 = \frac{8\mu L}{\pi} \left( R + \frac{R^2(1 - \nu^2)}{hE} \left[ (P_c + P_{0,CRV})/2 - IOP \right] \right)^{-4}, \qquad (36)$$

where $R$, $L$, and $h$ represent effective values for radius, length, and wall thickness of the venules compartment, $E$ and $\nu$ are Young's modulus and Poisson's ratio of the venules wall, respectively, and $P_c$ and $P_{0,CRV}$ are the pressure at the inlet and at the outlet of the venules, respectively. In [19] the results of the new model are compared

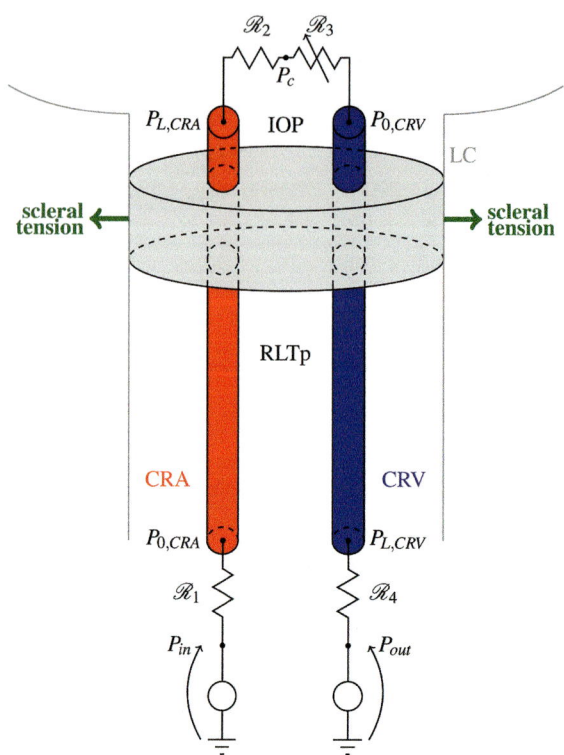

to experimental measurements to understand the mechanism behind the decrease in
CRA velocity induced by IOP elevation, measured in vivo in humans by Harris et
al. [63] and Findl et al. [42]. The mathematical model suggests that the increase in
vascular resistance caused by lamina cribrosa compression is not enough to explain
the velocity decrease observed in [42, 63]. Instead, the model suggests that this
decrease might be due to the IOP-induced increase in the vascular resistance $\mathscr{R}_3$ of
the retinal venules, see Fig. 18. Moreover, in [19, 20] a theoretical investigation is
performed to compare the influence of IOP and CSF on retinal hemodynamics. The
mathematical model results suggest that the IOP influence on retinal hemodynamics
might be direct via the resistance of the intraocular retinal vessels, while the CSF
influence on retinal hemodynamics might be mediated by associated changes in
mean arterial pressure.

Several models have been developed to investigate CRV hemodynamics [19, 20,
82, 84, 128]. These models mainly differ in four aspects: (1) the CRV is modeled
as a compliant vessel (assuming small deformations) or as a collapsible vessels
(assuming large deformations via a Starling-like resistor model, see Sect. 2.4); (2)
the CRV hemodynamics are coupled or not coupled to the lamina cribrosa deformations; (3) the CRV hemodynamics are coupled or not coupled to the upstream
and downstream vasculature; (4) the model is time-dependent or stationary. Note

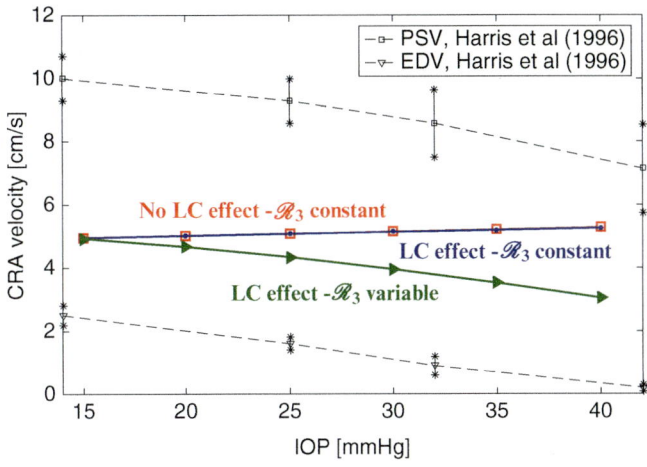

**Fig. 18** Comparison between in vivo measurements by Harris et al. [63] (black) and model predictions obtained by taking or not taking into account the lamina compression on the vessels, and by taking and not taking into account the effect of IOP on $\mathscr{R}_3$

that all of the CRV models considered here account for the effect of IOP and CSF pressure on CRV hemodynamics. In [19, 20], as described above, the CRV is modeled as a compliant vessel and the stationary CRV hemodynamics is coupled with the upstream/downstream circulation and the lamina cribrosa deformation.

The studies in [82, 84] assume that the CRV diameter changes only by a small percentage, i.e., compliant vessel assumption, hence the vessel resistance and capacitance are assumed to be constant. The models proposed in [82, 84] are also referred to as constant input variable output (CIVO) models. Here, a constant input flow is assumed from the capillaries into the CRV, which is equivalent to neglecting the effect of the vasculature upstream of the CRV. In particular, in [84], the CRV is modeled as two semi-infinite flexible cylinders representing the pre- and post-lamina cribrosa segments, connected by a rigid segment of finite length representing the translaminar segment, hence neglecting the effect of lamina cribrosa deformations on the CRV. The pre-laminar segment is subjected to IOP oscillations, and the post-laminar segment is subjected to oscillations in CSF pressure. Realistic IOP and CSF pressure waveforms are employed. The proposed CRV model is used to predict the phase relationship between oscillations in IOP, CSF pressure, and CRV diameter, since clinical experiments show spontaneous venous pulsations close to the lamina region [99]. The model shows that: (1) the driving forces of vein pulsation are the phase and amplitude difference between IOP and CSF pressure; and (2) phase delays between IOP oscillations and CRV pulsations are due to constriction of the vein close to the lamina region and to a dumping effect of the lamina region on the oscillations. This latter effect is proportional to the ratio of the CRV lamina segment resistance to the pre-or post-CRV lamina segment resistance. Levine et al. [84] conclude that CRV oscillations are dominated neither by IOP nor by CSF pressure, but instead are dominated by the difference in phase and amplitude between them.

In [128], Stewart et al. developed a two-dimensional collapsible compartmental model of the hemodynamics in the CRV, dividing the vessel into four compartments with different properties of the vessel walls: a Starling resistor in the intraocular segment subject to IOP, followed by a rigid segment in the optic nerve, followed by a Starling resistor in the nerve sheath subject to ICP, followed by rigid extra-neural segment, see Fig. 14. Note that the assumption of a rigid wall in the optic nerve segment is equivalent to neglecting the effect of the lamina deformations and the ICP effect in that segment. The model is time-dependent, the blood flow is modeled via incompressible Navier–Stokes equations, and the IOP and ICP (given as inputs of the model) oscillate in time as sinusoidal waves. ICP oscillations are assumed to be in phase with arterial pulsation, while IOP oscillations are offset of a fixed phase shift. For more details on the Starling resistor-like model used in [128] please refer to Sect. 2.4. The proposed CRV model is used to test if it is possible to estimate ICP non-invasively from an observation of retinal vein pulsation. The model results show a linear relation between IOP and ICP at the point of onset of CRV pulsations, in agreement with experiments on dogs [98]. Moreover, the results show that the onset of CRV pulsation is due to the collapse in the nerve sheath compartment. Stewart et al. conclude that, indeed, it might be possible to estimate ICP non-invasively in a clinical setting by varying IOP and measuring the onset of CRV pulsation via ophthalmodynamometry.

## 4 Other Vascular Beds

The previous sections showed that many theoretical and experimental studies have been focused on describing blood flow in the retina and in some major retrobulbar vessels. This is mainly due to the fact that many vascular, geometrical, and physical parameters pertaining to these vascular segments are accessible for measurement, including several non-invasive techniques that can be performed in a clinical setting. Despite being less accessible to imaging, the blood flow through the optic nerve head, the choroid, and the ciliary body also plays a very important role in supporting ocular functions in health and disease. The few theoretical models that have been developed to study choroidal, ciliary, and optic nerve head circulations are reviewed below.

### 4.1 Optic Nerve Head

Most of the modeling work related to the optic nerve head, to date, has focused on its biomechanical response to changes in IOP and/or CSF pressure [36, 51]. Remarkably, computational models have been capable of predicting the collagen fibril architecture in the optic nerve head and in corneo-scleral shells as a result of remodeling [49, 50]. The modeling of optic nerve head hemodynamics has only started in the recent years. The few contributions in the field, to date, are described below.

In the work by Chuangsuwanich et al. [29], two-dimensional capillary networks were artificially generated to identify and rank the morphologic factors influencing the hemodynamics and oxygen concentrations in the microvasculature of the lamina cribrosa. The basic assumptions underlying the artificial network generation are: (1) the lamina cribrosa is a curved surface with principal curvatures aligned along the inferior–superior and nasal–temporal directions; (2) each collagenous beam within the lamina cribrosa contains a microcapillary in its center that supplies nutrients and oxygen to adjacent axons; (3) the collagenous beams, and thus the microcapillaries, are parallel to the surface of the lamina cribrosa; (4) the microcapillaries have a uniform diameter of $8\,\mu m$; (5) oxygenated blood, originating from the Circle of Zinn-Haller or the short posterior arteries, feeds the lamina cribrosa uniformly at its periphery; and (6) venous drainage of the microcapillaries occurs through the central retinal vein that passes through the center of the lamina cribrosa. Under these general assumptions, a custom-written algorithm was used to generate artificial networks characterized by different sizes and arrangements of the pores within the collagenous beams as well as diameter and main curvatures of the lamina cribrosa. In the model, capillaries are modeled as rigid cylindrical tubes and therefore the blood flow in each capillary is computed via the Hagen–Poiseuille equation. Arterial and venous pressures are imposed as boundary conditions for the network and are driving the flow. Oxygen transport inside the capillaries is included along the axial direction of the vessels, whereas oxygen diffusion and consumption within the tissue is described by a Krogh cylinder-type model. Model simulations show that the imposed arterial and venous pressures and the diameter of the lamina cribrosa are the factors that influence hemodynamics and oxygenation the most. In particular, higher oxygen concentration across the network was predicted for smaller diameters of the lamina cribrosa, higher arterial pressure, and lower venous pressure.

In order to investigate the role of IOP on the perfusion of the lamina cribrosa, it is necessary to relax the rigid tube assumption and let the blood vessel deform under the action of transmural pressure differences. However, from the mathematical perspective, resolving blood flow and oxygen transport through a network of compliant and collapsible vessels is extremely challenging as it involves the coupling of highly nonlinear fluid–structure interaction problems, with the generation of spurious oscillations and reflected waves. Thus, in order to avoid the huge hurdles in accounting for the fine details of the microvascular network, Causin et al. [27] proposed a modeling approach based on the equations of poroelasticity. This approach was further extended by Bociu et al. [12] to include structural viscoelasticity. In this modeling approach, the lamina cribrosa is described as a deformable porous medium composed of a deformable solid (comprising collagen, elastin, extracellular matrix, and neural tissue) and an interconnected vascular porous space filled by blood, as depicted in Fig. 19. To this end, the concept of *vascular porosity* was introduced as the fluid volumetric fraction, namely

**Fig. 19** Schematic representation of the mathematical model based on deformable porous media describing the perfusion of the lamina cribrosa in the optic nerve head. Courtesy of Prada [111]

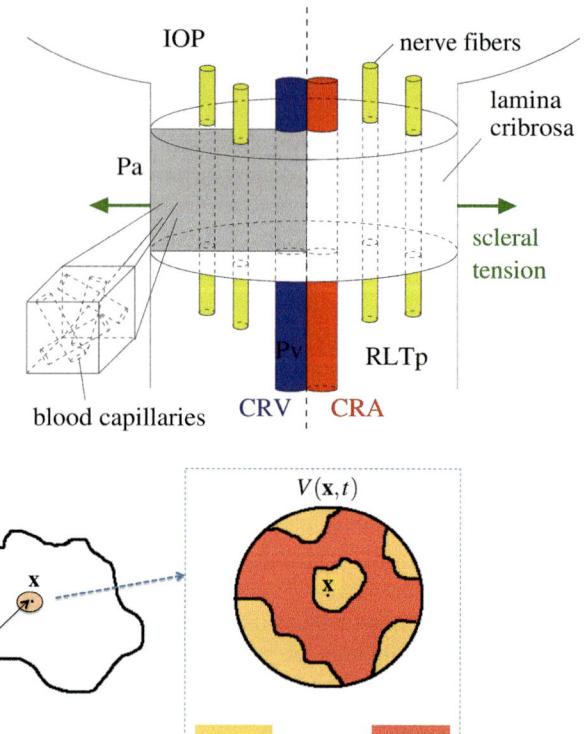

**Fig. 20** Schematic representation of a reference element volume $V(\mathbf{x}, t)$, whose barycenter is located at $\mathbf{x}$ at time $t$. $V(\mathbf{x}, t)$ is filled by two distinct phases, occupying the subvolumes $V_s(\mathbf{x}, t)$ and $V_f(\mathbf{x}, t)$

$$\phi(\mathbf{x}, t) = \frac{V_f(\mathbf{x}, t)}{V(\mathbf{x}, t)}, \tag{37}$$

where $V_f(\mathbf{x}, t)$ is the volume occupied by blood in any representative elementary volume $V(\mathbf{x}, t)$ located at $\mathbf{x}$ at time $t$, as depicted in Fig. 20. Under the assumption of a fully saturated mixture, the volumetric fraction $V_s(\mathbf{x}, t)$ of the solid component is given by $1 - \phi(\mathbf{x}, t)$. Moreover, under the assumptions of negligible inertia and small deformations [4, 44, 76, 81, 112], the motion of the poroelastic material is governed by the following equations for the balance of mass (of the fluid component) and linear momentum (for the fluid–solid mixture):

$$\frac{\partial \zeta}{\partial t} + \nabla \cdot \mathbf{v} = S(\mathbf{x}, t) \quad \text{and} \quad \nabla \cdot \mathbf{T} + \mathbf{F} = \mathbf{0} \quad \text{in } \Omega \times (0, T) \tag{38}$$

where $\zeta = \phi(\mathbf{x}, t) - \phi_0(\mathbf{x})$ is the fluid content, $\mathbf{v}$ is the discharge velocity, $\mathbf{T}$ is the stress tensor of the mixture (also known as total stress), $\mathbf{F}$ is a body force per unit of volume, $S$ is a net volumetric fluid production rate, and $\Omega$ is the region of space occupied by the mixture. Since biological tissues are almost incompressible, the fluid content can be written as $\zeta = \nabla \cdot \mathbf{u}$, where $\mathbf{u}$ is the solid displacement, implying that $\phi = \phi_0 + \nabla \cdot \mathbf{u}$. Darcy's law $\mathbf{v} = -\mathbf{K}\nabla p$ provides a relationship between velocity and pressure ($p$) via the permeability tensor $\mathbf{K}$ representing the geometrical architecture of the pores inside the matrix and the physical properties of the fluid flowing within them. The stress tensor $\mathbf{T}$ describes the combined mechanical behavior of all components in the mixture; in the case of incompressible mixture components and negligible viscous effects at the fluid–solid interface within the pores, one can write $\mathbf{T} = \boldsymbol{\sigma} - p\mathbf{I}$, where $\boldsymbol{\sigma}$ describes the mechanics of the solid component.

In Causin et al. [27], the permeability tensor was assumed to be a multiple of the identity, namely $\mathbf{K} = \kappa(\phi)\mathbf{I}$, with $\kappa(\phi) = \beta\phi^2$, where $\beta$ is a positive given constant, resulting from the assumptions that the capillaries change their volume at constant length and that blood flow within the pores is described by Poiseuille's law. A nonlinear constitutive equation was assumed for the structural stress tensor, namely $\boldsymbol{\sigma} = \lambda(\sigma_e)\nabla \cdot \mathbf{u}\,\mathbf{I} + \mu(\sigma_e)(\nabla\mathbf{u} + \nabla\mathbf{u}^T)$, where $\sigma_e$ is the effective stress as in Eq. (23) following [54]. The model was used to investigate how the distributions of stress, blood volume fraction (or vascular porosity), and blood velocity within the lamina cribrosa are influenced by different IOP levels and the enforcement of different mechanical constraints at the lamina's boundary. The model simulations predict different biomechanical and hemodynamic responses to IOP elevation depending on the region within the lamina and the degree of fixity of the scleral insertion. Specifically, when the boundary is mechanically clamped, IOP elevation leads to an increase in stress close to the lamina's boundary, making it more susceptible to tissue damage. On the other hand, when rotations are allowed at the boundary, the most vulnerable region appears to be located at the lamina's central axis, in proximity to the eye globe, where increased stress and reduced vascular porosity and blood velocity are predicted for increased levels of IOP. These results might help explain clinical and experimental observations showing that the response of the laminar tissue to changes in IOP is not uniform. For example, loss of collagen fibers was reported in the top-center of the lamina and recruitment of new collagen was detected at the lower surface, causing cupping in the lamina of animals suffering from prolonged IOP elevation [51].

In [12], Bociu et al. considered a different constitutive equation for the total stress tensor, namely

$$\mathbf{T} = \mathbf{T}_e + \mathbf{T}_v - p\mathbf{I}, \tag{39}$$

where $\mathbf{T}_e$ and $\mathbf{T}_v$ are the elastic and viscoelastic stress contributions, respectively, defined as

$$\mathbf{T}_e = 2\mu_e\epsilon(\mathbf{u}) + \lambda_e(\nabla \cdot \mathbf{u})\,\mathbf{I} \quad \text{and} \quad \mathbf{T}_v = 2\mu_v\epsilon(\mathbf{u}_t) + \lambda_v(\nabla \cdot \mathbf{u}_t)\,\mathbf{I}, \tag{40}$$

where $\epsilon(\mathbf{w})$ is the symmetric part of the gradient of the vector field $\mathbf{w}$, namely $\epsilon(\mathbf{w}) = (\nabla\mathbf{w} + \nabla\mathbf{w}^T)/2$, $\lambda_e$ and $\mu_e$ are the Lamé elastic parameters, and $\lambda_v$ and $\mu_v$ are the viscoelastic parameters. In this analysis, elastic and viscoelastic parameters are assumed to be given positive constants, whereas the permeability tensor is assumed to be a bounded function of the porosity, without specifying any particular functional form a priori. The mathematical analysis showed that the viscoelastic component of the stress tensor is a major determinant in the behavior of the solutions. From the theoretical viewpoint, existence of solutions to poro-viscoelastic models can be proved under less restrictive assumptions for data regularity when compared to the purely elastic case, suggesting that changes in viscoelastic tone due to aging or disease might significantly affect the perfusion in the optic nerve head. Specifically, the theoretical findings suggest that the lack of viscoelasticity may increase the susceptibility of the tissue to localized damage due to irregularities in the perfusion velocity as volumetric sources of linear momentum and/or boundary sources of traction experience sudden changes in time. Since IOP acts as a boundary source of traction on the optic nerve head tissue, see, for example, [27], the findings suggest that even physiological changes in IOP might induce pathological changes in the hemodynamics of the optic nerve head tissue if its viscoelasticity is not intact. Interestingly, viscoelastic changes in ocular tissues have been associated with glaucoma in experimental studies on rabbits and monkeys [37, 38]. On the other hand, sudden changes in gravitational acceleration, such as those experienced by astronauts during missions, translate into sudden changes in the volumetric source of linear momentum, which might increase tissue vulnerability to damage, as shown by the analysis in [12]. These considerations are particularly relevant for the optic nerve head tissue, whose pathological changes have been associated with the visual impairments affecting many crew members during and after long-duration space flights [86].

## 4.2   *Choroid*

A zero-dimensional model to study choroidal circulation was proposed by Kiel and Shepherd in [73]. The main goal of the model was to shed some light on the mechanisms giving rise to blood flow autoregulation in the choroid as MAP and IOP are varied. In [73], the authors performed experimental measurements on 8 anesthetized rabbits, whose ocular anatomy allows for direct access to the choroid. MAP was manipulated via an occluder in the thoracic vena cava, while IOP was held at 5, 15, and 25 mm Hg. IOP was controlled and manipulated via catheters inserted in the vitreous. Choroidal blood flow was measured via laser Doppler flowmetry. The experimental measurements indicated that the efficacy of autoregulation depended on the MAP and IOP levels in a very complex way, leading the authors to hypothesize that a myogenic response may be involved in choroidal autoregulation.

To test this hypothesis, the authors proposed a mathematical model composed of three vascular compartments (artery, capillary, and vein) encased in a pressurized chamber kept at pressure $P_{ch}$. Total blood flow is determined by the pressure difference $P_{in} - P_{out}$ at the ends of the network and by the hydraulic resistance across the network. Capillary resistance is assumed to be fixed, whereas arterial and venous resistances are assumed to vary according to active and passive mechanisms, respectively. More precisely, the arterial resistance varies with a myogenic response that is sensitive to the transmural pressure difference $P_a - P_{ch}$ via a phenomenological law based on experimental data. The venous resistance embodies a Starling resistor effect where vessel collapsibility depends on the transmural pressure difference $P_v - P_{ch}$ via a sigmoidal function. Model simulations showed similar results to those observed experimentally, thereby supporting the hypothesis that a myogenic mechanism may be responsible for the autoregulation of choroidal blood flow in the rabbit. Whether and to what extent these conclusions are applicable to the human eye remain open questions.

## 4.3   Ciliary Body

Blood flow in the ciliary body has been studied experimentally in the view of its relationship with aqueous humor production. The review article by Kiel et al. [74] indicates that this relationship is quite complex, as aqueous humor production depends on ciliary blood flow only below a critical level of perfusion and is independent of blood flow above it. The results discussed in the review also show that the oxygen supplied by the ciliary circulation is a critical factor in the relationship between ciliary blood flow and aqueous humor production. The authors further investigate these concepts by means of a lumped parameter model developed in STELLA (ISEE Systems, Lebanon, NH). The model describes: (1) blood flow in the choroidal and ciliary circulations; (2) aqueous humor production and drainage; (3) oxygen delivery and consumption in the ciliary circulation and in the ciliary processes. Model simulations allowed to infer quantitative interpretations of the Moses iconic 1981 graph [100] depicting steady state IOP as the intersection of aqueous humor inflow and outflow and show promise as effective tools to complement experimental data with virtual experiments.

## 5   Mathematics and Imaging

Doppler ultrasound is a consolidated technique to measure blood velocity waveform in some of the major ocular vessels, including the ophthalmic artery. Image processing of the OA Doppler waveform has been used to investigate the relevance of waveform parameters in ocular diseases, such as glaucoma, OA stenosis, and uveitis. Typical waveform parameters used in ophthalmology are maximum systolic

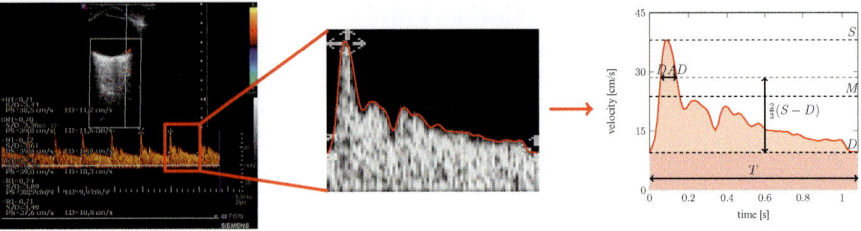

**Fig. 21** Schematic of the semi-automated image processing to detect the OA waveform parameters [22]

height of the waveform (S), end diastolic height of the waveform (D), mean height of the waveform (M), resistive index RI $=$ (S $-$ D)/S, and pulsatility index PI $=$ (S $-$ D)/M, see Fig. 21. Both RI and PI are a measure for the resistance of the vessel to blood flow.

Only recently, waveform parameters commonly used to study other vascular beds throughout the body, such as renal arteries, hepatic arteries, and lower limb arteries, have been used to gain new insight into the characterization of the OA velocity waveform in glaucoma patients [22, 109]. In particular, Carichino et al. [22] developed a computer-aided manipulation process of OA color Doppler images with the goal of extracting a novel set of waveform parameters and testing their relevance in glaucoma patients. One of the novel parameters considered is the normalized distance between the ascending and descending limb (DAD/T) of the wave at two thirds of the difference between S and D, see Fig. 21. This parameter has been already introduced in a previous study on patients with obliterating atherosclerosis in [104], where they found that DAD/T could identify the severity of obliterating atherosclerosis and the presence or absence of progression. Interestingly, glaucoma patients showed a significantly higher DAD/T compared to healthy individual [22]. Further studies are necessary to establish the correlation between this novel parameter and glaucoma and its potential use as a marker for progression.

Doppler ultrasound technology can be used to extract not only velocity profiles, but also Doppler power spectrums in the form of sonograms. Güler and Übeyli recently developed mathematical models to investigate how OA Doppler power spectrum might be utilized to classify patients with ocular diseases, such as OA stenosis and uveitis [57–60, 132]. They use different versions of artificial neural networks, such as a multilayer feedforward neural network [57, 58] and multilayer perceptron neural network [60]. These networks, given input parameters taken from the Doppler signal, are able to estimate if a subject is healthy or has developed an ocular disease. In particular, Güler and Übeyli focused on studying the impact of RI and PI as input parameters of OA waveforms in relation to OA stenosis. Although there is a statistical difference between RI and PI values of normal subjects and subjects with stenosis, it is difficult to separate healthy and stenosis subjects using them, because there is an overlap in RI and PI values among the two groups. In order to overcome the uncertainty in OA Doppler signal classification caused by the imprecise boundaries between RI and PI values of healthy and stenosis patients, an adaptive neuro-fuzzy inference system was introduced in the neural network [59, 132]. Thanks to the introduction of fuzzy logic, the authors are able to show that PI and RI have a considerable impact on the detection of OA stenosis, namely the neuro-fuzzy network is able to classify healthy individuals and stenosis patients with an accuracy of 97.22% [59].

# 6 Perspectives

In this chapter, we have reviewed the mathematical models that have been developed to date to study different aspects of ocular blood flow. We anticipate that the main challenge in the next decades will be to develop a unified framework to describe simultaneously:

1. *the interconnections between different vascular beds*, including retinal, choroidal, ciliary, and retrobulbar circulations;
2. *the coupling between phenomena occurring at different scales in space*, such as the motion of ions across cellular membranes to the macroscopic change of arteriole diameter, *and time*, such as the hemodynamic changes due to the circadian rhythm and slow changes due to tissue growth and remodeling; and
3. *the interplay between the deterministic laws of nature* governing the fluid dynamics of blood flow *and the intrinsic variability* of biophysical and geometrical characteristics among individuals.

Some preliminary contributions in this direction can be found in [13, 56, 120, 121]. In [56], Guidoboni et al. present a mathematical model that accounts for: (1) the flows of blood and aqueous humor in the eyes; (2) the flows of blood, cerebrospinal fluid, and interstitial fluid in the brain; and (3) their interactions. The flow is driven by the arterial blood pressure $P_A$, which is given as a variable input, while the venous pressure $P_V$ is kept constant. Model parameters have been calibrated on published data. The model is used to simulate changes in IOP, CSF pressure, flow, and pressure distributions across the whole system induced by changes in $P_A$. Model predicted relationships between IOP, $P_A$, and CSF pressure are within the same range as those reported by major clinical studies [34, 96, 117, 139]. In addition, choroidal venous pressure in the vortex veins computed by the model from flow and pressure distribution within the body is approximately equal to IOP over a wide range of values, confirming theoretically for the first time the findings presented by Bill in 1963 [11]. In [120], Sacco et al. present a mathematical model that simulates the macroscopic diameter change in retinal arterioles as a result of microscopic biochemical reactions stimulated by glutamate and nitric oxide and leading to the phosphorylation of myosin in the smooth muscle cells. The model predicts that the interaction between nitric oxide and endothelin plays a crucial role in the response of retinal arterioles to neural stimuli. In [121], Sala et al. present a mathematical model that accounts for the coupling between the blood flow within the lamina cribrosa, described as a three-dimensional porous medium, and the blood flow in the retinal, choroid, and retrobulbar circulation, described as a zero-dimensional network. The multiscale model is used to quantify the effect of IOP, CSF pressure, and blood pressure on the hemodynamic and biomechanic responses of the lamina cribrosa. Simulations suggest that similar levels of the translaminar pressure difference, defined as the difference between IOP and CSF pressure, lead to similar stress and strain distributions within the lamina but noticeably different perfusion velocities. Despite the encouraging results mentioned above, there is still a long way to go in order to attain a full multicomponent/multiscale/multiphysics mathematical description of ocular hemodynamics. Interestingly, the need for mathematical models capable of capturing the complex nature of the eye has motivated the study and development of novel mathematical techniques capable of posing and solving these problems correctly and accurately. For example, the need for solving the partial differential equations for deformable porous media flow in a bounded domain to model tissue perfusion has led to the proof of blow-up in finite time of the solution in the case of weak smoothness of the data in time [12, 133]. Another example is the development of numerical techniques that preserve the energy balance at the discrete level when decoupling three-dimensional and zero-dimensional problems [23]. Ultimately, combining the expertise of mathematicians, clinicians, scientists, engineers, and experimentalists will allow for the collection of data and development of models that provide a comprehensive understanding of the role of blood flow in ocular health and disease.

## Appendix 1: Poiseuille's Law

In 1840, Poiseuille (a physician) used water and mercury to impose a pressure gradient across a long narrow tube of length $L$ and diameter $D$. Empirically, he obtained the relationship that flow through the small tube was proportional to $\dfrac{D^4 \Delta P}{L}$, where $\Delta P$ was the pressure difference between the two ends ($\Delta P = P_0 - P_L$). The theory establishing this observed relationship was completed by Neumann and Hagenbach in 1858. First, considering a balance of forces on an inner cylindrical region within fluid yields

$$\pi r^2 P_0 - \pi r^2 P_L - 2\pi R L \tau = 0. \tag{41}$$

This gives shear stress $\tau = \dfrac{r \Delta P}{2L}$. Assuming a Newtonian fluid, $\tau = -\mu \dfrac{du}{dr}$ where $u(r)$ is the velocity in the radial direction. Integrating and applying the no-slip condition at the vessel wall ($r = a$) yields $u = \dfrac{\Delta P}{4L\mu}(a^2 - r^2)$. Assuming a non-uniform velocity, flow rate ($Q$) of the fluid can be calculated by integrating this velocity with respect to area:

$$Q = \int_0^a u(r)2\pi r dr = \frac{\pi \Delta P D^4}{128 L \mu} \tag{42}$$

where diameter $D = 2a$. Equation (42) is known as Poiseuille's law, even though Poiseuille did not formally derive it, and this type of flow is known as Poiseuille's flow. Importantly, Poiseuille's law emphasizes the dependence of flow on the fourth power of the diameter of the vessel.

## Appendix 2: Krogh Cylinder

When tissue oxygen consumption is calculated using a Krogh cylinder model, it is assumed that each oxygen-delivering vessel (e.g., capillary) provides oxygen via diffusion to only the nearest cylindrical region of tissue surrounding the vessel (as depicted in Fig. 7). In the Krogh model, oxygen diffusion is given by

$$D\alpha \frac{1}{r} \frac{d}{dr} \left( r \frac{dP}{dr} \right) = M \tag{43}$$

where $r$ is distance from the center of the vessel, $P(r, x)$ is the partial pressure of oxygen at radial distance $r$ within the tissue cylinder for a fixed position $x$ in the network, $D$ is the diffusion coefficient of oxygen in tissue, and $\alpha$ is the solubility of oxygen in tissue.

This diffusion equation is solved for the partial pressure of oxygen in tissue along the radial direction $r$ for a fixed position in the network $x$ assuming no flux across the tissue boundary $\left( \frac{dP(r_t)}{dt} = 0 \right)$ and assuming $PO_2$ at the edge of the vessel ($r = r_v$) is $P(r_v, x)$. Integrating the diffusion equation twice yields the solution, $P(r, x)$:

$$P(r, x) = P(r_v, x) + \frac{M}{D\alpha} \left[ \frac{r^2 - r_v^2}{4} + \frac{r_t^2}{2} \ln \frac{r_v}{r} \right]. \tag{44}$$

## Appendix 3: Modeling of Compressible and Collapsible Resistances

This section describes the main steps of the derivation of the formulas for the passive resistors of the model depicted in Fig. 3 that account for the Laplace/Startling mechanical response of the vessels to changes in the value of transmural pressure difference.

The vessel of interest is modeled as a straight cylinder $\Omega$ with cross-section $\Sigma$ and length $L$ in cross-section shape $(x, y, z)$, see Fig. 22. The cross-section shape is constraint-free, thus it might not be circular.

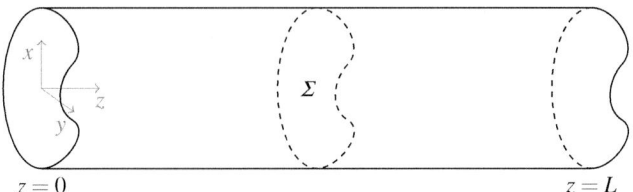

**Fig. 22** Representative cylinder $\Omega$ in the Cartesian coordinate system $(x, y, z)$ with cross-section, $\Sigma$, of length $L$ [24]

The motion of fluid inside the tube is initially described by the stationary Navier–Stokes equations. A model reduction technique [24, 113] is then adopted to obtain the following differential equation:

$$\frac{1}{K_r \rho} \frac{dp}{dz} + \frac{Q(z)}{A^2(z)} = 0, \tag{45}$$

where $Q$ is the volumetric average flow rate, $A$ is the cross-sectional area, $p$ is the pressure inside the tube, $\rho$ is the fluid (blood) density, and the value of $K_r$ depends on the profile that is chosen for the fluid motion ($K_r = \dfrac{8\pi\mu}{\rho}$ for the case of parabolic/Poiseuille's flow, where $\mu$ is the fluid viscosity) [17, 113]. It is worth mentioning that Eq. (45) contains the volumetric flow $Q$ and the pressure gradient along the tube $\dfrac{dp}{dz}$, therefore it also involves the vascular resistance R since these three quantities are related via Ohm's law (1). However, Eq. (45) does not yet contain the contribution of transmural pressure difference.

This problem is solved by introducing the tube law that relates the cross-sectional area to the transmural pressure difference [17, 52, 107, 126],

$$P(\alpha) = \frac{p(z) - p_e}{K_p}, \tag{46}$$

where $p_e$ represents the external pressure acting on the tube, $\alpha \stackrel{\text{def}}{=} \dfrac{A(z)}{A_{ref}}$ represents the ratio between the cross-sectional area $A(z)$ and the reference cross-sectional area $A_{ref}$ ($A_{ref} = A$ when $p - p_e = 0$), respectively. The constant $K_p$ is chosen in agreement with the value in [17, 24, 126] for a linear elastic tube. The explicit form of $P(\alpha)$ depends on the physiological and mechanical properties of the considered tube.

For a *compressible* tube, the cross-section is assumed to remain circular and the radial displacement due to changes in transmural pressure difference is computed using Laplace's law [24]. Through some algebraic manipulation it is possible to explicitly compute the tube law for a compressible tube as $P(\alpha) = k_L(\alpha^{1/2} - 1)$,

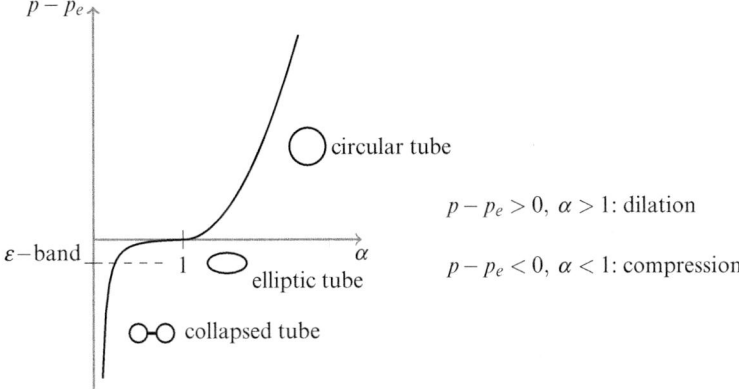

**Fig. 23** Graph of tube law experimental results for a collapsible tube, adapted from [17, 24, 52, 107, 126]. The cross-sectional shape changes from circular to elliptic to highly collapsed as the transmural pressure difference $p - p_e$ decreases

where the constant $k_L$ depends on the geometrical parameters of the considered tube, see [24].

For a *collapsible* tube, when $\alpha < 1$, the tube law is based on experimental results [17, 126], and it corresponds to $P(\alpha) = 1 - \alpha^{-3/2}$. This formula describes the phenomenon known as the Starling resistor effect, depicted in Fig. 23. For positive values of the transmural pressure difference, the vessel dilates to maintain a circular cross-section. For negative values of the transmural pressure difference, the cross-section becomes first elliptical and then highly collapsed. In the elliptical region ($\varepsilon$-band) small variations in transmural pressure difference produce large variations in the cross-sectional area (and in the value of the vascular resistance). In the highly collapsed region, large variations in the transmural pressure difference produce small variations in the cross-sectional area (and in the vascular resistance) since the specific shape of the collapsed cross-section yields a higher resistance than the elliptical shape.

The modeling choices adopted for $P(\alpha)$ for a compressible and collapsible tube are summarized below.

**Compliant tube** $\quad P(\alpha) = k_L(\alpha^{1/2} - 1),$

**Collapsible tube** $\quad P(\alpha) = \begin{cases} 1 - \alpha^{-3/2}, & \alpha \leq 1 \\ k_L(\alpha^{1/2} - 1), & \alpha > 1. \end{cases}$

Combining Eq. (45) and the tube law, the explicit formulation for the passive resistors in Fig. 3 is computed as:

**Compliant tube** $\quad R = \dfrac{K_r \rho L}{A_{ref}^2} \left[ \dfrac{\overline{p} - p_e}{K_p k_L} + 1 \right]^{-4}$

**Collapsible tube**    $R = \begin{cases} \dfrac{K_r \rho L}{A_{ref}^2} \left[ 1 - \dfrac{\overline{p} - p_e}{K_p} \right]^{4/3} , & \alpha \leq 1 \\[2em] \dfrac{K_r \rho L}{A_{ref}^2} \left[ \dfrac{\overline{p} - p_e}{K_p k_L} + 1 \right]^{-4} , & \alpha > 1, \end{cases}$

where $\overline{p}$ is the average pressure inside the tube. It is worth mentioning that the expressions obtained for the resistors are characterized by a strong nonlinear behavior and are able to model the collapse phenomenon (Starling resistor effect) characteristic of venous vessels.

Algebraic manipulation can be used to demonstrate that the modeling approach used in the compliant tube case is equivalent to Poiseuille's law

$$R = \frac{K_r \rho L}{A_{ref}^2} \left[ \frac{\overline{p} - p_{ext}}{K_p k_L} + 1 \right]^{-4} = \frac{128 \mu L}{\pi D^4}, \tag{47}$$

where $D$ represents the tube diameter.

The complete derivation of the vascular resistors presented in this appendix is available in [24].

# References

1. Aletti, M., Gerbeau, J., Lombardi, D.: Modeling autoregulation in three- dimensional simulations of retinal hemodynamics. Journal for Modeling in Ophthalmology **1** (2015)
2. Anderson, A.R.A., Chaplain, M.A.J.: A mathematical model for capillary network formation in the absence of endothelial cell proliferation. Applied mathematics letters **11**(3) 109–114 (1998)
3. Anderson, A.R.A., Chaplain, M.A.J.: Continuous and discrete mathematical models of tumorinduced angiogenesis. Bull. Math. Biol. **60**, 857–900 (1998)
4. Araujo, R.P., Sean McElwain, D.L.: A mixture theory for the genesis of residual stresses in growing tissues I: a general formulation. SIAM J. Appl. Math. **65**(4), 1261–1284 (2005)
5. Arciero, J.C., Carlson, B.E., Secomb, T.W.: Theoretical model of metabolic blood flow regulation: roles of ATP release by red blood cells and conducted responses. Am J Physiol Heart Circ Physiol **295** H1562–H1571 (2008)
6. Arciero, J., Harris, A., Siesky, B., et al.: Theoretical analysis of vascular regulatory mechanisms contributing to retinal blood flow autoregulation. Invest Ophthalmol Vis Sci. **54** 5584–5593 (2013)
7. Aubert, M., Chaplain, M.A.J., McDougall, S.R., Devlin, A., Mitchell, C.A.: A Continuum Mathematical Model of the Developing Murine Retinal Vasculature. Bull Math Biol **73** 2430–2451 (2011)
8. Balding, D., McElwain, D.: A mathematical model of tumour−induced capillary growth. J. Theor. Biol. 114, 53–73 (1985)
9. Bartha, K., Rieger, H.: Vascular network remodeling via vessel cooption, regression and growth in tumors. J Theor Biol. **241**(4) 903–18 (2006)
10. Bentley, K., Gerhardt, H., Bates, P.A.: Agent−based simulation of notch−mediated tip cell selection in angiogenic sprout initialisation. J Theor Biol. **250**(1) 25–36 (2008)

11. Bill, A.: The uveal venous pressure. Archives of Ophthalmology **69**(6) 780–782 (1963)
12. Bociu, L., Guidoboni, G., Sacco, R., Webster, J.T.: Analysis of nonlinear poro-elastic and poro-visco-elastic models. Archive for Rational Mechanics and Analysis **222**(3) 1445–1519 (2016)
13. Braun, R.J.: ARVO 2017 highlights in mathematical modeling. Journal for Modeling in Ophthalmology **2**(1) 4–6 (2018)
14. Buttery, R.G., Hinrichsen, C.F., Weller, W.L., Haight, J.R.: How thick should a retina be? A comparative study of mammalian species with and without intraretinal vasculature. Vis Res. **31**: 169–187 (1991)
15. Byrne, H.M., Chaplain, M.A.J.: Mathematical models for tumour angiogenesis: Numerical simulations and nonlinear wave solutions. Bull. Math. Biol. **57** 461–486 (1995)
16. Byrne, H.M., Chaplain, M.A.J.: Explicit solutions of a simplified model of capillary sprout growth during tumor angiogenesis. Appl. Math. Lett. **9** 69–74 (1996)
17. Cancelli, C., Pedley, T.J.: A separated-flow model for collapsible-tube oscillations. Journal of Fluid Mechanics. Cambridge Univ Press, **157** 375–404 (1985)
18. Carlson, B.E., Arciero, J.C., Secomb, T.W.: Theoretical model of blood flow autoregulation: roles of myogenic, shear-dependent, and metabolic responses. Am J Physiol Heart Circ Physiol. **295** H1572–H1579 (2008)
19. Carichino, L.: Multiscale mathematical modeling of ocular blood flow and oxygenation and their relevance to glaucoma. Ph.D. thesis, Purdue University (2016)
20. Carichino, L., Guidoboni, G., Siesky, B.A., Amireskandari, A., Januleviciene, I., Harris, A.: Effect of intraocular pressure and cerebrospinal fluid pressure on the blood flow in the central retinal vessels. In: Causin, P., Guidoboni, G., Sacco, R., Harris, A. (eds) Integrated Multidisciplinary Approaches in the Study and Care of the Human Eye. Kugler Publications, 59–66 (2014)
21. Carichino, L., Harris, A., Guidoboni, G., Siesky, B.A., Pinto, L., Vandewalle, E., Olafsdottir, O.B., Hardarson, S.H., Van Keer, K., Stalmans, I., Stefansson, E., Arciero, J.C.: A theoretical investigation of the increase in venous oxygen saturation levels in advanced glaucoma patients. Journal for Modeling in Ophthalmology **1** 64–87 (2016)
22. Carichino, L., Guidoboni, G., Verticchio Vercellin, A.C., Milano, G., Cutolo, C.A., Tinelli, C., De Silvestri, A., Lapin, S., Gross, J.C., Siesky, B.A., Harris, A.: Computer-aided identification of novel ophthalmic artery waveform parameters in healthy subjects and glaucoma patients. Journal for Modeling in Ophthalmology **1**(2), 59–69 (2016)
23. Carichino, L., Guidoboni, G., Szopos, M.: Energy-based operator splitting approach for the time discretization of coupled systems of partial and ordinary differential equations for fluid flows: The Stokes case. Journal of Computational Physics **364**, 235–256 (2018)
24. Cassani, S.: Blood circulation and aqueous humor flow in the eye: multi-scale modeling and clinical applications. Ph.D. thesis, Purdue University (2016)
25. Cassani, S., Harris, A., Siesky, B., Arciero, J: Theoretical analysis of the relationship between changes in retinal blood flow and ocular perfusion pressure J. Coupled Syst. Multiscale Dyn. **3**(1) (2015)
26. Cassani, S., Arciero, J., Guidoboni, G., Siesky, B., Harris, A.: Theoretical predictions of metabolic flow regulation in the retina Journal for Modeling in Ophthalmology. (2016)
27. Causin, P., Guidoboni, G., Harris, A., Prada, D., Sacco, R., Terragni, S.: A poroelastic model for the perfusion of the lamina cribrosa in the optic nerve head. Mathematical Biosciences **257** 33–41 (2014)
28. Causin, P., Guidoboni, G., Malgaroli, F., Sacco, R., Harris, A.: Blood flow mechanics and oxygen transport and delivery in the retinal microcirculation: multiscale mathematical modeling and numerical simulation. Biomech Model Mechanobiol. **15** 525–542 (2016)
29. Chuangsuwanich, T., Birgersson, K.E., Thiery, A., Thakku, S.G., Leo, H.L., Girard, M.J.A.: Factors Influencing Lamina Cribrosa Microcapillary Hemodynamics and Oxygen ConcentrationsModeling of Lamina Cribrosa Hemodynamics. Investigative Ophthalmology & Visual Science **57**(14) 6167–6179 (2016)

70. Jensen, O.E.: Flows through deformable airways. Centre for Mathematical Medicine, School of Mathematical Sciences, University of Nottingham, (2002)
71. Jeppesen, P., Aalkjaer, C., Bek, T.: Myogenic response in isolated porcine retinal arterioles. Curr Eye Res. **27** 217–222 (2003)
72. Katz, A.M.: Application of the Starling resistor concept to the lungs during CPPV. Critical Care Medicine. LWW, **5**(2), 67–72 (1977)
73. Kiel, J.W., Shepherd, A.P.: Autoregulation of choroidal blood flow in the rabbit. Investigative Ophthalmology & Visual Science **33**(8), 2399–2410 (1992)
74. Kiel, J.W., Hollingsworth, M., Rao, R., Chen, M., Reitsamer, H.A.: Ciliary blood flow and aqueous humor production. Progress in Retinal and Eye Research **30**(1), 1–17 (2011)
75. Kim, M., Lee, E.J., Seo, J.H., Kim, T.: Relationship of spontaneous retinal vein pulsation with ocular circulatory cycle. PLoS One. Public Library of Science, **9**(5), e97943 (2014)
76. Klisch, S.M.: Internally constrained mixtures of elastic continua. Math. Mech. Solids **4** 481–498 (1999)
77. Krogh, A.: The supply of oxygen to the tissues and the regulation of the capillary circulation. J Physiol **52** 457–474 (1919)
78. Kur, J., Newman, E.A., Chan–Ling, T.: Cellular and physiological mechanisms underlying blood flow regulation in the retina choroid in health disease. Prog Retin Eye Res **31**, 377–406 (2012)
79. Lakin, W.D., Stevens, S.A., Tranmer, B.I., Penar, P.L.: A whole-body mathematical model for intracranial pressure dynamics. Journal of Mathematical Biology **46**(4), 347–383 (2003)
80. Lau, J.C.M., Linsenmeier, R.A.: Oxygen consumption and distribution in the Long-Evans rat retina. Experimental Eye Research **102**, 50–58 (2012)
81. Lemon, G., King, J.R., Byrne, H.M., Jensen, O.E., Shakesheff,K.M.: Mathematical modelling of engineered tissue growth using a multiphase porous flow mixture theory. Journal of mathematical biology. J. Math. Biol. **52**, 571–594 (2006)
82. Levine, D.N.: Spontaneous pulsation of the retinal veins. Microvasc. Res. **56**(3), 154–165 (1998)
83. Levine, H.A., Tucker, A.L., Nilsen-Hamilton, M.: A Mathematical Model for the Role of Cell Signal Transduction in the Initiation and Inhibition of Angiogenesis. Growth Factors **20**(4) (2003)
84. Levine, D.N., Bebie, H.: Phase and amplitude of spontaneous retinal vein pulsations: an extended constant inflow and variable outflow model. Microvasc. Res. **106**, 67–79 (2016)
85. Liu, D., Wood, N. B., Witt, N., et al: Computational Analysis of Oxygen Transport in the Retinal Arterial Network. Current Eye Research **34**(11) 945–956 (2009)
86. Mader, T.H., Gibson, C.R., Pass, A.F., Kramer, L.A., Lee, A.G., Fogarty, J., Tarver, W.J., Dervay, J.P., Hamilton, D.R., Sargsyan, A. et al: Optic disc edema, globe flattening, choroidal folds, and hyperopic shifts observed in astronauts after long-duration space flight. Ophthalmology **118**(10) 2058–2069 (2011)
87. Maggelakis, S. A., Savakis, A. E.: A mathematical model of growth factor induced capillary growth in the retina. Math. Comput. Model. **24** 33–41. (1996)
88. Maggelakis, S. A., Savakis, A. E.: A mathematical model of retinal neovascularization. Math. Comput. Model. **29** 91–97 (1999)
89. Mandecka, A., Dawczynski, J., Blum, M., Muller, N., Kloos, C., Wolf, G., Vilser, W., Hoyer, H., Muller, U.A.: Influence of flickering light on the retinal vessels in diabetic patients. Diabetes Care. **30**(12) 3048–3052 (2007)
90. Mantzaris, N., Webb, S., Othmer, H.G.: Mathematical modeling of tumor-induced angiogenesis. J. Math. Biol. **49** 111–187 (2004)
91. McDougall S.R., Watson, M.G., Devlin, A.H., Mitchell, C.A., Chaplain, M.A.: A hybrid discrete-continuum mathematical model of pattern prediction in the developing retinal vasculature. Bull Math Biol. **74**(10) 2272–314 (2012)
92. McGuire, B.J., Secomb, T.W.: A theoretical model for oxygen transport in skeletal muscle under conditions of high oxygen demand. J Appl Physiol. **91**(5) 2255–65 (2001)

93. McGuire, B.J., Secomb, T.W.: Estimation of capillary density in human skeletal muscle based on maximal oxygen consumption rates. Am J Physiol Heart Circ Physiol **285** H2382–H2391 (2003)
94. Merks, R.M.H., Glazier, J.A.: Dynamic mechanisms of blood vessel growth Nonlinearity. **19**(1) C1-C10 (2006)
95. Misiulis, E., Džiugys, A., Navakas, R., Striūgas, N.: A fluid-structure interaction model of the internal carotid and ophthalmic arteries for the noninvasive intracranial pressure measurement method. Comput. Biol. Med. **84**, 79–88 (2017)
96. Mitchell, P. Lee, A.J., Wang, J.J., Rochtchina, E.: Intraocular pressure over the clinical range of blood pressure: blue mountains eye study findings. American Journal of Ophthalmology **140**(1), 131–132 (2005)
97. Morgan, W.H., Chauhan, B.C., Yu, D.Y., Cringle, S.J., Alder, V.A., House, P.H.: Optic disc movement with variations in intraocular and cerebrospinal fluid pressure. Invest. Ophthalmol. Vis. Sci. **43**(10), 1419–1428 (2002)
98. Morgan, W.H, Yu, D.Y., Balaratnasingam, C.: The role of cerebrospinal fluid pressure in glaucoma pathophysiology: the dark side of the optic disc. J. Glaucoma **17**(5), 408–413 (2008)
99. Morgan, W.H., Lind, C.R.P., Kain, S., Fatehee, N., Bala, A. Yu, D.Y.: Retinal Vein Pulsation Is in Phase with Intracranial Pressure and Not Intraocular PressureVein Pulsation Phase Relations. Invest. Ophthalmol. Vis. Sci. Eng. **53**(8), 4676–4681 (2012)
100. Moses R.A.: Intraocular Pressure. In *Adlers Physiology of the Eye: clinical application*, ed. by Moses, R.A. (C.V. Mosby Co, St Louis,1981), p. 227–254.
101. Nakabayashi, S., Nagaoka, T., Tani, T., Sogawa, K., Hein, T.W., Kuo, L., Yoshida, A.: Retinal arteriolar responses to acute severe elevation in systemic blood pressure in cats: role of endothelium-derived factors. Experimental Eye Research. Elsevier **103**, 63–70 (2012)
102. Neudorfer, M., Kessner, R., Goldenberg, D., Lavie, A., Kessler, A.: Retrobulbar blood flow changes in eyes with diabetic retinopathy: a 10-year follow-up study. Clinical Ophthalmology **8**, 2325–32 (2014)
103. Newson, T., El-Sheikh, A.: Mathematical modeling of the biomechanics of the lamina cribrosa under elevated intraocular pressures. J. Biomech. Eng. **128**(4), 496–504 (2006)
104. Oliva, I., Roztocil, K.: Toe pulse wave analysis in obliterating atherosclerosis. Angiology **34**(9), 610–619 (1983)
105. Olson, J.L., Asadi-Zeydabadi, M., Tagg, R.: Theoretical estimation of retinal oxygenation in chronic diabetic retinopathy. Comput. Biol. Med. **58**, 154–162 (2015)
106. Page, T.C., Light, W.R., Hellums, J.D.: Prediction of microcirculatory oxygen transport by erythrocyte/hemoglobin solution mixtures. Microvasc Res **56** 113–126 (1998)
107. Pedley, T.J., Brook, B.S., Seymour, R.S.: Blood pressure and flow rate in the giraffe jugular vein. Philosophical Transactions of the Royal Society of London B: Biological Sciences. The Royal Society, **351**(1342) 855–866 (1996)
108. Pettet, G.J., Byrne, H.M., McElwain, D.L.S., Norbury, J.: A model of wound-healing angiogenesis in soft tissue. Math. Biosci. **263**, 1487–1493 (1996)
109. Pinto, L.A., Vandewalle, E., DeClerck, E., Marques-Neves, C., Stalmans, I.: Ophthalmic artery Doppler waveform changes associated with increased damage in glaucoma patients. Invest. Ophthalmol. Vis. Sci. **53**(4), 2448–2453 (2012)
110. Popel, A.S.: Theory of oxygen transport to tissue. Crit Rev Biomed Eng **17** 257–321 (1989)
111. Prada, D.: A hybridizable discontinuous Galerkin method for nonlinear porous media viscoelasticity with applications in ophthalmology. Ph.D. thesis, Purdue University (2016)
112. Preziosi, L., Tosin, A.: Multiphase modelling of tumour growth and extracellular matrix interaction: mathematical tools and applications. J. Math. Biol. **58** 625–656 (2009)
113. Quarteroni, A., Formaggia, L., Ayache, N.: Mathematical modelling and numerical simulation of the cardiovascular system. EPFL (2002)
114. Formaggia, L., Quarteroni, A., Veneziani, A.: Cardiovascular Mathematics: Modeling and simulation of the circulatory system. Springer Science & Business Media (2010)

115. Ragauskas, A., Daubaris, G., Dziugys, A., Azelis, V., Gedrimas, V.: Innovative non-invasive method for absolute intracranial pressure measurement without calibration. Acta. Neurochir. **95**[Suppl.], 357–361 (2005)
116. Ragauskas, A., Matijosaitis, V., Zakelis, R., Petrikonis, K., Rastenyte, D., Piper, I., Daubaris, G.: Clinical assessment of noninvasive intracranial pressure absolute value measurement method. Neural. **78**(21), 1684–1691 (2012)
117. Ren, R., Jonas, J.B., Tian, G., Zhen, Y., Ma, K., Li, S., Wang, H., Li, B., Zhang, X., Wang, N.: Cerebrospinal fluid pressure in glaucoma: a prospective study. Ophthalmology **117**(2), 259–266 (2010)
118. Roos, M. W.: Theoretical estimation of retinal oxygenation during retinal artery occlusion. Physiological Measurement. IOP Publishing, **25**(6), 1523 (2004)
119. Saari, J.C.: Metabolism and photochemistry in the retina. In *Adler's Physiology of the Eye Clinical Application*, ed. by Moses, RA.; Hart, WM. (Mosby, St Louis,1987), p. 356–373.
120. Sacco, R., Mauri, A.G., Cardani, A., Siesky, B.A., Guidoboni, G., Harris, A.: Increased levels of nitric oxide may pathologically affect functional hyperemia in the retina: model and simulation. IOVS **58**(8) 214–214 (2017) (ARVO Abstract 214)
121. Sala, L., Prud'homme, C., Guidoboni, G., Szopos, M., Siesky, B.A., Harris, A.: Analysis of IOP and CSF alterations on ocular biomechanics and lamina cribrosa hemodynamics. IOVS (2018) (ARVO Abstract 4475)
122. Schugart, R.C., Friedman, A., Zhao, R., Sen, C.K.: Wound angiogenesis as a function of tissue oxygen tension: A mathematical model. PNAS **105** 2628–2633 (2008)
123. Secomb, T.W., Hsu, R., Park, E.Y., et al.: Greens function methods for analysis of oxygen delivery to tissue by microvascular networks. Ann Biomed Eng **32** 1519–1529 (2004)
124. Secomb, T.W.: Krogh-Cylinder and Infinite-Domain Models for Washout of an Inert Diffusible Solute from Tissue. Microcirculation **22**(1) 91–98 (2015)
125. Secomb, T.W.: A Greens function method for simulation of time-dependent solute transport and reaction in realistic microvascular geometries. Math Med Biol **33** 475–494 (2016)
126. Shapiro, A.H.: Steady flow in collapsible tubes. Journal of Biomechanical Engineering. American Society of Mechanical Engineers, **99**(3), 126–147 (1977)
127. Stathopoulos, N.A., Nair, P.K., Hellums, J.D.: Oxygen transport studies of normal and sickle red cell suspensions in artificial capillaries. Microvasc Res **34** 200–210 (1987)
128. Stewart, P.S., Jensen, O.E., Foss, A.JE.: A theoretical model to allow prediction of the csf pressure from observations of the retinal venous pulse. Invest. Ophthalmol. Vis. Sci. **55**(10), 6319–6323 (2014)
129. Takahashi, T., Nagaoka, T., Yanagida, H., Saitoh, T., Kamiya, A., Hein, T., Kuo, L., Yoshida, A.: A mathematical model for the distribution of hemodynamic parameters in the human retinal microvascular network. J. Biorheol. **23**(2), 77–86 (2009)
130. Tani,T., Nagaoka, T., Nakabayashi, S., Yoshioka, T., Yoshida, A.: Autoregulation of retinal blood flow in response to decreased ocular perfusion pressure in cats: Comparison of the effects of increased intraocular pressure and systemic hypotension. Investigative Ophthalmology and Visual Science. **55** 360 (2014).
131. Tham, Y.-C., Lim, S.-H., Gupta, P., et al: Inter-relationship between ocular perfusion pressure, blood pressure, intraocular pressure profiles and primary open-angle glaucoma: the Singapore Epidemiology of Eye Diseases study. B J Ophthalmol **0** 1–5 (2018)
132. Übeyli, E. D.: Adaptive neuro-fuzzy inference system employing wavelet coefficients for detection of ophthalmic arterial disorders. Expert. Syst. Appl. **34**(3), 2201–2209 (2008)
133. Verri, M., Guidoboni, G., Bociu, L., Sacco, R.: The role of structural viscoelasticity in deformable porous media with incompressible constituents: applications in biomechanics. Mathematical Biosciences and Engineering **154**(4), 933–959 (2018)
134. Wangsa-Wirawan, N.D., Linsenmeier, R.A.: Retinal oxygen: fundamental and clinical aspects. Archives of Ophthalmology **121**(4), 547–557 (2003)
135. Watson, M.G., McDougall, S.R., Chaplain, M. A. J., Devlin, A.H., Mitchell, C.A.: Dynamics of angiogenesis during murine retinal development: a coupled in vivo and in silico study J.R. Soc. Interface **9** 2351–2364 (2012)

136. Wellman, A., Genta, P.R., Owens, R.L., Edwards, B.A., Sands, S.A., Loring, S.H., White, D.P., Jackson, A.C., Pedersen, O.F., Butler, J.P.: Test of the Starling resistor model in the human upper airway during sleep. Journal of applied physiology. Am Physiological Soc, **117**(12), 1478–1485 (2014)
137. Whiteley, J.P., Gavaghan, D.J., Hahn, C.E.: Mathematical modelling of pulmonary gas transport. J Math Biol **47** 79–99 (2003)
138. Woo, S., Kobayashi, A.S., Schlegel, W.A., Lawrence, C.: Nonlinear material properties of intact cornea and sclera. Exp. Eye Res. **14**(1), 29–39 (1972)
139. Xu, L., Wang, H., Wang, Y., Jonas, J.B.: Intraocular pressure correlated with arterial blood pressure: the Beijing Eye Study. American Journal of Ophthalmology **144**(3), 461–462 (2007)
140. Ye, G.F., Moore, T.W., Buerk, D.G., Jaron, D.: A compartmental model for oxygen−carbon dioxide coupled transport in the microcirculation. Ann Biomed Eng. **22** 464–479 (1994)
141. Yu, D.Y., Cringle, S.J.: Oxygen distribution and consumption within the retina in vascularised and avascular retinas and in animal models of retinal disease. Progress in Retinal and Eye Research **20**(2) 175–208 (2001)
142. Zunino, P.: Mathematical and numerical modeling of mass transfer in the vascular system. Ph.D. thesis, École Oolytechnique Fédérale de Lausanne (2002)

# Part III
# Aqueous Humor

# 1 Introduction

Ocular aqueous humor dynamics (AHD) is fundamental to the maintenance of intraocular pressure (IOP). It is crucial to the health of the eye, especially to the avascular tissues of the anterior segment, and it is a key factor in some ocular pathologies related to IOP such as glaucoma. Characteristics of AHD are flow, facility, and pressure. The main parameters include aqueous humor production rate, usually measured as flow rate into the anterior chamber (Fa); the facility of outflow through the trabecular meshwork (Ctrab); the outflow from the anterior chamber angle into uveal tissues, the uveoscleral outflow (Fu); and the pressure in the vessels that drain aqueous humor from the trabecular outflow pathway, the episcleral venous pressure (EVP) (Fig. 1). These parameters are all interrelated as described by the modified Goldmann equation:

$$IOP = (Fa - Fu)/Ctrab + EVP.$$

This equation describes the steady-state IOP and not transient changes such as during a Valsalva maneuver.

Aqueous production starts in the ciliary process core of the ciliary body (Fig. 2). Fluid seeps from the capillaries within the core by ultrafiltration. Then by active transport it traverses the two epithelial layers lining the processes and is secreted into the posterior chamber (Fig. 1, arrow 1). Secretion requires movement of specific ions into the intercellular spaces between epithelial cells. These ions are mainly $Na^+$, $K^+$, $CO^2$, and $HCO_3^-$. Water follows by osmosis. Blocking transport of any of these ions can slow aqueous production rate. Some drugs lower IOP in this manner. Once in the posterior chamber, some fluid flows posteriorly across the vitreous cavity (Fig. 1, arrow 2). This flow is very hard to measure and is usually assumed to be unaffected by experimental manipulations. The flow that can be measured is the flow into the anterior chamber (aqueous flow, Fig. 1, arrow 3). "Aqueous production" is not synonymous with "aqueous flow." The aqueous humor in the anterior chamber feeds the avascular tissues of the cornea, lens, and trabecular meshwork and flushes away metabolic waste products.

The aqueous humor drains from the anterior chamber angle via two drainage pathways: trabecular (Fig. 1, arrow 5) and uveoscleral (Fig. 1, arrow 4). Trabecular outflow is the predominant, pressure-dependent pathway that provides the major resistance to drainage. The anatomy of this pathway is quite complex because it must provide the needed resistance to maintain IOP, yet it allows fluid to exit. The trabecular meshwork is composed of uveal, corneoscleral, and juxtacanalicular layers. These layered sheets of oval lacunae form 2–12 μm diameter openings [1]. The juxtacanalicular layer, which precedes Schlemm's canal, offers the most resistance to outflow in the trabecular meshwork system. The pathway distal to the TM also provides some outflow resistance. Distal flow starts at the collector channel entrances in the outer wall of Schlemm's canal, then moves into collector channels, the deep scleral plexus, the intrascleral venous plexus, and the episcleral

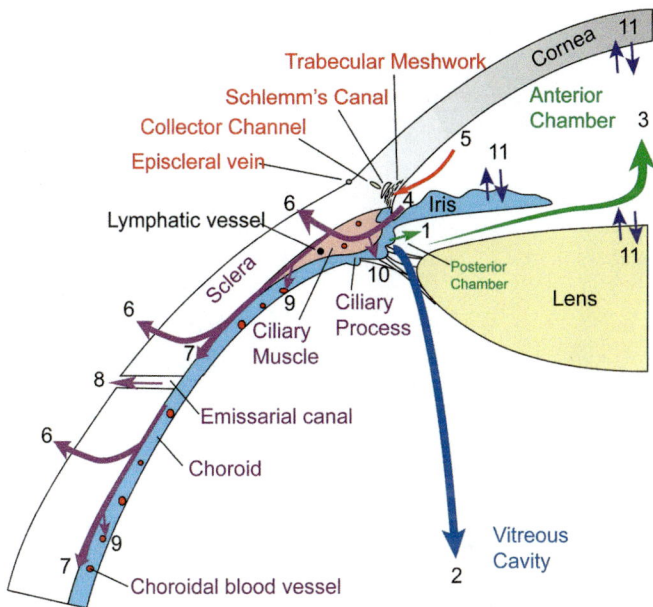

**Fig. 1** Circulation and drainage of aqueous humor. Ocular aqueous humor is secreted by the ciliary processes of the ciliary body into the posterior chamber (1). Some fluid drains across the vitreous cavity (2) and exits the eye across the retinal pigment epithelium. Most of the produced aqueous humor circulates between the lens and the iris through the pupil and into the anterior chamber (3). It drains from the anterior chamber by passive bulk flow via two pathways: the uveoscleral outflow pathway (4) and the trabecular outflow pathway (5). The trabecular outflow pathway includes the trabecular meshwork, Schlemm's canal, collector channels, deep scleral plexus, intrascleral venous plexus, episcleral veins, and anterior ciliary veins, and into the systemic circulation. In the uveoscleral outflow pathway, fluid first drains from the chamber angle into the tissue spaces of the ciliary muscle. From there, it drains in many directions, including across the sclera (6), along the supraciliary and suprachoroidal spaces (7), through emissarial canals (8), into choroidal vessels (9) and vortex veins (not drawn), or back into the ciliary processes (10) where it is secreted again. There is some exchange of aqueous humor with the iris, lens, and cornea (11), but this is not considered part of aqueous humor dynamics (Modified from Atlas of Glaucoma. Second Edition. Choplin N and Lundy C (eds). Informa UK Ltd., 2007. ISBN 9781841845180, with permission)

veins superficially. Fluid also drains into aqueous veins before reaching episcleral veins (Fig. 3). This pathway is affected by choroidal vascular volume, ocular pulse, and scleral rigidity [2, 3]. Improving drainage through or bypassing the trabecular meshwork altogether are effective therapeutic approaches to lower IOP.

The second outflow pathway, uveoscleral outflow (Fig. 1, arrows 4, 6–9), is the unconventional, pressure-independent pathway. This pathway drains fluid from the anterior chamber into the ciliary muscle. From there, fluid drains in multiple directions including the supraciliary and suprachoroidal spaces, across the sclera, through emmisarial canals, into choroidal vessels and into lymphatic vessels (Fig. 1). In other words, uveoscleral outflow encompasses the drainage of

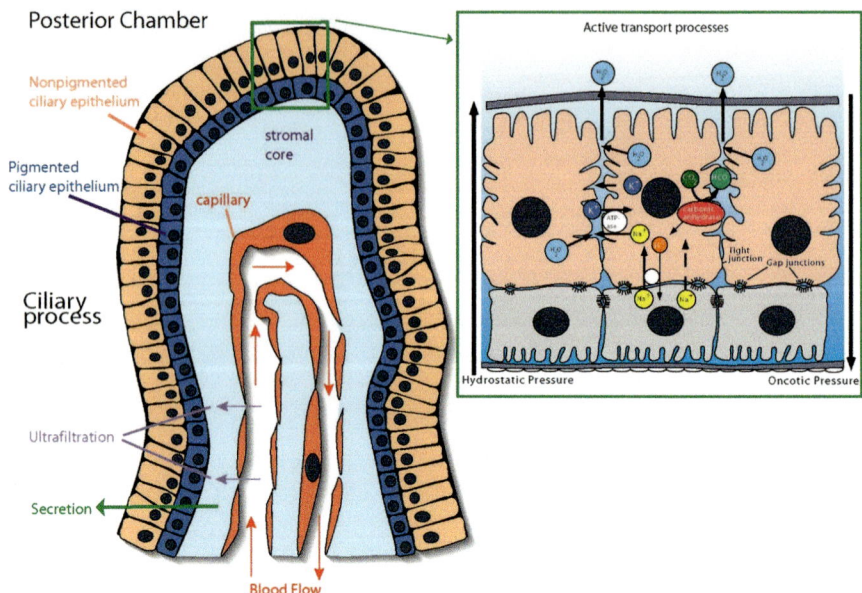

**Fig. 2** Production of aqueous humor. The ciliary process has a core containing capillaries and interstitial tissue surrounded by two epithelial layers, oriented apex to apex. The innermost layer consists of nonpigmented ciliary epithelial cells, which are connected to each other by tight junctions and desmosomes. The outer layer consists of pigmented ciliary epithelial cells connected to each other by gap junctions. The production of aqueous humor by the ciliary processes involves both ultrafiltration and active secretion. Plasma filtrate enters the interstitial space through capillary fenestrations (ultrafiltration). The capillary wall is a major barrier to plasma proteins, but the most significant barrier is in the nonpigmented ciliary epithelium where tight junctions occlude the apical region of the intercellular spaces. The outcome of this design is a high protein concentration in the tissue fluid. This causes a high oncotic pressure in the tissue fluid and a reduction in the transcapillary difference in the oncotic pressure. Under normal conditions, the movement of fluid into the posterior chamber requires secretion. (From Atlas of Glaucoma, with permission)

fluid from the anterior chamber angle other than through the trabecular meshwork. Uveoscleral outflow rates range from 10% [4] to 50% [5] of total outflow depending on health and age of the eye and time of day of the measurements. Relaxing the ciliary muscle enlarges spaces between muscle bundles, increases uveoscleral outflow, and lowers IOP. Additionally, biochemically modifying the extracellular matrix in the interstitial spaces of the ciliary muscle bundles improves drainage and lowers IOP. Contraction of the muscle enlarges the muscle bundles and narrows the interstitial spaces. Contraction also pulls on the scleral spur which is connected to the trabecular meshwork, thus stretching the tissue and enlarging the spaces in the meshwork. Thus, contraction decreases uveoscleral outflow yet improves trabecular outflow facility. The latter predominates over the former and IOP decreases.

The episcleral veins drain the aqueous humor that traverses the trabecular meshwork. Its pressure is important because were the pressure to increase, IOP

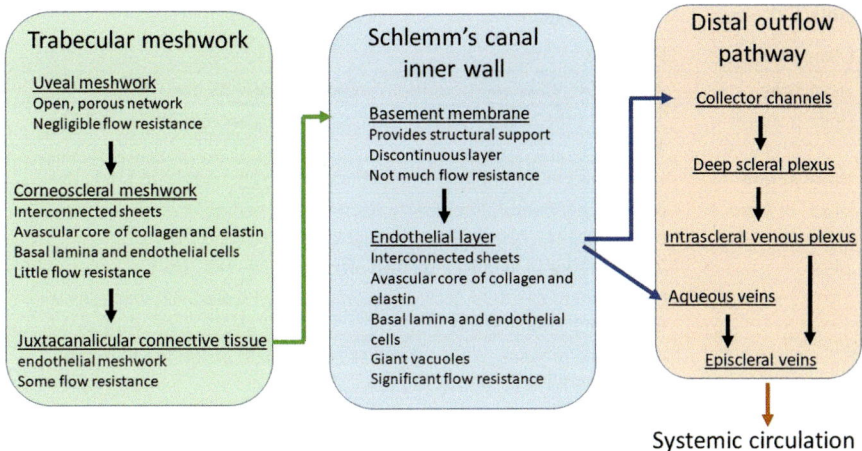

**Fig. 3** Details of trabecular and distal outflow. The arrows indicate the direction of fluid flow and the tissues along the pathway

would necessarily increase for aqueous humor to continue to drain. Conversely, if episcleral venous pressure were to decrease, IOP would decrease. Some treatments for glaucoma lower IOP in part by reducing the pressure in the episcleral veins.

The Goldmann equation is a key tool in the pursuit of understanding IOP changes during an active day, throughout life and in diseases such as glaucoma. The Goldmann equation was developed decades ago and is still of value today. However, with the wealth of new information obtained on each parameter since the equation was written, it may be time to consider a revision. This review summarizes the continually changing aspects of aqueous humor dynamics throughout life.

## 2   Fluctuations in IOP

Intraocular pressure varies predictably under numerous circumstances (Table 1). Some of these changes are transient, while others progress over a lifetime. Rapid changes usually are caused by fluctuations in episcleral venous pressure or changes in vascular volume, whereas longer term fluctuations or permanent changes are the result of alterations in inflow, outflow, or both.

### 2.1   24-h Changes

During the course of an active day, numerous situations can affect IOP including wearing tight clothing around the neck [6], playing a wind instrument [7], and

**Table 1** Normal fluctuations in intraocular pressure

| Changes | Outcome |
| --- | --- |
| Heart beat | Acute IOP peaks and troughs are synchronized with heart beat. |
| Body position | The lower the head relative to the heart, the higher the IOP [118, 144, 182]. |
| Body weight | Body mass index and IOP in adults are positively correlated [183, 184]. |
| Blood or venous pressure | IOP is higher with tight necktie (inconsistent finding [6, 143]), Valsalva maneuver [11], obstructive sleep apnea syndrome [185]. |
| 24-h | IOP is higher in the morning than afternoon [17, 186, 187]. |
| | Compared to day, IOP decreases at night when seated, but increases at night when supine [133]. |
| | Compared to young adults, the peak of 24-h IOP is delayed for older adults in both body positions [188]. |
| Seasonal | IOP is higher in winter than summer [16, 17, 189–191]. |
| Aging | IOP does not change with aging in healthy Chinese and Caucasian adults [5, 68]. |
| | IOP decreases in adults from Japan [65, 192–195]. |
| | IOP increases in adults from USA and Singapore [62, 196]. |
| Pregnancy | IOP decreases during pregnancy [72]. |

performing a Valsalva maneuver [8–11], to name a few. These activities alter the steady-state IOP by increasing systemic venous pressure which is transmitted through the jugular, orbital, and vortex veins to the choroid, causing vascular engorgement and increases in uveal volume. Once the activity ends, steady-state IOP returns. These non-steady-state situations that involve acute changes in blood pressure, flow, or volume are not described in the Goldmann equation.

Independent of physical activity, IOP follows a 24-h cycle. IOP at night is lower than during the day when a person is in the same body position for all measurements. However, IOP increases at night if the measurement is taken with the subject sitting during the day and supine at night [12, 13] which are the preferred body positions of most people. In healthy individuals, the peak IOP occurs at 8 AM and the trough IOP occurs at 8 PM. In the case of glaucoma, the 24-h cycle of IOP is shifted forward by about 8 h. Thus, in glaucomatous individuals, the peak IOP occurs at 4 PM and the trough IOP occurs at 4 AM [14, 15]. Whether this shift contributes to or is an outcome of glaucoma is unknown.

## 2.2 Seasonal Changes

The OHTS trial data across six climate zones in the USA found seasonal effects on IOP in all zones with IOP peaks between January and February [16, 17]. One hypothesis of several hypotheses for the circannual fluctuation in IOP may be related to the seasonal variation in sunlight hitting the eye [17]. This light affects the pineal gland, which is sensitive to the daily total amount of light signal reaching it by way of intrinsically photosensitive retinal ganglion cells in the retina. The pineal gland secretes more melatonin in the dark than in the light. Melatonin inhibits release of gonadotropin-releasing hormone in the anterior pituitary gland [18]. The downstream effect of this inhibition is decreased secretion of progesterone and estrogen. Both progesterone and estrogen lower IOP by increasing outflow facility [19, 20]. Therefore, during the winter months the decreased light levels lead to increased melatonin secretion and decreased progesterone and estrogen, resulting in a relative increase in IOP.

A clear understanding of melatonin and its IOP effects is muddied by the complexity of melatonin synthesis, secretion, and its various signaling pathways. Numerous studies of exogenous melatonin treatment have come to differing conclusions. While some studies have found that melatonin increases IOP [21–23], others have found that it decreases IOP. It is evident that seasonal changes in IOP cannot be explained by melatonin alone [24].

Another possible explanation for seasonal changes in IOP involves seasonal variation in sunlight and seasonal changes in expression of melatonin receptors. UVA radiation released by the sun upregulates expression of the NRF2 transcription factor [25]. NRF2 upregulates expression of NQO2, a melatonin receptor that decreases IOP [26]. Therefore, decreased light levels in the winter may result in decreased expression of NQO2 and increased IOP. To date, no studies on seasonal changes in NQO2 expression have been performed.

Just the secretion of melatonin alone illustrates its complexity. Although the pineal gland is the major site of melatonin synthesis in the body, pinealectomy had no effect on melatonin levels in the ciliary body of chickens; diurnal patterns continued as before the surgery [27]. The fact that no change was seen may be due to the fact that the retina itself synthesizes melatonin in a circadian rhythm [28]. Other tissues of the body, such as lymphocytes, also synthesize melatonin [29].

## 2.3   Recreational Substances

Recreational substances can affect IOP in numerous ways. Acute alcohol intake decreases IOP [30] by a mechanism unrelated to a change in outflow facility [31]. In acute alcohol intake, alcohol suppresses the release of antidiuretic hormone from the posterior pituitary gland. This suppression results in less reabsorption of water in the kidneys. Ultimately, there is a decrease in the total amount of fluid in the body, which decreases systemic blood pressure as well as IOP. Chronic alcohol consumption is associated with increased IOP [32, 33] by mechanisms unknown, but it may be related to effects on blood pressure. Chronic alcohol consumption is associated with systemic hypertension [34–38] which some (but not all) studies find to be correlated with ocular hypertension [39, 40].

Chronic cigarette use also is associated with increased IOP [41]. Nicotine-mediated activation of nicotinic acetylcholine receptors are upregulated by chronic

[5, 129]. An explanation for the slowing may be the thicker basement membrane of nonpigmented epithelium in eyes from patients between 60 and 70 years of age than patients younger than 50 years of age [130]. Between 4 and 9 years of age, aqueous flow appears to increase. This conclusion is based on a small study of ten children [117].

### 3.1.3 Pregnancy Changes in Aqueous Flow

Intraocular pressure is often found to be reduced during pregnancy. This is not a result of changes in aqueous flow [81, 131].

## 3.2 Uveoscleral Outflow

Rates of uveoscleral outflow slow from day to night and as one ages [132, 133]. In both circumstances, the reason can be explained in part by the concomitant slowing of aqueous humor production. Other factors exist that contribute to the rate of uveoscleral outflow and its fluctuations, one of which may be the ciliary muscle and its movement. This muscle serves three essential functions, all of which contribute to IOP in one way or another. (1) It is a key component in accommodation, contracting and relaxing to change the shape of the lens; (2) it provides a small resistance to uveoscleral drainage by nature of its extracellular matrix composition and the size of its muscle bundles; (3) it connects to the scleral spur and when contracted it maintains patency of the collapsible Schlemm's canal and modulates tension on its inner wall and juxtacanalicular tissue to directly affect trabecular outflow facility. As one ages, the ciliary muscle decreases in length, increases in width [134], and traps more pigment. The tendons of the ciliary muscle appear thickened, show increased amounts of associated microfibrils, and are surrounded by dense layers of thick collagen fibrils [135]. The morphological changes to the ciliary muscle and its diminished capacity to move when contracted may contribute to the age-related slowing of uveoscleral outflow.

By its anatomical connections to the trabecular meshwork via the scleral spur, contractions and relaxations of the ciliary muscle also affect the trabecular outflow pathway. It is possible, although not yet proven, that when the eye changes its focal distance from near to far and back, aqueous humor drainage alternates between the uveoscleral and trabecular outflow routes. At night when asleep, accommodative changes are not needed, the ciliary muscle remains at rest, and uveoscleral outflow slows. Interestingly, with aging, the contribution of uveoscleral drainage becomes smaller with greater dependency on the trabecular route for drainage of aqueous humor [68]. At the same time, aging appears to stiffen trabecular meshwork layers that keep the juxtacanalicular tissue and Schlemm's canal open even with fluctuating IOP [136]. Glaucomatous changes in the eye may lead to a reduction in the axial-tensile stiffness of the uveal and corneoscleral meshwork layers and an increase in compressive stiffness of the juxtacanalicular/inner wall tissues [137]. Under these conditions, the trabecular meshwork may be less capable of preventing Schlemm's canal collapse or the herniation of the trabecular meshwork into collector channels under elevated IOP [138], which may increase outflow resistance.

## 3.3   Episcleral Venous Pressure

Published values of episcleral venous pressure (EVP) range from 7 to 12 mmHg. These values were collected using a venomanometer [139], the only instrument designed for noninvasive measurement of EVP. The range in pressures is large because the technique has a significant subjective component. The venomanometer is clamped to a slit lamp, and the clear membrane of the venomanometer is positioned so that it can be seen through the oculars of the slit lamp. The membrane is placed on the sclera near the limbus where episcleral veins are visible. An episcleral vein is identified, and the pressure behind the membrane is slowly increased until the vein begins to collapse. The point of half collapse of the vein is considered the endpoint. This value is read off the dial on the venomanometer. Half collapse is the subjective endpoint but with sufficient experience consistency in the measurement is possible. Most reports list EVP values in the range of 8–10 mmHg. When a video imaging system is added to the venomanometer, the change in diameter of the vessel is recorded as the pressure is raised. The pressure level is stamped on each frame of the video. The images are saved and a frame-by-frame examination of the vein allows one to read the pressure at the beginning of vessel collapse instead of using half collapse as the endpoint. The method is labor intensive but more reproducible than the half collapse method. With this technique, EVP is approximately 7 mmHg [140].

The EVP increases by 1–3.6 mmHg at night compared to the day [12, 13]. This increase is due predominantly to a change in body position from upright during the day to recumbent at night. Effect of sleep itself on EVP has not been studied as the subject must be awake to make the measurement. EVP also can be elevated transiently by downward displacement [141] of the head or wearing a tight necktie [6]. Tight neckties may constrict the jugular vein, thereby increasing episcleral venous pressure, and in turn elevating IOP [142]. Not all studies have found this effect, however [143].

The difference in IOP between seated and supine postures is greater in patients with primary open angle glaucoma [144] than in healthy age-matched cohorts, and it is even greater in exfoliation glaucoma [145]. The IOP rise is thought to result from posture-related increases of EVP and choroidal blood volume. This posture effect on IOP is independent of age [144, 146]. EVP does not change with aging in healthy individuals [5].

Under some circumstances, elevated EVP may be associated with glaucoma. These conditions include idiopathic elevated episcleral venous pressure, episcleritis [147], Sturge Weber Syndrome [148, 149], spontaneous carotid cavernous fistulas, orbital tumors, and endocrine exophthalmos [150]. Patients without glaucoma but with obstructive sleep apnea (OSA) have normal IOP patterns at night [151], but patients with glaucoma and OSA seem to have increased IOP after therapy with continuous positive airway pressure [152].

## 3.4  Outflow Facility

Outflow facility is the facility of aqueous humor outflow from the anterior chamber angle. It describes predominantly the facility of trabecular outflow. The methods that measure facility cannot easily separate trabecular from uveoscleral outflow facility, but as the uveoscleral outflow facility is only a small percentage of trabecular outflow facility [153, 154], any detected changes are assumed to be from alterations in trabecular outflow facility. Outflow facility is the reverse of the resistance to outflow, i.e., when facility is high, resistance is low. Outflow facility is assessed by numerous methods, all of which determine a change in aqueous flow with a concurrent change in IOP. Tonographic and fluorophotometric methods are the only methods available for use in humans. With the tonographic method, IOP is measured continuously for 2 or 4 min with a weighted probe placed on the cornea of the supine subject. The weight is a 10 g donut that is attached to the shaft of the probe. The other eye fixates on a target on the ceiling. Minimal movement of the subject and the technician holding the probe yields best results. The change in pressure is assumed to be from a change in outflow rate of aqueous humor (determined by Friedenwald tables). The change in flow during the test divided by the simultaneous change in pressure yields outflow facility. The fluorophotometric method utilizes the same concepts as tonography. However, the method measures a change in flow rate following administration of an aqueous flow suppressant, such as acetazolamide. It slows flow and lowers pressure but does not affect any other parameter of the Goldmann equation. As with tonography, the ratio of the change in flow to change in pressure is outflow facility. Each of these methods has numerous assumptions and shortcomings that have been described elsewhere [155]. Nevertheless, important findings have been made using these methods.

Outflow facility is low in people with elevated IOP with or without several types of glaucoma. Outflow facility decreases at night in healthy humans [132, 133] but does not decrease at night in ocular hypertensive patients [121] presumably because it is already low during the day and cannot decrease further at night. Outflow facility decreases at night to maintain IOP within healthy levels despite the 50% nocturnal reduction in aqueous flow.

Outflow facility is regulated by very complex molecular signaling mechanisms, changes in trabecular cell shape, and anatomical changes in the lining of Schlemm's canal. Many laboratories are devoted to understanding these mechanisms. The more that is learned, the more one is awed by the complexity of the trabecular drainage system.

After crossing the trabecular meshwork, aqueous humor reaches the episcleral veins by flowing through collector channels or aqueous veins. Collector channels contribute some resistance to outflow and this should be taken into consideration in any mathematical model of aqueous humor dynamics. The first indication that collector channels have resistance was noted when complete removal of the trabecular meshwork (trabeculectomy), the site of highest resistance to outflow, did not drop IOP to the level of episcleral venous pressure. In eyes with and without primary open angle glaucoma [2, 156], a high percentage of aqueous humor outflow resistance in the distal parts of the outflow tissue was found at normal IOP which decreased as IOP rose. Morphological examination of outflow tissues in human eyes revealed some occluded collector channels that appear to open at a higher IOP. In eyes with primary open angle glaucoma, there is a decrease in the area of Schlemm's canal and a reduced number of open collector channels available for fluid movement [157], factors that explain, at least in part, the reduced outflow facility.

### 3.4.1 Changes in Outflow Facility During Pregnancy

The decrease in IOP that is found during pregnancy occurs primarily because of an increase in outflow facility [81]. One hypothesis for this effect is that excess progesterone levels during pregnancy block the ocular hypertensive effect of endogenous corticosteroids [81]. Nonpregnant women given progestin have increased outflow facility [19]. Relaxin also may play a role in IOP changes during pregnancy as discussed above. Since multiple factors are changing during pregnancy, additional explanations for the reduced IOP cannot be ruled out.

## 4 Drug Efficacy Varies with AHD Fluctuations

### 4.1 Inflow Drugs

Drugs that slow aqueous inflow into the anterior chamber (inflow drugs) can significantly and rapidly lower IOP. Most of the early drugs used to treat glaucoma worked by this mechanism. Three classes of drugs that have this therapeutically beneficial characteristic are beta-adrenergic antagonists (beta-blockers), carbonic anhydrase inhibitors (CAIs), and alpha$_2$ adrenergic agonists. Beta-blockers (such as timolol) reduce aqueous flow by blocking beta-adrenoceptors located in the ciliary processes [158], reduce the production of cAMP, and slow the secretion of aqueous humor by the ciliary epithelia. Timolol has potent effects on aqueous flow during the day but no effect on aqueous flow or IOP at night. It cannot improve on the physiological factors regulating normal 24-h fluctuations of IOP and nightly slowing of aqueous flow. Although timolol is ineffective at lowering IOP at night, it does dampen the normal IOP increase upon awakening the following morning [159].

Carbonic anhydrase inhibitors (CAIs) slow aqueous production in a way completely different from beta-blockers. Several different isoforms of carbonic anhydrase enzymes are located in the pigmented and nonpigmented ciliary epithelia of the ciliary processes (Fig. 2). The coordinated function of these enzymes, along with ion cotransporters and exchangers, contributes to aqueous humor formation by implementing bicarbonate transport across the ciliary epithelium [160]. Bicarbonate, sodium, and chloride are transported into the intercellular spaces between nonpigmented ciliary epithelial cells, drawing water by osmosis. Inhibitors of carbonic anhydrase appear to block aqueous humor formation by inhibiting NaCl uptake from the stroma [161], and slowing the rate of water movement into the posterior chamber. Aqueous flow rate into the anterior chamber decreases, a change that is detectable by fluorophotometry. The outcome is a reduction in IOP.

CAIs are short acting drugs that have inconsistent effects at night. Some studies reported reductions in nocturnal IOP and aqueous flow beyond normal 24-h fluctuations. Other studies did not [159, 162–164]. Unlike timolol, the morning rise in IOP upon awakening was not inhibited by the CAI, dorzolamide [159].

Alpha$_2$ agonists reduce aqueous flow on the short term but with continued treatment, the aqueous flow effect fades [165]. The strong vasoconstrictive effect by brimonidine on uveal blood vessels reduces blood flow and hence ciliary body volume [166]. Insufficient blood supply causes insufficient delivery of oxygen and other energy sources to the ciliary processes, and as a consequence, aqueous humor production slows. Aqueous production should return to normal when the vasoconstrictive effect fades, and blood flow again delivers adequate energy sources to the ciliary processes. With continued dosing of brimonidine for a month, the aqueous flow effect is lost, replaced by an increase in uveoscleral outflow [165]. The cause of the latter effect is unclear but may result from release of endogenous prostaglandins and molecular and cellular changes in the ciliary muscle. When given three times daily for 6 weeks, brimonidine did not slow aqueous flow or lower IOP at night. Any nocturnal drug effect on inflow or outflow was not sufficient for a nocturnal change in IOP to be detectable [12].

## 4.2 Outflow Drugs

Approved outflow drugs for the treatment of elevated IOP include sympathomimetics, parasympathomimetics, prostaglandin analogs, and rho kinase inhibitors. The sympathomimetics and parasympathomimetics are rarely used for chronic lowering of IOP anymore and their IOP effects at night are unknown.

Prostaglandin analogs (PGAs) reduce IOP mainly by improving uveoscleral outflow [167, 168] although improvements in outflow facility also have been reported [169, 170]. One mechanism by which uveoscleral outflow is increased may include widening of the spaces among the ciliary muscle bundles [171, 172] as a result of relaxation of the ciliary muscle [173, 174], or remodeling of the extracellular matrix of the ciliary muscle [175–177]. Another possibility is the

42. Govind, A.P., P. Vezina, and W.N. Green, *Nicotine-induced upregulation of nicotinic receptors: underlying mechanisms and relevance to nicotine addiction.* Biochem Pharmacol, 2009. **78**(7): p. 756-65.
43. Jiwani, A.Z., et al., *Effects of caffeinated coffee consumption on intraocular pressure, ocular perfusion pressure, and ocular pulse amplitude: a randomized controlled trial.* Eye (Lond), 2012. **26**(8): p. 1122-30.
44. Opremcak, E.M. and P.A. Weber, *Interaction of timolol and caffeine on intraocular pressure.* J Ocul Pharmacol, 1985. **1**(3): p. 227-34.
45. Tomida, I., R.G. Pertwee, and A. Azuara-Blanco, *Cannabinoids and glaucoma.* Br J Ophthalmol, 2004. **88**(5): p. 708-13.
46. Zhan, G.L., et al., *Effects of marijuana on aqueous humor dynamics in a glaucoma patient.* J Glaucoma, 2005. **14**(2): p. 175-7.
47. Chien, F.Y., et al., *Effect of WIN 55212-2, a cannabinoid receptor agonist, on aqueous humor dynamics in monkeys.* Arch Ophthalmol, 2003. **121**(1): p. 87-90.
48. Green, K. and J.E. Pederson, *Effect of 1 -tetrahydrocannabinol on aqueous dynamics and ciliary body permeability in the rabbit.* Exp Eye Res, 1973. **15**(4): p. 499-507.
49. Merritt, J.C., et al., *Topical delta 9-tetrahydrocannabinol and aqueous dynamics in glaucoma.* J Clin Pharmacol, 1981. **21**(8-9 Suppl): p. 467S-471S.
50. *American Academy of Ophthalmology Complementary Therapy Task Force, Hoskins Center for Quality Eye Care. Marijuana in the Treatment of Glaucoma CTA - 2014 Complementary Therapy Assessments. San Francisco: American Academy of Ophthalmology.* 2014.
51. Jaafar, M.S. and G.A. Kazi, *Normal intraocular pressure in children: a comparative study of the Perkins applanation tonometer and the pneumatonometer.* J Pediatr Ophthalmol Strabismus, 1993. **30**(5): p. 284-7.
52. Jiang, W.J., et al., *Intraocular pressure and associated factors in children: the Shandong children eye study.* Invest Ophthalmol Vis Sci, 2014. **55**(7): p. 4128-34.
53. Dusek, W.A., B.K. Pierscionek, and J.F. McClelland, *Age variations in intraocular pressure in a cohort of healthy Austrian school children.* Eye (Lond), 2012. **26**(6): p. 841-5.
54. Tint, N.L., et al., *Hormone therapy and intraocular pressure in nonglaucomatous eyes.* Menopause, 2010. **17**(1): p. 157-60.
55. Feldman, F., J. Bain, and A.R. Matuk, *Daily assessment of ocular and hormonal variables throughout the menstrual cycle.* Arch Ophthalmol, 1978. **96**(10): p. 1835-8.
56. Becker, B. and C.K. Ramsey, *Plasma cortisol and the intraocular pressure response to topical corticosteroids.* Am J Ophthalmol, 1970. **69**(6): p. 999-1003.
57. Becker, B., *Intraocular pressure response to topical corticosteroids.* Invest Ophthalmol, 1965. **4**: p. 198-205.
58. Otte, C., et al., *A meta-analysis of cortisol response to challenge in human aging: importance of gender.* Psychoneuroendocrinology, 2005. **30**(1): p. 80-91.
59. Zhou, J.N., et al., *Alterations in the circadian rhythm of salivary melatonin begin during middle-age.* Journal of Pineal Research, 2003. **34**(1): p. 11-16.
60. Ohashi, Y., et al., *Differential pattern of the circadian rhythm of serum melatonin in young and elderly healthy subjects.* Biological Signals, 1997. **6**(4-6): p. 301-306.
61. Wetterberg, L., et al., *Normative melatonin excretion: a multinational study.* Psychoneuroendocrinology, 1999. **24**(2): p. 209-226.
62. Klein, B.E., R. Klein, and K.L. Linton, *Intraocular pressure in an American community. The Beaver Dam Eye Study.* Invest Ophthalmol Vis Sci, 1992. **33**(7): p. 2224-8.
63. Leske, M.C., et al., *Distribution of intraocular pressure. The Barbados Eye Study.* Arch Ophthalmol, 1997. **115**(8): p. 1051-7.
64. Memarzadeh, F., et al., *Associations with intraocular pressure in Latinos: the Los Angeles Latino Eye Study.* Am J Ophthalmol, 2008. **146**(1): p. 69-76.
65. Nakano, T., et al., *Long-term physiologic changes of intraocular pressure: a 10-year longitudinal analysis in young and middle-aged Japanese men.* Ophthalmology, 2005. **112**(4): p. 609-16.

66. Lin, H.Y., et al., *Intraocular pressure measured with a noncontact tonometer in an elderly Chinese population: the Shihpai Eye Study.* Arch Ophthalmol, 2005. **123**(3): p. 381-6.

67. Lee, J.S., et al., *Relationship between intraocular pressure and systemic health parameters in a Korean population.* Clinical and Experimental Ophthalmology, 2002. **30**(4): p. 237-241.

68. Guo, T., et al., *Aqueous humour dynamics and biometrics in the ageing Chinese eye.* Br J Ophthalmol, 2017. **101**(9): p. 1290-1296.

69. Khawaja, A.P., et al., *Associations with intraocular pressure across Europe: The European Eye Epidemiology (E3) Consortium.* European Journal of Epidemiology, 2016. **31**(11): p. 1101-1111.

70. Cohen, E., et al., *Relationship Between Body Mass Index and Intraocular Pressure in Men and Women: A Population-based Study.* J Glaucoma, 2016. **25**(5): p. e509-13.

71. Kocak, N., et al., *Evaluation of the intraocular pressure in obese adolescents.* Minerva Pediatr, 2015. **67**(5): p. 413-8.

72. Wang, C., et al., *Changes in intraocular pressure and central corneal thickness during pregnancy: a systematic review and Meta-analysis.* Int J Ophthalmol, 2017. **10**(10): p. 1573-1579.

73. Akar, Y., et al., *Effect of pregnancy on intraobserver and intertechnique agreement in intraocular pressure measurements.* Ophthalmologica, 2005. **219**(1): p. 36-42.

74. Ibraheem, W.A., et al., *Tear Film Functions and Intraocular Pressure Changes in Pregnancy.* Afr J Reprod Health, 2015. **19**(4): p. 118-22.

75. Qureshi, I.A., X.R. Xi, and T. Yaqob, *The ocular hypotensive effect of late pregnancy is higher in multigravidae than in primigravidae.* Graefes Arch Clin Exp Ophthalmol, 2000. **238**(1): p. 64-7.

76. Qureshi, I.A., et al., *Effect of third trimester of pregnancy on diurnal variation of ocular pressure.* Chin Med Sci J, 1997. **12**(4): p. 240-3.

77. Qureshi, I.A., *Measurements of intraocular pressure throughout the pregnancy in Pakistani women.* Chin Med Sci J, 1997. **12**(1): p. 53-6.

78. Qureshi, I.A., *Intraocular pressure and pregnancy: a comparison between normal and ocular hypertensive subjects.* Arch Med Res, 1997. **28**(3): p. 397-400.

79. Qureshi, I.A., *Intraocular pressure: association with menstrual cycle, pregnancy and menopause in apparently healthy women.* Chin J Physiol, 1995. **38**(4): p. 229-34.

80. Saylik, M. and S.A. Saylik, *Not only pregnancy but also the number of fetuses in the uterus affects intraocular pressure.* Indian J Ophthalmol, 2014. **62**(6): p. 680-2.

81. Ziai, N., et al., *Beta-human chorionic gonadotropin, progesterone, and aqueous dynamics during pregnancy.* Arch Ophthalmol, 1994. **112**(6): p. 801-6.

82. Kass, M.A. and M.L. Sears, *Hormonal regulation of intraocular pressure.* Surv Ophthalmol, 1977. **22**(3): p. 153-76.

83. Saylik, M. and S.A. Saylık, *Not only pregnancy but also the number of fetuses in the uterus affects intraocular pressure.* Indian journal of ophthalmology, 2014. **62**(6): p. 680-2.

84. Hisaw, F.L., *Experimental relaxation of the pubic ligament of the guinea pig.* Proc Soc Exp Biol Med 1926. **23**(8): p. 661-663.

85. Sherwood, O.D., *Relaxin's physiological roles and other diverse actions.* Endocr Rev, 2004. **25**(2): p. 205-34.

86. Bathgate, R.A., et al., *Relaxin family peptides and their receptors.* Physiol Rev, 2013. **93**(1): p. 405-80.

87. Dehghan, F., et al., *The effect of relaxin on the musculoskeletal system.* Scand J Med Sci Sports, 2014. **24**(4): p. e220-9.

88. Weiss, G., E.M. O'Byrne, and B.G. Steinetz, *Relaxin: a product of the human corpus luteum of pregnancy.* Science, 1976. **194**(4268): p. 948-9.

89. Quagliarello, J., et al., *Serial relaxin concentrations in human pregnancy.* Am J Obstet Gynecol, 1979. **135**(1): p. 43-4.

90. Goldsmith, L.T. and G. Weiss, *Relaxin in human pregnancy.* Ann N Y Acad Sci, 2009. **1160**: p. 130-5.

91. Palejwala, S., et al., *Relaxin gene and protein expression and its regulation of procollagenase and vascular endothelial growth factor in human endometrial cells.* Biol Reprod, 2002. **66**(6): p. 1743-8.

92. Goldsmith, L.T., et al., *Relaxin regulation of endometrial structure and function in the rhesus monkey.* Proc Natl Acad Sci U S A, 2004. **101**(13): p. 4685-9.

93. Lenhart, J.A., et al., *Relaxin increases secretion of tissue inhibitor of matrix metalloproteinase-1 and -2 during uterine and cervical growth and remodeling in the pig.* Endocrinology, 2002. **143**(1): p. 91-8.

94. De Groef, L., et al., *MMPs in the trabecular meshwork: promising targets for future glaucoma therapies?* Invest Ophthalmol Vis Sci, 2013. **54**(12): p. 7756-63.

95. Borras, T., L.K. Buie, and M.G. Spiga, *Inducible scAAV2.GRE.MMP1 lowers IOP long-term in a large animal model for steroid-induced glaucoma gene therapy.* Gene Ther, 2016.

96. O'Callaghan, J., et al., *Therapeutic potential of AAV-mediated MMP-3 secretion from corneal endothelium in treating glaucoma.* Hum Mol Genet, 2017. **26**(7): p. 1230-1246.

97. Pattabiraman, P.P. and C.B. Toris, *The exit strategy: Pharmacological modulation of extracellular matrix production and deposition for better aqueous humor drainage.* Eur J Pharmacol, 2016. **787**: p. 32-42.

98. Weinreb, R., E. Cotlier, and B.Y.J.T. Yue, *The extracellular matrix and its modulation in the trabecular meshwork.* Survey of Ophthalmology, 1996. **40**(5): p. 379-390.

99. Tane, N., et al., *Effect of excess synthesis of extracellular matrix components by trabecular meshwork cells: possible consequence on aqueous outflow.* Exp Eye Res, 2007. **84**(5): p. 832-42.

100. Acott, T.S. and M.J. Kelley, *Extracellular matrix in the trabecular meshwork.* Exp Eye Res, 2008. **86**(4): p. 543-61.

101. Fuchshofer, R. and E.R. Tamm, *Modulation of extracellular matrix turnover in the trabecular meshwork.* Experimental Eye Research, 2009. **88**(4): p. 683-688.

102. Keller, K.E., et al., *Extracellular matrix turnover and outflow resistance.* Exp Eye Res, 2009. **88**(4): p. 676-82.

103. Samuel, C.S., R.J. Summers, and T.D. Hewitson, *Antifibrotic Actions of Serelaxin - New Roles for an Old Player.* Trends Pharmacol Sci, 2016. **37**(6): p. 485-97.

104. Bennett, R.G., *Relaxin and its role in the development and treatment of fibrosis.* Transl Res, 2009. **154**(1): p. 1-6.

105. Bennett, R.G., et al., *Relaxin decreases the severity of established hepatic fibrosis in mice.* Liver Int, 2014. **34**(3): p. 416-26.

106. Samuel, C.S., et al., *Anti-fibrotic actions of relaxin.* Br J Pharmacol, 2017. **174**(10): p. 962-976.

107. Fuchshofer, R. and E.R. Tamm, *The role of TGF-beta in the pathogenesis of primary open-angle glaucoma.* Cell Tissue Res, 2012. **347**(1): p. 279-90.

108. Heeg, M.H., et al., *The antifibrotic effects of relaxin in human renal fibroblasts are mediated in part by inhibition of the Smad2 pathway.* Kidney Int, 2005. **68**(1): p. 96-109.

109. Sassoli, C., et al., *Relaxin prevents cardiac fibroblast-myofibroblast transition via notch-1-mediated inhibition of TGF-beta/Smad3 signaling.* PLoS One, 2013. **8**(5): p. e63896.

110. Masterson, R., et al., *Relaxin down-regulates renal fibroblast function and promotes matrix remodelling in vitro.* Nephrol Dial Transplant, 2004. **19**(3): p. 544-52.

111. Huang, X., et al., *Relaxin regulates myofibroblast contractility and protects against lung fibrosis.* Am J Pathol, 2011. **179**(6): p. 2751-65.

112. Chow, B.S., et al., *Relaxin signals through a RXFP1-pERK-nNOS-NO-cGMP-dependent pathway to up-regulate matrix metalloproteinases: the additional involvement of iNOS.* PLoS One, 2012. **7**(8): p. e42714.

113. Wang, C., et al., *The Anti-fibrotic Actions of Relaxin Are Mediated Through a NO-sGC-cGMP-Dependent Pathway in Renal Myofibroblasts In Vitro and Enhanced by the NO Donor, Diethylamine NONOate.* Front Pharmacol, 2016. **7**: p. 91.

114. Chow, B.S., et al., *Relaxin requires the angiotensin II type 2 receptor to abrogate renal interstitial fibrosis.* Kidney Int, 2014. **86**(1): p. 75-85.

115. Singh, S., R.L. Simpson, and R.G. Bennett, *Relaxin activates peroxisome proliferator-activated receptor gamma (PPARgamma) through a pathway involving PPARgamma coactivator 1alpha (PGC1alpha).* J Biol Chem, 2015. **290**(2): p. 950-9.
116. Singh, S. and R.G. Bennett, *Relaxin family peptide receptor 1 activation stimulates peroxisome proliferator-activated receptor gamma.* Ann N Y Acad Sci, 2009. **1160**: p. 112-6.
117. Brubaker, R.F., *Flow of aqueous humor in humans [The Friedenwald Lecture].* Invest Ophthalmol Vis Sci, 1991. **32**(13): p. 3145-66.
118. Carlson, K.H., et al., *Effect of body position on intraocular pressure and aqueous flow.* Invest Ophthalmol Vis Sci, 1987. **28**(8): p. 1346-52.
119. Gharagozloo, N.Z., R.H. Baker, and R.F. Brubaker, *Aqueous dynamics in exfoliation syndrome.* Am J Ophthalmol, 1992. **114**(4): p. 473-8.
120. Larsson, L.I., E.S. Rettig, and R.F. Brubaker, *Aqueous flow in open-angle glaucoma.* Arch Ophthalmol, 1995. **113**(3): p. 283-6.
121. Fan, S., et al., *Aqueous humor dynamics during the day and night in volunteers with ocular hypertension.* Arch Ophthalmol, 2011. **129**(9): p. 1162-6.
122. Koskela, T. and R.F. Brubaker, *The nocturnal suppression of aqueous humor flow in humans is not blocked by bright light.* Invest Ophthalmol Vis Sci, 1991. **32**(9): p. 2504-6.
123. McLaren, J.W., R.F. Brubaker, and J.S. FitzSimon, *Continuous measurement of intraocular pressure in rabbits by telemetry.* Invest Ophthalmol Vis Sci, 1996. **37**(6): p. 966-75.
124. Dortch-Carnes, J. and G. Tosini, *Melatonin receptor agonist-induced reduction of SNP-released nitric oxide and cGMP production in isolated human non-pigmented ciliary epithelial cells.* Exp Eye Res, 2013. **107**: p. 1-10.
125. Wiechmann, A.F. and C.R. Wirsig-Wiechmann, *Melatonin receptor mRNA and protein expression in Xenopus laevis nonpigmented ciliary epithelial cells.* Exp Eye Res, 2001. **73**(5): p. 617-23.
126. Osborne, N.N. and G. Chidlow, *The presence of functional melatonin receptors in the iris-ciliary processes of the rabbit eye.* Exp Eye Res, 1994. **59**(1): p. 3-9.
127. Viggiano, S.R., et al., *The effect of melatonin on aqueous humor flow in humans during the day.* Ophthalmology, 1994. **101**(2): p. 326-31.
128. Bill, A., *Blood circulation and fluid dynamics in the eye.* Physiol Rev, 1975. **55**(3): p. 383-417.
129. Zhao, M., et al., *Aqueous humor dynamics during the day and night in juvenile and adult rabbits.* Invest Ophthalmol Vis Sci, 2010. **51**(6): p. 3145-51.
130. Okuyama, M., S. Okisaka, and Y. Kadota, *[Histological analysis of aging ciliary body].* Nippon Ganka Gakkai Zasshi, 1993. **97**(11): p. 1265-73.
131. Green, K., et al., *Aqueous humor flow rate and intraocular pressure during and after pregnancy.* Ophthalmic Res, 1988. **20**(6): p. 353-7.
132. Nau, C.B., et al., *Circadian Variation of Aqueous Humor Dynamics in Older Healthy Adults.* Investigative Ophthalmology & Visual Science, 2013. **54**(12): p. 7623-7629.
133. Liu, H., et al., *Aqueous humor dynamics during the day and night in healthy mature volunteers.* Arch Ophthalmol, 2011. **129**(3): p. 269-75.
134. Sheppard, A.L. and L.N. Davies, *The effect of ageing on in vivo human ciliary muscle morphology and contractility.* Invest Ophthalmol Vis Sci, 2011. **52**(3): p. 1809-16.
135. Tamm, E., et al., *Posterior attachment of ciliary muscle in young, accommodating old, presbyopic monkeys.* Invest Ophthalmol Vis Sci, 1991. **32**(5): p. 1678-92.
136. Camras, L.J., et al., *Differential effects of trabecular meshwork stiffness on outflow facility in normal human and porcine eyes.* Invest Ophthalmol Vis Sci, 2012. **53**(9): p. 5242-50.
137. Camras, L.J., et al., *Circumferential tensile stiffness of glaucomatous trabecular meshwork.* Invest Ophthalmol Vis Sci, 2014. **55**(2): p. 814-23.
138. Battista, S.A., et al., *Reduction of the available area for aqueous humor outflow and increase in meshwork herniations into collector channels following acute IOP elevation in bovine eyes.* Invest Ophthalmol Vis Sci, 2008. **49**(12): p. 5346-52.
139. Zeimer, R.C., et al., *A practical venomanometer. Measurement of episcleral venous pressure and assessment of the normal range.* Arch Ophthalmol, 1983. **101**(9): p. 1447-9.

140. Sit, A.J., et al., *A novel method for computerized measurement of episcleral venous pressure in humans.* Exp Eye Res, 2011. **92**(6): p. 537-44.
141. Lavery, W.J. and J.W. Kiel, *Effects of Head Down Tilt on Episcleral Venous Pressure in a Rabbit Model.* Experimental eye research, 2013. **111**: p. 88-94.
142. Bigger, J.F., *Glaucoma with elevated episcleral venous pressure.* South Med J, 1975. **68**(11): p. 1444-8.
143. Theelen, T., et al., *Impact factors on intraocular pressure measurements in healthy subjects.* Br J Ophthalmol, 2004. **88**(12): p. 1510-1.
144. Katsanos, A., et al., *The Effect of Posture on Intraocular Pressure and Systemic Hemodynamic Parameters in Treated and Untreated Patients with Primary Open-Angle Glaucoma.* J Ocul Pharmacol Ther, 2017. **33**(8): p. 598-603.
145. Ozkok, A., et al., *Posture-induced changes in intraocular pressure: comparison of pseudoexfoliation glaucoma and primary open-angle glaucoma.* Jpn J Ophthalmol, 2014. **58**(3): p. 261-6.
146. Cymbor, M., E. Knapp, and F. Carlin, *Idiopathic elevated episcleral venous pressure with secondary glaucoma.* Optom Vis Sci, 2013. **90**(7): p. e213-7.
147. Pikkel, J., et al., *Is Episcleritis Associated to Glaucoma?* J Glaucoma, 2015. **24**(9): p. 669-71.
148. Jorgensen, J.S. and R. Guthoff, *[Sturge-Weber syndrome: glaucoma with elevated episcleral venous pressure].* Klin Monbl Augenheilkd, 1987. **191**(4): p. 275-8.
149. Shiau, T., et al., *The role of episcleral venous pressure in glaucoma associated with Sturge-Weber syndrome.* J AAPOS, 2012. **16**(1): p. 61-4.
150. Jorgensen, J.S. and R. Guthoff, *[The role of episcleral venous pressure in the development of secondary glaucomas].* Klin Monbl Augenheilkd, 1988. **193**(5): p. 471-5.
151. Kiekens, S., et al., *Continuous positive airway pressure therapy is associated with an increase in intraocular pressure in obstructive sleep apnea.* Invest Ophthalmol Vis Sci, 2008. **49**(3): p. 934-40.
152. Alvarez-Sala, R., et al., *Nasal CPAP during wakefulness increases intraocular pressure in glaucoma.* Monaldi Arch Chest Dis, 1994. **49**(5): p. 394-5.
153. Toris, C.B. and J.E. Pederson, *Effect of intraocular pressure on uveoscleral outflow following cyclodialysis in the monkey eye.* Invest Ophthalmol Vis Sci, 1985. **26**(12): p. 1745-9.
154. Bill, A., *Conventional and uveo-scleral drainage of aqueous humour in the cynomolgus monkey (Macaca irus) at normal and high intraocular pressures.* Exp Eye Res, 1966. **5**(1): p. 45-54.
155. Toris, C.B., *Chapter 7 Aqueous Humor Dynamics I. Measurement Methods and Animal Studies* Current Topics in Membranes, 2008. **62**: p. 193-229.
156. Rosenquist, R., et al., *Outflow resistance of enucleated human eyes at two different perfusion pressures and different extents of trabeculotomy.* Curr Eye Res, 1989. **8**(12): p. 1233-40.
157. Hann, C.R., et al., *Anatomic changes in Schlemm's canal and collector channels in normal and primary open-angle glaucoma eyes using low and high perfusion pressures.* Invest Ophthalmol Vis Sci, 2014. **55**(9): p. 5834-41.
158. Bartels, S.P., et al., *Pharmacological effects of topical timolol in the rabbit eye.* Invest Ophthalmol Vis Sci, 1980. **19**(10): p. 1189-97.
159. Gulati, V., et al., *Diurnal and nocturnal variations in aqueous humor dynamics of patients with ocular hypertension undergoing medical therapy.* Arch Ophthalmol, 2012. **130**(6): p. 677-84.
160. Shahidullah, M., et al., *Studies on bicarbonate transporters and carbonic anhydrase in porcine nonpigmented ciliary epithelium.* Invest Ophthalmol Vis Sci, 2009. **50**(4): p. 1791-800.
161. Civan, M.M., *The Eye's Aqueous humor*, in *Current Topics in Membranes*, M.M. Civan, Editor. 2008, Elsevier, Inc: San Diego. p. 231-272.
162. McCannel, C.A., S.R. Heinrich, and R.F. Brubaker, *Acetazolamide but not timolol lowers aqueous humor flow in sleeping humans.* Graefes Arch Clin Exp Ophthalmol, 1992. **230**(6): p. 518-20.

163. Topper, J.E. and R.F. Brubaker, *Effects of timolol, epinephrine, and acetazolamide on aqueous flow during sleep.* Invest Ophthalmol Vis Sci, 1985. **26**(10): p. 1315-9.
164. Vanlandingham, B.D., T.L. Maus, and R.F. Brubaker, *The effect of dorzolamide on aqueous humor dynamics in normal human subjects during sleep.* Ophthalmology, 1998. **105**(8): p. 1537-40.
165. Toris, C.B., C.B. Camras, and M.E. Yablonski, *Acute versus chronic effects of brimonidine on aqueous humor dynamics in ocular hypertensive patients.* Am J Ophthalmol, 1999. **128**(1): p. 8-14.
166. Reitsamer, H.A., M. Posey, and J.W. Kiel, *Effects of a topical alpha2 adrenergic agonist on ciliary blood flow and aqueous production in rabbits.* Exp Eye Res, 2006. **82**(3): p. 405-15.
167. Toris, C.B., C.B. Camras, and M.E. Yablonski, *Effects of PhXA41, a new prostaglandin F2 alpha analog, on aqueous humor dynamics in human eyes.* Ophthalmology, 1993. **100**(9): p. 1297-304.
168. Ziai, N., et al., *The effects on aqueous dynamics of PhXA41, a new prostaglandin F2 alpha analogue, after topical application in normal and ocular hypertensive human eyes.* Arch Ophthalmol, 1993. **111**(10): p. 1351-8.
169. Lim, K.S., et al., *Mechanism of Action of Bimatoprost, Latanoprost, and Travoprost in Healthy Subjects.* Ophthalmology. **115**(5): p. 790-795.e4.
170. Bahler, C.K., et al., *Prostaglandins increase trabecular meshwork outflow facility in cultured human anterior segments.* Am J Ophthalmol, 2008. **145**(1): p. 114-9.
171. Lutjen-Drecoll, E. and E. Tamm, *Morphological study of the anterior segment of cynomolgus monkey eyes following treatment with prostaglandin F2 alpha.* Exp Eye Res, 1988. **47**(5): p. 761-9.
172. Tamm, E., M. Rittig, and E. Lutjen-Drecoll, *[Electron microscopy and immunohistochemical studies of the intraocular pressure lowering effect of prostaglandin F2 alpha].* Fortschr Ophthalmol, 1990. **87**(6): p. 623-9.
173. Poyer, J.F., C. Millar, and P.L. Kaufman, *Prostaglandin F2 alpha effects on isolated rhesus monkey ciliary muscle.* Invest Ophthalmol Vis Sci, 1995. **36**(12): p. 2461-5.
174. Alphen, G.W., P.B. Wilhelm, and P.W. Elsenfeld, *The effect of prostaglandins on the isolated internal muscles of the mammalian eye, including man.* Doc Ophthalmol, 1977. **42**(2): p. 397-415.
175. Sagara, T., et al., *Topical prostaglandin F2alpha treatment reduces collagen types I, III, and IV in the monkey uveoscleral outflow pathway.* Arch Ophthalmol, 1999. **117**(6): p. 794-801.
176. Weinreb, R.N., et al., *Prostaglandins increase matrix metalloproteinase release from human ciliary smooth muscle cells.* Invest Ophthalmol Vis Sci, 1997. **38**(13): p. 2772-80.
177. Ocklind, A., *Effect of latanoprost on the extracellular matrix of the ciliary muscle. A study on cultured cells and tissue sections.* Exp Eye Res, 1998. **67**(2): p. 179-91.
178. Stjernschantz, J., et al., *Uveoscleral Outflow Biology and Clinical Aspects.* 1998, London, UK: Mosby International Limited.
179. Kazemi, A., et al., *The Effects of Netarsudil Ophthalmic Solution on Aqueous Humor Dynamics in a Randomized Study in Humans.* J Ocul Pharmacol Ther, 2018. **34**(5): p. 380-386.
180. Kiel, J.W. and C.C. Kopczynski, *Effect of AR-13324 on episcleral venous pressure in Dutch belted rabbits.* J Ocul Pharmacol Ther, 2015. **31**(3): p. 146-51.
181. Inoue, T. and H. Tanihara, *Rho-associated kinase inhibitors: a novel glaucoma therapy.* Prog Retin Eye Res, 2013. **37**: p. 1-12.
182. Marshall-Goebel, K., et al., *Intracranial and Intraocular Pressure During Various Degrees of Head-Down Tilt.* Aerosp Med Hum Perform, 2017. **88**(1): p. 10-16.
183. Lam, C.T., G.E. Trope, and Y.M. Buys, *Effect of Head Position and Weight Loss on Intraocular Pressure in Obese Subjects.* J Glaucoma, 2017. **26**(2): p. 107-112.

184. Geloneck, M.M., et al., *Correlation between intraocular pressure and body mass index in the seated and supine positions.* J Glaucoma, 2015. **24**(2): p. 130-4.
185. Cohen, Y., et al., *The effect of nocturnal CPAP therapy on the intraocular pressure of patients with sleep apnea syndrome.* Graefes Arch Clin Exp Ophthalmol, 2015. **253**(12): p. 2263-71.
186. David, R., et al., *Diurnal intraocular pressure variations: an analysis of 690 diurnal curves.* The British Journal of Ophthalmology, 1992. **76**(5): p. 280-283.
187. Zeimer, R.C., J.T. Wilensky, and D.K. Gieser, *Presence and Rapid Decline of Early Morning Intraocular Pressure Peaks in Glaucoma Patients.* Ophthalmology, 1990. **97**(5): p. 547-550.
188. Mansouri, K., R.N. Weinreb, and J.H.K. Liu, *Effects of Aging on 24-Hour Intraocular Pressure Measurements in Sitting and Supine Body Positions.* Investigative Ophthalmology & Visual Science, 2012. **53**(1): p. 112-116.
189. Cheng, J., et al., *Seasonal changes of 24-hour intraocular pressure rhythm in healthy Shanghai population.* Medicine (Baltimore), 2016. **95**(31): p. e4453.
190. Bengtsson, B.O., *Some factors affecting the distribution of intraocular pressures in a population.* Acta Ophthalmologica, 1972. **50**(1): p. 33-46.
191. Blumenthal, M., et al., *Seasonal Variation in Intraocular Pressure.* American Journal of Ophthalmology, 1970. **69**(4): p. 608-610.
192. Fukuoka, S., et al., *Intraocular pressure in an ophthalmologically normal Japanese population.* Acta Ophthalmol, 2008. **86**(4): p. 434-9.
193. Kashiwagi, K., T. Shibuya, and S. Tsukahara, *De novo age-related retinal disease and intraocular-pressure changes during a 10-year period in a Japanese adult population.* Jpn J Ophthalmol, 2005. **49**(1): p. 36-40.
194. Nomura, H., et al., *Age-related changes in intraocular pressure in a large Japanese population: a cross-sectional and longitudinal study.* Ophthalmology, 1999. **106**(10): p. 2016-22.
195. Nomura, H., et al., *The relationship between age and intraocular pressure in a Japanese population: The influence of central corneal thickness.* Current Eye Research, 2002. **24**(2): p. 81-85.
196. Wong, T.T., et al., *The relationship of intraocular pressure with age, systolic blood pressure, and central corneal thickness in an Asian population.* Invest Ophthalmol Vis Sci, 2009. **50**(9): p. 4097-102.
197. Sit, A.J., et al., *Circadian Variation of Aqueous Dynamics in Young Healthy Adults.* Investigative Ophthalmology & Visual Science, 2008. **49**(4): p. 1473-1479.
198. Becker, B., *The Decline in Aqueous Secretion and Outflow Facility with Age[*].* American Journal of Ophthalmology. **46**(5): p. 731-736.
199. Gabelt, B.A.T. and P.L. Kaufman, *Changes in aqueous humor dynamics with age and glaucoma.* Progress in Retinal and Eye Research, 2005. **24**(5): p. 612-637.
200. Coakes, R.L. and R.F. Brubaker, *The mechanism of timolol in lowering intraocular pressure. In the normal eye.* Arch Ophthalmol, 1978. **96**(11): p. 2045-8.
201. Yablonski, M.E., et al., *A fluorophotometric study of the effect of topical timolol on aqueous humor dynamics.* Exp Eye Res, 1978. **27**(2): p. 135-42.
202. Ingram, C.J. and R.F. Brubaker, *Effect of brinzolamide and dorzolamide on aqueous humor flow in human eyes.* Am J Ophthalmol, 1999. **128**(3): p. 292-6.
203. Christiansen, G.A., et al., *Mechanism of ocular hypotensive action of bimatoprost (Lumigan) in patients with ocular hypertension or glaucoma.* Ophthalmology. **111**(9): p. 1658-1662.
204. Brubaker, R.F., et al., *Effects of AGN 192024, a new ocular hypotensive agent, on aqueous dynamics[1].* American Journal of Ophthalmology. **131**(1): p. 19-24.
205. Dinslage, S., et al., *The influence of Latanoprost 0.005% on aqueous humor flow and outflow facility in glaucoma patients: a double-masked placebo-controlled clinical study.* Graefe's Archive for Clinical and Experimental Ophthalmology, 2004. **242**(8): p. 654-660.
206. Toris, C.B., et al., *Effects of Travoprost on Aqueous Humor Dynamics in Patients With Elevated Intraocular Pressure.* Journal of Glaucoma, 2007. **16**(2): p. 189-195.

207. Toris, C.B., et al., *Effects of brimonidine on aqueous humor dynamics in human eyes.* Arch Ophthalmol, 1995. **113**(12): p. 1514-7.
208. Maus, T.L., C. Nau, and R.F. Brubaker, *Comparison of the early effects of brimonidine and apraclonidine as topical ocular hypotensive agents.* Arch Ophthalmol, 1999. **117**(5): p.586-91.
209. Larsson, L.I., *Aqueous humor flow in normal human eyes treated with brimonidine and timolol, alone and in combination.* Arch Ophthalmol, 2001. **119**(4): p. 492-5.

# Aqueous Humor Dynamics and Its Influence on Glaucoma

Frances Meier-Gibbons and Marc Töteberg-Harms

**Abstract** The chapter describes the anatomical and functional features of the aqueous humor (AH) dynamics with special focus on pathological changes in glaucoma. The main therapeutic approaches to medically and surgically regulate AH production and outflow are discussed.

## 1 Introduction

Aqueous humor (AH) plays an important role in the homeostasis of the eye. A disturbed regulation of the AH dynamics leads to an elevated or decreased intraocular pressure (IOP). An Elevation of IOP is an important risk factor for the development of glaucomatous disease [1]. The entity "glaucoma" comprises various diseases with different etiological factors, but all lead to progressive neuropathy of the optic nerve and ultimately blindness [2].

AH is produced within the pigmented epithelium of the ciliary body and passes from the posterior to the anterior chamber [3]. AH exits the eye through two different pathways, primarily through the trabecular ("conventional") outflow system (composed of the trabecular meshwork (TM), Schlemm's canal (SC), and the collector channels), and culminates in the episcleral venous system [3]. To a lesser extent, the outflow uses the uveoscleral ("unconventional") pathway [3]. This complex process is regulated by various molecules, and many mechanisms within the process remain unknown.

The main therapeutic approaches to the regulation of AH are as follows: (1) a reduction of AH production and (2) an enhancement of AH outflow. Both approaches are feasible with antiglaucomatous medication and surgical methods [2].

F. Meier-Gibbons (✉)
Eye Center Rapperswil, Rapperswil, Switzerland

M. Töteberg-Harms
Department of Ophthalmology, University Hospital Zurich, Zurich, Switzerland

© Springer Nature Switzerland AG 2019        191
G. Guidoboni et al. (eds.), *Ocular Fluid Dynamics*, Modeling and Simulation in Science, Engineering and Technology, https://doi.org/10.1007/978-3-030-25886-3_7

To target directly the primary cause of the resistance, which leads to an elevation of IOP, would be physiological, i.e., the juxtacanalicular TM and inner wall of Schlemm's canal. New molecules which address such structures have been investigated in recent years, and have yielded promising results not only for lowering IOP but also for protecting the endangered fibers of the optic nerve.

## 2  History

Galen first described AH during the second century AD. He indicated that the eye contains two fluids, namely the AH and the vitreous [4]. His descriptions of the anatomy of the eye were used by ophthalmologists until the seventeenth/eighteenth century [4]. For a long time, particularly during the medieval Islamic time period, the crystalline lens was presumed to be the center of the eye. In this position, the lens together with the surrounding liquids was the center of discussions related to the activity of the eye in the vision process [4]. In the eighteenth century, AH was described as a circulating liquid in a capsule, surrounded by a membrane [4]. By the end of the nineteenth century, the general principles of AH dynamics were accepted. Later, in the twentieth century, such principles were complemented with the discovery of the episcleral vessel plexus by Seidel in 1921 [5].

## 3  Clinical Importance of AH

The importance of AH for the eye is based on various factors: (1) AH helps stabilize the globe [5], (2) its content provides nutrition for the avascular cornea and crystalline lens [5], (3) locally and systemically applied drugs can circulate and be distributed within different ocular structures [5], (4) mediators and anti-inflammatory cells can circulate between the posterior and the anterior chamber in pathologic conditions [3], and (5) a disturbance of the equilibrium between AH production and AH drainage can lead to an elevation of IOP and plays an important role in the pathogenesis of glaucoma [1].

## 4  AH Production and AH Composition: Known Facts and New Knowledge

The formation of AH is a complex process which involves a series of steps and takes place in the ciliary body. The primary three steps in the formation of AH are ultrafiltration, active secretion, and diffusion.

## 4.1  Ultrafiltration

The purpose of the first step is to transport fluid from the capillaries into the stroma. Studies have shown that approximately 4% of blood plasma passing through the ciliary processes is filtered into the interstitial space between the capillaries and the ciliary epithelium [5]. This process is guided by differences in both hydrostatic and oncotic pressures [5].

## 4.2  Active Transport

Concurrently, approximately a million non-pigmented ciliary epithelium cells actively secrete ions into the intercellular clefts which attract a significant amount of water via osmotic forces [4]. Na-K-ATPases, Cl channels, and $NaK_2Cl$ co-transporters are involved in this transport [2, 3, 5]. The process is mediated by aquaporins (AcPs), which are active water channels that aid the transport of fluids against an osmotic gradient [3]. Such active transport accounts for approximately 80–90% of AH formation [3]. The anatomical structure of the intercellular spaces helps guide the flow of liquids directly into the posterior chamber.

## 4.3  Diffusion

Nutrients and other substances join AH by diffusion. As AH passes from the posterior to the anterior chamber, AH increasingly resembles actual plasma. The nutrients (mainly glucose, carbohydrates, urea, glutathione, amino acids, oxygen, and potassium) are exchanged against carbon dioxide, lactate, and pyruvate from the surrounding tissue [3, 5].

Until present day, no direct measurements of the production of AH have been possible, which is a noteworthy point. The production rate can only be estimated by measuring the amount of AH passing from the posterior to the anterior chamber [2]. The production of AH shows a circadian rhythm, which is influenced by several factors. The most important factors are age, sex, blood flow through the ciliary body, neuronal control, hormonal effects, and intracellular regulators [2]. The majority of studies find a production rate of 2–3 $\mu l$/min [5].

## 5  Outflow Dynamics

## 5.1  Trabecular and Uveoscleral Outflow

The main outflow system is the trabecular or conventional outflow. AH passes through the TM into the Schlemm's canal and subsequently directly into the episcleral veins [5]. Recent studies have demonstrated that the outflow system can additionally act as a biomechanical pump, which partially depends on the IOP [5, 6]. The conventional outflow system has several critical functions [5]: (1) In order to retain the optical functionality of the eye, the TM should allow AH to leave the anterior chamber, yet blocks blood reflux. It is an important part of the blood–aqueous barrier [5]. (2) The outflow system should control the IOP and account for different rates of AH formation, IOP levels, and changing tones of the ciliary muscle [5]. (3) TM has phagocytizing functions to remove foreign material [5]. The second outflow system is the uveoscleral or unconventional outflow. In contrast to the conventional outflow, the structures are less defined: the AH passes the ciliary body and the surrounding tissues of the uvea and leaves the eye through the sclera. This uveoscleral outflow is age-dependent and accounts for approximately 25–57% of the total flow in young humans, which is reduced with increasing age [2]. Studies have indicated that the uveoscleral outflow increases in inflamed eyes [5, 7, 8]. Uveoscleral outflow is mainly IOP independent. However, uveoscleral outflow reacts to the tone of the ciliary muscle (e.g., during accommodation) in a depressed lens or to applied medications (parasympaticomimetics, prostaglandin agonists) [3].

## 5.2 The Normal Structure of the Outflow System: TM, Schlemm's Canal, Collector Channels, and the Venous System: Known Facts and New Knowledge

The TM is composed of three layers, namely the uveal meshwork, the corneoscleral meshwork, and the juxtacanalicular meshwork [2–5]. The inner layer of the TM (the uveoscleral meshwork) is adjacent to the anterior chamber. It consists of four layers of interconnecting bands of collagen, elastin, and large holes. The second layer—the corneoscleral meshwork—shows a similar structure [2, 3, 5]. The innermost layer—the juxtacanalicular meshwork—is adjacent to the inner wall of Schlemm's canal and consists of an extracellular matrix, glycosaminoglycans, and juxtacanalicular cells [2, 5]. Recent studies identified new structures in the TM, including specialized filopodias which are called "tunneling nanotubes" and communicate signals and organelles directly between the cells of TM [9]. One study confirms the existence of the three-dimensional structure of the TM, which consists of collagens and glycosaminoglycans [10]. A separate study highlights the importance of collagen in the regulation of the resistance [11]. Other studies emphasize the influence of oxidative stress which results in damage of the TM cells [12–15]. A reaction of malfunctioning TM cells is the production of proteins which can alter AH and the secretion of vascular endothelial growth factor (VEGF) [16]. VEGF can interfere at the level of the endothelium of Schlemm's canal and enhance outflow. This phenomenon may explain the decreased outflow facility in patients receiving anti-VEGF therapy [17]. Many other factors that influence TM cells and the structure of TM have been identified in prior years, including that CLANS (cross-linked actin networks) reorganize the cytoskeletal arrangement in TM and stiffen the tissue [18]. TgFB alters the extracellular matrix homeostasis and the contractility of the cells by interacting with other proteins [19–24]. Elevated levels of TgFB 2 are found not only in the AH but also in samples of reactive astrocytes of the optic nerve head of glaucoma patients. The presence of such levels indicates its role both in AH outflow, and directly in the optic nerve head among glaucoma patients, leading to an accelerated degeneration of nerve fibers [19]. Various factors influence the TM cells, including, for example, the multifunctional transmembrane protein anoctamin, which is coded by the gene Ano6 and can induce a swelling of TM cells [25]. Matrix metalloproteinases (MMPs), a group of enzymes (proteases part of the metzincin superfamily), are capable of degrading various types of proteins and collagen in the extracellular matrix and enhancing TM outflow by constant remodeling of the tissue. TIMPS (tissue inhibitors of MMP), on the contrary, counteract the actions of MMPs and lead to increased outflow resistance [26]. Nitric oxide (NO) and endothelin 1 influence the vascular smooth muscle contractility and therein the function of TM [27]. NO has been detected in many cells of the ciliary body and the TM [3]. In TM, endogenous NO regulates the steady state of cell volumes and leads by way of a multifunctional pathway to a reduction of cell volumes. Such regulation simultaneously enhances the outflow capacities [28]. AH continues to pass through the TM into Schlemm's canal, which is a vascular sinus bordered by an inner and outer wall [2, 5]. The inner wall of Schlemm's canal plays an important role in the resistance, and therefore acts as a barrier which AH must overcome. The endothelial cells of the inner wall, which are not continuously arranged, can change their configuration and material properties in response to IOP changes [29]. AH seemingly passes the inner wall by way of microsized transendothelial pores. Such pores are inconsistently distributed and may be altered in glaucoma [30]. More recent studies confirm the importance of the pores of the Schlemm's canal as it relates to the regulation of the AH outflow [31, 32]. Gene models are used to demonstrate that 113 genes are involved in the regulation of Schlemm's canal cell stiffness and pore formation [33]. However, the outer wall of Schlemm's canal consists of a continuous single layer of epithelium cells and does not react to IOP changes [5]. The AH continues to flow through 20–30 collector channels, which arise from the outer wall of Schlemm's canal and drain into a system of intrascleral, episcleral, and subconjunctival venous vessels [5]. The fusion of AH and blood flow in the episcleral veins is visible at the slit lamp and reveals a typical laminar flow. The intensity of the outflow of AH is determined by the difference between the IOP and the pressure in the veins; a rise in the outflow pressure (IOP less episcleral venous pressure) leads to a rise in the trabecular outflow [2]. The episcleral pressure is, among other factors, influenced by the body position [2]. It is well-known that IOP fluctuates in a circadian rhythm; IOP may also modulate with changing body positions and is normally higher in the supine position at night [2]. However, in sitting position, IOP does not change significantly during the night despite a reduction of 50% of AH flow [34, 35]. Nau showed through a study that nocturnal AH flow reduction in healthy persons is compensated by a small decrease in conventional outflow and a larger decrease in uveoscleral outflow [34]. However, IOP change seems to mainly derive from changes in episcleral venous pressure and less often from changes in the AH outflow facility [36].

## 5.3   Sites of Resistance in the AH Outflow System

Two specific sites are considered to be the place of main resistance [2, 3, 5]: (1) the juxtacanalicular space and (2) the endothelium of Schlemm's canal. In order to maintain equilibrium between production and outflow, a certain amount of resistance in the outflow system is necessary [2, 5]. Studies have shown that the consistency of the extracellular matrix and the trabecular meshwork cells is altered in patients with glaucoma [2, 37, 38]. Considering that the juxtacanalicular space is described as a sponge with a fiber network and filled with proteins, cells, and fibrils, changes of this tissue likely account for approximately 75% of resistance [5, 39]. Many factors influence the resistance, yet one of the main components is glycosaminoglycans, which are large molecules able to attract a significant amount of water and therefore reduce the diameter of the outflow channels [5]. Changes in the structure of glycosaminoglycans may also lead to a deposition of fibrous material and therefore an increased electron density of the TM structures among aging patients and patients with glaucoma [40]. The process of aging additionally leads to a senescence of the TM with autophagy of the cells and an augmentation of the extracellular matrix within the juxtacanalicular region [41]. External factors can also influence the structure of the TM and consequently the outflow. Approximately one-third of the population reacts to locally applied steroid therapy with an elevation in IOP [42]. The reason for IOP elevation seems to lie in the fourfold stiffening of the extracellular matrix (ECM) within the TM and coupled with the elevated expression of matrix proteins by steroids (i.e., in this study—dexamethasone). Such changes eventually result in increased outflow resistance and an elevation of IOP [43]. The endothelium of the inner wall of Schlemm's canal can be altered by an accumulation of bulk particles, which block the outflow and biochemically changing endothelial cells [5]. Changes in the state of the ciliary muscle tone and the innervation of the ciliary blood vessels are additional factors which influence the trabecular outflow via a direct connection of the ciliary muscle within the juxtacanalicular region [44].

## 6   Alterations of the Outflow Structures in Glaucoma

A healthy eye has a balance between AH production and AH outflow. An imbalance induces an increase or decrease in IOP. Such increases in IOP consequently give rise to the progressive damage of the retinal nerve fibers, destruction of retinal ganglion cells, and eventually blindness within the patient [45]. The changes in the aging eye will be discussed elsewhere. In glaucoma, the rise in IOP is predominantly caused by reduced outflow at the level of the TM (i.e., conventional outflow), which is responsible for approximately 75% of the outflow capacity [2, 5]. Studies have documented anatomical changes in the TM, particularly including an increase of extracellular matrix which leads to a tissue stiffness [2]. Other changes, among which primarily include cell stiffness and changes in the pore size and consistency within the Schlemm's canal, have been documented by other studies.

# 7 Therapeutic Approaches in Glaucoma: From the Past to the Future

The entity of the disease "glaucoma" was first used in Ophthalmology by Hippocrates (in the literature "aphorisms"). He defined the word "glaucosis" which he associated with "dimness in the vision of elderly people." For a significant part of history, the etiology of the disease was not clear. In the nineteenth century, specifically with the help of new optical methods (e.g., invention of the ophthalmoscope in 1851 by Helmholtz), glaucoma was classified into different subtypes and the first treatment options (i.e., the iridectomy by Albrecht von Graefe in 1856) were developed [46]. The first medical treatment of elevated IOP was performed in 1877 by Weber. Weber's treatment consisted of a cholinergic agonist, which remained for many decades the only medication used as glaucoma treatment [2]. Almost a century later, in 1967, a local beta-adrenergic antagonist was introduced into glaucoma therapy, followed by other antiglaucomatous drugs [2]. Until present times, the only proven method to inhibit the progression of glaucoma is, remarkably, a reduction in IOP [1]. Various factors influence the progression of glaucoma; however, perhaps most well-known is that an elevated IOP is a major risk factor for the progression of the disease [1]. Other treatment options (e.g., neuroprotection) have not yet indicated promising results. An additional common acknowledgement is that an imbalance in the production of AH and its outflow via the conventional and the unconventional route lead to an elevation of IOP [1]. Therefore, IOP can be successfully reduced in two ways: (1) reduce the production of AH ("inflow") and (2) enlarge the outflow via the two known routes [2]. Both therapeutic approaches are possible with medication (inflow, outflow, and combined inflow/outflow drugs) and surgery. Surgical options are presented later.

# 8 Inflow Medications

Apart from beta-adrenergic antagonists (introduced in 1967), adrenergic agonists have been used in the treatment of elevated IOP. Adrenergic agonists work with a dual action (inflow and outflow); however, the precise mechanism of their action continues to be discussed [2]. The third drug class which reduces AH production is the class of carbonic anhydrase inhibitors (CAI). CAIs have been used as a systemic medication since 1952; however, the first locally applied CAI has been available since the mid-1990s [47].


# 9   Outflow Medications

## 9.1   Cholinergic Agonists

Cholinergic agonists were the first antiglaucomatous drugs described in history [2]. They act either directly (by stimulating nicotinic or muscarinic receptors) or indirectly (by inhibiting cholinesterase or enhancing acetylcholine release) [2]. Pilocarpine was the first drug used in the treatment of glaucoma and continues to be the most often used cholinergic agonist [2]. The mechanism of action is an enhancement of the TM outflow through the stimulation of postsynaptic muscarinic receptors, which act on the ciliary muscle and on the scleral spur [2]. The reduction of uveoscleral outflow is significant, although this mechanism interestingly does not counteract the opposite mechanism of action of prostaglandin agonist (PG). Both drugs, if applied jointly, show an additive effect in lowering IOP. The main problems of cholinergic agonists are represented by the short time of action and the side effects (i.e., the dosing is 3–4 times per day and the side effect of a miosis reduces therapy adherence) [2].


## 9.2   Prostaglandin Analogues

Recent studies related to the mechanisms of AH dynamics led to the development of the most important antiglaucomatous drug class. In 1996, the first locally applied prostaglandin (PG) F 2alpha analogue entered the market (Latanoprost). Four other drugs (namely Unoprostone, Travoprost, Tafluprost, and Bimatoprost) followed shortly thereafter. Unoprostone was used in few countries and was less effective as it relates to IOP reduction [2]. Latanoprost, Travoprost, and Tafluprost are prostanoids, while Bimatoprost is a prostamide; prostanoids and prostamides are prodrugs of PGF 2 alpha [48]. There are different mechanisms of action of PGs: (1) they reduce IOP through change in the outflow system, primarily through an increase in uveoscleral outflow and, less pronounced, an increase in the trabecular outflow facility which therefore increases aqueous flow. (2) They influence unconventional outflow: PG bind to receptors in the ciliary muscle, which leads to a change in the structure of the ciliary cells, widening of the spaces between the bundles of the muscle, change in the extracellular matrix, and a relaxation of the ciliary muscle [48]. The different types of prostanoids variably bind to the nine identified subtypes of PG receptors, and therefore, no distinct prostamide receptor has been identified until now [2, 48]. The change in the extracellular matrix is derived from the direct reaction of PG with matrix metalloproteinases (MMPs) which reduce outflow resistance through the degradation of extracellular matrix. The action of MMPs is regulated by their inhibitors, tissue inhibitors of metalloproteinases (TIMPs), an interaction which may be targeted as a future therapeutic option [49]. (3) Influence on the conventional outflow: studies show structural changes and regional loss of endothelial cells of Schlemm's canal as well as a loss of extracellular matrix of the juxtacanalicular region in patients treated with Latanoprost [49]. Such changes, especially the focal loss of extracellular matrix in the juxtacanalicular region, may enhance the outflow of AH. (4) PG moreover affects the molecular level, which leads to rapid changes in the signaling of cell cultures. (5) The mechanism of action of the prostamide agonist, Bimatoprost, is not yet fully understood [48]. A recent study shows that Bimatoprost reacts with cells of TM and Schlemm's canal, which leads to decreased cell contractibility and an enhancement of the outflow system. The effect could be overcome by the application of a prostamide antagonist, AGN211334 [50]. In summary, PGs reduce IOP by changing the outflow system—primarily by increasing the uveoscleral outflow and, less pronounced, by increasing the trabecular outflow facility which thereby increases aqueous flow [2]. The clinical advantages of PG therapy include the once-daily application form, the effective IOP-lowering (around 30%) efficacy, and the few systemic side effects (caution should be specifically be exercised as it relates to patients with a severe asthma) [51, 52]. The primary disadvantages are the local side effects. PG can lead to a marked pigmentation of the iris and the periocular skin and a hyperemia of the conjunctiva [51]. To enhance IOP reduction, PGs are available in a combined drug together with a beta-blocker.

## 10 Combined Inflow–Outflow Medication

### 10.1 Alpha-Adreno Receptor Agonists

In the 1920s, Adrenaline was used to lower IOP as a topical agent or provided subconjunctivally [53]. The use of adrenaline was discontinued due to its systemic sympatheticomimetic side effects. Locally applied Brimonidine was finally introduced in 1996 following two generations of less effective selective adrenergic agents [2]. Alpha-adrenoreceptor agonists have a combined action on the inflow and outflow of AH [54]. They reduce AH production by approximately 20% and enhance uveoscleral outflow, which leads to an IOP reduction of 20–25% [55]. Both actions are mediated through the stimulation of alpha 2-adrenergic receptors within different structures of the eye. Brimonidine may have an eventual neuroprotective action. The results of the studies are controversial; however, an explanation could be related to the upregulation of antiapoptotic genes [56]. Clinical observations largely maintain that the drug is effective, but has local and systemic side effects. Patients may complain about allergic reactions and sympatimimetic systemic side effects (e.g., changes in the level of the blood pressure and cerebrovascular signs including headache) [2]. To enhance Brimonidine's effects, it is available in a combined drug together with a beta-blocker. Interestingly, the combined drugs reveal less local side effects.

## 11  Future Therapeutic Approaches

Until present day, drugs that have been used in the treatment of glaucoma have either influenced the inflow, the outflow, or have had a combined mode of action. Interestingly, the main sites of resistance in the tissue (and therefore the main sites leading to an elevation of IOP) have not been addressed in therapy. However, in December 2017, the first drug acting directly on the trabecular cells was approved by FDA: a rhokinase inhibitor [57]. Netarsudil ophthalmic solution 0.2% has a triple action: first, it inhibits the rhokinase pathway (reducing fibrosis of trabecular meshwork); second, it inhibits the norephedrine pathway (decreasing AH production); and third, it lowers episcleral venous pressure (enhancing outflow). The important action in the rhokinase pathway is its suppression of the activity of profibrotic proteins (especially TGF-beta 2) and its regulatory influence on the actomyosin dynamics of various cell types [57]. Through a once-daily application, its IOP-lowering capacities are comparable to PG [57]. The main (local) side effect is a hyperemia. An additive effect might exist with PG; studies for a combination with latanoprost and with a beta-blocker are ongoing [57]. A separate distinct rhokinase inhibitor, ripasudil, has been approved in Japan, and other molecules are the subject of ongoing studies. Latanoprostene bunod ophthalmic solution 0.024% is a novel PG F2 alpha agonist which was approved by FDA in November 2017. It has a dual action and efficiently reduces IOP, with demonstration of the same side effects as the PG in current clinical use [2, 58, 59]. The new molecule is the first PG which donates nitric oxide (NO) while metabolizing. It metabolizes to Latanoprost acid (increasing uveoscleral outflow) and to butanediol mononitrate (which releases NO and increases conventional outflow by a relaxation and change in the size of the smooth muscle cells and TM cells) [2, 58, 59]. Other molecules exist in different phases of clinical evaluation and might soon enter the market. Such molecules essentially act on the tissues of the outflow systems. Trabodenoson (INO-8875) is an adenosine alpha 1 receptor agonist which increases outflow by upregulation of MMPs. First clinical studies are published [60]. Soluble guanylate cyclase stimulator (sGC) IWP 953 increases AH outflow capacity in monkeys via a regulation of the sCG/cGMP (cyclic guanosin-$3',5'$-monophosphate) pathway [61]. Oxidative stress may lead to a reduced TM outflow by a change in the molecular structure and functional actions of TM [12]. Transcription factor Hairy and Enhancer of split-1 (HES 1) is involved in the extracellular matrix system and may rescue the functional damage caused by oxidative stress through an inhibition of the proliferative and migratory changes of the TM [12]. Prostaglandin receptor EPs 3/FP agonist ONO

9054 has a dual action—essentially, it increases trabecular and uveoscleral outflow by latching onto the according receptors. It is interesting to observe that molecules stimulating EP and FP receptors receive the same clinical reactions (enhancement of uveoscleral outflow), yet it is known that the receptors differ structurally and functionally. Another newly used molecule is Butaprost, an EP 2 agonist, which has an IOP-lowering effect as studied in monkeys [10]. PGE(2) analogues act as agonists on the receptors EP 2 and EP 4 [2]. The IOP-lowering capacity is higher, but they are less stable in aqueous solution and cause a greater number of side effects than the known molecules [2]. Chromokalin (CKLP1) is a water-soluble ATP-sensitive potassium channel opener which enhances distal conventional outflow by decreasing outflow resistance. Contrarily, it lowers IOP through the reduction of episcleral venous pressure [10, 62, 63]. Anoctamin, especially Ano6, influences the swelling and therefore the resistance of TM cells in cell cultures. Tie 2 agonists (acting on the signaling process between angiopoietin (Angtpt) and its Tie 2 receptor system) influence SC cell integrity in mice models. The aging process is induced by a malfunctioning with the related changes in the SC cells which could be slowed down by an activation of the signaling process [64]. Antagonists of sphingosine-1-phosphate (S1P1) receptor molecules may reduce TM resistance [65]. Cannabinoids may lead to a migration of TM cells and therefore to a reduction in IOP [66]. In studies on monkeys, it was found that serotonin 2 receptor agonists lower IOP by binding on 5-HT receptors [67]. The method of action is an enhanced uveoscleral outflow [67]. Serotonin, an important neurotransmitter, together with serotonergic nerves and serotonin receptors, is moreover found throughout the eye [2]. Furthermore, transforming growth factor B (TGF B) is elevated in AH of glaucoma patients [12]. The molecule alters the extracellular matrix homeostasis and the contractility of the TM cells and further influences the resistance of TM [12, 24]. Bone morphogenetic protein (BMP) counteracts the profibrotic actions of TGF B and could be used as a future therapeutic option [68]. Latrunculins, a family of natural products found in sponges, increase the outflow facility by enlarging the extracellular space and changing the integrity of the TM [2, 69, 70]. MMPs constantly modulate the TM independently of the IOP and, hence, enhance conventional outflow. Conversely, TIMPs counteract this enhancement [26]. Studies have demonstrated that patients with glaucoma show a MMP/TIMP imbalance with an elevated level of TIMPs in their AH composition [26, 71, 72]. Therefore, a drug-induced change of the imbalance in favor of MMPs could be a therapeutic option in glaucoma patients. The role of exogenous NO in the conventional outflow physiology has been earlier discussed. Knowing that endogenous NO is involved in the physiological regulation of AH outflow facility at the level of endothelial cells of SC, it might be considered a novel treatment option [58, 73]. One new drug using the regulatory action of NO was recently approved by FDA [57]. Many studies have been performed and are ongoing which use the growing knowledge about genetic processes; a mutation of the gene MYOC which encodes the protein myocilin is involved in certain types of glaucoma, especially the juvenile onset glaucoma and approximately 4% of POAG patients [74]. Studies found out that mutant myocilin induces abnormal extracellular matrix accumulation in the reticulum of TM which eventually leads to a reduced outflow. An influence on the mutation of the MYOC gene could further be used in glaucoma therapy [75]. The gene regulator micro RNA is important in the regulation of TM cells and may act as a proapoptotic agent in TM cells (especially among glaucoma patients). Stem cells may replace missing or malfunctioning TM cells in the future [76].

## 12  Surgical Approaches to Influence AH Dynamics and to Treat Glaucoma

In most cases of glaucoma management, topical medications are capable of decreasing IOP to an extent, which sufficiently slows down the progression of glaucoma and prevents blindness during one's lifetime. However, over time, medications can become inefficient to sufficiently lower IOP or IOP can suddenly decompensate and thus no longer be controlled by topical medication. In such cases when an accelerated rate of progression jeopardizes the preservation of vision, surgical interventions become necessary.

There are generally two potential approaches to influence AH dynamics surgically, i.e., (1) to increase AH outflow and/or (2) to reduce AH productions.

## 13  Increase in AH Outflow

There are various approaches to surgically increase AH outflow. The two known outflow systems, trabecular and uveoscleral outflow, can be the target of surgical approaches. In addition, all physiological outflow systems can be bypassed and AH can be guided outside of the globe and into a subconjunctival bleb.

### 13.1  Increase in Trabecular Outflow

The trabecular outflow system is the target of most minimally invasive glaucoma surgery (MIGS) procedures [77]. One of the first invented techniques, excimer laser trabeculostomy (or trabeculotomy, ELT) uses an excimer laser (MLase AG, Germering, Germany) to create approximately ten laser channels, which connect the anterior chamber with Schlemm's canal [78–85]. Therefore, ELT bypasses the highest outflow resistance, i.e., the juxtacanalicular trabecular meshwork and inner wall of Schlemm's canal. The same result is achieved by the iStent and the iStent Incect (Glaukos Corp, San Clemente, CA), with an implant while ELT does not require an implant [86, 87]. The removal of large parts of the juxtacanalicular trabecular meshwork and inner wall of Schlemm's canal is achieved by trabeculotomy ab interno with the trabectome (Neomedix Corp., Tustin, CA) [88–90].

## 13.2 Increase in Uveoscleral Outflow

Currently, there are two devices which increase uveoscleral outflow, namely the CyPass uveoscleral shunt (Alcon Novartis, Fort Worth, TX) and the iStent Supra (Glaukos Corp, San Clemente, CA) [91–94]. Each of these MIGS procedures connects the anterior chamber with the uveoscleral space and enhances uveoscleral outflow. As of August 29, 2018, Alcon announced voluntary market withdrawal of CyPass Micro-Stent for surgical glaucoma due to concern of enhanced endothelial cell loss in some cases.

A different approach is MicroPulse MP3 cyclophotocoagulation (Iridex Corp, Mountain View, CA) [95–100]. The mechanism of action is unknown. Initial laboratory studies by Murray Johnstone (presented as a poster at the annual meeting of the American Glaucoma Society 2017 in San Diego, CSA) suggest an increase in uveoscleral outflow. The procedure is repeatable and success rates are approximately 74% [98].

## 13.3 Bypassing Trabecular and Uveoscleral Outflow

The most commonly performed glaucoma surgeries are likely trabeculectomy and implantation of glaucoma drainage devices (e.g., the Ahmed valve, the Baerveldt, or Molteno implants). Success rates for trabeculectomy are around 66%, while success rates for the glaucoma drainage devices are around 60% [101, 102].

A less invasive approach is the XEN Gel stent (Allergan Inc., Madison, NJ), a small porcine collagen tube that is implanted ab interno through a clear-cornea incision [103, 104]. Data from large randomized controlled trials is lacking.

The IDE trial was published in 2019 [136]. The study showed a reduction in IOP and IOP-lowering medication. Target pressure was 15 mmHg on 1 medication at 24 months. However, 41% of the eyes required needlings in the postoperative course.

A second device is the PRESERFLO (Santen, Osaka, Japan; prior name: InnFocus MicroShunt). The PRESERFLO is made out of poly(styrene-block-isobutylene-block-styrene) or SIBS [137, 138]. First published results with a 3-year follow-up show a lower target pressure (11 mmHg at 3 years postop on less than 1 medication) and fewer needlings required after implantation (approx. 4%) [139]. A major difference between the XEN Gel stent and the PRESERFLO is the approach of implantation. The Xen Gel stent is inserted ab internally from the anterior chamber through a clear cornea incision while the PRESERFLO is implanted from ab externo after conjunctival dissection.

# 14   Reduction in AH Production

Destruction of the ciliary body epithelium, i.e., by cyclodestructive procedures, results in a decrease in AH formation assuming that a constant outflow IOP will be reduced as a result of a reduced AH production. Therefore, cyclodestructive procedures are used to treat glaucoma. However, AH production typically does not change significantly during one's lifetime. A surgical reduction of AH formation seems to be a non-physiological approach with regard to glaucoma management. Thus, indications for cyclodestructive procedures are reserved either for refractory cases of glaucoma after outflow surgeries have failed or in eyes with little to no visual potential. Generally, the ciliary body epithelium can either be destroyed by heat, i.e., 810 nm diode trans-scleral cyclophotocoagulation (ts-CPC) [105–113] and endo-cyclophotocoagulation (ECP) [114–119], or by freezing, i.e., trans-scleral cryocoagulation [120–124]. ECP is usually performed on pseudophakic eyes or in combination with cataract surgery. Generally, it is possible to perform ECP in phakic eyes; however, a pars plana approach with prior anterior vitrectomy is required. The advantage of ECP is its allowance for direct visualization of the target tissue. On the contrary, ECP is a more invasive procedure compared to ts-CPC and cryocoagulation, which are both performed from the outside through the intact sclera and globe. The ECP hand-piece (Endo Optiks/Beaver Visitec, Waltham, MA) contains a light fiber, video fiber, and a laser fiber. The ECP system is available with an 18- to 25-gauge system. Success rates for ts-CPC are generally favorable with an IOP $\leq 21$ mmHg ranging between 54 and 93% [125]. Success rates appear to be correlated with the amount of treatment energy [125]. Influencing factors are baseline IOP, patient's age, types of refractory glaucoma, history of previous surgery, male gender, and pigmentation [125]. The reported success rates for ECP are between 12 and 91% [126].

# 15   Other Laser Procedures to Lower Intraocular Pressure

Argon laser trabeculoplasty (ALT) [127] and selective laser trabeculoplasty (SLT) [128] are office laser procedures commonly used to lower IOP. Their exact mode of action is unknown [129]. In theory, both increase trabecular outflow [130]. ALT and SLT typically lower IOP only temporarily [131]. The extent to which ALT and SLT lower IOP is comparable [132]. SLT is repeatable and is possible to perform after failed ALT [133, 134]. Contrarily, SLT is not successful after a substantial amount of trabecular meshwork has been removed, e.g., after failed trabeculectomy ab interno with the trabectome [134]. ALT and SLT are both laser interventions with a favorable risk profile [132]. The IOP-lowering effectiveness of 360° SLT is comparable to that of latanoprost [135].

# 16 Future Outlook

The surgical armamentarium of MIGS procedures is rapidly evolving. More MIGS procedures have been under investigation and will be approved to use by authorities in the near future. The lack of long-term follow-up, randomized control trials (RCTs), and comparative trials is their major flaw. As long as we do not have a standard for RCTs, we cannot compare outcomes, efficacy, and safety of different MIGS procedures. A standard could be a certain follow-up time (e.g., 2 or 5 years), washed-out IOP before surgery and at the end of the follow-up, a standardized protocol on how to reestablish IOP-lowering topical treatment during follow-up. A consensus meeting between the US Food and Drug Administration and the American glaucoma society made a first attempt toward this goal. The result was a minimum of 2 years follow-up and washed-out IOP at baseline and after 1 and 2 years. However, their favorable risk profile compared to standard glaucoma procedures, e.g., trabeculectomy, and their short downtime and fast visual recovery for the patient are their major advantages.

# 17 Summary

For the first time since 1996 (when the PG latanoprost was placed on the market), new classes of antiglaucomatous drugs have been developed which act on the main structures involved in the AH dynamics. Many promising new molecules and therapeutic options are the subject of ongoing studies, not only involving a reduction in IOP but hopefully also in the field of neuroprotection. Currently, the main goal of glaucoma therapy is to stabilize the disease; however, in the future, the goal may consist of using genetic alternations to prevent the optic nerve damage which leads to onset of the disease.

# References

1. Kotecha A, Lim S, Garway-Heath D (2009) Tonometry and Intraocular Pressure Fluctuation. In: Shaarawy TM, Sherwood MB, Hitchings RG, Crowston JG (eds) Glaucoma, Medical Diagnosis & Therapy. Saunders Elsevier Limited, Amsterdam, Niederlande, pp. 103-113.
2. Toris CB (2010) Pharmacotherapies for glaucoma. Curr Mol Med 10: 824-840
3. Goel M, Picciani RG, Lee RK, Bhattacharya SK (2010) Aqueous humor dynamics: a review. Open Ophthalmol J 4: 52-59 DOI https://doi.org/10.2174/1874364101004010052
4. Mark HH (2010) Aqueous humor dynamics in historical perspective. Surv Ophthalmol 55: 89-100 DOI https://doi.org/10.1016/j.survophthal.2009.06.005
5. Stamper RL, Lieberman MF, Drake MV (2009) Aqueous Humor Dynamics. In: Stamper RL, Lieberman MF, Drake MV (eds) Becker-Shaffer's Diagnosis and Therapy of the Glaucomas. MOSBY Elsevier, Maryland Heights, MO, pp. 8-67.

6. Wang W, Qian X, Song H, Zhang M, Liu Z (2016) Fluid and structure coupling analysis of the interaction between aqueous humor and iris. Biomed Eng Online 15: 133 DOI https://doi.org/10.1186/s12938-016-0261-3
7. Toris CB, Gregerson DS, Pederson JE (1987) Uveoscleral outflow using different-sized fluorescent tracers in normal and inflamed eyes. Exp Eye Res 45: 525-532
8. Toris CB, Pederson JE (1987) Aqueous humor dynamics in experimental iridocyclitis. Invest Ophthalmol Vis Sci 28: 477-481
9. Keller KE, Bradley JM, Sun YY, Yang YF, Acott TS (2017) Tunneling Nanotubes are Novel Cellular Structures That Communicate Signals Between Trabecular Meshwork Cells. Invest Ophthalmol Vis Sci 58: 5298-5307 DOI https://doi.org/10.1167/iovs.17-22732
10. Osmond M, Bernier SM, Pantcheva MB, Krebs MD (2017) Collagen and collagen-chondroitin sulfate scaffolds with uniaxially aligned pores for the biomimetic, three dimensional culture of trabecular meshwork cells. Biotechnol Bioeng 114: 915-923 DOI https://doi.org/10.1002/bit.26206
11. Huang W, Fan Q, Wang W, Zhou M, Laties AM, Zhang X (2013) Collagen: a potential factor involved in the pathogenesis of glaucoma. Med Sci Monit Basic Res 19: 237-240 DOI 10.12659/MSMBR.889061
12. Xu L, Zhang Y, Guo R, Shen W, Qi Y, Wang Q, Guo Z, Qi C, Yin H, Wang J (2017) HES1 promotes extracellular matrix protein expression and inhibits proliferation and migration in human trabecular meshwork cells under oxidative stress. Oncotarget 8: 21818-21833 DOI 10.18632/oncotarget.15631
13. Zhao J, Wang S, Zhong W, Yang B, Sun L, Zheng Y (2016) Oxidative stress in the trabecular meshwork (Review). Int J Mol Med 38: 995-1002 DOI https://doi.org/10.3892/ijmm.2016.2714
14. Babizhayev MA, Yegorov YE (2011) Senescent phenotype of trabecular meshwork cells displays biomarkers in primary open-angle glaucoma. Curr Mol Med 11: 528-552
15. Izzotti A, Longobardi M, Cartiglia C, Sacca SC (2011) Mitochondrial damage in the trabecular meshwork occurs only in primary open-angle glaucoma and in pseudoexfoliative glaucoma. PLoS One 6: e14567 DOI https://doi.org/10.1371/journal.pone.0014567
16. Sacca SC, Izzotti A (2014) Focus on molecular events in the anterior chamber leading to glaucoma. Cell Mol Life Sci 71: 2197-2218 DOI https://doi.org/10.1007/s00018-013-1493-z
17. Reina-Torres E, Wen JC, Liu KC, Li G, Sherwood JM, Chang JY, Challa P, Flugel-Koch CM, Stamer WD, Allingham RR, Overby DR (2017) VEGF as a Paracrine Regulator of Conventional Outflow Facility. Invest Ophthalmol Vis Sci 58: 1899-1908 DOI https://doi.org/10.1167/iovs.16-20779
18. Bermudez JY, Montecchi-Palmer M, Mao W, Clark AF (2017) Cross-linked actin networks (CLANs) in glaucoma. Exp Eye Res 159: 16-22 DOI https://doi.org/10.1016/j.exer.2017.02.010
19. Fuchshofer R, Tamm ER (2012) The role of TGF-beta in the pathogenesis of primary open-angle glaucoma. Cell Tissue Res 347: 279-290 DOI https://doi.org/10.1007/s00441-011-1274-7
20. Junglas B, Kuespert S, Seleem AA, Struller T, Ullmann S, Bosl M, Bosserhoff A, Kostler J, Wagner R, Tamm ER, Fuchshofer R (2012) Connective tissue growth factor causes glaucoma by modifying the actin cytoskeleton of the trabecular meshwork. Am J Pathol 180: 2386-2403 DOI https://doi.org/10.1016/j.ajpath.2012.02.030
21. Muralidharan AR, Maddala R, Skiba NP, Rao PV (2016) Growth Differentiation Factor-15-Induced Contractile Activity and Extracellular Matrix Production in Human Trabecular Meshwork Cells. Invest Ophthalmol Vis Sci 57: 6482-6495 DOI https://doi.org/10.1167/iovs.16-20671
22. Su Y, Yang CY, Li Z, Xu F, Zhang L, Wang F, Zhao S (2012) Smad7 siRNA inhibit expression of extracellular matrix in trabecular meshwork cells treated with TGF-beta2. Mol Vis 18: 1881-1884

23. Wang J, Harris A, Prendes MA, Alshawa L, Gross JC, Wentz SM, Rao AB, Kim NJ, Synder A, Siesky B (2017) Targeting Transforming Growth Factor-beta Signaling in Primary Open-Angle Glaucoma. J Glaucoma 26: 390-395 DOI https://doi.org/10.1097/IJG.0000000000000627

24. Wordinger RJ, Clark AF (2014) Lysyl oxidases in the trabecular meshwork. J Glaucoma 23: S55-58 DOI https://doi.org/10.1097/IJG.0000000000000127

25. Banerjee J, Leung CT, Li A, Peterson-Yantorno K, Ouyang H, Stamer WD, Civan MM (2017) Regulatory Roles of Anoctamin-6 in Human Trabecular Meshwork Cells. Invest Ophthalmol Vis Sci 58: 492-501 DOI https://doi.org/10.1167/iovs.16-20188

26. Ashworth Briggs EL, Toh T, Eri R, Hewitt AW, Cook AL (2015) TIMP1, TIMP2, and TIMP4 are increased in aqueous humor from primary open angle glaucoma patients. Mol Vis 21: 1162-1172

27. Dismuke WM, Liang J, Overby DR, Stamer WD (2014) Concentration-related effects of nitric oxide and endothelin-1 on human trabecular meshwork cell contractility. Exp Eye Res 120: 28-35 DOI https://doi.org/10.1016/j.exer.2013.12.012

28. Ellis DZ, Sharif NA, Dismuke WM (2010) Endogenous regulation of human Schlemm's canal cell volume by nitric oxide signaling. Invest Ophthalmol Vis Sci 51: 5817-5824 DOI https://doi.org/10.1167/iovs.09-5072

29. Stamer WD, Braakman ST, Zhou EH, Ethier CR, Fredberg JJ, Overby DR, Johnson M (2015) Biomechanics of Schlemm's canal endothelium and intraocular pressure reduction. Prog Retin Eye Res 44: 86-98 DOI https://doi.org/10.1016/j.preteyeres.2014.08.002

30. Braakman ST, Read AT, Chan DW, Ethier CR, Overby DR (2015) Colocalization of outflow segmentation and pores along the inner wall of Schlemm's canal. Exp Eye Res 130: 87-96 DOI https://doi.org/10.1016/j.exer.2014.11.008

31. Braakman ST, Pedrigi RM, Read AT, Smith JA, Stamer WD, Ethier CR, Overby DR (2014) Biomechanical strain as a trigger for pore formation in Schlemm's canal endothelial cells. Exp Eye Res 127: 224-235 DOI https://doi.org/10.1016/j.exer.2014.08.003

32. Braakman ST, Moore JE, Jr., Ethier CR, Overby DR (2016) Transport across Schlemm's canal endothelium and the blood-aqueous barrier. Exp Eye Res 146: 17-21 DOI https://doi.org/10.1016/j.exer.2015.11.026

33. Cai J, Perkumas KM, Qin X, Hauser MA, Stamer WD, Liu Y (2015) Expression Profiling of Human Schlemm's Canal Endothelial Cells From Eyes With and Without Glaucoma. Invest Ophthalmol Vis Sci 56: 6747-6753 DOI https://doi.org/10.1167/iovs.15-17720

34. Nau CB, Malihi M, McLaren JW, Hodge DO, Sit AJ (2013) Circadian variation of aqueous humor dynamics in older healthy adults. Invest Ophthalmol Vis Sci 54: 7623-7629 DOI https://doi.org/10.1167/iovs.12-12690

35. Liu H, Fan S, Gulati V, Camras LJ, Zhan G, Ghate D, Camras CB, Toris CB (2011) Aqueous humor dynamics during the day and night in healthy mature volunteers. Arch Ophthalmol 129: 269-275 DOI https://doi.org/10.1001/archophthalmol.2011.4

36. Selvadurai D, Hodge D, Sit AJ (2010) Aqueous humor outflow facility by tonography does not change with body position. Invest Ophthalmol Vis Sci 51: 1453-1457 DOI https://doi.org/10.1167/iovs.09-4058

37. Sihota R, Lakshmaiah NC, Walia KB, Sharma S, Pailoor J, Agarwal HC (2001) The trabecular meshwork in acute and chronic angle closure glaucoma. Indian J Ophthalmol 49: 255-259

38. Read AT, Chan DW, Ethier CR (2007) Actin structure in the outflow tract of normal and glaucomatous eyes. Exp Eye Res 84: 214-226

39. Maepea O, Bill A (1989) The pressures in the episcleral veins, Schlemm's canal and the trabecular meshwork in monkeys: effects of changes in intraocular pressure. Exp Eye Res 49: 645-663

40. Pescosolido N, Cavallotti C, Rusciano D, Nebbioso M (2012) Trabecular meshwork in normal and pathological eyes. Ultrastruct Pathol 36: 102-107 DOI https://doi.org/10.3109/01913123.2011.634090

41. Pulliero A, Seydel A, Camoirano A, Sacca SC, Sandri M, Izzotti A (2014) Oxidative damage and autophagy in the human trabecular meshwork as related with ageing. PLoS One 9: e98106 DOI https://doi.org/10.1371/journal.pone.0098106
42. Kersey JP, Broadway DC (2006) Corticosteroid-induced glaucoma: a review of the literature. Eye (Lond) 20: 407-416 DOI https://doi.org/10.1038/sj.eye.6701895
43. Raghunathan VK, Morgan JT, Park SA, Weber D, Phinney BS, Murphy CJ, Russell P (2015) Dexamethasone Stiffens Trabecular Meshwork, Trabecular Meshwork Cells, and Matrix. Invest Ophthalmol Vis Sci 56: 4447-4459 DOI https://doi.org/10.1167/iovs.15-16739
44. McDougal DH, Gamlin PD (2015) Autonomic control of the eye. Compr Physiol 5: 439-473 DOI https://doi.org/10.1002/cphy.c140014
45. Donegan RK, Lieberman RL (2016) Discovery of Molecular Therapeutics for Glaucoma: Challenges, Successes, and Promising Directions. J Med Chem 59: 788-809 DOI https://doi.org/10.1021/acs.jmedchem.5b00828
46. Wikipedia tfe Alfred Carl Graefe.
47. Pfeiffer N (1997) Dorzolamide: development and clinical application of a topical carbonic anhydrase inhibitor. Surv Ophthalmol 42: 137-151
48. Winkler NS, Fautsch MP (2014) Effects of prostaglandin analogues on aqueous humor outflow pathways. J Ocul Pharmacol Ther 30: 102-109 DOI https://doi.org/10.1089/jop.2013.0179
49. Sagara T, Gaton DD, Lindsey JD, Gabelt BT, Kaufman PL, Weinreb RN (1999) Reduction of collagen type I in the ciliary muscle of inflamed monkey eyes. Invest Ophthalmol Vis Sci 40: 2568-2576
50. Stamer WD, Piwnica D, Jolas T, Carling RW, Cornell CL, Fliri H, Martos J, Pettit SN, Wang JW, Woodward DF (2010) Cellular basis for bimatoprost effects on human conventional outflow. Invest Ophthalmol Vis Sci 51: 5176-5181 DOI https://doi.org/10.1167/iovs.09-4955
51. Inoue K (2014) Managing adverse effects of glaucoma medications. Clin Ophthalmol 8: 903-913 DOI https://doi.org/10.2147/OPTH.S44708
52. Toris CB, Gabelt BT, Kaufman PL (2008) Update on the mechanism of action of topical prostaglandins for intraocular pressure reduction. Surv Ophthalmol 53 Suppl1: S107-120 DOI https://doi.org/10.1016/j.survophthal.2008.08.010
53. Wang YL, Hayashi M, Yablonski ME, Toris CB (2002) Effects of multiple dosing of epinephrine on aqueous humor dynamics in human eyes. J Ocul Pharmacol Ther 18: 53-63 DOI https://doi.org/10.1089/108076802317233216
54. Giovannitti JA, Jr., Thoms SM, Crawford JJ (2015) Alpha-2 adrenergic receptor agonists: a review of current clinical applications. Anesth Prog 62: 31-39 DOI https://doi.org/10.2344/0003-3006-62.1.31
55. Cantor LB (2000) The evolving pharmacotherapeutic profile of brimonidine, an alpha 2-adrenergic agonist, after four years of continuous use. Expert Opin Pharmacother 1: 815-834 DOI https://doi.org/10.1517/14656566.1.4.815
56. Wheeler LA, Lai R, Woldemussie E (1999) From the lab to the clinic: activation of an alpha-2 agonist pathway is neuroprotective in models of retinal and optic nerve injury. Eur J Ophthalmol 9 Suppl 1: S17-21
57. Rao PV, Pattabiraman PP, Kopczynski C (2017) Role of the Rho GTPase/Rho kinase signaling pathway in pathogenesis and treatment of glaucoma: Bench to bedside research. Exp Eye Res 158: 23-32 DOI https://doi.org/10.1016/j.exer.2016.08.023
58. Aliancy J, Stamer WD, Wirostko B (2017) A Review of Nitric Oxide for the Treatment of Glaucomatous Disease. Ophthalmol Ther 6: 221-232 DOI https://doi.org/10.1007/s40123-017-0094-6
59. Garcia GA, Ngai P, Mosaed S, Lin KY (2016) Critical evaluation of latanoprostene bunod in the treatment of glaucoma. Clin Ophthalmol 10: 2035-2050 DOI https://doi.org/10.2147/OPTH.S103985

60. Myers JS, Sall KN, DuBiner H, Slomowitz N, McVicar W, Rich CC, Baumgartner RA (2016) A Dose-Escalation Study to Evaluate the Safety, Tolerability, Pharmacokinetics, and Efficacy of 2 and 4 Weeks of Twice-Daily Ocular Trabodenoson in Adults with Ocular Hypertension or Primary Open-Angle Glaucoma. J Ocul Pharmacol Ther 32: 555-562 DOI https://doi.org/10.1089/jop.2015.0148

61. Ge P, Navarro ID, Kessler MM, Bernier SG, Perl NR, Sarno R, Masferrer J, Hannig G, Stamer WD (2016) The Soluble Guanylate Cyclase Stimulator IWP-953 Increases Conventional Outflow Facility in Mouse Eyes. Invest Ophthalmol Vis Sci 57: 1317-1326 DOI https://doi.org/10.1167/iovs.15-18958

62. Roy Chowdhury U, Viker KB, Stoltz KL, Holman BH, Fautsch MP, Dosa PI (2016) Analogs of the ATP-Sensitive Potassium (KATP) Channel Opener Cromakalim with in Vivo Ocular Hypotensive Activity. J Med Chem 59: 6221-6231 DOI https://doi.org/10.1021/acs.jmedchem.6b00406

63. Roy Chowdhury U, Rinkoski TA, Bahler CK, Millar JC, Bertrand JA, Holman BH, Sherwood JM, Overby DR, Stoltz KL, Dosa PI, Fautsch MP (2017) Effect of Cromakalim Prodrug 1 (CKLP1) on Aqueous Humor Dynamics and Feasibility of Combination Therapy With Existing Ocular Hypotensive Agents. Invest Ophthalmol Vis Sci 58: 5731-5742 DOI https://doi.org/10.1167/iovs.17-22538

64. Kim J, Park DY, Bae H, Park DY, Kim D, Lee CK, Song S, Chung TY, Lim DH, Kubota Y, Hong YK, He Y, Augustin HG, Oliver G, Koh GY (2017) Impaired angiopoietin/Tie2 signaling compromises Schlemm's canal integrity and induces glaucoma. J Clin Invest 127: 3877-3896 DOI https://doi.org/10.1172/JCI94668

65. Sumida GM, Stamer WD (2011) S1P(2) receptor regulation of sphingosine-1-phosphate effects on conventional outflow physiology. Am J Physiol Cell Physiol 300: C1164-1171 DOI https://doi.org/10.1152/ajpcell.00437.2010

66. Ramer R, Hinz B (2010) Cyclooxygenase-2 and tissue inhibitor of matrix metalloproteinases-1 confer the antimigratory effect of cannabinoids on human trabecular meshwork cells. Biochem Pharmacol 80: 846-857 DOI https://doi.org/10.1016/j.bcp.2010.05.010

67. Gabelt BT, Okka M, Dean TR, Kaufman PL (2005) Aqueous humor dynamics in monkeys after topical R-DOI. Invest Ophthalmol Vis Sci 46: 4691-4696 DOI https://doi.org/10.1167/iovs.05-0647

68. Tovar-Vidales T, Fitzgerald AM, Clark AF (2016) Human trabecular meshwork cells express BMP antagonist mRNAs and proteins. Exp Eye Res 147: 156-160 DOI https://doi.org/10.1016/j.exer.2016.05.004

69. Sabanay I, Tian B, Gabelt BT, Geiger B, Kaufman PL (2006) Latrunculin B effects on trabecular meshwork and corneal endothelial morphology in monkeys. Exp Eye Res 82: 236-246 DOI https://doi.org/10.1016/j.exer.2005.06.017

70. Ethier CR, Read AT, Chan DW (2006) Effects of latrunculin-B on outflow facility and trabecular meshwork structure in human eyes. Invest Ophthalmol Vis Sci 47: 1991-1998 DOI https://doi.org/10.1167/iovs.05-0327

71. De Groef L, Van Hove I, Dekeyster E, Stalmans I, Moons L (2013) MMPs in the trabecular meshwork: promising targets for future glaucoma therapies? Invest Ophthalmol Vis Sci 54: 7756-7763 DOI https://doi.org/10.1167/iovs.13-13088

72. De Groef L, Van Hove I, Dekeyster E, Stalmans I, Moons L (2014) MMPs in the neuroretina and optic nerve: modulators of glaucoma pathogenesis and repair? Invest Ophthalmol Vis Sci 55: 1953-1964 DOI https://doi.org/10.1167/iovs.13-13630

73. Chang JY, Stamer WD, Bertrand J, Read AT, Marando CM, Ethier CR, Overby DR (2015) Role of nitric oxide in murine conventional outflow physiology. Am J Physiol Cell Physiol 309: C205-214 DOI https://doi.org/10.1152/ajpcell.00347.2014

74. Tamm ER (2002) Myocilin and glaucoma: facts and ideas. Prog Retin Eye Res 21: 395-428

75. Kasetti RB, Phan TN, Millar JC, Zode GS (2016) Expression of Mutant Myocilin Induces Abnormal Intracellular Accumulation of Selected Extracellular Matrix Proteins in the Trabecular Meshwork. Invest Ophthalmol Vis Sci 57: 6058-6069 DOI https://doi.org/10.1167/iovs.16-19610
76. Paulaviciute-Baikstiene D, Barsauskaite R, Januleviciene I (2013) New insights into pathophysiological mechanisms regulating conventional aqueous humor outflow. Medicina (Kaunas) 49: 165-169
77. Saheb H, Ahmed, II (2012) Micro-invasive glaucoma surgery: current perspectives and future directions. Curr Opin Ophthalmol 23: 96-104 DOI https://doi.org/10.1097/ICU.0b013e32834ff1e7
78. Herdener S, Pache M (2007) [Excimer laser trabeculotomy: minimally invasive glaucoma surgery]. Ophthalmologe 104: 730-732 DOI https://doi.org/10.1007/s00347-007-1598-6
79. Pache M, Wilmsmeyer S, Funk J (2006) [Laser surgery for glaucoma: excimer-laser trabeculotomy]. Klin Monbl Augenheilkd 223: 303-307 DOI https://doi.org/10.1055/s-2005-858861
80. Toteberg-Harms M, Ciechanowski PP, Hirn C, Funk J (2011) [One-year results after combined cataract surgery and excimer laser trabeculotomy for elevated intraocular pressure]. Ophthalmologe 108: 733-738 DOI https://doi.org/10.1007/s00347-011-2337-6
81. Toteberg-Harms M, Hanson JV, Funk J (2013) Cataract surgery combined with excimer laser trabeculotomy to lower intraocular pressure: effectiveness dependent on preoperative IOP. BMC Ophthalmol 13: 24 DOI https://doi.org/10.1186/1471-2415-13-24
82. Toteberg-Harms M, Wachtl J, Schweier C, Funk J, Kniestedt C (2017) Long-term efficacy of combined phacoemulsification plus trabeculectomy versus phacoemulsification plus excimer laser trabeculotomy. Klin Monbl Augenheilkd 234: 457-463 DOI https://doi.org/10.1055/s-0043-100291
83. Walker R, Specht H (2002) [Theoretical and physical aspects of excimer laser trabeculotomy (ELT) ab interno with the AIDA laser with a wave length of 308 mm]. Biomed Tech (Berl) 47: 106-110
84. Wilmsmeyer S, Philippin H, Funk J (2006) Excimer laser trabeculotomy: a new, minimally invasive procedure for patients with glaucoma. Graefes Arch Clin Exp Ophthalmol 244: 670-676 DOI https://doi.org/10.1007/s00417-005-0136-y
85. Berlin MS, Rajacich G, Duffy M, Grundfest W, Goldenberg T (1987) Excimer laser photoablation in glaucoma filtering surgery. Am J Ophthalmol 103: 713-714
86. Spiegel D, Garcia-Feijoo J, Garcia-Sanchez J, Lamielle H (2008) Coexistent primary open-angle glaucoma and cataract: preliminary analysis of treatment by cataract surgery and the iStent trabecular micro-bypass stent. Adv Ther 25: 453-464 DOI https://doi.org/10.1007/s12325-008-0062-6
87. Bahler CK, Hann CR, Fjield T, Haffner D, Heitzmann H, Fautsch MP (2012) Second-generation trabecular meshwork bypass stent (iStent inject) increases outflow facility in cultured human anterior segments. Am J Ophthalmol 153: 1206-1213 DOI https://doi.org/10.1016/j.ajo.2011.12.017
88. Akil H, Chopra V, Huang AS, Swamy R, Francis BA (2017) Short-Term Clinical Results of Ab Interno Trabeculotomy Using the Trabectome with or without Cataract Surgery for Open-Angle Glaucoma Patients of High Intraocular Pressure. J Ophthalmol 2017: 8248710 DOI https://doi.org/10.1155/2017/8248710
89. Gunderson E (2008) Trabeculotomy Ab Interno, using the Trabectome: a promising treatment for patients with open-angle glaucoma. Insight 33: 13-15
90. Minckler D, Mosaed S, Francis B, Loewen N, Weinreb RN (2014) Clinical results of ab interno trabeculotomy using the Trabectome for open-angle glaucoma: the mayo clinic series in Rochester, Minnesota. Am J Ophthalmol 157: 1325-1326 DOI https://doi.org/10.1016/j.ajo.2014.02.030
91. Hoeh H, Vold SD, Ahmed IK, Anton A, Rau M, Singh K, Chang DF, Shingleton BJ, Ianchulev T (2016) Initial Clinical Experience With the CyPass Micro-Stent: Safety and Surgical Outcomes of a Novel Supraciliary Microstent. J Glaucoma 25: 106-112 DOI https://doi.org/10.1097/IJG.0000000000000134

92. Hoh H, Grisanti S, Grisanti S, Rau M, Ianchulev S (2014) Two-year clinical experience with the CyPass micro-stent: safety and surgical outcomes of a novel supraciliary micro-stent. Klin Monbl Augenheilkd 231: 377-381 DOI https://doi.org/10.1055/s-0034-1368214

93. Huisingh C, McGwin G (2014) Response: Optical coherence tomography of the suprachoroid after CyPass Micro-Stent implantation for the treatment of open angle glaucoma. Br J Ophthalmol 98: 847 DOI https://doi.org/10.1136/bjophthalmol-2013-304806

94. Saheb H, Ianchulev T, Ahmed, II (2014) Optical coherence tomography of the suprachoroid after CyPass Micro-Stent implantation for the treatment of open-angle glaucoma. Br J Ophthalmol 98: 19-23 DOI https://doi.org/10.1136/bjophthalmol-2012-302951

95. Gavris MM, Olteanu I, Kantor E, Mateescu R, Belicioiu R (2017) IRIDEX MicroPulse P3: innovative cyclophotocoagulation. Rom J Ophthalmol 61: 107-111

96. Lee JH, Shi Y, Amoozgar B, Aderman C, De Alba Campomanes A, Lin S, Han Y (2017) Outcome of Micropulse Laser Transscleral Cyclophotocoagulation on Pediatric Versus Adult Glaucoma Patients. J Glaucoma 26: 936-939 DOI https://doi.org/10.1097/IJG.0000000000000757

97. Emanuel ME, Grover DS, Fellman RL, Godfrey DG, Smith O, Butler MR, Kornmann HL, Feuer WJ, Goyal S (2017) Micropulse Cyclophotocoagulation: Initial Results in Refractory Glaucoma. J Glaucoma 26: 726-729 DOI https://doi.org/10.1097/IJG.0000000000000715

98. Kuchar S, Moster MR, Reamer CB, Waisbourd M (2016) Treatment outcomes of micropulse transscleral cyclophotocoagulation in advanced glaucoma. Lasers Med Sci 31: 393-396 DOI https://doi.org/10.1007/s10103-015-1856-9

99. Aquino MC, Barton K, Tan AM, Sng C, Li X, Loon SC, Chew PT (2015) Micropulse versus continuous wave transscleral diode cyclophotocoagulation in refractory glaucoma: a randomized exploratory study. Clin Exp Ophthalmol 43: 40-46 DOI https://doi.org/10.1111/ceo.12360

100. Tan AM, Chockalingam M, Aquino MC, Lim ZI, See JL, Chew PT (2010) Micropulse transscleral diode laser cyclophotocoagulation in the treatment of refractory glaucoma. Clin Exp Ophthalmol 38: 266-272 DOI https://doi.org/10.1111/j.1442-9071.2010.02238.x

101. Budenz DL, Barton K, Gedde SJ, Feuer WJ, Schiffman J, Costa VP, Godfrey DG, Buys YM, Ahmed Baerveldt Comparison Study G (2015) Five-year treatment outcomes in the Ahmed Baerveldt comparison study. Ophthalmology 122: 308-316 DOI https://doi.org/10.1016/j.ophtha.2014.08.043

102. Gedde SJ, Schiffman JC, Feuer WJ, Herndon LW, Brandt JD, Budenz DL, Tube Versus Trabeculectomy Study G (2009) Three-year follow-up of the tube versus trabeculectomy study. Am J Ophthalmol 148: 670-684 DOI https://doi.org/10.1016/j.ajo.2009.06.018

103. Hohberger B, Welge-Lussen UC, Lammer R (2018) MIGS: therapeutic success of combined Xen Gel Stent implantation with cataract surgery. Graefes Arch Clin Exp Ophthalmol DOI https://doi.org/10.1007/s00417-017-3895-3

104. Dupont G, Collignon N (2016) [New Surgical Approach in Primary Open-Angle Glaucoma: Xen Gel Stent a Minimally Invasive Technique]. Rev Med Liege 71: 90-93

105. Winkler NF, Funk J (2013) [Transscleral cyclophotocoagulation as primary surgical intervention in glaucoma]. Klin Monbl Augenheilkd 230: 353-357 DOI https://doi.org/10.1055/s-0032-1328359

106. Spencer AF, Vernon SA (1999) "Cyclodiode": results of a standard protocol. Br J Ophthalmol 83: 311-316

107. Schlote T, Grub M, Kynigopoulos M (2008) Long-term results after transscleral diode laser cyclophotocoagulation in refractory posttraumatic glaucoma and glaucoma in aphakia. Graefes Arch Clin Exp Ophthalmol 246: 405-410 DOI https://doi.org/10.1007/s00417-007-0708-0

108. Rotchford AP, Jayasawal R, Madhusudhan S, Ho S, King AJ, Vernon SA (2010) Transscleral diode laser cycloablation in patients with good vision. Br J Ophthalmol 94: 1180-1183 DOI https://doi.org/10.1136/bjo.2008.145565

109. Pucci V, Tappainer F, Borin S, Bellucci R (2003) Long-term follow-up after transscleral diode laser photocoagulation in refractory glaucoma. Ophthalmologica 217: 279-283 DOI https://doi.org/10.1159/000070635
110. Pucci V, Marchini G, Pedrotti E, Morselli S, Bonomi L (2001) Transscleral diode laser photocoagulation in refractory glaucoma. Ophthalmologica 215: 263-266 DOI https://doi.org/10.1159/000050871
111. Leszczynski R, Gierek-Lapinska A, Forminska - Kapuscik M (2004) Transscleral cyclophotocoagulation in the treatment of secondary glaucoma. Med Sci Monit 10: CR542-548
112. Kosoko O, Gaasterland DE, Pollack IP, Enger CL (1996) Long-term outcome of initial ciliary ablation with contact diode laser transscleral cyclophotocoagulation for severe glaucoma. The Diode Laser Ciliary Ablation Study Group. Ophthalmology 103: 1294-1302
113. Hauber FA, Scherer WJ (2002) Influence of total energy delivery on success rate after contact diode laser transscleral cyclophotocoagulation: a retrospective case review and meta-analysis. J Glaucoma 11: 329-333
114. Chen J, Cohn RA, Lin SC, Cortes AE, Alvarado JA (1997) Endoscopic photocoagulation of the ciliary body for treatment of refractory glaucomas. Am J Ophthalmol 124: 787-796
115. Uram M (1992) Ophthalmic laser microendoscope endophotocoagulation. Ophthalmology 99: 1829-1832
116. Uram M (1992) Ophthalmic laser microendoscope ciliary process ablation in the management of neovascular glaucoma. Ophthalmology 99: 1823-1828
117. Uram M (1995) Combined phacoemulsification, endoscopic ciliary process photocoagulation, and intraocular lens implantation in glaucoma management. Ophthalmic Surg 26: 346-352
118. Lima FE, Magacho L, Carvalho DM, Susanna R, Jr., Avila MP (2004) A prospective, comparative study between endoscopic cyclophotocoagulation and the Ahmed drainage implant in refractory glaucoma. J Glaucoma 13: 233-237
119. Lindfield D, Ritchie RW, Griffiths MF (2012) 'Phaco-ECP': combined endoscopic cyclophotocoagulation and cataract surgery to augment medical control of glaucoma. BMJ Open 2 DOI https://doi.org/10.1136/bmjopen-2011-000578
120. Bellows AR, Grant WM (1973) Cyclocryotherapy in advanced inadequately controlled glaucoma. Am J Ophthalmol 75: 679-684
121. Bellows AR, Grant WM (1978) Cyclocryotherapy of chronic open-angle glaucoma in aphakic eyes. Am J Ophthalmol 85: 615-621
122. Caprioli J, Strang SL, Spaeth GL, Poryzees EH (1985) Cyclocryotherapy in the treatment of advanced glaucoma. Ophthalmology 92: 947-954
123. De Roetth A, Jr. (1968) Cryosurgery for the treatment of advanced chronic simple glaucoma. Am J Ophthalmol 66: 1034-1041
124. De Roetth A, Jr. (1968) Cryosurgery for the treatment of advanced chronic simple glaucoma. Trans Am Ophthalmol Soc 66: 45-61
125. Ishida K (2013) Update on results and complications of cyclophotocoagulation. Curr Opin Ophthalmol 24: 102-110 DOI https://doi.org/10.1097/ICU.0b013e32835d9335
126. Cohen A, Wong SH, Patel S, Tsai JC (2017) Endoscopic cyclophotocoagulation for the treatment of glaucoma. Surv Ophthalmol 62: 357-365 DOI https://doi.org/10.1016/j.survophthal.2016.09.004
127. Englert JA, Cox TA, Allingham RR, Shields MB (1997) Argon vs diode laser trabeculoplasty. Am J Ophthalmol 124: 627-631
128. Latina MA, Park C (1995) Selective targeting of trabecular meshwork cells: in vitro studies of pulsed and CW laser interactions. Exp Eye Res 60: 359-371
129. Amon M, Menapace R, Radax U, Wedrich A, Skorpik C (1990) Long-term follow-up of argon laser trabeculoplasty in uncontrolled primary open-angle glaucoma. A study with standardized extensive preoperative treatment. Ophthalmologica 200: 181-188
130. Wise JB, Witter SL (1979) Argon laser therapy for open-angle glaucoma. A pilot study. Arch Ophthalmol 97: 319-322

131. Juzych MS, Chopra V, Banitt MR, Hughes BA, Kim C, Goulas MT, Shin DH (2004) Comparison of long-term outcomes of selective laser trabeculoplasty versus argon laser trabeculoplasty in open-angle glaucoma. Ophthalmology 111: 1853-1859 DOI https://doi.org/10.1016/j.ophtha.2004.04.030

132. Martinez-de-la-Casa JM, Garcia-Feijoo J, Castillo A, Matilla M, Macias JM, Benitez-del-Castillo JM, Garcia-Sanchez J (2004) Selective vs argon laser trabeculoplasty: hypotensive efficacy, anterior chamber inflammation, and postoperative pain. Eye (Lond) 18: 498-502 DOI https://doi.org/10.1038/sj.eye.6700695

133. Polat    J,    Grantham    L,    Mitchell    K,    Realini    T    (2016)    Repeatability    of selective    laser    trabeculoplasty.    Br    J    Ophthalmol    100:    1437-1441    DOI https://doi.org/10.1136/bjophthalmol-2015-307486

134. Toteberg-Harms M, Rhee DJ (2013) Selective laser trabeculoplasty following failed combined phacoemulsification cataract extraction and ab interno trabeculectomy. Am J Ophthalmol 156: 936-940 e932 DOI https://doi.org/10.1016/j.ajo.2013.05.044

135. Nagar M, Ogunyomade A, O'Brart DP, Howes F, Marshall J (2005) A randomised, prospective study comparing selective laser trabeculoplasty with latanoprost for the control of intraocular pressure in ocular hypertension and open angle glaucoma. Br J Ophthalmol 89: 1413-1417 DOI https://doi.org/10.1136/bjo.2004.052795

136. Reitsamer H, Sng C, Vera V, Lenzhofer M, Barton K, Stalmans I; Apex Study Group. Two-year results of a multicenter study of the ab interno gelatin implant in medically uncontrolled primary open-angle glaucoma. Graefes Arch Clin Exp Ophthalmol. 2019 May;257(5):983-996.

137. Pinchuk L, Riss I, Batlle JF, Kato YP, Martin JB, Arrieta E, Palmberg P, Parrish RK, Weber BA, Kwon Y, Parel JM. The use of poly(styrene-block-isobutylene-block-styrene) as a microshunt to treat glaucoma. Regen Biomater. 2016 Jun;3(2):137-42.

138. Pinchuk L, Riss I, Batlle JF, Kato YP, Martin JB, Arrieta E, Palmberg P, Parrish RK 2nd, Weber BA, Kwon Y, Parel JM. The development of a micro-shunt made from poly(styrene-block-isobutylene-block-styrene) to treat glaucoma. J Biomed Mater Res B Appl Biomater. 2017 Jan;105(1):211-221.

139. Batlle JF, Fantes F, Riss I, Pinchuk L, Alburquerque R, Kato YP, Arrieta E, Peralta AC, Palmberg P, Parrish RK 2nd, Weber BA, Parel JM. Three-Year Follow-up of a Novel Aqueous Humor MicroShunt. J Glaucoma. 2016 Feb;25(2):e58-65.

# Approaches to Aqueous Humor Outflow Imaging

Jenna Tauber and Larry Kagemann

**Abstract** The aqueous humor outflow (AHO) tract is the pathway by which fluid travels from the eye's anterior chamber to drain into the venous circulation. This tract plays an important role in the regulation of eye pressure and is likely key in the pathogenesis of glaucoma. As such, the outflow tract is a target for alteration by medical and surgical treatment approaches. To understand this pathway's anatomy and physiology, numerous structural and functional studies have been conducted. While the ability to conduct such studies has advanced along with modern imaging approaches, many challenges to accurately assessing its structure and function remain.

Aqueous humor outflow (AHO) is a subject of great interest in Ophthalmology, especially in a time of minimally invasive glaucoma surgeries (MIGS) designed to enhance outflow. Whereas the trabecular meshwork was once the major focus of anterior segment outflow tract research, the entire conventional AHO pathway can now be studied. Aqueous humor from the anterior chamber traverses multiple layers of trabecular meshwork, enters Schlemm's canal, passes into collector channels, spreads into an intrascleral venous plexus, and empties into aqueous and episcleral veins, ultimately entering the venous circulation. Existing research can be organized as either structural or functional assessments: the former focusing on the physical arrangement of the AHO tract, and the latter as a means to document dynamic fluid flow.

J. Tauber (✉) · L. Kagemann
Department of Ophthalmology, NYU Langone Medical Center, NYU School of Medicine, New York, NY, USA
e-mail: Jenna.Tauber@nyulangone.org; Lawrence.Kagemann@nyumc.org; Lawrence.Kagemann@fda.hhs.gov

© Springer Nature Switzerland AG 2019
G. Guidoboni et al. (eds.), *Ocular Fluid Dynamics*, Modeling and Simulation in Science, Engineering and Technology, https://doi.org/10.1007/978-3-030-25886-3_8

# 1 Structural Assessment of the AHO Pathway

Early attempts to characterize the 3D structure of the AHO pathway included *casting studies*. Agents, such as neoprene latex and vulcanizing silicone, would be injected into outflow structures. After agent polymerization and tissue digestion, a 3D cast representation of the structure would be left [1, 2]. In conjunction with gross anatomy, histology and electron microscopy studies, casting allowed researchers to better delineate and understand AHO anatomy [3–5]. One important finding of this work was the understanding that AHO structures vary along their 360-degree circumference.

While extremely informative, casting has important limitations: because early injection studies using neoprene rubber required supra-physiological pressures, representations of proximal structures, including Schlemm's canal, may not reflect physiologically normative conditions. In addition, even the most precise and careful work when dealing with post-mortem specimens is prone to inaccuracies. There are challenges associated with tissue procurement, fixation artifacts, and inconsistencies in sample preparations [3, 5, 6].

In 1963, Morton Grant described *physiologic experimental studies* using an aqueous humor perfusion ring [7]. He applied a basic formula of fluid dynamics to quantify outflow in enucleated eyes. Using $C = F/P$, he could calculate the facility of aqueous outflow (C). In his experiment, he controlled the rate of saline solution injection (F) to create steady in and outflow and established steady-state moderate pressure (P) conditions in the eye. Using such methods, Grant and others were able to show that the trabecular meshwork acts as a primary resistor, accounting for 70–75% of AHO resistance [8, 9]. These techniques have also been used to show that there is ongoing resistance distal to the trabecular meshwork and that AHO resistance is higher in eyes with glaucoma. Building on this model to investigate theories of disease, others have studied features of the AHO pathway including extracellular matrix and cytoskeleton, pro-fibrotic factor secretion, funneling theories, and biomechanics [5, 6, 10–12]. This work has been meaningful, contributing to the development of commonly used minimally invasive glaucoma surgery (MIGS) techniques; however, it does not address the critically important structure–function relationship of the outflow tract.

## 1.1 Imaging Techniques

Postmortem trabecular meshwork imaging has been performed using *two-photon microscopy (TPM)* [13] (Fig. 1). TPM is an imaging technique that uses near-infrared laser to visualize whole fixed or live tissue. It is superior to single photon microscopy in that it can penetrate deeper to develop high-resolution optical sectioning with less scatter, absorption, and phototoxicity [14–17]. TPM can use multiple modalities, including autofluorescence, direct-labeled fluorescence,

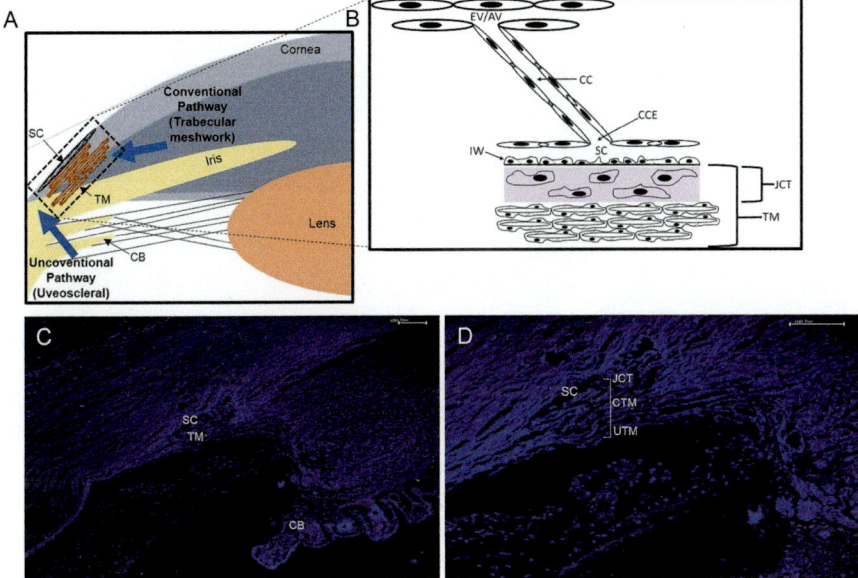

**Fig. 1** (**a**) Schematic diagram summarizing the conventional and uveoscleral aqueous humor outflow pathways in the eye's anterior chamber. (**b**) Detailed schematic illustrating the trabecular meshwork (TM), collector channel entrances (CCE), collector channels (CC), episcleral vein (EV), and aqueous vein (AV). (**c**) Section from a fluoresce-stained anterior chamber in which the ciliary body (CB), Schlemm's canal (SC), and TM are labeled. (**d**) Magnified view of the TM in which juxtacanalicular tissue (JCT), uveoscleral (UTM), and conventional (CTM) can be visualized. (Image courtesy of Carreon et al. [54])

indirect anti-body-labeled epifluorescence, and second harmonic generation. Some of these techniques have been used to isolate components of extracellular matrix, including collagen and elastin [18–20]. While it can be used to examine live cell dynamics, it has yet to be applied to human subjects in vivo.

*Three-dimensional micro-computed tomography (3D micro-CT)* can be used to image the entire anterior segment, including the AHO pathway beyond the trabecular meshwork [21]. In 3D micro-CT, a molybdenum X-ray source is used to scan the tissue. Rays are converted to visible light when projected onto a crystal scintillator plate and reconstructed in 3D. In enucleated eyes, 360° of outflow structure can be seen, including Schlemm's canal, collector channels, and large intrascleral vessels [22].

Optical coherence tomography (OCT) uses near-infrared low-coherence laser interferometry to capture 3D in vivo images of biologic tissue [23] (Fig. 2). After many years of use in retina and optic nerve head structure analyses, *anterior segment (AS-)OCT* was finally developed. This technology can identify regions of low reflectivity corresponding to the AHO pathway structures. It is especially simple to image Schlemm's canal because fluid in this space provides a clear hypo-reflective

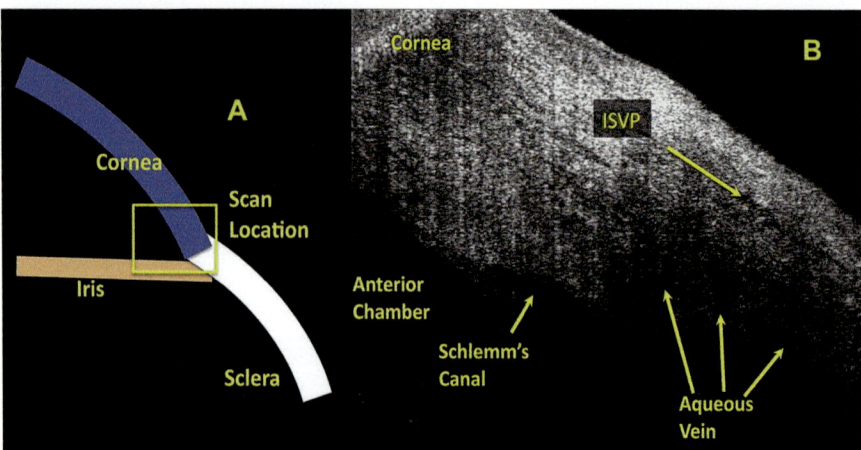

**Fig. 2** (**a**) Cartoon demonstrating the region of the eye captured using this anterior chamber spectral-domain optical coherence tomography imaging technique. (**b**) The scan is labeled to illustrate visualization of Schlemm's canal, aqueous veins, and the intrascleral venous plexus (ISVP). (Figure courtesy of Kagemann et al. [25])

signal [24–26]. With image processing techniques, 3D reconstruction and even in silico casting are possible [24]. Automated segmentation methods are an ongoing subject of research in which information derived from one limbal segment can be extrapolated to the entire circumference [27]. A clinical and surgical focus has also been placed on AS-OCT AHO pathway research in studies that have looked at relationships between visible structures and medical and surgical therapies [28–30].

*Phase-sensitive (PhS-)OCT* is a variation of OCT technology that is capable of quantifying nanometer scale tissue motion with velocities as slow as 2.6 nm/s [31]. This has been applicable to studying trabecular meshwork motion, which occurs in the 1 Hz range. PhS-OCT trabecular meshwork imaging has shown pulse-dependent motion, which correlates with the cardiac pulse [32].

In spite of all this promising work, a major limitation still exists to using OCT in the anterior segment: there has yet to be any histological confirmation of the signals being seen. Until a ground truth is established, caution must be taken not to draw hurried conclusions from these OCT findings.

In addition, there are still some barriers to successful segmentation. Because Schlemm's canal and collector channels both appear distinctly hypo-reflective on OCT, they can be distinguished from their surroundings; however, it can be difficult to delineate one from the other. Similarly, other hypo-reflective structures, such as vessels and lymphatics, may be incorrectly deemed collector channels. In the past, researchers have made assumptions to decide how segmentations should be performed, but this may mean leaving out or misrepresenting the true number of collector channels in the AHO tract [27].

A third limitation is that an OCT cannot capture the entire limbal circumference at once. The average limbal radius is 11.5 mm—a circumference of 36.11 mm. Clinical OCT macular imaging covers approximately 5–6 mm [33]. In one study, more than 5000 B-scans with overlapping volumes were needed to achieve good quality 360-degree coverage [27]. Scans must overlap to account for the 35 μm distance between B-scans, which could otherwise easily overlook collector channels [33]. Finally, until an adequate eye-tracking function has been implemented into the technology, it will be difficult to compare images taken at different time periods.

# 2 Functional Assessment of AHO

## 2.1 Static Techniques

With static testing, *tracers* are introduced into the anterior chamber, and then the eye is examined microscopically or histologically. Tracer position can be recorded considering factors like time of tracer flow and particle size. Much of this work has been performed using gold particles. With electron microscopes, it is possible to observe gold uptake segmentally [34]. Fluorescent microspheres (0.2–20 μm) or quantum dots (0.01 μm) have also been used as tracers [33]. In live or enucleated eyes, confocal or fluorescent microscopes can be used with these tracers for global or segmental AHO assessment.

It is important to note that the majority of aqueous is composed of water with a molecular weight of 18 daltons (Da). Because tracer molecules are often larger than water molecules, particles can accumulate at AHO filtration points, thus creating a static visualization of trabecular meshwork and collector channel flow. When using tracers, multiple kinds with different characteristics can be used to reflect the flow of different aqueous molecules, which include vitamin C, DNA, RNA, proteins, and ions.

With a tracer technique, quantitative methods are also possible. For example, percent filtration length (PEFL), a measure of tracer utilization in the trabecular meshwork, can be calculated from a histological segmentation. PEFL = L/TL, where L represents the total distance traversed by the tracer and TL represents the total length of the observable outflow pathways [35].

These concepts can be applied in biochemical and genetic studies to see how chemical agents, like ROCK inhibitors, or intraocular pressure variations affect PEFL values [36, 37]. Regions seen to have elevated or reduced tracer accumulation can be probed to evaluate DNA and RNA expression as well as protein levels in animal models and enucleated human eyes [38, 39]. Unfortunately, tracers have not been deemed safe in humans, so at this point they cannot be used for measurements in vivo.

## 2.2   Real-Time

In the absence of tissue processing requirements, real-time AHO investigations can be performed in living humans. If there is no lag time between obtaining a tissue sample and analyzing it, findings may be more accurate. With the application of small safe tracers in more natural clinical and surgical environments, real-time AHO imaging techniques have the potential to be extremely valuable.

### 2.2.1   Episcleral Fluid Waves

The first of such techniques involves the study of *episcleral venous fluid waves*. A phacoemulsification unit is a tool used routinely in cataract surgery to emulsify the natural lens as well as to aspirate and irrigate balance salt solution into the anterior chamber. The tool is also used for phaco-trabectome MIGS procedures, which ablate the trabecular meshwork to decrease intraocular pressure in glaucoma. During this procedure, Fellman had the idea to position his microscope over the nasal limbus to view the perilimbal vasculature at the moment of irrigation [40]. In some cases, a wave could be seen in post-limbal episcleral veins, indicating patency of the distal AHO collector system.

One benefit of this technique includes the ability to use it during routine surgeries with equipment that is familiar to surgeons. A limitation, however, is that anterior chamber pressure is elevated during this procedure, which may result in non-physiologic results. Even if this pressure were calculated in an attempt to quantify fluid flow, salt solution might leak around the surgical site wounds during the procedure, and it would be difficult to measure this loss. In addition, there is an inherent challenge to this approach in that the observer is required to measure the flushing of blood by a clear solution: a loss-of-signal test. By contrast, it is less complicated to observe outflow tracts being filled with a substance than being emptied.

### 2.2.2   Canalography

The invention of *canalograms* also began with another MIGS approach: the canalo-plasty. During this procedure, Schlemm's canal is exposed using a microcatheter and stented with a tube intended to improve drainage. As the probe tool is reversed out of the canal, a viscoelastic fluorescein tracer can be injected and the pathway can be visualized [41–44]. The resulting images are very good, but there are clear imperfections to this method. By introducing tracer directly into Schlemm's canal with this *ab-externo* approach, the trabecular meshwork is bypassed entirely, and its contribution to the outflow tract is overlooked. Also, there are technical limitations. It would be impossible to remove the probe quickly enough to ensure filling along the entire limbal circumference at once. Therefore, the region closest to the injection might appear artificially to have more flow.

### 2.2.3 Aqueous Angiography

Built further upon the idea of canalography is *aqueous angiography*. Here tracers (indocyanine green, fluorescein, and others) are introduced into the anterior chamber using an *ab-interno* approach, and then images are taken externally. By filling the anterior chamber, the trabecular meshwork influences outflow in a more physiologic pattern.

Though intended to be used on humans in vivo, study began on post-mortem porcine, bovine, and human eyes using experimental setups [45–49]. In these experiments, constant-pressure gravity-driven tracer was delivered from a reservoir above the eye using routine operating room anterior chamber maintainers and tubing [33]. Spectralis HRA + OCT [Heidelberg Engineering, Heidelberg, Germany] was used to capture axial images on fluorescein or indocyanine green capture mode. Traction sutures in post-mortem eyes and spontaneous eye movement in living subjects were required to eliminate signal interference from the eyelids.

Multi-model imaging has been used to validate angiographic findings: intrascleral lumens can be correlated with AS-OCT findings, and fixable fluorescent dextran tracers trapped in the trabecular meshwork can be seen as angiographically positive regions [45, 47–49]. Quantitative analysis is possible by applying imaging algorithms to automatically compute focal and global flow and to track the way these patterns change over time [50].

Of course to perform aqueous angiography on living subjects, a non-laboratory setup is required. This involves positioning the patient in the routine supine operating room posture and using a Specralis FLEX module—a flexible arm with multiple pivot joints and HRA + OCT installed onto it [48, 49]. Studies applying these methods have been able to identify dynamic changes in flow, with signals arising and diminishing within the same region through an undetermined mechanism. Major appeals of this approach relate to its potential application in MIGS. It could help to answer such questions as, "should trabecular MIGS target areas of higher or lower flow for maximum benefit to the patient"? Aqueous angiography does not require any wounds beyond those created during routine cataract surgery, so perhaps this analysis could become a completely routine step that is personalized to the individual patient in surgical planning.

In spite of the many benefits to using aqueous angiography, it still is not an entirely physiological method. Though tracer delivery into the anterior chamber allows for trabecular meshwork involvement, it does not consider that aqueous is naturally delivered posterior to the iris and migrates anteriorly before entering the AHO tract. By bypassing this mechanism, the chamber may be artificially deepened. In addition, using lid specula or traction sutures could theoretically alter the ocular surface, exerting forces that are not present naturally. Finally, the routine use of antimuscarinic drops to dilate live human eyes could very likely impact angiography results by altering trabecular meshwork capacitance [48].

Using mathematical modeling, these techniques can be considered in combination to dig deeper into aspects of the AHO pathway. For example, as aqueous angiography begins at the trabecular meshwork and canalography begins at

Schlemm's canal, their difference represents the outflow at the trabecular meshwork (Canalogram − aqueous angiogram = trabecular meshwork) [33].

# 3 Noninvasive Techniques to Assess AHO

The goal of noninvasive imaging would be to allow for individualized surgical AHO planning prior to MIGS in an approach that has little to no risk to the patient. At this point, there are no noninvasive options, though future research is looking into OCT angiography (OCTA) for the anterior segment. OCTA works by separating moving signal scatters from static background tissue to build angiograms [51]. A challenge is that the clear non-particulate nature of aqueous fluid makes it difficult to visualize using this imaging technology. Efforts have been made to discover contrast agents that can be used with OCTA in the anterior segment, such as gold [52]. Of course, however, once a contrast agent is required, the approach can no longer be considered truly noninvasive.

With this limitation in mind, there is a desire to find and exploit a chemical that is native to the aqueous and present in greater concentrations compared to serum. A front-runner candidate in this search so far has been Vitamin C [53].

Over the years, many innovative approaches have been taken to visualizing AHO. With each advancement, limitations are considered, which facilitates the growth of new ideas. As this work continues, noninvasive functional approaches might be developed, which would have the potential to revolutionize patient care.

# References

1. Ashton N. Anatomical study of Schlemm's canal and aqueous veins by means of neoprene casts. Part I. Aqueous veins. *The British journal of ophthalmology*. 1951;35(5):291–303.
2. Van Buskirk EM. The canine eye: the vessels of aqueous drainage. *Investigative ophthalmology & visual science*. 1979;18(3):223–230.
3. Dvorak T. Further Studies on the Canal of Schlemm: Its Anastomoses and Anatomic Relations Georgiana. *American Journal of Ophthalmology*. 1955;39(4, Part 2):65–89.
4. Ujiie K, Bill A. The drainage routes for aqueous humor in monkeys as revealed by scanning electron microscopy of corrosion casts. *Scanning electron microscopy*. 1984(Pt 2):849–856.
5. Johnson M. 'What controls aqueous humour outflow resistance?'. *Exp Eye Res*. 2006;82(4):545–557.
6. Lutjen-Drecoll E, Shimizu T, Rohrbach M, Rohen JW. Quantitative analysis of 'plaque material' in the inner- and outer wall of Schlemm's canal in normal- and glaucomatous eyes. *Exp Eye Res*. 1986;42(5):443–455.
7. Grant WM. Experimental aqueous perfusion in enucleated human eyes. *Archives of ophthalmology (Chicago, Ill : 1960)*. 1963;69:783–801.
8. Van Buskirk EM. Trabeculotomy in the immature, enucleated human eye. *Investigative ophthalmology & visual science*. 1977;16(1):63–66.

9. Rosenquist R, Epstein D, Melamed S, Johnson M, Grant WM. Outflow resistance of enucleated human eyes at two different perfusion pressures and different extents of trabeculotomy. *Current eye research.* 1989;8(12):1233–1240.

10. Swaminathan SS, Oh DJ, Kang MH, et al. Secreted protein acidic and rich in cysteine (SPARC)-null mice exhibit more uniform outflow. *Investigative ophthalmology & visual science.* 2013;54(3):2035–2047.

11. Tripathi RC, Li J, Chan WF, Tripathi BJ. Aqueous humor in glaucomatous eyes contains an increased level of TGF-beta 2. *Exp Eye Res.* 1994;59(6):723–727.

12. Overby DR, Stamer WD, Johnson M. The changing paradigm of outflow resistance generation: towards synergistic models of the JCT and inner wall endothelium. *Exp Eye Res.* 2009;88(4):656–670.

13. Chu ER, Gonzalez JM, Jr., Tan JC. Tissue-based imaging model of human trabecular meshwork. *Journal of ocular pharmacology and therapeutics : the official journal of the Association for Ocular Pharmacology and Therapeutics.* 2014;30(2–3):191–201.

14. Aptel F, Olivier N, Deniset-Besseau A, et al. Multimodal nonlinear imaging of the human cornea. *Investigative ophthalmology & visual science.* 2010;51(5):2459–2465.

15. Johnson AW, Ammar DA, Kahook MY. Two-photon imaging of the mouse eye. *Investigative ophthalmology & visual science.* 2011;52(7):4098–4105.

16. Schenke-Layland K. Non-invasive multiphoton imaging of extracellular matrix structures. *Journal of biophotonics.* 2008;1(6):451–462.

17. Zipfel WR, Williams RM, Webb WW. Nonlinear magic: multiphoton microscopy in the biosciences. *Nature biotechnology.* 2003;21(11):1369–1377.

18. Gonzalez JM, Jr., Heur M, Tan JC. Two-photon immunofluorescence characterization of the trabecular meshwork in situ. *Investigative ophthalmology & visual science.* 2012;53(7):3395–3404.

19. Huang AS, Gonzalez JM, Jr., Le PV, Heur M, Tan JC. Sources of structural autofluorescence in the human trabecular meshwork. *Investigative ophthalmology & visual science.* 2013;54(7):4813–4820.

20. Tan JC, Gonzalez JM, Jr., Hamm-Alvarez S, Song J. In situ autofluorescence visualization of human trabecular meshwork structure. *Investigative ophthalmology & visual science.* 2012;53(4):2080–2088.

21. Jorgensen SM, Demirkaya O, Ritman EL. Three-dimensional imaging of vasculature and parenchyma in intact rodent organs with X-ray micro-CT. *The American journal of physiology.* 1998;275(3 Pt 2):H1103–1114.

22. Hann CR, Bentley MD, Vercnocke A, Ritman EL, Fautsch MP. Imaging the aqueous humor outflow pathway in human eyes by three-dimensional micro-computed tomography (3D micro-CT). *Experimental Eye Research.* 2011;92(2):104–111.

23. Huang D, Swanson EA, Lin CP, et al. Optical coherence tomography. *Science.* 1991;254(5035):1178–1181.

24. Kagemann L, Wollstein G, Ishikawa H, et al. Visualization of the conventional outflow pathway in the living human eye. *Ophthalmology.* 2012;119(8):1563–1568.

25. Kagemann L, Wollstein G, Ishikawa H, et al. 3D visualization of aqueous humor outflow structures in-situ in humans. *Exp Eye Res.* 2011;93(3):308–315.

26. Li P, Butt A, Chien JL, et al. Characteristics and variations of in vivo Schlemm's canal and collector channel microstructures in enhanced-depth imaging optical coherence tomography. *The British journal of ophthalmology.* 2017;101(6):808–813.

27. Huang AS, Belghith A, Dastiridou A, Chopra V, Zangwill LM, Weinreb RN. Automated circumferential construction of first-order aqueous humor outflow pathways using spectral-domain optical coherence tomography. *Journal of biomedical optics.* 2017;22(6):66010.

28. Li G, Farsiu S, Chiu SJ, et al. Pilocarpine-induced dilation of Schlemm's canal and prevention of lumen collapse at elevated intraocular pressures in living mice visualized by OCT. *Investigative ophthalmology & visual science.* 2014;55(6):3737–3746.

29. Li G, Mukherjee D, Navarro I, et al. Visualization of conventional outflow tissue responses to netarsudil in living mouse eyes. *European journal of pharmacology.* 2016;787:20–31.

30. Skaat A, Rosman MS, Chien JL, et al. Microarchitecture of Schlemm Canal Before and After Selective Laser Trabeculoplasty in Enhanced Depth Imaging Optical Coherence Tomography. *Journal of glaucoma.* 2017;26(4):361–366.
31. Wang RK, Kirkpatrick S, Hinds M. Phase-sensitive optical coherence elastography for mapping tissue microstrains in real time. *Applied Physics Letters.* 2007;90(16):164105.
32. Li P, Shen TT, Johnstone M, Wang RK. Pulsatile motion of the trabecular meshwork in healthy human subjects quantified by phase-sensitive optical coherence tomography. *Biomedical optics express.* 2013;4(10):2051–2065.
33. Huang AS, Francis BA, Weinreb RN. Structural and functional imaging of aqueous humour outflow: a review. *Clinical & Experimental Ophthalmology.* 2017:1–11.
34. Sabanay I, Gabelt BT, Tian B, Kaufman PL, Geiger B. H-7 effects on the structure and fluid conductance of monkey trabecular meshwork. *Archives of ophthalmology (Chicago, Ill : 1960).* 2000;118(7):955–962.
35. Battista SA, Lu Z, Hofmann S, Freddo T, Overby DR, Gong H. Reduction of the available area for aqueous humor outflow and increase in meshwork herniations into collector channels following acute IOP elevation in bovine eyes. *Investigative ophthalmology & visual science.* 2008;49(12):5346–5352.
36. Ren R, Li G, Le TD, Kopczynski C, Stamer WD, Gong H. Netarsudil Increases Outflow Facility in Human Eyes Through Multiple Mechanisms. *Investigative ophthalmology & visual science.* 2016;57(14):6197–6209.
37. Zhu JY, Ye W, Gong HY. Development of a novel two color tracer perfusion technique for the hydrodynamic study of aqueous outflow in bovine eyes. *Chinese medical journal.* 2010;123(5):599–605.
38. Keller KE, Bradley JM, Vranka JA, Acott TS. Segmental versican expression in the trabecular meshwork and involvement in outflow facility. *Investigative ophthalmology & visual science.* 2011;52(8):5049–5057.
39. Vranka JA, Bradley JM, Yang YF, Keller KE, Acott TS. Mapping molecular differences and extracellular matrix gene expression in segmental outflow pathways of the human ocular trabecular meshwork. *PloS one.* 2015;10(3):e0122483.
40. Fellman RL, Grover DS. Episcleral venous fluid wave: intraoperative evidence for patency of the conventional outflow system. *Journal of glaucoma.* 2014;23(6):347–350.
41. Aktas Z, Tian B, McDonald J, et al. Application of canaloplasty in glaucoma gene therapy. where are we? *Journal of ocular pharmacology and therapeutics : the official journal of the Association for Ocular Pharmacology and Therapeutics.* 2014;30(2–3):277–282.
42. Grieshaber MC. Ab externo Schlemm's canal surgery: viscocanalostomy and canaloplasty. *Developments in ophthalmology.* 2012;50:109–124.
43. Grieshaber MC, Pienaar A, Olivier J, Stegmann R. Clinical evaluation of the aqueous outflow system in primary open-angle glaucoma for canaloplasty. *Investigative ophthalmology & visual science.* 2010;51(3):1498–1504.
44. Zeppa L, Ambrosone L, Guerra G, Fortunato M, Costagliola C. Using canalography to visual-ize the in vivo aqueous humor outflow conventional pathway in humans. *JAMA ophthalmology.* 2014;132(11):1281.
45. Huang AS, Saraswathy S, Dastiridou A, et al. Aqueous Angiography with Fluorescein and Indocyanine Green in Bovine Eyes. *Translational vision science & technology.* 2016;5(6):5.
46. Huang AS, Saraswathy S, Dastiridou A, et al. Aqueous Angiography-Mediated Guidance of Trabecular Bypass Improves Angiographic Outflow in Human Enucleated Eyes. *Investigative ophthalmology & visual science.* 2016;57(11):4558–4565.
47. Saraswathy S, Tan JC, Yu F, et al. Aqueous Angiography: Real-Time and Physiologic Aqueous Humor Outflow Imaging. *PloS one.* 2016;11(1):e0147176.
48. Huang AS, Camp A, Xu BY, Penteado RC, Weinreb RN. Aqueous Angiography: Aqueous Humor Outflow Imaging in Live Human Subjects. *Ophthalmology.* 2017;124(8):1249–1251.
49. Huang AS, Li M, Yang D, Wang H, Wang N, Weinreb RN. Aqueous Angiography in Living Nonhuman Primates Shows Segmental, Pulsatile, and Dynamic Angiographic Aqueous Humor Outflow. *Ophthalmology.* 2017;124(6):793–803.

50. Loewen RT, Brown EN, Roy P, Schuman JS, Sigal IA, Loewen NA. Regionally Discrete Aqueous Humor Outflow Quantification Using Fluorescein Canalograms. *PloS one.* 2016;11(3):e0151754.
51. Wang RK, Jacques SL, Ma Z, Hurst S, Hanson SR, Gruber A. Three dimensional optical angiography. *Optics express.* 2007;15(7):4083–4097.
52. Wang B, Kagemann L, Schuman JS, et al. Gold nanorods as a contrast agent for Doppler optical coherence tomography. *PloS one.* 2014;9(3):e90690.
53. Purcell EF, Lerner LH, Kinsey VE. Ascorbic acid in aqueous humor and serum of patients with and without cataract; physiologic significance of relative concentrations. *AMA archives of ophthalmology.* 1954;51(1):1–6.
54. Carreo T, van der Merwe E, Fellman RL, Johnstone M, Bhattacharya SK. Aqueous outflow — A continuum from trabecular meshwork to episcleral veins. *Prog Retin Eye Res.* 2017;57:108–133.

# Mathematical Models of Aqueous Production, Flow and Drainage

Mariia Dvoriashyna, Jan O. Pralits, Jennifer H. Tweedy, and Rodolfo Repetto

**Abstract** The aqueous humour (AH) is a transparent fluid with water-like properties that fills the anterior chamber (AC, the region between the cornea and the iris) and the posterior chamber (PC, the region between the iris and the lens) of the eye, which are connected at the pupil. AH is produced at ciliary processes, and it flows from the PC to the AC, where it is drained in the trabecular meshwork. AH flow is important physiologically, as it governs intraocular pressure and delivers nutrients to avascular ocular tissues. Disruption of AH flow may lead to multiple pathological conditions, such as glaucoma and nutrient depletion. Studying aqueous production, flow and drainage is thus relevant to understand eye physiology and pathophysiology.

Mathematical modelling has proven to be a very useful tool for studying AH, as it allows one to understand the mechanisms of the flow by studying them separately. In this chapter we outline the mathematical models of AH production, different AH flow mechanisms and drainage, subsequently. We focus on analytical works and briefly mention the main conclusions of numerical ones.

## 1 Introduction

The anterior segment of the eye consists of the anterior (AC) and posterior (PC) chambers, which are connected to each other by the pupil aperture (see Fig. 1). The AC is delimited anteriorly by the cornea and posteriorly by the iris and pupil, while the PC is the space between the iris and the lens. These two chambers contain a clear

M. Dvoriashyna
Department of Applied Mathematics and Theoretical Physics,
University of Cambridge, Cambridge, UK

J. O. Pralits · R. Repetto (✉)
Department of Civil, Chemical and Environment Engineering, University of Genoa, Genova, Italy
e-mail: rodolfo.repetto@unige.it

J. H. Tweedy
Department of Bioengineering, Imperial College London, London, UK

© Springer Nature Switzerland AG 2019
G. Guidoboni et al. (eds.), *Ocular Fluid Dynamics*, Modeling and Simulation in
Science, Engineering and Technology, https://doi.org/10.1007/978-3-030-25886-3_9

**Fig. 1** Sketch of the anterior
segment of the eye. Drawing
by Federica Grillo
(University of Genoa, Italy)

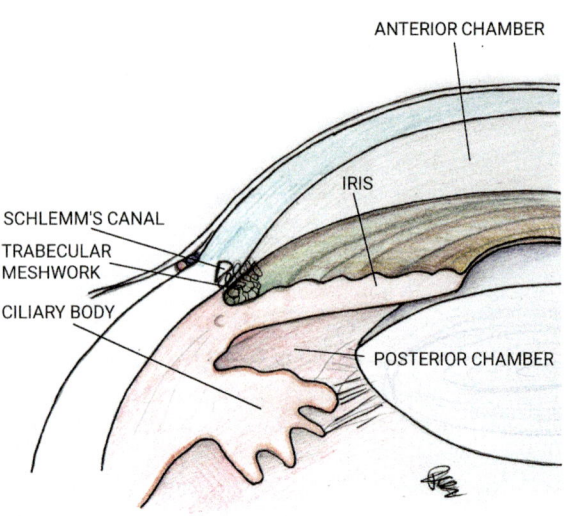

ANTERIOR CHAMBER

IRIS

SCHLEMM'S CANAL

TRABECULAR
MESHWORK

CILIARY BODY

POSTERIOR CHAMBER

fluid named aqueous humour (AH). AH is an aqueous solution containing a mixture of electrolytes, organic solutes, growth factors and other proteins [23]. From the mechanical point of view it can be modelled as a Newtonian fluid with properties similar to water, i.e. a density $\rho \approx 10^3$ kg/m$^3$ and a kinematic viscosity $\nu \approx 10^{-6}$ m$^2$/s (see Table 1).

AH is produced by the ciliary body in the PC at a rate of approximately 3 μl/min ($\approx 5 \times 10^{-11}$ m$^3$/s) [11, 12] and flows through the pupil into the AC, from where it exits the eye via the conventional and uveoscleral pathways. In the former pathway, which in humans accounts for the majority of the drainage [2], AH percolates through the trabecular meshwork (TM), enters Schlemm's canal and is finally drained into the venous system through small, roughly evenly spaced, channels. In the uveoscleral (or unconventional) outflow pathway, AH passes from the AC into the ciliary muscle and then into the supraciliary and suprachoroidal spaces, and finally exits the eye through the sclera.

The flow of AH supplies nutrients to the avascular tissues of the anterior chamber, in particular the TM, the lens and the corneal endothelium. Moreover, the balance between the rate of production and resistance to drainage of AH governs the intraocular pressure (IOP), an important clinical parameter, which ranges between 12 and 22 mmHg (1660–2933 Pa) in healthy subjects. An elevated IOP is correlated with the occurrence of open-angle glaucoma (OAG) [45, 77], one of the most common causes of blindness worldwide [40, 63].

Mechanical phenomena are heavily involved in the physiology of AH, and mathematical models have proved to be extremely useful tools, both for understanding eye physiology and also for studying pathophysiology. Mathematical models can also potentially help diagnosis of pathology and improve medical and surgical treatments and the design of prosthetic devices.

**Table 1** Parameter values used for the simulations

| Aqueous properties | | | |
|---|---|---|---|
| Density | $\rho$ | $1000$ kg/m$^3$ | |
| Flux | $Q$ | $3\,\mu\text{l/min} =$ $5 \cdot 10^{-11}$ m$^3$/s | [11, 12] |
| Kinematic viscosity | $\nu$ | $7.5 \cdot 10^{-7}$ m$^2$/s | [6] |
| Specific heat at constant pressure | $c_p$ | $4.178 \cdot 10^3$ J kg$^{-1}$K$^{-1}$ (water at $37°$) | [5] |
| Thermal conductivity | $k$ | $0.578$ Wm$^{-1}$K$^{-1}$ | [57] |
| Thermal expansion coefficient | $\alpha$ | $3 \cdot 10^{-4}$ K$^{-1}$ (water at $30°$) | [5] |
| *Geometrical characteristics of the anterior chamber* | | | |
| Diameter | $L_{AC}$ | $\approx 13$ mm | |
| Maximum height | $H_{AC}$ | $2.63$ mm | ISO-11979-3 [30] |
| Average height | $\bar{H}_{AC}$ | $1.3$ mm | [21] |
| Minimum radius of curvature of the cornea | | $6.8$ mm | ISO-11979-3 [30] |
| Radius of curvature of the lens | | $\approx 10$ mm | ISO-11979-3 [30] |
| *Geometrical characteristics of the posterior chamber* | | | |
| Length of the bPC | $L_{bPC}$ | $2.57$ mm | [20] |
| Maximum height of the bPC | $H_{bPC}$ | $0.94$ mm | [20] |
| Average height of the bPC | $\bar{H}_{bPC}$ | $0.44$ mm | |
| Length of the ILC | $L_{ILC}$ | $1.14$ mm | [20] |
| Minimum height of the ILC | $H_{ILC}$ | $7\,\mu$m | [68] |
| Average height of the ILC | $\bar{H}_{ILC}$ | $11\,\mu$m | [20] |

The chapter is organised as follows: In Sect. 2, we consider models of AH production at the ciliary body epithelium, the mechanism of which is not fully understood, as both fluid mechanics and cellular electrophysiology are involved. In Sect. 3, we discuss mathematical modelling of the flow of AH, beginning with general characteristics and a mathematical framework for studying it in Sect. 3.1. In Sect. 3.2 onward, we cover the different physical mechanisms that drive the flow in the PC and AC. Finally in Sect. 4, we describe models of AH drainage from the eye, including both the well-studied conventional outflow and also the uveoscleral outflow.

Although we mention the most recent and up-to-date thinking on each of the phenomena we describe, our deliberate focus is on relatively simple analytical works, and we typically consider simplified and idealised geometries and flow conditions that are amenable to analytical treatment. The advantage of this approach is that it helps us to isolate the characteristic and key ingredients of each of the mechanisms, and thereby elucidate the underlying physics more effectively than would numerical results.

## 2   Models of Aqueous Production

### 2.1   Anatomical Structure of the Ciliary Body

As pointed out in the introduction, the rate of AH production is a key factor, together with the resistance to aqueous drainage, for controlling IOP. AH is produced continuously by the ciliary body at an average rate of approximately $3\,\mu l/min$ ($\approx 5 \times 10^{-11}\ m^3/s$) [11, 12]. The production rate is known to vary significantly with circadian rhythm (see Part 3, Sect. 6 of the present volume).

Understanding the mechanisms of AH production and their relative importance is relevant for many reasons. In particular, lowering the production rate could prevent glaucoma or be used to treat it. In the present section, we discuss the physical mechanisms that govern AH production and shortly review the literature concerning the related mathematical models.

In order to understand the physics behind AH production, we first need to briefly review the anatomy of the ciliary body. This tissue has annular shape, is located behind the iris and separates the PC from the vitreous cavity. Its base is home to the ciliary muscle, which controls lens accommodation. The surface of the ciliary body has a series of wrinkles called ciliary processes, which are oriented radially so as to point towards the pupil, and have the role of increasing the surface area available for fluid production [15].

The ciliary processes are linen by a double layer of epithelial cells, which is a peculiar feature of this tissue (see Fig. 2). The inner layer consists of pigmented cells and the outer one of non-pigmented ones. The epithelial cells within each layer and between the layers are connected by gap junctions, which allow the cells to be electrically coupled. This arrangement is likely to promote a cooperative work of these two cell layers in the AH production. The non-pigmented epithelial cells are connected to each other by tight junctions that block the passage of large molecules and maintain a transepithelial potential difference. The non-pigmented epithelial cells of the ciliary processes, create the "blood-aqueous barrier", which is effectively a separation between the AH and blood circulation. On the side of the PC, the non-pigmented ciliary epithelial cells present a very high degree of membrane infolding, which is a further mean for increasing the capacity of the tissue to secrete fluid.

The ciliary body is a highly vascularised tissue with a complicated arrangement of small blood vessels. Capillaries in the ciliary processes are fenestrated, and this permits passage of plasma proteins into the stroma, the tissue within the ciliary body [61]. Such large molecules, however, cannot move from the stroma into the AH, owing to the existence of the tight junctions in the epithelium. For this reason, the chemical composition of the AH differs significantly from that of blood plasma, since, the former is almost entirely free of proteins [72]. This is essential for the AH to be a perfectly transparent liquid.

The tight junctions connecting the non-pigmented cells of the ciliary epithelium also separate the cell membrane into two different parts: the membrane on the side

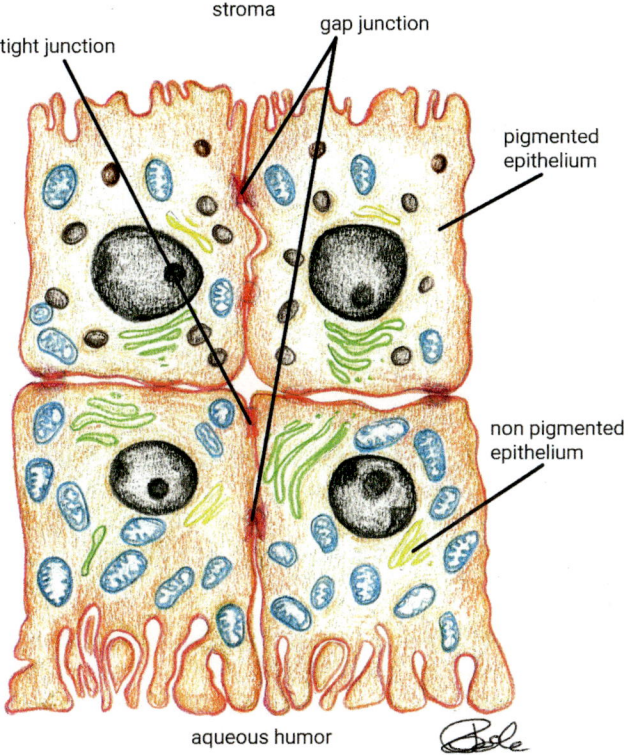

**Fig. 2** Sketch of the cell arrangement in the ciliary epithelium. Drawing by Federica Grillo (University of Genoa, Italy)

of the AH and the membrane on the side of the stroma. The distribution of ion channels and exchangers is different in the membranes, which allows the cells to create a net flux of ions across the cell layer. It is generally accepted that this uneven distribution of ion transporters on the cell membrane generates an ion flux directed from the stroma into the PC [72].

## 2.2 Mechanisms of Aqueous Production

Aqueous production is a complex phenomenon that involves various different physical mechanisms, which are described below.

*Mechanical pressure*   A standing pressure drop of approximately 15 mmHg exists between the stroma and the AH humour in the PC. This mechanical pressure jump across the double cell layer of the ciliary epithelium originates from the blood pressure in the capillaries and tends to pump fluid into the PC.

*Oncotic pressure*    As mentioned in the previous section, the tight junctions connecting the non-pigmented ciliary epithelial cells act as a sieve for large molecules, which cannot flow across the double cell layer into the PC. As a result of this, the stroma has a higher protein concentration than the AH (which is almost protein free). This establishes an oncotic pressure difference across the ciliary epithelium, with the oncotic pressure being higher on the stromal side. This, in turn, drives water flux, which is directed from the PC back into the stroma, and thus is opposite to that produced by mechanical pressure.

*Osmotic pressure*    Ions are actively transported across the ciliary epithelium. When a cell layer transports ions, this typically establishes an osmotic pressure difference across the layer due to different ion concentrations on the two sides. Many well-documented examples are present in the literature, see, for instance, [78] for transporting the Necturus gallbladder, [42] in the proximal tubules and [19] in the retinal pigment epithelium. The typical picture is that "water follows the ions", i.e. water motion occurs in the main direction of ion transport. This is also the case for the ciliary epithelium: the osmotic pressure difference is such as to drive water flux from the stroma into the PC. We note that this is not a "passive flow" in the sense that it requires energy expenditure for the "active" ion transport.

The general view [15] is that the first two mechanisms approximately balance each other (they approximately have the same magnitude and opposite sign) and, therefore, active ion transport across the ciliary epithelium plays a major role in AH production.

## 2.3   *Mathematical Models of Aqueous Production*

The actual geometry of ciliary epithelium is very complex and, therefore, the most obvious approach to model the problem of aqueous production is to treat the whole epithelium as a semipermeable membrane. Most of the mathematical models of AH production are based on this approach [39, 47, 70]. The volumetric flux per unit surface through the ciliary epithelium $F$ (positive if the flux is from the stroma to the PC) can be written as follows:

$$F = L_p \left( \Delta p - \sigma_p \Delta \Pi_p - \sigma_s \Delta \Pi_s \right), \tag{1}$$

where $L_p$ is the membrane hydrodynamic conductivity (measured in the international system in $m^2 s/kg$), $p$ denotes the mechanical pressure, $\Pi_p$ the oncotic pressure and $\Pi_s$ the osmotic one. The operator $\Delta$ denotes the difference between the value of the quantity it is applied to in the stroma and in the AH. Finally, $\sigma_p$ and $\sigma_s$ are the reflection coefficients for proteins and low-molecular components. We recall that the osmotic pressure can be related to the concentration of a species through van't Hoff's law [58]

$$\Pi = RTc, \tag{2}$$

where $R$ is the universal gas constant (8.314 J/mol K), $T$ is the absolute temperature and $c$ the molar concentration of the considered species.

In the expression (1), all mechanisms described in the above section are accounted for. In particular, the first term represents the effect of a mechanical pressure drop. When $\Delta p > 0$, which is normally the case in the ciliary processes, this term describes fluid flux from the stroma to the PC, proportional to the pressure jump. The second term accounts for the oncotic pressure difference (osmotic effect due to proteins). Since large molecules cannot pass through the tight junctions between non-pigmented epithelial cells, the reflection coefficient $\sigma_p$ is close to 1. The concentration of blood proteins is larger in the stroma than in the AH, which implies $\Delta \Pi_p > 0$. Thus, the second term in (1) provides a negative contribution to $F$. Finally, the last term in (1) accounts for the osmotic flux due to a jump in ion concentration across the epithelium. This term is typically positive and contributes to increase fluid flux towards the PC. Estimates of the values of the various parameters that appear in Eq. (1) are provided in [47] and [70].

Kiel et al. [39] improved the above simple model by also accounting for the role of ciliary blood flow on AH production. The authors propose a lumped parameter (0-dimensional) mathematical model that couples blood flow in the choroid and ciliary processes (including autoregulation mechanisms), oxygen delivery to and consumption by ocular tissues and AH production (which is treated in a similar way as discussed above). Their results, which are in good agreement with experimental observations, suggest that there is an interplay between ciliary blood flow and AH production. In particular, AH production grows significantly with increasing ciliary blood flow at low values of perfusion and progressively becomes blood flow independent at high perfusion rates.

The above models, based on a description of the ciliary epithelium as a semipermeable membrane, cannot capture the details of AH production at the level of a single cell. Many works in the field of water transport across epithelial and endothelial layers have shown that, in particular, the cleft gap between adjacent cells can play a significant role. The underlying physical mechanism was very clearly explained by Diamond and Bossert [16], who proposed a mathematical model of what is known as standing gradient osmotic flow. The idea behind this mechanism is explained in Fig. 3, where cells are schematised as squares separated by a gap (the

**Fig. 3** Diagram of the standing gradient system. Blue arrows indicate ion fluxes and black arrows water motion

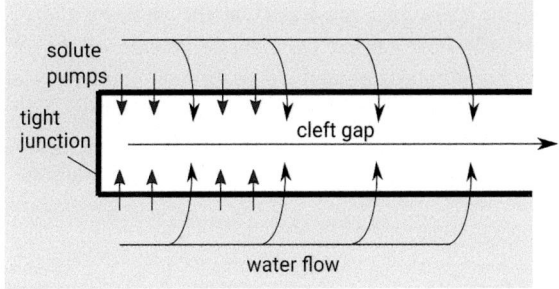

cleft gap). On one side, we have a tight junction and, on the other, the cleft opens into the PC. Ions are actively pumped into the cleft owing to the presence of Na/K-ATPase channels on the basolateral membrane of the cell. This creates an excess ion concentration in the cleft, which drives water flux from the cell into the cleft. Ions in turn are transported along the cleft to the PC by advection and electrodiffusion. This effectively couples fluid and ion transport. The problem was solved by Diamond and Bossert [16], who took advantage of the slender shape of the cleft to reduce the governing system of PDEs to a set of ODEs. A similar approach was adopted more recently by Avtar et al. [4], who specifically considered the problem of AH production by the ciliary epithelium. We note that the standing gradient osmotic flow hypothesis can explain also the occurrence of isotonic transport, i.e. in the absence of transepithelial jump in osmolarity.

Mathematical modelling can also help understanding more fundamental aspects of the process of AH production. A recent example is the study by Mauri et al. [48], who theoretically investigated the role of the bicarbonate ion ($HCO_3^-$) on the transepithelial potential difference and the sodium potassium pump (Na/K-ATPase) in the non-pigmented epithelial cell in the ciliary processes, in connection with their role in the active secretion of AH. The model suggests that $HCO_3^-$ inhibition may prevent physiologically correct baseline values of the non-pigmented transepithelial potential difference and Na/K-ATPase function.

# 3   Aqueous Humour Flow

In this section, we will outline mathematical models of aqueous flow produced by different mechanisms. We will first introduce lubrication theory, an approximation technique utilised for modelling flows in thin domains and then apply it to describe the flow in the AC or PC.

## 3.1   General Characteristics of Fluid Flow in the Anterior and Posterior Chambers

Flow of AH through the AC and PC is driven by various mechanisms, in particular: (1) a pressure drop between the ciliary processes where it is produced (see Sect. 2) and the trabecular meshwork at the angle of the eye, leading to pressure-driven flow; (2) flow caused by a shape change of the AC and/or PC, such as due to lens accommodation and iris motion; (3) buoyancy-driven flow due to temperature differences between the anterior and posterior of the AC (only when eyelids are open and subject is in an environment that is significantly below (or above) body temperature); (4) AH motion caused by eye rotations.

As will be described in Sect. 3.1.1, both the AC and PC are relatively shallow in the antero–posterior direction, compared with their typical widths in the superior–inferior and lateral directions. Denoting a characteristic length in the anterior–posterior direction as $H$ and that in the orthogonal directions as $L$, we define an aspect ratio of the domain $\epsilon = H/L \ll 1$. The Reynolds number of the flow is given by $Re = UL/\nu$, where $U$ is characteristic velocity and $\nu$ kinematic viscosity, which is typically small enough, indicating that there is no turbulence; in addition, $\epsilon^2 Re$ is also small, meaning that the flow of AH can be effectively described via the lubrication approximation of the governing nonlinear Navier–Stokes equations, and this is described in detail in Sect. 3.1.2. Since the resulting system of lubrication equations governing the motion is linear, the different flow mechanisms can be studied separately and then superposed to find their effects in combination. We consider each of the four mechanisms (1)–(4) listed above in turn in Sects. 3.2–3.5.

### 3.1.1 Geometrical Properties of the Anterior and Posterior Chambers Based on Measurements

In Fig. 4 and Table 1, we report the characteristic dimensions of the AC and PC. The AC can be visualised very accurately using high-resolution optical imaging techniques such as OCT, and thus reliable measurements are available for its geometry. There are significant differences in the geometry between subjects, and also between races, particularly in the height of the AC. The values given in Table 1 are taken from the standard ISO-11979-3 [30]; note that the radius of the lens varies during lens accommodation.

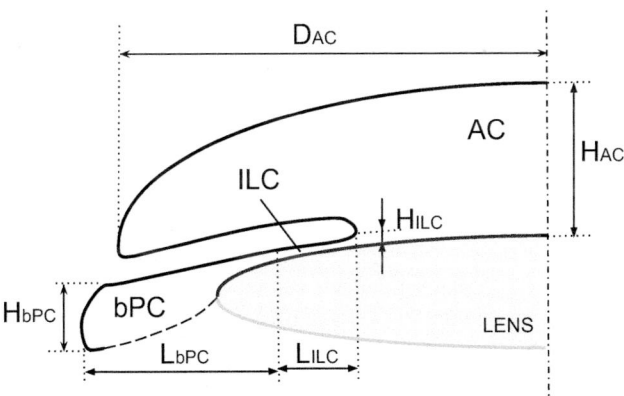

**Fig. 4** Characteristic dimensions of the AC and PC. Non-permeable walls are shown by solid lines and a fictitious wall delimiting the bPC is given by a dashed line. The grey zone indicates the crystalline lens and the whole geometry is symmetric around the dashed-dotted line. The meaning and values of the characteristic dimensions are reported in Table 1, with $D_{\mathrm{AC}} = L_{\mathrm{AC}}/2$

Visualising the PC is more challenging owing to its small size and to the fact that optical measurements cannot be taken behind the iris. The values reported in Table 1 are taken from [20], who analysed an image of the PC obtained using a high-frequency ultrasound scan on a healthy human subject. The scan had a resolution of 20–30 μm (axial–lateral), which was sufficient to measure the geometrical characteristics away from the pupil, but not the thickness of the iris–lens channel (ILC), the region where the iris gets very close to the lens, which is extremely thin. The height of the ILC reported in the table is thus not directly measured but it is taken from estimates by Silver and Quigley [68]. Following [20], it is convenient for the arguments that follow to consider the PC as subdivided into two regions: the ILC (close to the pupil) and the rest of the domain that we denote as "bulk" posterior chamber (bPC).

We evaluate the aspect ratio of each region of the domain as $\epsilon_i = \bar{H}_i/L_i$, with $i = $ AC, bPC, ILC, and where $\bar{H}_i$ is an estimate of the average depth of region $i$ and $L_i$ the corresponding length. With the values from Table 1, we obtain the following estimates: $\epsilon_{AC} = 0.1$, $\epsilon_{bPC} = 0.17$ and $\epsilon_{ILC} = 0.01$.

### 3.1.2 Lubrication Approximation for Flow in a Thin Domain

The analysis here considers flow in a thin domain, with characteristic thickness $H$ and length $L$, such as the AC and PC, with a small aspect ratio $\epsilon = H/L$. In what follows, $x$ and $y$ denote the streamwise and out of the plane coordinates, while $z$ is the direction of the depth, normal to the $x-y$ plane. The corresponding velocity vector is $\mathbf{u} = (u, v, w)$ and $p$ is the pressure. AH flow is governed by Navier–Stokes equations for incompressible fluid

$$\nabla \cdot \mathbf{u} = 0, \tag{3}$$

$$\frac{\partial \mathbf{u}}{\partial t} + (\mathbf{u} \cdot \nabla)\mathbf{u} + \frac{1}{\rho}\nabla p = \nu\nabla^2\mathbf{u}. \tag{4}$$

For the sake of discussion, we introduce the following scales and non-dimensional variables, denoted by primes,

$$t = Tt', \ x = Lx', \ y = Ly', \ z = Hz', \ u = Uu', \ v = Uv', \ w = \epsilon Uw', \ p = Pp', \tag{5}$$

where $U$, $T$ and $P$ are characteristic streamwise velocity, time and pressure scales, respectively. Note that the depth coordinate $z$ is scaled by $H$, while the other two directions are scaled by $L$. The continuity equation suggests to scale $w$ with $\epsilon U$.

The dimensionless continuity equation (3) can then be written

$$\frac{\partial u'}{\partial x'} + \frac{\partial v'}{\partial y'} + \frac{\partial w'}{\partial z'} = 0, \tag{6}$$

with all terms being equally important. The momentum equation (4) in the $x$-direction, for an incompressible fluid, in dimensionless form can now be written using (5) as

$$\frac{U}{T}\frac{\partial u'}{\partial t'} + \frac{U^2}{L}\left(u'\frac{\partial u'}{\partial x'} + v'\frac{\partial u'}{\partial y'} + w'\frac{\partial u'}{\partial z'}\right) + \frac{P}{\rho L}\frac{\partial p'}{\partial x'}$$
$$= \frac{\nu U}{H^2}\left(\frac{H^2}{L^2}\frac{\partial^2 u'}{\partial x'^2} + \frac{H^2}{L^2}\frac{\partial^2 u'}{\partial y'^2} + \frac{\partial^2 u'}{\partial z'^2}.\right) \tag{7}$$

Dividing Eq. (7) by $\nu U/H^2$, and using the definitions of $\epsilon$ and $Re$, we obtain

$$\frac{H^2}{\nu T}\frac{\partial u'}{\partial t'} + \epsilon^2 Re\left(u'\frac{\partial u'}{\partial x'} + v'\frac{\partial u'}{\partial y'} + w'\frac{\partial u'}{\partial z'}\right) + \frac{H^2 P}{\mu U L}\frac{\partial p'}{\partial x'} = \epsilon^2\left(\frac{\partial^2 u'}{\partial x'^2} + \frac{\partial^2 u'}{\partial y'^2}\right) + \frac{\partial^2 u'}{\partial z'^2}, \tag{8}$$

where $\mu$ is dynamic viscosity, $\mu = \nu\rho$. In the absence of any forced unsteady motion of the surrounding surfaces, the natural time scale is $T = L/U$. Consequently, the first term on the left-hand side of Eq. (8) is also multiplied by $\epsilon^2 Re$, which is normally referred to as reduced Reynolds number. Omitting terms of order $\epsilon^2$ and $\epsilon^2 Re$ or smaller, and balancing the pressure gradient with the dominant viscous term, Eq. (8) can be approximated to leading order as

$$\frac{\partial p'}{\partial x'} = \frac{\partial^2 u'}{\partial z'^2}, \tag{9}$$

where we have set $P = \mu LU/H^2$. It is important to note that, differently from what happens in the case of low Reynolds number Stokes flows, in the context of lubrication theory, inertial terms are subdominant even if the Reynolds number is of order one.

The approximation of the momentum equation in the $y$-direction gives a balance between the pressure and viscous terms, similar to Eq. (9). If the considered scaling is introduced in the momentum equation (4) in the $z$-direction, we get

$$\epsilon^4 Re\frac{\partial w'}{\partial t'} + \epsilon^4 Re\left(u'\frac{\partial w'}{\partial x'} + v'\frac{\partial w'}{\partial y'} + w'\frac{\partial w'}{\partial z'}\right) + \frac{\partial p'}{\partial z'} = \epsilon^4\frac{\partial^2 w'}{\partial x'^2} + \epsilon^4\frac{\partial^2 w'}{\partial y'^2} + \epsilon^2\frac{\partial^2 w'}{\partial z'^2}. \tag{10}$$

Clearly, at leading order one obtains

$$\frac{\partial p'}{\partial z'} = 0. \tag{11}$$

To summarise, the lubrication approximation for thin layers in dimensional form leads to the following linear system:

$$\frac{\partial p}{\partial x} - \mu \frac{\partial^2 u}{\partial z^2} = 0, \tag{12a}$$

$$\frac{\partial p}{\partial y} - \mu \frac{\partial^2 v}{\partial z^2} = 0, \tag{12b}$$

$$\frac{\partial p}{\partial z} = 0, \tag{12c}$$

$$\nabla \cdot \mathbf{u} = 0. \tag{12d}$$

The system of equations (12) is accompanied by adequate boundary conditions. For applications such as the AC and PC, it is reasonable to impose no-slip conditions on the velocity components on non-permeable surfaces.

In the following sections, lubrication theory is applied in spherical and cylindrical coordinates. Although equations (12) are different in other coordinate systems, the simplification procedure effectively follows the same steps.

## 3.2 Pressure-Driven Flow

In this section, we focus on the characteristics of the flow in the PC and AC, neglecting the problem of drainage that will be considered later, in Sect. 4, and imposing a fixed value of the pressure at the TM. Moreover, AH inflow will be treated as a boundary condition at the ciliary processes, where a given incoming flow rate will be prescribed. The flow considered in this section is thus driven by a pressure drop between the ciliary processes and the TM.

The aim of this section is to investigate how the pressure varies from the PC to the AC. This is relevant because the pressure drop between the two chambers pushes forward the iris, which is a compliant tissue. In cases when the pressure difference between PC and AC is too large the iris can be displayed forward so much as to close the angle it forms with the cornea, where the TM is located. In turn, this produces an increase of AH outflow resistance (possibly substantially) and can lead to angle-closure glaucoma, a very serious and acute condition that requires immediate surgical treatment. Angle-closure glaucoma occurs in about 0.5% of white and black people and about 1.5% of Chinese and Indian individuals over the age of 40 [26]. Even though angle-closure glaucoma is less prevalent than open-angle glaucoma, it may be responsible for as much blindness worldwide [60].

Since the AC is thicker than the PC (see Table 1) most of the resistance to AH flow from the ciliary processes to the TM, and therefore most of the pressure drop, occurs in correspondence of the PC and, in particular, of the ILC. The resistance to AH flow can increase substantially in the case of pupillary block, which is a condition occurring when the iris and lens come in contact to each other. Pupillary block can be complete or partial, depending on whether the whole pupil is blocked or only part of it [46].

For the above reasons, we shall focus in the section on the flow in the PC and assume that the pressure in the AC is constant. The validity of this assumption is corroborated by numerical simulations of the flow in the AC by Repetto et al. [62], who showed that, in normal conditions, the pressure drop across the AC is as small as $7 \times 10^{-7}$ mmHg.

The first analytical model of AH flow in the PC was proposed by Friedland [25]. The model was based on the assumption that the PC has a constant depth. A similar model was also developed by Silver and Quigley [68], who, however, just considered the flow through the ILC (still assumed to be a gap of constant height). In other words, the authors assumed that AH pressure in the PC before reaching the ILC was constant. The authors' model shows that the pressure drop between the PC and AC is highly dependent on the thickness and (to a lesser extent) the length of the ILC.

In the following, we present a simplified version of the model proposed by Dvoriashyna et al. [20], which extends the above works by accounting for a realistic shape of the PC. In order to describe the geometry of the PC, the authors adopt a system of spherical coordinates $(r, \theta, \phi)$, as shown in Fig. 5a. The axis $\theta = 0, \pi$ is an antero-posterior axis, passing through the centre of the pupil, and the PC is assumed to be an axisymmetric domain. The lens is assumed to be a spherical surface with radius $R$, described by the equation $r = R$. The depth of the PC is described by a function $h(\theta)$ that is obtained from an ultrasound scan image of a human eye (see Table 1 for the characteristic dimensions of the domain). As anticipated in Sect. 3.1, the PC is divided into two regions, bPC ($\theta_{pc} \leq \theta < \theta_{ch}$) and ILC ($\theta_{ch} \leq \theta \leq \theta_p$), which are smoothly connected to each other, see Fig. 5b.

Since, as discussed in Sect. 3.1, both the bPC and the ILC regions are thin (the corresponding aspect ratio $\epsilon$ being significantly smaller than 1), lubrication theory can be adopted. Owing to the axial symmetry of the domain, and in the absence of pupillary block, the flow is also assumed to be axisymmetric. Moreover, the

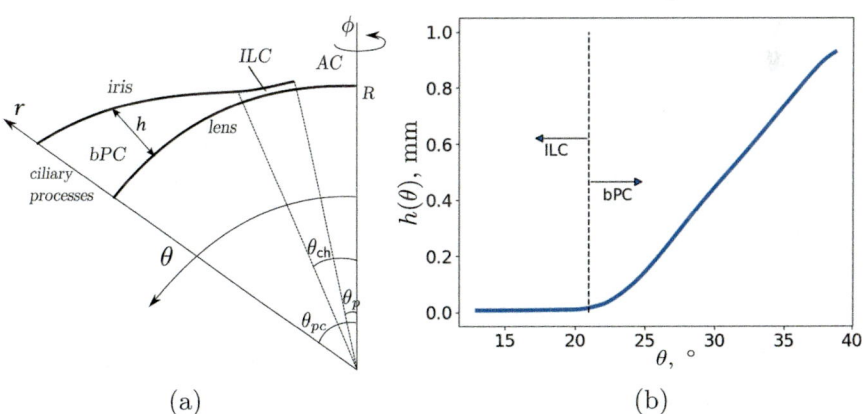

(a)                                                            (b)

**Fig. 5** (a) Sketch of the notation and coordinate system. (b) Function $h(\theta)$ that describes the height of the PC

flow is considered steady, being induced by a steady pressure gradient. Thus, the velocity vector can be written as $\mathbf{u} = (u_r(r, \theta), u_\theta(r, \theta), 0)$, with $u_r$ and $u_\theta$ denoting the radial and zenithal velocity component, respectively. Moreover, in agreement with the standard lubrication theory approach, the pressure $p$ is only function of $\theta$, being constant throughout the thickness of the gap in the limit of thin film flow, see Eq. (12c). Following a procedure analogous to that described in Sect. 3.1 (but now adopting spherical coordinates), the dimensional Navier–Stokes equation in the $\theta$-direction and the continuity equation simplify to

$$\frac{1}{r}\frac{dp}{d\theta} = \mu\frac{1}{r^2}\frac{\partial}{\partial r}\left(r^2\frac{\partial u_\theta}{\partial r}\right), \tag{13a}$$

$$\frac{1}{r\sin\theta}\frac{\partial(\sin\theta\,u_\theta)}{\partial\theta} + \frac{1}{r^2}\frac{\partial(r^2 u_r)}{\partial r} = 0, \tag{13b}$$

The above equations have to be solved subjected to no-slip conditions at the boundaries of the domain ($r = R$ and $r = R + h$). A fixed AH flux $Q$ is assumed to enter the domain at the ciliary body, which leads to the following condition (the derivation is based on the use of Eq. (16)):

$$\frac{dp}{d\theta} = \frac{6\mu Q}{\pi h^3 \sin\theta} \qquad \text{at } \theta = \theta_{pc}. \tag{14}$$

Finally, in correspondence of the pupil, a given value of the pressure is assumed (taken equal to zero without loss of generality)

$$p = 0 \qquad \text{at } \theta = \theta_p. \tag{15}$$

By integrating Eq. (13a) twice with respect to $r$ and imposing no-slip conditions at the lens and the iris, one obtains the following expression for the velocity:

$$u_\theta = \frac{1}{2\mu r}\frac{dp}{d\theta}\left(r^2 + R(R + h) - r(2R + h)\right), \tag{16}$$

where the pressure still has to be determined. This is done by integrating the continuity equation (13b) with respect to $r$ from $r = R$ to $r = R + h$, and imposing again the no-slip condition, which leads to the following equation for $p$:

$$\frac{d}{d\theta}\left(h^3 \sin\theta\frac{dp}{d\theta}\right) = 0. \tag{17}$$

The above equation has to be solved subjected to the conditions (14) and (15).

The solution of Eq. (17) is readily found numerically. An analytical solution can also be found for specific values of the function $h(\theta)$ and, in particular, for the case $h = \text{const}$. In this case, the solution for the pressure reads

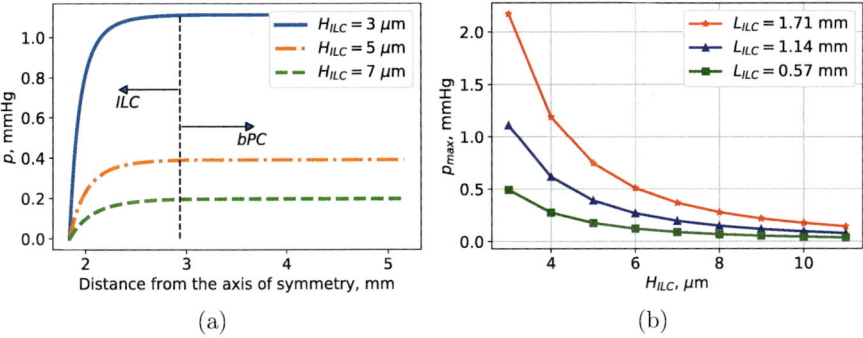

**Fig. 6** (**a**) Pressure distribution along the PC. The ciliary processes are on the right of the plot, the pupil on the left. (**b**) Maximum value of the pressure attained in the PC as a function of the height of the ILC, for different values of the ILC length

$$p = \frac{6\mu Q}{\pi h^3} \ln \left( \frac{\tan \theta/2}{\tan \theta_p/2} \right). \tag{18}$$

In Fig. 6, the results of the model are shown. The plots are obtained adopting the shape function $h(\theta)$ shown in Fig. 5b. In the left panel, the pressure distribution along the PC, from the ciliary processes (right) to the pupil (left), is reported for different values of the ILC height. It appears that most of the pressure drop occurs in the ILC and the pressure experiences little variation in the bPC. This confirms the validity of the approach proposed by Silver and Quigley [68], who neglected pressure variation in the bulk of the PC. In Fig. 6b, the maximum value of the pressure in the bPC (i.e. in correspondence of the ciliary processes) is plotted vs the height of the ILC, for three different values of the ILC length. From inspection of both plots, it clearly appears that the height of the ILC plays a fundamental role and that the pressure drop from PC to AC can reach relatively high values if the ILC is very thin.

The model described above can be easily extended to study the effect of a partial pupillary block. In this case, it is assumed that outflow from the PC to the AC can occur only through a certain region of the pupil. Clearly, this breaks the axial symmetry of the flow and the equations described above need to be suitably modified. The approach based on the lubrication theory, however, remains valid and the solution procedure is identical. For the non-axisymmetric case, the equation for the pressure (the non-axisymmetric analogue of Eq. (17)) reads

$$\frac{\partial}{\partial \theta} \left( \sin \theta h^3 \frac{\partial p}{\partial \theta} \right) + \frac{h^3}{\sin \theta} \frac{\partial^2 p}{\partial \phi^2} = 0. \tag{19}$$

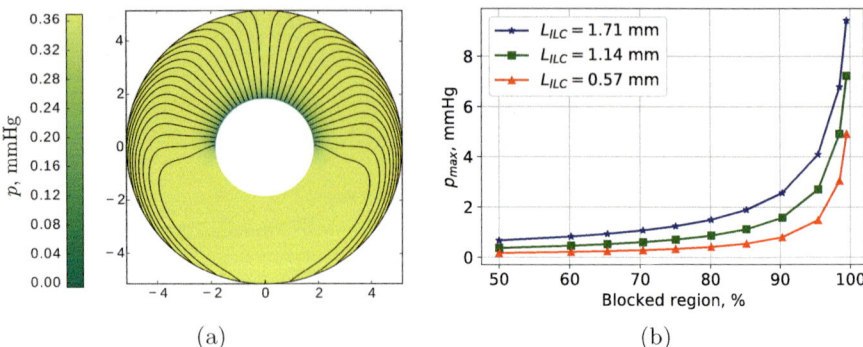

**Fig. 7** (**a**) Pressure distribution (shading) and streamlines of the depth-averaged velocity in the PC in the case of partial pupillary block. The lower half of the pupil blocked. (**b**) Maximum pressure in the PC as a function of the percentage of blocked region of the pupil for different lengths of iris–lens channel

In order to model the existence of a partial pupillary block the boundary condition (15) has to be modified, to account for the fact that the outflow does not take place over the whole length of the pupil, and can be written as

$$p = 0 \text{ in the open region,} \tag{20a}$$

$$\frac{\partial p}{\partial \theta} = 0 \text{ in the blocked region.} \tag{20b}$$

Equation (19) with the boundary conditions (20a,b) can be solved numerically and some results are shown in Fig. 7. Figure 7a is relative to the case in which half of the pupil is blocked (the lower part in the figure). Colours represent the pressure and in black streamlines of the depth-averaged velocity are also shown. The break of the axial symmetry is evident: a fluid particle entering the domain from the bottom part has to go all the way round to exit from the upper part of the pupil. This obviously implies that the resistance to flow increases with respect to the normal case as it appears from Fig. 7b. In the figure, the maximum pressure in the PC is shown as a function of the degree of closure of the pupil (percentage of blocked region). The pressure grows to infinity as the blockage tends to be complete, as there is no outflow in this limit. However, the figure shows that for the pressure in the PC to grow to relatively large values the percentage of blocked pupil has to be quite large.

A more comprehensive, fully numerical, model of the flow in the PC and AC was proposed by Heys et al. [32], who combined AH flow with passive iris deformation. In this fluid structure interaction model, the authors treated the AH as a Newtonian fluid and the iris as an incompressible linear elastic solid. Their results are in general agreement with those described above. In particular, the model predicts that in normal conditions the minimum distance between iris and lens (i.e. the minimum depth of the ILC) is 4.4 μm, consistent with the values adopted in Fig. 6. Moreover, the pressure drop between the PC and the AC in normal conditions is predicted to be 0.23 mmHg, which also agrees well with the values reported in Fig. 6. The authors' model shows that decreasing the pupil diameter the pressure drop between PC and AC increases, the thickness of the ILC also slightly increases and its length increases. The predictions of the model by Heys et al. [32] of iris contour and IOP in normal eyes are also consistent with clinical observation.

## 3.3    Flow Induced by Lens Accommodation and Iris Movement

Iris motion during miosis (pupil contraction) and mydriasis (pupil dilation) and also lens accommodation can cause transient flows from the PC to the AC. We will focus here on iris motion but a similar approach as the one described below would also apply to the case of flow generated by lens accommodation.

Milton and Longtin [52] report that pupil constriction typically lasts less than 1 s, whereas pupil dilation takes a couple of seconds. This implies that velocities and pressure generated during miosis are larger than during mydriasis and thus only miosis will be considered in the following. Over the time scale of a second, it is reasonable to assume that the total volume of AH contained in the AC and PC remains constant. Therefore, flow from one chamber to the other can occur if the ratio between the volumes of the two regions changes. Dorairaj et al. [17] performed experiments in which they estimated the PC volume from ultrasound images before and after pupil diameter changes. They report that, in the case of pupil dilation, changes in the PC volume $\Delta V$ during iris motion are small (in the order of some percent of the total PC volume $V$).

In order to study fluid flow generated by miosis, we follow here the analysis proposed by Dvoriashyna et al. [20]. In agreement with the observations by Dorairaj et al. [17], the authors assumed the $\Delta V/V \ll 1$, with $V$ denoting the volume of the PC before miosis. The duration of pupil contraction is denoted by $T$. With $\Delta V = 0.05\,V$ and $T = 1$ s, a flux of 106 µl/min from the PC to the AC is predicted, which is significantly larger than the flux due to production/drainage, described in the previous section (3 µl/min). The corresponding values of the Reynolds number in the bPC and ILC are $Re_{bPC} \approx 0.51$ and $Re_{ILC} \approx 12$, which means that $\epsilon_{bPC}^2 Re_{bPC} \approx 1.5 \times 10^{-2}$ and $\epsilon_{ILC}^2 Re_{ILC} \approx 1.2 \times 10^{-3}$. Thus, lubrication theory applies also in this case and, notably, the time derivative in the equations of motion can still be neglected, even in the present case of an unsteady flow. This means that the velocity field at a given time is fully determined by the iris velocity at that particular time, which we denote by $\mathbf{v}(\theta) = \big(v_r(\theta), v_\theta(\theta), 0\big)$ and which is assumed to be axisymmetric.

We follow the same approach and adopt the same notation as in Sect. 3.2, but we now have to account for the fact that the iris moves. Thus the boundary condition at $r = R + h$ has to be modified to $\mathbf{u}(r, \theta) = \mathbf{v}(r, \theta)$, with $\mathbf{v}$ denoting the iris velocity at time $t$ that needs to be specified.

With the use of this new boundary condition, Eq. (16) takes the following modified form:

$$u_\theta = \frac{1}{2\mu r}\frac{dp}{d\theta}\left(r^2 + R(R+h) - r(2R+h)\right) + v_\theta\frac{R+h}{h}\left(1 - \frac{R}{r}\right). \qquad (21)$$

Moreover, the equation for the pressure (17) now reads

$$\frac{1}{12\mu\sin\theta}\frac{d}{d\theta}\left(h^3\sin\theta\frac{dp}{d\theta}\right) = (R+h)^2 v_r + \frac{h}{2\sin\theta}\frac{d}{d\theta}\left((R+h)\sin\theta v_\theta\right) - \frac{R+h}{2}\frac{dh}{d\theta}v_\theta$$
$$(22)$$

In order to close the problem, the velocity of the iris $\mathbf{v}$ has to be specified. Dvoriashyna et al. [20] have proposed a method to estimate the functions $v_r(\theta)$ and $v_\theta(\theta)$, which we do not report in detail here for the sake of saving space. However, we note that one of the key assumptions made by the authors is that the minimum distance between the iris and the lens remains constant during iris motion. In the writers view, the only possible alternative approach, which would allow one to overcome the need of assuming how the thickness of the ILC changes during the transient flow generated by miosis, would be to develop a fluid structure interaction model of iris motion during miosis. This was done by Huang and Barocas [33], who developed an axisymmetric model of the flow of AH in the PC and AC, with the iris modelled as an elastic solid with additional active elastic terms. However, the authors' model does not account for the dynamical behaviour of the system during miosis and only considers steady-state solutions.

Equation (22) can be easily solved numerically and some results are shown in Fig. 8, obtained using the function $\mathbf{v}(\theta)$ proposed by Dvoriashyna et al. [20]. In the figure, the maximum pressure in the PC during miosis is plotted as a function of the imposed ILC depth for different values of volume changes in the PC. Comparing this figure with Fig. 6b, which is analogous but obtained for the case of the production/drainage flow, it appears that miosis can generate much larger pressures in the PC. We note that these large values of the pressure last for a very short time. Moreover, the model is likely to overestimate the maximum value of the pressure in the PC during miosis, since it is based on the assumption that the height of the iris–lens channel remains unchanged.

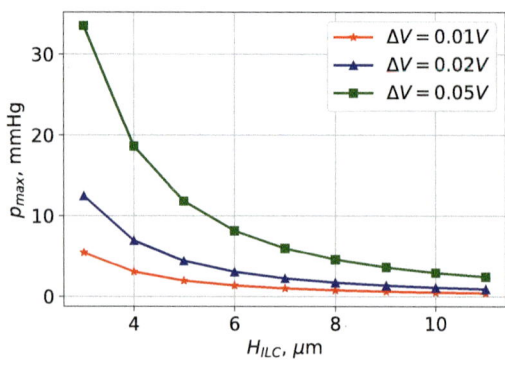

**Fig. 8** Maximum value of the pressure attained in the PC during miosis as a function of the height of the ILC. The three curves correspond to different changes of volume in the PC, $\Delta V$, after miosis

## 3.4   Flow Induced by Thermal Effects

Fluid flow in the anterior chamber was observed clinically already a century ago and it was first suggested by Türk [74] that this might result from the temperature gradient across the anterior chamber. It is by now known [13, 24, 31, 62] that the flow appears in a single convection cell, rising (opposing gravity) in the back of the AC and falling in the front (along the cornea). Moreover, the lateral movement, in the nasal-temporal direction, is weak. The temperature gradient exists since the temperature in the back of the AC is close to body temperature (37°) while the cornea is exposed to a surrounding condition (for instance, 25° during the day when the eyelids are open). The temperature difference inside the AC is however not that large. The thickness of the cornea is less than 1 mm and its thermal properties close to that of water. It is therefore likely that the temperature on the inside of the cornea is close to body temperature. Nevertheless, what will be shown in this section is that even small temperature differences within the AC cause fluid motion with maximum velocities larger than those created by the pressure-driven flow (see Sect. 3.2).

Aqueous motion in the AC induced by thermal effects is important as it is one of the primary mechanisms inducing fluid mixing, at least during daytime. One of the key roles of AH is to deliver nutrients to the avascular tissues of lens and cornea. The existence of relatively effective mixing processes in the AC avoids the generation of local regions of nutrient depletion in the AH, which in turn would decrease the efficiency of transport to the target tissues.

Moreover, the thermal flow in the AC being more intense than that due to production and drainage, also has an important role when blood cells are present in the AC. Such cells tend to settle in the bottom part of the AC when standing in upright position, a condition known as hyphema [65]. The thermally induced flow is likely to have a role in the resuspension of such particles and eventually help their clearance from the AC. As discussed in [13] this concerns red (erythrocytes) and white (leukocytes) blood cells as well as pigment particles. The presence of white blood cells (with size ranging from 6 to 20 μm) normally indicates that the patient is suffering from uveitis which might lead to damage within the eye and sight loss. It is known that most of the pigment cells present in the AC originate from the back of the iris and enter through the pupil aperture. Accumulation of pigment cells might occur on the inner surface of the cornea forming a so-called Krukenberg spindle [43].

For the above reasons, it is important to study the flow patterns, velocity magnitude and corresponding wall shear stresses in the AC induced by thermal effects.

### 3.4.1   An Analytical Model for Thermally Driven Flow in the Anterior Chamber

In this section, an analytical solution, based on the work by Canning et al. [13], of the fluid flow due to thermal effects is derived and discussed. A schematic drawing

There are three governing equations, describing the mechanics of the wall, mass conservation and momentum conservation. The governing equations are made dimensionless in the following way. The lengths along the channel are non-dimensionalised with respect to $s$, the half-distance between neighbouring channels, the height of the channel with respect to $h_0$, the difference between the pressure in the canal and that in the collecting channels ($P(x) - P_{cc}$) with respect to the difference between the intraocular pressure and the pressure in the collecting channels, $IOP - P_{cc}$. The velocity scale is found using scalings appropriate to low-Reynolds-number flows, and balance the pressure gradient and viscous terms. We assume that Schlemm's canal is long and thin, that is $s \gg h_0$, which is the main assumption of lubrication theory. This means that the $x$- and $y$-velocity components $u$ and $v$ are expected to scale in proportion to the aspect ratio, so the scale of $v$ is $h_0/s$ times that of $u$ (thus $v$ is much smaller than $u$). Using the aforementioned scaling in the $x$-component of the Navier–Stokes equations, we obtain the velocity scales as $h_0^2(IOP - P_{cc})/(s\mu)$ and $h_0^3(IOP - P_{cc})/(s^2\mu)$, in the $x$- and $y$-directions, respectively. Moreover, the flux $Q$ along Schlemm's canal scales as the cross-sectional area times the velocity scale in the $x$-direction.

Using the linear spring equation, the height of the canal can be related to the pressure drop across the inner wall as

$$\frac{h_0 - h(x)}{h_0} = \frac{IOP - P(x)}{E}, \tag{38}$$

where $P(x)$ is the pressure inside Schlemm's canal. Fluid flows through the wall of Schlemm's canal at a rate equal to the pressure difference across the wall, $IOP-P$, multiplied by the conductance per unit length, $1/R_w$, where $R_w$ is the wall resistance. Mass conservation is expressed by the following equation:

$$\frac{dQ}{dx} = \frac{IOP - P}{R_w}. \tag{39}$$

Using lubrication theory, the following dimensionless system is obtained:

$$1 - h' = \frac{IOP - P_{cc}}{E}(1 - P'), \tag{40a}$$

$$\frac{dQ'}{dx'} = \frac{s^2\mu}{h_0^3 w R_w}(1 - P'), \tag{40b}$$

$$Q' = \frac{-h'^3}{12}\frac{dP'}{dx'}, \tag{40c}$$

where $\{\}'$ denotes dimensionless variables. Combining Eqs. (40a)–(40c), we obtain an ordinary differential equation for $h'$

$$\frac{d}{dx'}\left(h'^3\frac{dh'}{dx'}\right) + \beta^2(1 - h') = 0, \tag{41}$$

where

$$\beta = \sqrt{\frac{12s^2\mu}{h_0^3 w R_w}}, \tag{42}$$

is the ratio between the resistance in the undeformed canal and the resistance of the inner wall. Equation (41) is solved imposing the following two boundary conditions:

$$\frac{dh'}{dx'} = 0 \quad \text{at} \quad x' = 0, \text{ and} \quad h' = 1 - \mathcal{P} \quad \text{at} \quad x' = 1, \tag{43}$$

where $\mathcal{P} = (\text{IOP} - P_{cc})/E$. With the first condition in Eq. (43), we demand that the solution is symmetric. This means that $h'$ must be an even function about $x' = 0$. In the second condition, we assume that the pressure in the collecting channels equals that in Schlemm's canal at $x' = 1$ which means that we know the displacement of the inner wall at that point.

Results obtained by Johnson and Kamm [37] solving Eq. (41) show that the flow rate increases as the pressure drop $\mathcal{P}$ increases. For small pressure drops, $Q$ is a linear function of $\mathcal{P}$ indicating that the resistance to flow is approximately constant, which, in turn, means that the majority of the resistance comes from traversing the inner wall of Schlemm's canal. When the pressure drop is large, the resistance to flow in the canal increases, since the canal becomes narrower. Without any further considerations, the model predicts total collapse of Schlemm's canal when $\mathcal{P}$ reaches 1. However, Schlemm's canal has small protrusions called septae from its outer wall, which prevent complete collapse because they support the wall. Johnson and Kamm [37] incorporated this effect into the model by assuming that the height of Schlemm's canal cannot collapse to less than the height of the septae. The results, including the septae in the model, show that three regimes exist, with threshold values of the pressure $\mathcal{P}_C$ and $\mathcal{P}_R$, which are schematically reported in Fig. 15:

1. Regime I ($\mathcal{P} < \mathcal{P}_C$): The canal height is greater than the septae everywhere, and thus the canal behaves as a porous compliant wall.
2. Regime II ($\mathcal{P}_C < \mathcal{P} < \mathcal{P}_R$): In this case, the canal can be divided into two regions: one, extending from $x' = 0$ to $x' = x'_c$, where the canal behaves as a porous compliant wall; and the second, extending from $x' = x'_c$ to $x' = 1$, where the canal is collapsed to the septae and behaves as a porous rigid channel.
3. Regime III ($\mathcal{P} > \mathcal{P}_R$): The wall touches the septae all along its length and behaves as a porous rigid channel.

**Fig. 15** Sketch of typical pressure-flow curve for the rigid septae model. Sketches show the channel geometry for each of the three regimes. Reproduced from [37]. Here $\Delta p = IOP - P_{cc}$ and $\bar{Q} = \mu Q(s)s/wh_0^3 E$

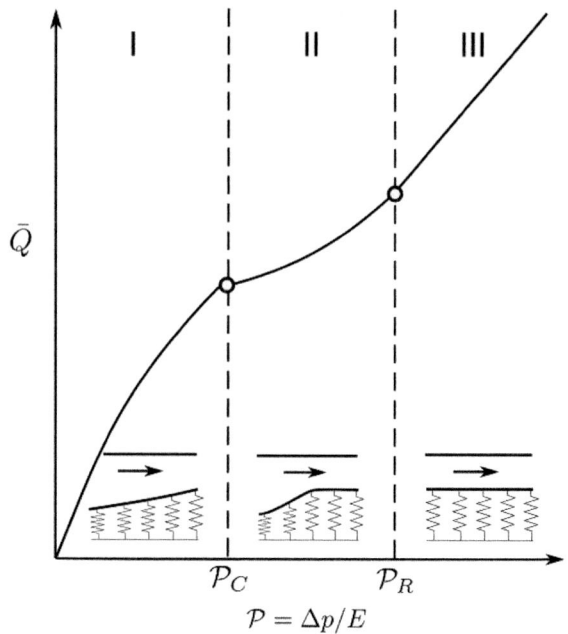

More recent studies have extended this model by including a better geometrical representation of Schlemm's canal and adding a poroelastic model of the trabecular meshwork, whose porosity changes under strain as Schlemm's canal collapses [3].

Ethier et al. [22] analysed transmission electron micrographs of the juxtacanalicular tissue to estimate its Darcy permeability, and they found that the resistance of the juxtacanalicular tissue alone is too small, by a factor of 10–100, to explain the pressure drop across the tissue. They proposed that the excess resistance is generated by the existence of an extracellular matrix gel between the tissues. A numerical model was developed by Kumar et al. [44] to solve the problem for the flow and pressure in the AC, using a value of the Darcy permeability of the trabecular meshwork estimated from the pore size. The authors concluded that the majority of the pressure drop occurs across the trabecular meshwork as the AH exits the eye.

Johnson et al. [36] suggested that, in addition to the extracellular matrix, the fact that the high-resistance juxtacanalicular tissue abuts the inner wall of Schlemm's canal means the AH has to converge through the juxtacanalicular tissue towards each of the holes in the inner wall. This leads to the so-called funneling hypothesis, in which the total resistance of the combined tissues is greater than the sum of the resistances of the tissues separately because the flow is not parallel in the juxtacanalicular tissue, see Fig. 16a. In addition, the pressure gradients that exist within the flowing AH generate mechanical loads that change the morphology of the tissue, making it harder to estimate the exact resistance. A possible treatment for glaucoma is therefore pharmacological agents that induce breaks in the structure of the inner wall of Schlemm's canal as these reduce the funneling and therefore

**Fig. 16** Schematic diagram showing the funneling phenomenon. (**a**) Flow distribution with inner wall cells attached to substratum; (**b**) uniform flow that results when these attachments are broken, for example, by a pharmacological agent. JCT: juxtacanalicular tissue

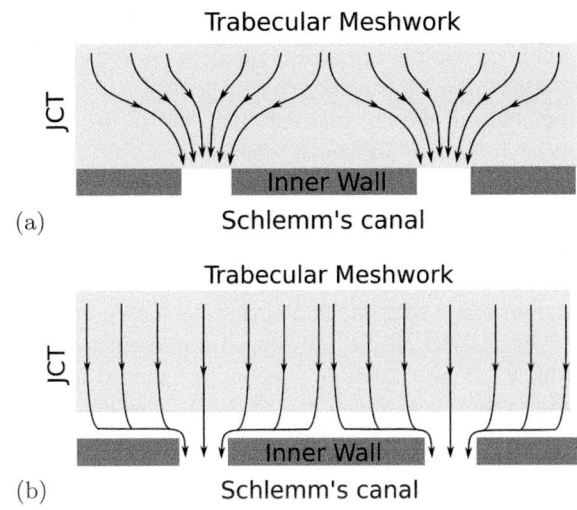

the resistance, an example of which is shown in Fig. 16b. This is thought to explain the "washout phenomenon" that is seen in a variety of non-human animal eyes in which the outflow resistance diminishes with increasing volumetric flux of AH [28], and the effectiveness of surgeries that aim to enlarge Schlemm's canal is probably explained by inadvertent rupture of the inner wall of the canal during the surgery, which in turn greatly reduces the resistance to outflow [34].

Johnson et al. [36] used a mathematical model to study the combined resistance of a porous medium when it abuts a fenestrated surface. In the limit of small pore radius, they were able to obtain an analytical solution for the effective resistance of the combined tissue. Merchant and Heys [51] showed that allowing for spatially heterogeneous permeability of the juxtacanalicular tissue leads to an increased tissue resistance, and furthermore that adding a model of the pores in the endothelial cells of the inner wall of Schlemm's canal significantly increases the resistance of the combined tissue.

## 4.4 Analytical and Computational Studies of the Treatments for Glaucoma

A number of modelling studies have addressed the response of AH drainage to various glaucoma treatments.

Kotliar et al. [41] developed a mathematical model to calculate the IOP after either a non-penetrating deep sclerectomy or a trabeculectomy. Villamarin et al. [76] developed a finite element model of the flow of AH, including that through the trabecular meshwork, for both normal and glaucomatous eyes, and used it to investigate the effectiveness of a scleral implant to treat glaucoma. To determine the permeability of the trabecular meshwork, they used the fact that the IOP should be within physiological ranges suitable for normal and glaucomatous eyes.

One common treatment for glaucoma is to insert a small shunt with one end in the angle of the eye and the other end in the subconjunctival space. Fluid flows through the low-resistance shunt to create a bleb, a small raised area of the sclera over a pocket of fluid in the subconjunctival space. Numerical models of the fluid flow with such a shunt have been studied by Gardiner et al. [27], Niederer et al. [54], Pan et al. [56], Stay et al. [69], Siewert et al. [66, 67], Schmidt et al. [64]. Other models have considered microstents in Schlemm's canal to keep it patent [79] and the effect of artificial drainage channels created by laser application [14].

Kapnisis et al. [38] developed a numerical model of the aspiration of AH, in which a small volume of aqueous is removed from the AC to be tested. Their aim was to find from which part of the AC the aspirated AH fluid was taken, and they showed that it was localised around the tip of the needle used for aspirating, and that it is particularly difficult to aspirate from the angle of the eye, which is unfortunate as the fluid in the angle contains proteins that would not be present in the AH from the bulk, and therefore the aspirated AH is not representative of the AH entering the trabecular meshwork.

**Acknowledgements** The authors thank Prof. Federica Grillo, University of Genoa (Italy), for drawing Figs. 1 and 2. Mariia Dvoriashyna acknowledges the Department of Civil, Chemical and Environmental Engineering of the University of Genoa (Italy), where she worked as a PhD student when the original version of this chapter was written.

# References

1. O. Abouali, A. Modareszadeh, A. Ghaffarieh, and J. Tu. Investigation of saccadic eye movement effects on the fluid dynamic in the anterior chamber. *J. Biomech. Eng.*, 134(2):021002, Feb. 2012. ISSN 1528-8951. https://doi.org/10.1115/1.4005762. PMID: 22482669.
2. A. Alm and S. F. E. Nilsson. Uveoscleral outflow – A review. *Exp. Eye Res.*, 88:760–768, 2009.
3. R. Avtar and R. Srivastava. Modelling aqueous humor outflow through trabecular meshwork. *Appl. Math. Comput.*, 189:734–745, 2007.
4. R. Avtar, R. Srivastava, and D. Nigam. A mathematical model for solute coupled water transport in the production of aqueous humor. *Applied Mathematical Modelling*, 32(7): 1350–1369, 2008.
5. G. K. Batchelor. *An Introduction to Fluid Dynamics.* Cambridge University Press, 1967.
6. J. A. Beswick and C. McCulloch. Effect of hyaluronidase on the viscosity of the aqueous humour. *Br. J. Ophthamol.*, 40:545–548, 1956.
7. A. Bill. Conventional and uveo-scleral drainage of aqueous humour in the cynomolgus monkey (Macaca irus) at normal and high intraocular pressures. *Exp. Eye Res.*, 5:45–54, 1966a.
8. A. Bill. The routes for bulk drainage of aqueous humour in the vervet monkey (Cercopithecus ethiops). *Exp. Eye Res.*, 5:55–57, 1966b.
9. A. Bill and K. Hellsing. Production and drainage of aqueous humor in the cynomolgus money (Macaca irus). *Invest. Ophthalmol.*, 4:920–926, 1965.
10. A. Bill and C. I. Phillips. Uveoscleral drainage of aqueous humour in human eyes. *Exp. Eye Res.*, 12:275–281, 1971.

11. R. F. Brubaker. Measurement of aqueous flow by fluorophotometry. In *The Glaucomas*. Mosby (St. Louis), 1989.

12. R. F. Brubaker. Flow of aqueous humor in humans [the friedenwald lecture]. *Investigative Ophthalmology and Visual Science*, 32 (13):3145–3166, 1991. ISSN, 1552-5783. PMID: 1748546.

13. C. R. Canning, M. J. Greaney, J. N. Dewynne, and A. Fitt. Fluid flow in the anterior chamber of a human eye. *IMA J. Math. Appl. Med. Biol.*, 19:31–60, 2002.

14. D. Chai, G. Chaudhary, E. Mikula, H. Sun, and T. Juhasz. 3D finite element model of aqueous outflow to predict the effect of femtosecond laser created partial thickness drainage channels. *Laser. Surg. Med.*, 40:188–195, 2008.

15. N. A. Delamere. Ciliary body and ciliary epithelium. *Advances in organ biology*, 10:127–148, 2005.

16. J. M. Diamond and W. H. Bossert. Standing-gradient osmotic flow: A mechanism for coupling of water and solute transport in epithelia. *The Journal of general physiology*, 50(8): 2061–2083, 1967.

17. S. Dorairaj, J. M. Liebmann, C. Tello, V. Barocas, and R. Ritch. Posterior chamber volume does not change significantly during dilation. *Br. J. Ophthalmol.*, 93:1514–1517, 2009.

18. P. G. Drazin and W. H. Reid. *Hydrodynamic Stability*. Cambridge University Press, 1981.

19. M. Dvoriashyna, A. J. Foss, E. A. Gaffney, O. E. Jensen, and R. Repetto. Osmotic and electroosmotic fluid transport across the retinal pigment epithelium: A mathematical model. *Journal of theoretical biology*, 456:233–248, 2018a.

20. M. Dvoriashyna, R. Repetto, M. Romano, and J. Tweedy. Aqueous humour flow in the posterior chamber of the eye and its modifications due to pupillary block and iridotomy. *Mathematical medicine and biology: a journal of the IMA*, 35(4):447–467, 2018b.

21. M. Dvoriashyna, R. Repetto, and J. Tweedy. Oscillatory and steady streaming flow in the anterior chamber of the moving eye. *Journal of Fluid Mechanics*, 863:904–926, 2019.

22. C. R. Ethier, R. D. Kamm, B. A. Palaszewski, M. C. Johnson, and T. M. Richardson. Calculation of flow resistance in the juxtacanalicular meshwork. *Invest. Ophthalmol. Visual Sci.*, 27:1741–1750, 1986.

23. M. P. Fautsch and D. H. Johnson. Aqueous humor outflow: what do we know? where will it lead us? *Investigative ophthalmology & visual science*, 47 (10):4181–4187, 2006.

24. A. D. Fitt and G. Gonzalez. Fluid mechanics of the human eye: Aqueous humour flow in the anterior chamber. *Bull. Math. Biol.*, 68(1):53–71, 2006.

25. A. B. Friedland. A hydrodynamic model of acqueous flow in the posterior chamber of the eye. *Bull. Math. Biol.*, 40:223–235, 1978.

26. D. S. Friedman. Epidemiology of angle-closure glaucoma. *Journal of Current Glaucoma Practice*, 1(1): 1–3, 2007.

27. B. S. Gardiner, D. W. Smith, M. Coote, and J. G. Crowston. Computational modeling of fluid flow and intra-ocular pressure following glaucoma surgery. *PLoS ONE*, 5:e13178, 2010.

28. H. Gong and T. F. Freddo. The washout phenomenon in aqueous outflow – Why does it matter? *Exp. Eye Res.*, 88:729–737, 2009.

29. H. Y. Gong, R. C. Tripathi, and B. J. Tripathi. Morphology of the aqueous outflow pathway. *Microsc. Res. Tech.*, 33:336–367, 1996.

30. H. v. Helmholtz. *Handbuch der physiologischen Optik*. University of Michigan Library, third edition, 1909.

31. J. J. Heys and V. H. Barocas. A Boussinesq model of natural convection in the human eye and formation of krunberg's spindle. *Ann. Biomed. Eng.*, 30:392–401, 2002.

32. J. J. Heys, V. H. Barocas, and M. J. Taravella. Modeling passive mechanical interaction between aqueous humor and iris. *Transactions of the ASME*, 123:540–547, December 2001.

33. E. C. Huang and V. H. Barocas. Active iris mechanics and pupillary block: steady-state analysis and comparison with anatomical risk factors. *Annals of biomedical engineering*, 32(9): 1276–1285, 2004.

34. D. H. Johnson and M. Johnson. How does nonpenetrating glaucoma surgery work? Aqueous outflow resistance and glaucoma surgery. *J. Glaucoma*, 10:55–67, 2001.

35. M. Johnson. What controls aqueous humor outflow resistance? *Exp. Eye Res.*, 82(4):545–557, 2006.

36. M. Johnson, A. Shapiro, C. R. Ethier, and R. D. Kamm. Modulation of outflow resistance by the pores of the inner wall endothelium. *Invest. Ophthalmo. Vis. Sci.*, 33:1670–1675, 1992.

37. M. C. Johnson and R. D. Kamm. The role of schlemm's canal in aqueous outflow from the human eye. *Invest. Ophthalmol. Visual Sci.*, 24(3): 320–325, Mar. 1983. ISSN 0146-0404. PMID: 6832907.

38. K. Kapnisis, M. van Doormaal, and C. R. Ethier. Modeling aqueous humor collection from the human eye. *J. Biomech.*, 42:2454–2457, 2009.

39. J. Kiel, M. Hollingsworth, R. Rao, M. Chen, and H. Reitsamer. Ciliary blood flow and aqueous humor production. *Prog. Retin. Eye Res.*, 30(1):1–17, 2011.

40. C. C. Klaver, R. C. Wolfs, J. R. Vingerling, A. Hofman, and P. T. de Jong. Age-specific prevalence and causes of blindness and visual impairment in an older population: the rotterdam study. *Archives of ophthalmology*, 116(5): 653–658, 1998.

41. K. E. Kotliar, T. V. Kozlova, and I. M. Lanzl. Postoperative aqueous outflow in the human eye after glaucoma filtration surgery: biofluidmechanical considerations. *Biomed. Tech.*, 54:14–22, 2009.

42. T. Krahn and A. M. Weinstein. Acid/base transport in a model of the proximal tubule brush border: impact of carbonic anhydrase. *American Journal of Physiology-Renal Physiology*, 270 (2):F344–F355, 1996.

43. F. Krukenberg. Beiderseitige angeborene melanose de hornhaut. *Klin. Mbl. Augenheilk.*, 37:254–258, 1899.

44. S. Kumar, S. Acharya, R. Beuerman, and A. Palkama. Numerical solution of ocular fluid dynamics in a rabbit eye: Parametric effects. *Ann. Biomed. Eng.*, 34:530–544, 2006.

45. Y. H. Kwon, J. H. Fingert, M. H. Kuehn, and W. L. Alward. Primary open-angle glaucoma. *New England Journal of Medicine*, 360(11): 1113–1124, 2009.

46. J. M. Liebmann and R. Ritch. Laser surgery for angle closure glaucoma. In *Seminars in ophthalmology*, volume 17, pages 84–91. Taylor & Francis, 2002.

47. G. Lyubimov, I. Moiseeva, and A. Stein. Dynamics of the intraocular fluid: Mathematical model and its main consequences. *Fluid Dynamics*, 42(5):684–694, 2007.

48. A. G. Mauri, L. Sala, P. Airoldi, G. Novielli, R. Sacco, S. Cassani, G. Guidoboni, B. Siesky, and A. Harris. Electro-fluid dynamics of aqueous humor production. simulations and new directions. *Journal for Modeling in Ophthalmology*, 1 (2):48–58, 2016.

49. D. M. Maurice. The von sallmann lecture 1996: an ophthalmological explanation of rem sleep. *Experimental eye research*, 66(2):139–145, 1998.

50. J. W. McLaren. Measurement of aqueous humor flow. *Exp. Eye Res.*, 88:641–647, 2009.

51. B. M. Merchant and J. J. Heys. Effects of variable permeability on aqueous humor outflow. *Appl. Math. Comput.*, 196:371–380, 2008.

52. J. G. Milton and A. Longtin. Evaluation of pupil constriction and dilation from cycling measurements. *Vision research*, 30(4):515–525, 1990.

53. S. Modarreszadeh, O. Abouali, A. Ghaffarieh, and G. Ahmadi. Physiology of aqueous humor dynamic in the anterior chamber due to rapid eye movement. *Physiol Behav*, 2014.

54. P. Niederer, F. Fankhauser, and S. Kwasniewska. Hydrodynamics of aqueous humor in chronic simple glaucoma. *Ophthalmologe*, 109:30–36, 2012.

55. OpenFOAM, the Open Source CFD Toolbox by OpenCFD Ltd. http://openfoam.com.

56. T. Pan, M. S. Stay, V. H. Barocas, J. D. Brown, and B. Ziaie. Modeling and characterization of a valved glaucoma drainage device with implications for enhanced therapeutic efficacy. *IEEE Trans. Biomed. Eng.*, 52:948–951, 2005.

57. H. F. Poppendiek, R. Randall, J. A. Breeden, J. E. Chambers, and J. R. Murphy. Thermal conductivity measurements and predictions for biological fluids and tissues. *Cryobiology*, 3(4):318–327, 1967. ISSN 0011-2240. doi: 10.1016/S0011-2240(67)80005-1.

58. R. F. Probstein. *Physicochemical hydrodynamics: an introduction*. John Wiley & Sons, 2005.

59. D. Purves, G. J. Augustine, D. Fitzpatrick, L. C. Katz, A.-S. LaMantia, J. O. McNamara, S. M. Williams, et al. Types of eye movements and their functions. *Neuroscience*, pages 361–390, 2001.

60. H. A. Quigley and A. T. Broman. The number of people with glaucoma worldwide in 2010 and 2020. *British journal of ophthalmology*, 90(3): 262–267, 2006.
61. G. Raviola. The structural basis of the blood-ocular barriers. *Experimental eye research*, 25:27–63, 1977.
62. R. Repetto, J. O. Pralits, J. H. Siggers, and P. Soleri. Phakic iris-fixated intraocular lens placement in the anterior chamber: effects on aqueous flow. *Invest. Ophthalmol. Visual Sci.*, 56(5): 3061–3068, 2015.
63. A. R. Rudnicka, S. Mt-Isa, C. G. Owen, D. G. Cook, and D. Ashby. Variations in primary open-angle glaucoma prevalence by age, gender, and race: a bayesian meta-analysis. *Investigative ophthalmology & visual science*, 47 (10):4254–4261, 2006.
64. W. Schmidt, C. Schultze, O. Stachs, R. Allemann, M. Lobler, K. Sternberg, U. Hinze, B. N. Chichkov, R. Guthoff, and K. P. Schmitz. Concept of a pressure-controlled microstent for glaucoma therapy. *Klin. Monatsbl. Augenh.*, 227:946–952, 2010.
65. R. R. Seeley, T. D. Stephens, and P. Tate. *Anatomy & Physiology*. McGraw-Hill, New York, 5th edition, 2000.
66. S. Siewert, C. Schultze, W. Schmidt, U. Hinze, B. Chichkov, A. Wree, K. Sternberg, R. Allemann, R. Guthoff, and K. P. Schmitz. Development of a micro-mechanical valve in a novel glaucoma implant. *Biomed. Microdevices*, 14:907–920, 2012.
67. S. Siewert, M. Saemann, W. Schmidt, M. Stiehm, K. Falke, N. Grabow, R. Guthoff, and K. P. Schmitz. Coupled analysis of fluid–structure interaction of a micro-mechanical valve for glaucoma drainage devices. *Klin. Monatsbl., Augenh.*, 232:1374–1380, 2015.
68. D. M. Silver and H. A. Quigley. Aqueous flow through the iris-lens channel: estimates o the differential pressure between the anterior and the posterior chambers. *J. Glaucoma*, 13(2):100–107, April 2004.
69. M. S. Stay, T. Pan, J. D. Brown, B. Ziaie, and V. H. Barocas. Thin-film coupled fluid–solid analysis of flow through the Ahmed$^{tm}$ glaucoma drainage device. *J. Biomech. Eng.*, 127:776–781, 2005.
70. M. Szopos, S. Cassani, G. Guidoboni, C. Prud'homme, R. Sacco, B. Siesky, and A. Harris. Mathematical modeling of aqueous humor flow and intraocular pressure under uncertainty: towards individualized glaucoma management. *Journal for Modeling in Ophthalmology*, 1 (2):29–39, 2016.
71. E. R. Tamm. The trabecular meshwork outflow pathways: Structural and functional aspects. *Exp. Eye Res.*, 88:648–655, 2009.
72. C. To, C. Kong, C. Chan, M. Shahidullah, and C. Do. The mechanism of aqueous humour formation. *Clinical and Experimental Optometry*, 85(6): 335–349, 2002.
73. C. B. Toris, M. E. Yablonski, and Y.-L. W. ad C. B. Camras. Aqueous humor dynamics in the aging human eye. *Am. J, Ophthalmol.*, 127(4):407–412, 1999.
74. S. Türk. Untersuchungen über eine strömung in der vorderen augenkammer. *Gaefes Arch. Ophtalmol.*, 64:481–501, 1906.
75. J. H. Tweedy, J. O. Pralits, R. Repetto, and P. Soleri. Flow in the anterior chamber of the eye with an implanted iris-fixated artificial lens. *Mathematical Medicine and Biology: A Journal of the IMA*, page dqx007, 2017.
76. A. Villamarin, S. Roy, R. Hasballa, O. Vardoulis, P. Reymond, and N. Stergiopulos. 3D simulation of the aqueous flow in the human eye. *Med. Eng. Phys.*, 34:1462–1470, 2012.
77. R. N. Weinreb and P. T. Khaw. Primary open-angle glaucoma. *The Lancet*, 363(9422):1711–1720, 2004.
78. A. Weinstein and J. Stephenson. Electrolyte transport across a simple epithelium. steady-state and transient analysis. *Biophysical journal*, 27(2):165–186, 1979.
79. F. Yuan, A. T. Schieber, L. J. Camras, P. J. Harasymowycz, L. W. Herndon, and R. R. Allingham. Mathematical modeling of outflow facility increase with trabecular meshwork bypass and Schlemm canal dilation. *J. Glaucoma*, 25:355–364, 2016.

# Part IV
# Vitreous Humor

# Vitreous Physiology

Gian Paolo Giuliari, Peter Bracha, A. Bailey Sperry, and Thomas Ciulla

**Abstract** The translucency of the vitreous makes its full structure and composition challenging to completely elucidate, but what is known about the anatomy and biochemistry of this body and its impact on optic function is fundamental to understanding ocular health and disorders, as it carries out several important functions within the eye. The particular makeup of structural protein fibers is known to play a pivotal role in stabilizing the vitreous and maintaining its morphological integrity, and disruptions in this network, due to genetics, disease, or environmental changes, may result in certain conditions and ocular pathologies. Research has recently shown that the concentrations of ions, nutrients, and other proteins and small molecules in the vitreous can also be affected by disease. Age-related changes to the vitreous are predominantly due to changes in the density and increasing liquefaction, which weaken its structural integrity and adhesion to the internal limiting membrane and may result in posterior vitreous detachment or collapse of the vitreous body. Familiarity with the anatomy, biochemistry, and development of, and changes to the vitreous facilitates an increased knowledge of its role in maintaining overall ocular health and may also further the understanding of certain conditions and ocular pathologies.

The vitreous body is a clear gel of approximately 4 mL that occupies about 80% of the globe. The role of the vitreous in the development of eye disorders is an area of intense research. Advancements in imaging of the vitreous in recent years have suggested roles in the pathogenesis of retinal tears and detachment, vitreomacular traction (VMT), macular holes, macular edema, proliferative retinopathies, retinal vein occlusion, and exudative age-related macular degeneration [1, 2]. To better

G. P. Giuliari · P. Bracha
Department of Ophthalmology, Indiana University School of Medicine, Indianapolis, IN, USA

A. B. Sperry
Tufts University, Medford, MA, USA

T. Ciulla (✉)
Midwest Eye Institute, Indianapolis, IN, USA

© Springer Nature Switzerland AG 2019                                          267
G. Guidoboni et al. (eds.), *Ocular Fluid Dynamics*, Modeling and Simulation in
Science, Engineering and Technology, https://doi.org/10.1007/978-3-030-25886-3_10

understand the role of the vitreous in these disorders, it is essential to be familiar with vitreous embryology, anatomy and physiology, composition, and normal aging changes.

# 1   Vitreous Embryology

Embryologic development of the eye occurs between gestational weeks 3 and 10, and involves ectoderm, neural crest cells, and mesenchyme. Optic vesicles derive from the neural tube, extending from the forebrain through mesenchyme toward the surface ectoderm. The optic vesicles invaginate to form double-layered structures, known as the optic cups, while also interacting with the surface ectoderm to induce lens precursor formation. The optic cups remain attached to the forebrain via optic stalks, precursors of the optic nerves.

The double layer structure of the optic cup forms the precursor of the neuroretina and RPE layers, respectively. The vitreous body, derived from mesenchymal cells of neural crest origin, forms in the center of the optic cup posterior to the lens. Blood vessels extend through the vitreous and supply the lens precursor. The hyaloid artery, which spans the embryologic vitreous from the precursor optic nerve to the developing lens, ultimately regresses, but can persist in cases of persistent fetal vasculature; a Bergmeister's papilla is a hyaloid artery remnant attached at the optic disk.

# 2   Gross Anatomy of the Vitreous

The appearance of the vitreous body varies depending upon the method of examination. For example, the vitreous in a young healthy eye may appear homogenous and clear when viewed with diffuse illumination, while the same vitreous will show fibrils with associated areas of condensation using the slit lamp biomicroscopy [3].

Light microscopy reveals a complex structure with multiple fibrils. Electron microscopy demonstrates a meshwork of collagen fibrils as well as a solution of salts, proteins, and hyaluronic acid within a meshwork of insoluble protein fibers. Recently, the laminar configuration of the peripheral vitreous has been demonstrated.

## 2.1   Vitreous Landmarks

The anterior surface of the vitreous, the anterior hyaloid membrane, is not a true membrane but rather an increase in the density of collagen fibrils. It has a concave configuration, and as it attaches to the posterior lens capsule, which is convex, forms the patella fossa. Also, the anterior vitreous attaches to the posterior lens capsule along a circular zone called the hyaloideocapsular ligament of Wieger. A potential space between these two structures, Berger's space, represents the anterior extension of Cloquet's canal, which is a remnant of the primary vitreous. Cloquet's canal arises at the level of the optic disc, forming a funnel called the space of Martegiani. As it projects anteriorly toward the vitreous cavity, it becomes narrower.

Occasionally, a failure in the complete regression of the hyaloid artery may show a Mittendorft's dot, classically located nasally in the posterior lens capsule, or a Bergmeister's papilla, which consists of a small tuft of fibrous tissue at the level of the optic nerve, as previously described.

Another important landmark is the vitreous base, the most important area of vitreous attachment; it is a circumferential zone where the vitreous attaches to the epithelium of the pars plana and to the peripheral retina. It extends about 2–6 mm wide. The anterior margin is about 5 mm posterior to the limbus. It extends posterior to the ora serrata about 2 mm in the temporal quadrants and 3 mm in the nasal quadrants.

### 2.1.1 Vitreous Biochemistry

Up to 98% of the vitreous is water, while the remaining 2% is composed of structural proteins, extracellular matrix components, and miscellaneous compounds, the transparent nature of which makes its study challenging. The structure of the vitreous is maintained by a thin network of unbranched heterotopic collagen fibrils, most of which are comprised of collagen types II, V/XI, and IX in a molar ratio of 75:10:15, respectively (Table 1).

## 2.2 Collagen

Collagen type II is the major structural protein of the vitreous core, comprising up to 75% of the total collagen in the human vitreous. The filaments are thin with a length of 8–16 nm with occasional bindings. The average interfibrillar distance is 1–2 μm, which may be responsible for the soft gel features. As in cartilage, it consists of

**Table 1** Vitreous biochemistry

| Component | Type | Function |
|---|---|---|
| Collagen | II, V, IX, XI, XVIII | Structural protein |
| | | Role in inhibiting angiogenesis |
| Hyaluronan | | Most important structural element of the vitreous |
| Chondroitin sulfate | Versican | Molecular morphology of the vitreous |
| Fibrills | | Defects may be associated with vitreous liquefaction |
| Opticin | | Prevents collagen aggregation and helps to stabilize the vitreous |

heterotypic fibrils. Reoperation after vitrectomy has demonstrated that the vitreous does not reform; however, a type II procollagen is secreted [4]. Similarities between the collagen of the vitreous and cartilage may explain why some inborn errors of type II collagen metabolism share similar phenotypic expression in joints and the vitreous [5].

The second most common collagen is type IX, accounting for up to 15%. It always contains a chondroitin sulfate glycosaminoglycan chain linked to the alpha2 chain at the NC3 domain, and is located outside of the fibrils, while collagens type V and XI are located in the core of the fibrils [6]. Recently, it was demonstrated that collagen type XVIII may play a role in inhibiting angiogenesis [7, 8]. The spaces left by these collagen fibers are mostly filled by glycosaminoglycans (GAGs), in particular hyaluronan [9].

## 2.3   Hyaluronan

Hyaluronan (HA) is the most important structural element of the vitreous gel, but it is not visible clinically or by electron microscopy. It appears shortly after birth and is thought to be synthesized by hyalocytes, Muller cells, and/or the ciliary body. HA links to type IX collagen, which plays a role in bridging adjacent collagen fibrils while at the same time spacing them apart to minimize light scattering and maintain vitreous transparency [10]. Enlargement of this molecule within the collagen fibril of the vitreous matrix, induced by fluctuations in the ionic balance and hydration of the vitreous, as may be seen in a condition such as diabetes, may produce a mechanical force that can be transmitted to the optic nerve, retina, and neovascular complexes [11]. Additionally, it may influence the diffusion of drugs through the vitreous [12].

## 2.4   Chondroitin Sulfate

The chondroitin sulfate (CS) found in the vitreous is mainly in the form of versican [13]. It is thought to form complexes with HA as well as with microfibrillar proteins, which may play a pivotal role in keeping the molecular morphology of the vitreous [14]. The excess vitreous liquefaction noted in patients with Wagner syndrome is thought to be related to mutations in the versican gene.

## 2.5   Fibrills

Mutations in the gene that encodes fibrillin-1 (*FBN1* on chromosome 15q21) are associated with vitreous liquefaction in patients with Marfan syndrome, and are significantly associated with the development of retinal detachments. They are also related to ectopia lentis [9].

## 2.6 Opticin

Opticin attaches to the collagen fibrils by binding to different GAGs, preventing aggregation with adjacent collagen [9, 15, 16]. It also binds to CS and heparan, suggesting a possible role in vitreoretinal adhesion [16, 17]. It may also play a role in stabilizing the vitreous by binding to CS and proteoglycans [18].

## 2.7 Vitreoretinal Interface

It still remains unknown precisely how the retina and the vitreous interact. The vitreous is firmly attached at places where the internal limiting membrane is thinnest, that is, the vitreous base, optic disc and macula, and over retinal blood vessels [1]. At the base area, the vitreous and the retina are connected by vitreous collagen fibers passing though the internal limiting membrane and intertwining with retinal collagen [19]. The collagen fibrils of the posterior vitreous cortex are indirectly attached to the internal limiting membrane and to the retina via laminin and fibronectin [20]. The attachment between the vitreous and the macula is called the vitreomacular interface.

The vitreoretinal interface consists of a protein matrix, which includes collagen IV, fibronectin, and laminin, that serves as an adhesive sheet facilitating the connection of the posterior vitreous to the internal limiting membrane [10, 21, 22].

## 2.8 The Role of the Vitreous in Ocular Health

Although vitreoretinal surgeons are able to remove the vitreous without significant consequences, the healthy vitreous has an important role and carries out several functions. As noted above, embryologically, it occupies a space around which the optic cup develops, and also hosts a blood supply to the developing lens.

The vitreous transmits up to 90% of light wavelengths between 300 nm and 1400 nm, stabilizes the globe, and is a pathway for nutrients to reach the avascular lens and retina. Also, small water-soluble substances in the retina may diffuse into the vitreous across the blood–retinal barrier. Moreover, it may act as a reservoir for substances such as oxygen, glucose, and ascorbic acid, which support the metabolism of adjacent structures, and as a possible pump for metabolic waste products.

However, determining the biochemical composition of the living human vitreous has mostly only been possible in animal models [23]. Variations in the biochemistry, anatomy, and protein composition have been noted in different mammals [24, 25].

The most accepted concentrations of sodium, potassium, chloride, phosphate, ascorbate, glucose, and lactate are taken from data in rabbit studies, a model that has its own unique characteristics [23, 26–28].

More recent studies have raised the need for a better understanding of and baseline data on these concentrations, as they may be affected by diseases.

Deficiency of trace elements such as iron, zinc, selenium, and copper has been associated with glaucoma, pseudoexfoliation syndrome, Behcet's disease, and age-related macular degeneration [29–32].

Another example of the role of the vitreous in the development of ocular disease comes with the comparison of diabetic and non-diabetic subgroups, which have revealed significant differences in the mean concentrations of a number of analytes. In this regard, the vitreous of diabetic patients has greater concentrations of glucose, lactate, copper, iron, and transferrin but lower concentrations of magnesium [33, 34]. Furthermore, there may be a relationship between vitreous and serum glucose which suggests that an equilibration of glucose occurs across the blood–retinal barrier [35].

### 2.8.1 Vitreous Aging Changes

Most of the structural changes of the anatomy of the vitreous are secondary to changes in its density. In younger eyes, the vitreous appears uniform when visualized macroscopically with diffuse illumination. However, there are already subtle differences in density between regions of the vitreous. This difference become more evident with aging as the central liquefaction progresses, and up to 25% of people may show some degree of vitreous liquefaction between 20 and 29 years of age [36].

The posterior vitreous cortex is the only zone of optical density in young eyes, and has a thickness of about 2–3 mm. Studies have demonstrated the presence of holes over the optic disc, fovea, and blood vessels [3]. Over time, liquid may enter through these holes, accessing the space between the cortical vitreous and the retina during posterior vitreous detachment. This differentiation and degeneration will continue with aging; however, the posterior vitreous may remain unchanged until late adulthood, while the center suffers more extend liquefaction. Changes may occur earlier in life in cases of myopia, intraocular inflammation, trauma, or surgery [37].

With aging, weakening of the adhesion between the posterior vitreous cortex and the internal limiting membrane occurs, possibly due to biochemical alterations such as changes in the galactose β (1,3)-N-acetylglucosamine [38–40].

After the age of 60 years, weakening of the posterior vitreous cortex/internal limiting membrane adhesion at the posterior pole allows liquid vitreous to enter the retrocortical space. Volume displacement from the central vitreous to the preretinal space causes the collapse of the vitreous body [21].

The most important age-related change is the posterior vitreous detachment, which is defined as a separation of the posterior vitreous cortex and the internal limiting membrane [36, 37]. Studies suggest that the vitreous is only able to tolerate a certain amount of liquefaction and instability before a posterior vitreous detachment occurs [36].

# 3 Conclusion

Although the vitreous has historically been difficult to study due to its "transparent by design" nature, it plays key roles in the embryologic development of the eye, stabilization of the globe, transmission of light, and circulation of nutrients within the eye. Familiarity with the vitreous micro- and macro-anatomy, physiology, and age-related changes facilitates an understanding of the vitreous' role in a variety of important ocular pathologies.

# References

1. Johnson MW. Posterior vitreous detachment: evolution and complications of its early stages. *American journal of ophthalmology*. 2010;149(3):371-382.e371.
2. Krebs I, Brannath W, Glittenberg C, Zeiler F, Sebag J, Binder S. Posterior vitreomacular adhesion: a potential risk factor for exudative age-related macular degeneration? *American journal of ophthalmology*. 2007;144(5):741-746.
3. Sebag J. Age-related changes in human vitreous structure. *Graefe's archive for clinical and experimental ophthalmology = Albrecht von Graefes Archiv fur klinische und experimentelle Ophthalmologie*. 1987;225(2):89-93.
4. Itakura H, Kishi S, Kotajima N, Murakami M. Vitreous collagen metabolism before and after vitrectomy. *Graefe's archive for clinical and experimental ophthalmology = Albrecht von Graefes Archiv fur klinische und experimentelle Ophthalmologie*. 2005;243(10):994-998.
5. Maumenee IH. Vitreoretinal degeneration as a sign of generalized connective tissue diseases. *American journal of ophthalmology*. 1979;88(3 Pt 1):432-449.
6. Bishop PN, Reardon AJ, McLeod D, Ayad S. Identification of alternatively spliced variants of type II procollagen in vitreous. *Biochemical and biophysical research communications*. 1994;203(1):289-295.
7. Bhutto IA, Kim SY, McLeod DS, et al. Localization of collagen XVIII and the endostatin portion of collagen XVIII in aged human control eyes and eyes with age-related macular degeneration. *Investigative ophthalmology & visual science*. 2004;45(5):1544-1552.
8. Ohlmann AV, Ohlmann A, Welge-Lussen U, May CA. Localization of collagen XVIII and endostatin in the human eye. *Current eye research*. 2005;30(1):27-34.
9. Bishop PN. Structural macromolecules and supramolecular organisation of the vitreous gel. *Progress in retinal and eye research*. 2000;19(3):323-344.
10. Scott JE, Chen Y, Brass A. Secondary and tertiary structures involving chondroitin and chondroitin sulphates in solution, investigated by rotary shadowing/electron microscopy and computer simulation. *European journal of biochemistry*. 1992;209(2):675-680.

11. Sebag J. Diabetic vitreopathy. *Ophthalmology.* 1996;103(2):205-206.
12. Kim H, Robinson SB, Csaky KG. Investigating the movement of intravitreal human serum albumin nanoparticles in the vitreous and retina. *Pharmaceutical research.* 2009;26(2):329-337.
13. Reardon A, Heinegard D, McLeod D, Sheehan JK, Bishop PN. The large chondroitin sulphate proteoglycan versican in mammalian vitreous. *Matrix biology: journal of the International Society for Matrix Biology.* 1998;17(5):325-333.
14. Cain SA, Morgan A, Sherratt MJ, Ball SG, Shuttleworth CA, Kielty CM. Proteomic analysis of fibrillin-rich microfibrils. *Proteomics.* 2006;6(1):111-122.
15. Reardon AJ, Le Goff M, Briggs MD, et al. Identification in vitreous and molecular cloning of opticin, a novel member of the family of leucine-rich repeat proteins of the extracellular matrix. *The Journal of biological chemistry.* 2000;275(3):2123-2129.
16. Hindson VJ, Gallagher JT, Halfter W, Bishop PN. Opticin binds to heparan and chondroitin sulfate proteoglycans. *Investigative ophthalmology & visual science.* 2005;46(12):4417-4423.
17. Sanders EJ, Walter MA, Parker E, Aramburo C, Harvey S. Opticin binds retinal growth hormone in the embryonic vitreous. *Investigative ophthalmology & visual science.* 2003;44(12):5404-5409.
18. Monfort J, Tardif G, Roughley P, et al. Identification of opticin, a member of the small leucine-rich repeat proteoglycan family, in human articular tissues: a novel target for MMP-13 in osteoarthritis. *Osteoarthritis and cartilage.* 2008;16(7):749-755.
19. Wang J, McLeod D, Henson DB, Bishop PN. Age-dependent changes in the basal retinovitreous adhesion. *Investigative ophthalmology & visual science.* 2003;44(5):1793-1800.
20. Matsumoto B, Blanks JC, Ryan SJ. Topographic variations in the rabbit and primate internal limiting membrane. *Investigative ophthalmology & visual science.* 1984;25(1):71-82.
21. Sebag J. Molecular biology of pharmacologic vitreolysis. *Transactions of the American Ophthalmological Society.* 2005;103:473-494.
22. Le Goff MM, Bishop PN. Adult vitreous structure and postnatal changes. *Eye (London, England).* 2008;22(10):1214-1222.
23. Andersen MV. Changes in the vitreous body pH of pigs after retinal xenon photocoagulation. *Acta ophthalmologica.* 1991;69(2):193-199.
24. Lee B, Litt M, Buchsbaum G. Rheology of the vitreous body: part 3. Concentration of electrolytes, collagen and hyaluronic acid. *Biorheology.* 1994;31(4):339-351.
25. Noulas AV, Skandalis SS, Feretis E, Theocharis DA, Karamanos NK. Variations in content and structure of glycosaminoglycans of the vitreous gel from different mammalian species. *Biomedical chromatography: BMC.* 2004;18(7):457-461.
26. Los LI, van Luyn MJ, Nieuwenhuis P. Vascular remnants in the rabbit vitreous body. I. Morphological characteristics and relationship to vitreous embryonic development. *Experimental eye research.* 2000;71(2):143-151.
27. Los LI, van Luyn MJ, Nieuwenhuis P. Organization of the rabbit vitreous body: lamellae, Cloquet's channel and a novel structure, the 'alae canalis Cloqueti'. *Experimental eye research.* 1999;69(3):343-350.
28. Los LI, van Luyn MJ, Eggli PS, Dijk F, Nieuwenhuis P. Vascular remnants in the rabbit vitreous body. II. Enzyme digestion and immunohistochemical studies. *Experimental eye research.* 2000;71(2):153-165.
29. Bruhn RL, Stamer WD, Herrygers LA, Levine JM, Noecker RJ. Relationship between glaucoma and selenium levels in plasma and aqueous humour. *The British journal of ophthalmology.* 2009;93(9):1155-1158.
30. Esalatmanesh K, Jamshidi A, Shahram F, et al. Study of the correlation of serum selenium level with Behcet's disease. *International journal of rheumatic diseases.* 2011;14(4):375-378.
31. Yilmaz A, Ayaz L, Tamer L. Selenium and pseudoexfoliation syndrome. *American journal of ophthalmology.* 2011;151(2):272-276.e271.
32. Ugarte M, Osborne NN, Brown LA, Bishop PN. Iron, zinc, and copper in retinal physiology and disease. *Survey of ophthalmology.* 2013;58(6):585-609.

33. Viktorinova A, Toserova E, Krizko M, Durackova Z. Altered metabolism of copper, zinc, and magnesium is associated with increased levels of glycated hemoglobin in patients with diabetes mellitus. *Metabolism: clinical and experimental.* 2009;58(10):1477-1482.

34. Simo-Servat O, Hernandez C, Simo R. Usefulness of the vitreous fluid analysis in the translational research of diabetic retinopathy. *Mediators of inflammation.* 2012;2012:872978.

35. Kokavec J, Min SH, Tan MH, et al. Biochemical analysis of the living human vitreous. *Clinical & experimental ophthalmology.* 2016;44(7):597-609.

36. Foos RY, Wheeler NC. Vitreoretinal juncture. Synchysis senilis and posterior vitreous detachment. *Ophthalmology.* 1982;89(12):1502-1512.

37. Sebag J. Ageing of the vitreous. *Eye (London, England).* 1987;1 (Pt 2):254-262.

38. Larsson L, Osterlin S. Posterior vitreous detachment. A combined clinical and physiochemical study. *Graefe's archive for clinical and experimental ophthalmology = Albrecht von Graefes Archiv fur klinische und experimentelle Ophthalmologie.* 1985;223(2):92-95.

39. Vaughan-Thomas A, Gilbert SJ, Duance VC. Elevated levels of proteolytic enzymes in the aging human vitreous. *Investigative ophthalmology & visual science.* 2000;41(11):3299-3304.

40. Russell SR, Shepherd JD, Hageman GS. Distribution of glycoconjugates in the human retinal internal limiting membrane. *Investigative ophthalmology & visual science.* 1991;32(7):1986-1995.

# 1   Developmental Abnormalities

## 1.1   Stickler Syndrome

The vitreous collagen forms the fibrillary component of the vitreous, and consists of types V and XI in the core, type II surrounding this core, and type IX on its outermost surface. The vitreoretinal interface includes the retina's internal limiting membrane, which is composed of type IV collagen [1]. Stickler syndrome is a genetic connective tissue disorder with mutations in genes encoding collagen subunits, resulting in malformation of primarily type II collagen, but also types IX and XI. Autosomal dominant variants can be caused by mutations in COL2A1 (80–90% of cases), COL11A1 (10–20% of cases), or COL11A2. Autosomal recessive variants can be due to mutations in COL9A1, COL9A2, or COL9A3. Clinically, patients have an optically empty vitreous with a thin layer of cortical vitreous immediately posterior to the lens (membranous subtype) or threadlike, avascular membranes that adhere circumferentially to the retina (beaded congenital vitreous anomaly). Patients classically have high myopia and equatorial and perivascular lattice degeneration, placing them at high risk of rhegmatogenous retinal detachment. Patients are at increased risk of open-angle glaucoma and premature cataract formation.

Stickler syndrome can be classified into syndromic and ocular-only subtypes, determined by the underlying genetic mutation. Genetic testing is available and can help with both diagnosis and prognosis.

Syndromic variants, which are more common than the ocular-only subtypes, are associated with midfacial flattening, oral clefting, and Pierre Robin sequence, features which can help clue practitioners to the right diagnosis. Musculoskeletal manifestations are common and include femoral head failure, precocious arthritis, and mild spondyloepiphyseal dysplasia. Both conductive and sensorineural hearing loss are common.

Management of the associated systemic conditions is important, as craniofacial and hearing issues can be amenable to treatment. Anesthesiologists should be informed of the diagnosis so that they can be prepared for variations during intubation. Amblyopic refractive errors can be treated with spectacles or contacts, and retinal detachment should be treated in standard fashion. Genetic counseling can be offered and screening of potentially affected family members should be considered [2, 3].

## 1.2   Wagner Syndrome

As noted above, the vitreous glycoaminoglycosides (GAGs) include hyaluronan and chondroitin sulfate, with the former being more abundant [1]. Wagner syndrome is an autosomal dominant disease caused by a mutation in the chondroitin sulfate proteoglycan 2 gene (CSPG2), and is clinically and pathologically distinct from ocular-only variants of Stickler's syndrome, despite a similarly optically empty vitreous with myopia. The CSPG2 gene encodes a core protein of the proteoglycan versican, a key vitreous structural protein. Classical features include vitreous syneresis, posterior strands and veils in the vitreous cavity, and avascular membranes. Patients may also have subnormal rod and cone responses on ERG, early cataract formation, age-dependent chorioretinal atrophy, and peripheral tractional retinal detachment from a progressive and extensive vitreoretinopathy. Unlike in classical Stickler syndrome, Wagner syndrome rarely has associated systemic manifestations [4].

## 1.3 Persistent Fetal Vasculature

Persistent fetal vasculature, also known as persistent hyperplastic primary vitreous, is a unilateral developmental anomaly that typically presents as a white retrolental mass at birth. It is caused by the persistence of the anterior hyaloid artery, which is the first embryological source of nutrients to the anterior eye. Typically it is associated with the ocular findings of microphthalmos, elongated ciliary processes, and lenticular opacification. Other associated findings include a shallow anterior chamber, glaucoma, and vitreous hemorrhage; pathologic neuroophthalmic developmental sequelae include amblyopia and strabismus. The precise pathogenesis is poorly understood. Most cases are sporadic, but autosomal dominant and recessive inheritance patterns have been described [5].

## 2 Normal Aging

The normal vitreous undergoes changes with age that can result in symptomatic floaters. Proteins can aggregate into visible masses that can result in light scattering and shadowing, most noticeable in brighter light. Liquefaction of the vitreous body occurs initially in a patchy fashion. The resulting lacunae can cause light scatter, contributing to the sensation of floaters. The liquefaction of vitreous typically starts around the age of four, increases in volume with age, and results in approximately 50% liquefaction of the gel in septuagenarians [6]. Syneresis and synchysis are controversial terms used to describe the process of vitreous liquefaction and protein aggregation; the precise terminology in literature is conflicting, with some authors describing syneresis as liquefaction and other authors describing syneresis as protein aggregation [7, 8]. Due to the conflicting terminology, this book chapter uses descriptive terms, vitreous liquefaction, and vitreous protein aggregation.

## 3  Posterior Vitreous Detachment

Posterior vitreous detachment (PVD) is a very common process where the typically adherent vitreous cortex separates from the neurosensory retina. Aging and pathologic processes lead to liquefaction and condensation of proteins within the body of the vitreous and weakening of adhesions between the posterior vitreous cortex and the internal limiting membrane (ILM). These biochemical changes combined with mechanical forces of a rotating globe lead to a separation of the posterior hyaloid from the ILM (Fig. 1). This typically starts in the macula, and separation planes between the vitreous cortex and neurosensory retina allow chronic or acute progression of the detachment anteriorly. Firm points of attachment between the cortex and the ILM include the fovea, optic nerve, and vitreous base, as well as major blood vessels, border of lattice degeneration, and chorioretinal scars. These firm points of attachment punctuate the four stages of PVD. Stage 1 PVD involves a partial PVD in the macula, with persistent cortical attachment to the fovea, optic nerve, and midperipheral retina. Stage 2 PVD involves a partial PVD progressing perifoveally, with persistent cortical attachment to the fovea, optic nerve, and midperipheral retina. Stage 3 PVD involves further progression through the fovea, with persistent cortical attachment to the optic nerve and midperipheral retina. Stage 4 PVD is considered a total PVD, with a visible Weiss Ring and persistent cortical attachment to the vitreous base only [9].

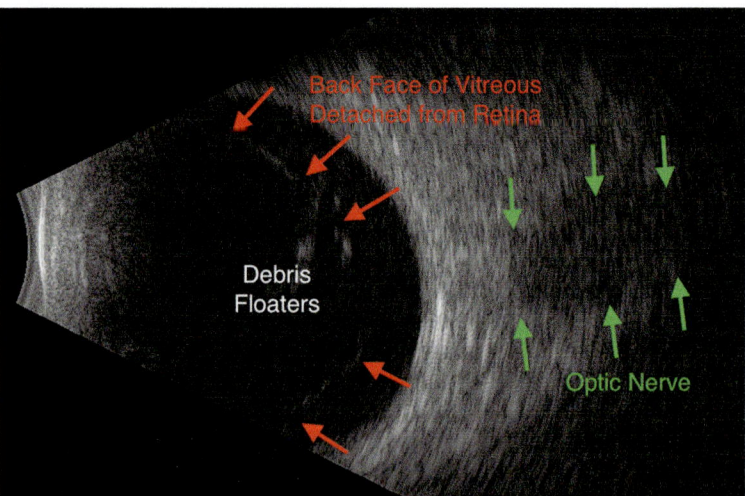

**Fig. 1** Posterior vitreous detachment. This B-scan ultrasound image clearly reveals the posterior hyaloid, detached from the retina. Vitreous debris are noted

Firm points of attachment between the vitreous cortex and retinal structures also contribute to vitreoretinal disorders, when accelerated vitreous liquefaction outpaces the age-related decline in vitreoretinal adhesion, resulting in pathologic separation of cortex from retina. When the vitreous cortex separates, the mechanical forces can result in symptomatic flashes and condensation of cortex can result in floaters. If these large floaters are very symptomatic, a pars plana vitrectomy (PPV) or, more recently, as an investigational option, YAG vitreolysis can be considered [9].

# 4  Anomalous Posterior Vitreous Detachment

## 4.1  Vitreomacular Traction

Vitreomacular traction (VMT) is a spectrum of disease where the vitreous cortex does not separate normally from the macula. Vitreomacular adhesion (VMA) is the least severe stage of disease, where a partial PVD is present but foveal attachment persists without distortion of the normal retinal contour (i.e., a stage 2 PVD, part of normal PVD evolution). In one observational, cross-sectional study of 335 eyes in 271 patients examined by OCT and ultrasonography, vitreofoveal adhesion was present in 19% of eyes, as assessed by OCT. As expected, the prevalence of vitreofoveal adhesion decreased with age, affecting 26% of patients <70 years of age compared to 9% of patients >80 years of age [10].

In VMT, as traction progresses in the setting of a firm vitreous cortex-foveal adhesion, the foveal anatomy becomes increasingly distorted by centripetal vitreoretinal traction. Patients typically present during the sixth to eighth decade of life and experience metamorphopsia and central visual blur when forces result in distortion of the normal foveal contour. In early stages, the anomalous adhesion spontaneously resolves in 50% of cases. If symptomatic and chronic, practitioners can consider ocriplasmin, a protease, to degrade proteins, including laminin which is a key component of the vitreomacular interface. Patients with focal adhesions tend to respond better to this chemical vitreolysis [11]. Alternatively, surgical lysis of the vitreomacular adhesion can be considered.

## 4.2  Idiopathic Macular Hole

An idiopathic macular hole is hypothesized to be a sequelae of vitreomacular traction, resulting from avulsion of the vitreous cortex from the macula and the formation of a partial- to full-thickness neurosensory retinal break [12]. Patients are variably symptomatic, from a mild central blur and metamorphopsia to a large central scotoma. Macular holes can be also staged according to their appearance, with stage 0 being an anomalous PVD with persistent foveal attachment. A stage

1 hole is either a foveal detachment (1A) or an isolated break in the outer fovea (1B). A stage 1A macular hole appears ophthalmoscopically as a central yellow spot with loss of foveolar depression and represents early serous detachment of foveolar retina. It can sometimes be mistaken for a solitary foveal soft druse on clinical exam. A stage 1B macular hole appears ophthalmoscopically as a 250-μm yellow ring forming an irregular slightly elevated ridge and results from centrifugal accumulation of foveolar xanthophyll [13]. A stage 2 hole is an eccentric or small, round full-thickness defect <400 μm in diameter. In many of these cases, an eccentric tear forms in the contracted prefoveolar vitreous cortex, producing a pseudo-operculum [13]. In the past, these tears have been thought to represent full-thickness retinal flap tears, or true opercula. A stage 3 hole is a full-thickness hole >400 μm in diameter and is typically surrounded by a small rim of thickened and localized detached perifoveal retina. It can be accompanied by an overlying, free pseudo-operculum. Yellow, clumped RPE deposits can be noted in the base of the hole [13]. Finally, a stage 4 hole is a full-thickness hole with a complete PVD.

Macular holes are treated based upon their symptomology, stage, chronicity, and etiology. Stage 1 holes have a 50% chance of spontaneous resolution following vitreofoveolar separation. Stage 2 through 4 holes have less than a 10% chance of spontaneous resolution, therefore a PPV with gas tamponade and ILM peel can be considered for symptomatic patients. ILM peeling is thought to reduce the risk of hole recurrence through elimination of a scaffolding for cellular proliferation and subsequent mechanical tangential traction [14].

## 4.3 Epiretinal Membrane

An epiretinal membrane (ERM) is a fibrocellular proliferation at the vitreoretinal interface that results in distortion of the underlying retina and can cause visual symptoms ranging from asymptomatic to debilitating central metamorphopsia and reduced acuity [15]. The membrane is composed of retinal and extraretinal cells and extracellular matrix components. The cellular source for the membrane includes glial cells, hyalocytes, macrophages, retinal pigment epithelial cells, fibroblasts, and myofibroblasts. Epithelial mesenchymal transition (EMT) of retinal pigment epithelium (RPE) cells that escape through retinal breaks is one potential source that may play a role in ERM [16]. The transdifferentiation of various cell types into myofibroblasts makes the precise origin of the cellular component difficult to determine. The cellular proliferation produces excessive and disorganized collagen and other extracellular matrix components, and contraction of the fibrocellular membrane results in distortion of the underlying retina [17]. Therapeutically, an ERM can be surgically removed through PPV, with or without the removal of the underlying ILM.

## *4.4 Vitreous Hemorrhage*

Vitreous cortex has increased adherence to retinal vessels and as the vitreous detaches, tractional forces can cause blood vessels to shear and then bleed into the vitreous. Smaller vitreous hemorrhages tend to resolve spontaneously, but larger hemorrhages often need to be cleared through PPV. Ocular ultrasonography is indicated for vitreous hemorrhage to rule out retinal tear or detachment, either of which would prompt urgent PPV and repair. A unique sequela of vitreous hemorrhage, particularly in eyes with severe and chronic pathology, is cholersterolosis bulbi (alternatively known as synchysis scintillans), in which cholesterol crystals accumulate subretinally, within the vitreous body or anterior chamber following resolution of the hemorrhage [18].

## *4.5 Retinal Tears and Detachments*

Strong adhesion between the vitreous and peripheral retina can result in a retinal tear with the progression of a posterior vitreous detachment. Tears are typically symptomatic with flashes and floaters, and vision can significantly decline if vitreous hemorrhage or retinal detachment occurs. Early recognition and timely treatment with laser barricade is essential to prevent the development of a retinal detachment. If retinal detachment occurs, the status of macular attachment and chronicity dictate the urgency of repair. When the macula is attached and the onset is acute, urgent repair is indicated; following macular detachment, the urgency of immediate repair diminishes.

## 5 Macular Edema

Vitreous cortex may play a role in macular edema from various etiologies, including pseudophakic cystoid macular edema (Irvine-Gass syndrome), diabetic macular edema, retinal vein occlusion, uveitis, and vitreomacular traction syndrome [19]. It is thought that traction on macular vasculature can compromise vessel integrity, leading to increased permeability and macular edema. Another hypothetical mechanism for vascular instability includes the sequestration of permeability factors in the gelatinous, native vitreous. Several authors have reported case series suggesting benefit of PPV with posterior hyaloid excision in the treatment of diabetic macular edema, especially those cases in which OCT suggests traction [20]. However, medical treatment with anti-vascular endothelial growth factor (anti-VEGF) therapy or steroids has a more favorable risk–benefit profile, and PPV plays a role only in recalcitrant cases and instances of vitreomacular traction.

# 6   The Role of Vitreous in Retinal Neovascularization

## 6.1   Proliferative Diabetic Retinopathy

The vitreous acts as a scaffold for retinal neovascularization (NV), which grows into the vitreous, instead of in the plane of the retina, in ischemic retinopathies, most commonly in proliferative diabetic retinopathy (PDR). The vitreous cortex can shift during PVD, causing traction on the NV, which shears and hemorrhages into the vitreous. In addition, NV in the vitreous cortex can fibrose, contract, and lead to tractional retinal detachment. In more severe cases of PDR, specifically those with tractional retinal detachment or severe non-clearing vitreous hemorrhage, PPV is indicated. Pars plana vitrectomy is clearly beneficial for the treatment of advanced active PDR, not only excising the vitreous scaffold on which NV grows but also excising the vitreous cortex which plays a role in NV traction and vitreous hemorrhage.

## 6.2   Retinopathy of Prematurity

Retinopathy of prematurity (ROP) is a developmental pathology of retinal vasculogenesis, where peripheral retinal ischemia results in neovascularization and in severe cases retinal detachment. Retinal vascular development begins at the optic disc at 16 weeks gestation and normally reaches the nasal and temporal ora serrata at 36 and 40 weeks, respectively. Suboptimal oxygenation and comorbid conditions, such as sepsis, result in premature closure of peripheral capillaries and prevent normal vascular development. The pathogenesis is complex, but principal factors include suboptimal postnatal retinal oxygenation and abnormal systemic growth factor production, both of which are a result of extreme prematurity, multi-organ insufficiency, and resuscitative efforts. This peripheral retinal ischemia can result in characteristic vascular changes, the early stages of which are a retinal demarcation line (Stage 1 ROP) that can develop into a ridge, with height and width (Stage 2 ROP). As disease progresses, neovascularization develops at this ridge and blood vessels extend into the vitreous (Stage 3 ROP). The blood vessels can then fibrose and contract, which results in a subtotal retinal detachment (Stage 4 ROP) and can culminate in a total retinal detachment (Stage 5 ROP). Early identification of ROP is crucial for early treatment and a reduction in the risk of development of retinal detachment. Peripheral retinal ablation of ischemic retina, either with laser or cryotherapy, increases the chances of favorable outcomes in select patients [21]. Anti-VEGF therapy has also demonstrated efficacy, but the ideal dosage, injection frequency, and follow-up remain unclear [22].

## 6.3   Familial Exudative Vitreoretinopathy

Familial exudative vitreoretinopathy is a genetic condition caused by mutations in genes that promote retinal vasculogenesis. The hallmark of FEVR is the failure of the temporal retina to properly vascularize, resulting in peripheral ischemia with ensuing VEGF-driven exudation and fibrovascular proliferation. As in PDR, this abnormal tissue can result in vitreous hemorrhage, retinal folds, tractional and exudative retinal detachment as well as a large positive angle kappa due to dragging of the macula. The phenotype is similar to retinopathy of prematurity (ROP), and the two can typically be distinguished by the premature birth and early critical care found in ROP. It is typically inherited in an autosomal dominant manner, but autosomal recessive and X-linked recessive forms have also been described. Several causative genes have been identified, but remain unidentified in numerous other cases, and the function of the known genes has not been fully elucidated. Four of the five implicated genes affect the Norrin/Frizzled4 signaling pathway, suggesting an important role in retinal vascular development.

Diagnosis is facilitated by wide field fluorescein angiography, which can visualize capillary loss typical of ischemic retina. Treatment depends on the stage of disease. Mild disease, without neovascularization or exudation, is typically observed. Neovascular disease is typically treated with peripheral retinal photocoagulation in the area of ischemic retina. Fibrovascular traction, resulting in detachment or significant distortion of the retina, can be treated surgically. Referral to a geneticist should be considered for possible genetic testing and counseling of family members. If the *LRP5* mutation is identified, dual energy X-ray absorptiometry (DEXA) scans should be performed to assess bone mineral density due to the high rate of associated osteopenia and osteoporosis [23].

# 7   Metabolic Disorders

## 7.1   Asteroid Hyalosis

Asteroid hyalosis is a benign and frequently asymptomatic accumulation of small 10–100 nm clumps of fat, phospholipids, and calcium in the vitreous. The pathogenesis is poorly understood. The vitreous aggregates, if in large enough quantity, can obscure visualization of the posterior pole, and fluorescein angiography can provide superior visualization of the retina over indirect ophthalmoscopy due to the lack of reflectance of the fluorescent wavelength coming from the retinal vasculature. In contrast, the wavelengths of light from the indirect headset reflect off of the calcified deposits, significantly reducing image contrast of underlying retina. If significantly affecting quality of vision or obscuring underlying retina from effective and necessary monitoring, PPV can remove asteroid hyalosis.

## 7.2 Amyloidosis

Amyloid deposits are aggregates of low-molecular-weight proteins with a beta-pleated sheet conformation in which the adjacent strands of polypeptides run in opposite directions ("anti-parallel"). Deposition of amyloid can occur in almost any organ to result in a myriad of diseases, each possessing a constellation of clinical findings. Clinicians have classified amyloidoses as follows: (1) primary amyloidosis, with no coexisting systemic disease; (2) multiple myeloma-associated amyloidosis; (3) secondary or reactive amyloidosis associated with chronic infectious disease or chronic inflammatory disease; (4) heredofamilial amyloidosis associated with some cardiovascular, renal, or neuropathic syndromes, or with familial Mediterranean fever; (5) local amyloidosis, resulting in single organ system deposition without systemic disease; (6) aging associated amyloidosis, often in the brain and heart.

Amyloid can affect the eye and adjacent structures in many ways. A hereditary form of systemic amyloidosis, designated as familial amyloidotic polyneuropathy (FAP), involves the vitreous and the peripheral nerves. FAP had been traditionally subdivided into four types. Type I FAP includes vitreous amyloidosis with an autonomic and peripheral neuropathy affecting the lower extremities most frequently. Type II FAP also develops vitreous amyloidosis and peripheral neuropathy, but the upper extremities, instead of the lower extremities, are affected first, and there is often an associated cardiomyopathy. Carpal tunnel syndrome can be present. FAP types I and II are caused by accumulation of mutant transthyretin. Patients with type III and IV FAP do not develop vitreous opacities.

The vitreous opacities can mimic infectious, inflammatory, and neoplastic etiologies. Pars plana vitrectomy can be employed diagnostically and therapeutically for vitreous amyloidosis. It is indicated when the vitreous deposits substantially reduce visual acuity.

## References

1. Schneider, E.W. and M.W. Johnson, *Emerging nonsurgical methods for the treatment of vitreomacular adhesion: a review.* Clin Ophthalmol, 2011. **5**: p. 1151-65.
2. Snead, M.P., et al., *Stickler syndrome, ocular-only variants and a key diagnostic role for the ophthalmologist.* Eye (Lond), 2011. **25**(11): p. 1389-400.
3. Robin, N.H., R.T. Moran, and L. Ala-Kokko, *Stickler Syndrome*, in *GeneReviews(R)*, M.P. Adam, et al., Editors. 1993, University of Washington, Seattle
4. Meredith, S.P., et al., *Clinical characterisation and molecular analysis of Wagner syndrome.* Br J Ophthalmol, 2007. **91**(5): p. 655-9.
5. Shastry, B.S., *Persistent hyperplastic primary vitreous: congenital malformation of the eye.* Clin Exp Ophthalmol, 2009. **37**(9): p. 884-90.
6. Milston, R., M.C. Madigan, and J. Sebag, *Vitreous floaters: Etiology, diagnostics, and management.* Surv Ophthalmol, 2016. **61**(2): p. 211-27.
7. Sebag, J., *Age-related changes in human vitreous structure.* Graefes Arch Clin Exp Ophthalmol, 1987. **225**(2): p. 89-93.

8. Le Goff, M.M. and P.N. Bishop, *Adult vitreous structure and postnatal changes.* Eye (Lond), 2008. **22**(10): p. 1214-22.

9. Johnson, M.W., *Posterior vitreous detachment: evolution and complications of its early stages.* Am J Ophthalmol, 2010. **149**(3): p. 371-82.e1.

10. Schwab, C., et al., *Prevalence of early and late stages of physiologic PVD in emmetropic elderly population.* Acta Ophthalmol, 2012. **90**(3): p. e179-84.

11. Neffendorf, J.E., et al., *Ocriplasmin for symptomatic vitreomacular adhesion.* Cochrane Database Syst Rev, 2017. **10**: p. Cd011874.

12. Gass, J.D., *Idiopathic senile macular hole. Its early stages and pathogenesis.* Arch Ophthalmol, 1988. **106**(5): p. 629-39.

13. Krzystolik, M.G., T.A. Ciulla, and A.R. Frederick, Jr., *Evolution of a full-thickness macular hole.* Am J Ophthalmol, 1998. **125**(2): p. 245-7.

14. Bainbridge, J., E. Herbert, and Z. Gregor, *Macular holes: vitreoretinal relationships and surgical approaches.* Eye (Lond), 2008. **22**(10): p. 1301-9.

15. Ciulla, T.A. and R.D. Pesavento, *Epiretinal fibrosis.* Ophthalmic Surg Lasers, 1997. **28**(8): p. 670-9.

16. Shu, D.Y. and F.J. Lovicu, *Myofibroblast transdifferentiation: The dark force in ocular wound healing and fibrosis.* Prog Retin Eye Res, 2017.

17. Bu, S.C., et al., *Idiopathic epiretinal membrane.* Retina, 2014. **34**(12): p. 2317-35.

18. Wand, M., T.R. Smith, and D.G. Cogan, *Cholesterosis bulbi: the ocular abnormality nown as synchysis scintillans.* Am J Ophthalmol, 1975. **80**(2): p. 177-83.

19. Golan, S. and A. Loewenstein, *Surgical treatment for macular edema.* Semin Ophthalmol, 2014. **29**(4): p. 242-56.

20. Haller, J.A., et al., *Vitrectomy outcomes in eyes with diabetic macular edema and vitreomacular traction.* Ophthalmology, 2010. **117**(6): p. 1087-1093.e3.

21. *Revised indications for the treatment of retinopathy of prematurity: results of the early treatment for retinopathy of prematurity randomized trial.* Arch Ophthalmol, 2003. **121**(12): p. 1684-94.

22. Mintz-Hittner, H.A., K.A. Kennedy, and A.Z. Chuang, *Efficacy of intravitreal bevacizumab for stage 3+ retinopathy of prematurity.* N Engl J Med, 2011. **364**(7): p. 603-15.

23. Gilmour, D.F., *Familial exudative vitreoretinopathy and related retinopathies.* Eye (Lond), 2015. **29**(1): p. 1-14.

# Vitreous Imaging

**Adam T. Chin and Caroline R. Baumal**

**Abstract** The anatomy of the vitreous and the vitreoretinal interface is important to understand its role in various disease states. However, the transparent nature of the vitreous presents a unique challenge for characterization of its anatomy. This chapter explores the techniques utilized to image the vitreous including slit lamp biomicroscopy, optical coherence tomography (OCT), B-scan ultrasonography, magnetic resonance imaging (MRI), and in vitro techniques. An overview of these technologies with their clinical applications is highlighted.

## 1 Introduction

Despite being the largest anatomic structure of the human eye, the vitreous has historically been difficult to assess clinically and histopathologically. The vitreous is gel-like extracellular matrix composed primarily of water with collagen fibrils and hyaluronan (also referred to as hyaluronic acid) [1]. As such, traditional means of tissue preparation lead to dehydration and introduce artifacts that have led anatomists to varied conclusion regarding the structure of the vitreous. The molecular organization of collagen and hyaluronan is exquisitely evolved to be almost completely optically transparent allowing 90% of incident light transmission to the retina. Therefore, in vivo examination with visible light has also challenged those who study the vitreous. Fortunately, the last few decades have ushered in new imaging technologies that enable more routine study of the normal vitreous anatomy and the vitreoretinal interface in vivo.

The transparency and apparent inactivity of the vitreous belie a complex, dynamic architecture which is important in ocular pathophysiology. The vitreous' role in the pathophysiology of numerous ocular conditions including vitreomacular

A. T. Chin · C. R. Baumal (✉)
Department of Ophthalmology/Vitreoretinal Surgery, New England Eye Center, Tufts University School of Medicine, Boston, MA, USA
e-mail: cbaumal@tuftsmedicalcenter.org

© Springer Nature Switzerland AG 2019                                        289
G. Guidoboni et al. (eds.), *Ocular Fluid Dynamics*, Modeling and Simulation in
Science, Engineering and Technology, https://doi.org/10.1007/978-3-030-25886-3_12

traction, macular hole, diabetic macular edema, proliferative diabetic retinopathy, proliferative vitreoretinopathy, and retinal detachment has been recognized [1]. Further, the vitreous has become an important target for pharmacotherapy. With the expanding role of intravitreal injections in the treatment of neovascularization and inflammation, it is important to understand the vitreous humor as a depot for therapeutics. Finally, as real-time imaging of the vitreous and vitreoretinal interface advances, there is a potential for intraoperative imaging to guide vitreoretinal surgeons [2].

## 2   Anatomy and Physiology of the Vitreous

The vitreous is a gel-like structure occupying the space between the internal limiting membrane (ILM) on the surface of the retina and the posterior lens capsule [3] It establishes an optically clear visual axis, and its biochemical properties inhibit cellular migration and proliferation [4]. The peripheral vitreous, or cortex, is composed of densely packed collagenous fibers running parallel to the surface of the retina and hyalocytes. The peripheral cortex attaches firmly to the posterior lens capsule, optic nerve, retinal vasculature, macula, and vitreous base (Fig. 1a). In the core or central vitreous, collagen fibers are less densely arranged and tend to run antero-posteriorly. A space known as *Berger's space* exists posterior to the lens and is bordered by the cortical vitreous adherence to the posterior lens capsule. Berger's space connects to a central canal, called the *hyaloid* or *Cloquet's canal*, which is a small, transparent canal running through the vitreous body from the posterior lens to the optic nerve. During fetal development, the hyaloid canal contains the hyaloid artery which is an extension of the central retinal artery and nourishes the developing lens. The *area of Martegiani* is a space that lies anterior to the optic nerve head and represents the posterior terminus of the embryologic system of canals and spaces in the vitreous [5].

With increasing age, the vitreous undergoes a process of progressive liquefaction in which the organized interaction between collagen and hyaluronan breaks down, and collagen fibrils aggregate. Within the first decade of life, a liquid pocket known a posterior premacular vitreous pocket, or *bursa premacularis*, can already be identified (Fig. 1b). By the second decade, one-fifth of the vitreous is liquid, and by the ninth decade, this fraction has increased to approximately one half [3]. Simultaneously, the adherence between the ILM of the retina and the posterior cortical vitreous weakens. With age, the collagen aggregates at the vitreous base as it widens [3]. Gradually, the vitreous gel liquefies in a process known as synchisis, and shrinks (syneresis) leading to separation or detachment of the posterior vitreous cortex from the ILM allowing liquefied vitreous fluid into this space. Typically, the posterior vitreous first detaches at the perifoveal macula while tight adherence at the fovea and peripapillary region remains. Next, the foveal vitreous is released. Ultimately, the vitreous separates from the peripapillary region and the entire posterior vitreous detaches. These degenerative vitreous changes are normal with aging. By 70 years

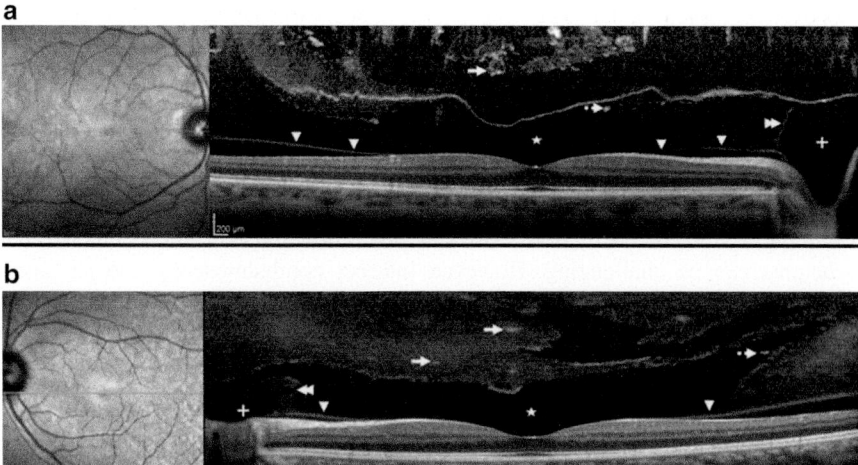

**Fig. 1** (**a**) Enface scanning laser ophthalmoscopy (SLO) and SD-OCT (Spectralis® HRA+OCT, Heidelberg Engineering, Germany) line scan through the fovea and optic nerve of a healthy 40 year-old male demonstrates normal anatomic structures of the posterior vitreous, vitreoretinal interface and retina. Anterior to the fovea, the bursa premacularis (star) appears as a hyporeflective space. Scattered bursa premacularis granular opacities (dotted arrow) and cortical vitreous granular opacities (arrow) can be seen. Adjacent to the bursa premacularis and anterior to the optic nerve, the area of Martegiani (plus) also appears as a hyporeflective space and is separated from the bursa premacularis by a thin septum (double arrowhead). The hyperreflective band attached to the inner retina is the posterior hyaloid of the vitreous cortex (arrowheads). In this individual, the peripapillary and perifoveal vitreous attachments to the retina can be visualized. (**b**) Enface scanning laser ophthalmoscopy (SLO) and SD-OCT (Spectralis HRA+OCT, Heidelberg Engineering, Germany) line scan through the fovea and optic nerve of a 50-year-old female. The bursa premacularis (star) appears as a hyporeflective space anterior to the fovea. Scattered bursa premacularis granular opacities (dotted arrow) and cortical vitreous granular opacities (arrow) can be seen. Adjacent to the bursa premacularis and anterior to the optic nerve, the area of Martegiani (plus) also appears as a hyporeflective space separated from the bursa premacularis by a thin septum (double arrowhead). The hyperreflective band that separates from the inner retina is the posterior hyaloid of the vitreous cortex (arrowheads). The peripapillary and perifoveal vitreous remains adherent around the fovea and edge of the optic nerve

of age, two-thirds of people have a complete posterior vitreous detachment (PVD), in which only the vitreous base remains attached to the peripheral cortical vitreous [6]. As the vitreous detaches, traction induced by persistent or excessive physiologic attachments to the fovea, retinal vasculature, peripheral retina, or optic nerve may lead to the following disorders: vitreomacular traction, macular hole, epiretinal membrane, vitreous hemorrhage, or tear of the retina. There also may be abnormally adherent attachments of the vitreous to the margins of peripheral lattice degeneration or to chorioretinal scars, and PVD may lead to retinal tear or retinal detachment in these eyes. Fortunately, in the majority of people, PVD occurs without any sequelae. The process of PVD may be acute, subacute, or prolonged requiring months for the process to complete. It may be asymptomatic or accompanied by transient photopsia and/or vitreous floaters.

## 3  Ophthalmoscopy/Biomicroscopy

The reduced optical density of the vitreous means direct visualization of its internal structures can be challenging. However, indirect ophthalmoscopy and slit lamp biomicroscopy remain integral techniques for clinical evaluation of the vitreous and vitreoretinal interface. In order to improve visualization of vitreous structures, adequate pupillary dilation, dark adaptation of the examiner, and maximizing the angle between the observer and incident illumination, as in the case of slit lamp biomicroscopy, are helpful [7].

Indirect ophthalmoscopy provides a binocular view of the vitreous via a head-piece and handheld condensing lens. A light source in the headset of the indirect ophthalmoscope produces diffuse light that illuminates the eye. A condensing lens, typically powered between +14 and +30 diopters, gathers divergent light leaving the eye to produce a magnified horizontally and vertically inverted image. Binocularity improves depth perception allowing the viewer to appreciate features such as the elevation of retinal tear or proliferative vitreoretinopathy. Indirect ophthalmoscopy, however, has limited utility in delineating fine features of vitreous structure, given low magnification and low angle of separation between the illumination source and the examiner. This technique can be combined with digital photography—especially with handheld smartphone or similar devices—to produce a 2-dimensional image [8]. Depression of the sclera while examining peripheral retina and vitreous allows for dynamic evaluation of vitreoretinal attachments.

Slit lamp ophthalmoscopy also provides a binocular view of the vitreous and can be readily combined with digital photography. The high magnification and adjustable slit beam allow for detailed examination of the vitreous. Using a high intensity slit beam of light and focusing the instrument just posterior to the lens, the anterior vitreous is readily visible without the use of additional lenses (Fig. 2). The anterior vitreous examination is clinically important and may be diagnostic. Examples of abnormal findings upon anterior vitreous examination include inflammatory white cells in uveitis and endophthalmitis, suspended retinal pigment epithelial cells in retinal tears, and an optically empty or membranous vitreous in Sticklers syndrome. These findings should be differentiated from normal condensed vitreous fibers of the aging vitreous [9]. Examination of the vitreous in the posterior and peripheral retina requires utilization of the slit lamp in conjunction with a double aspheric handheld lens, Hruby lens, or contact lens. Handheld +60 to +90 diopter lenses provide a horizontally and vertically inverted view of the retina and vitreous, while the −55 diopter Hruby lens lies coaxially when attached to the slit lamp and produces an upright virtual image. With the handheld double aspheric lens, dehiscence of the posterior cortical vitreous can be identified as a subtle hole in the posterior vitreous cortex, gray linear structure in front of the

**Fig. 2** Slit lamp photograph of anterior vitreous of a 39-year-old female with high myopia. With age, the vitreous liquefies (synchysis) and collagen fibers aggregate (syneresis). In this slit lamp photograph, an opaque membrane of aggregated collagen fibers can be seen immediately posterior to the retrolental (Berger's) space

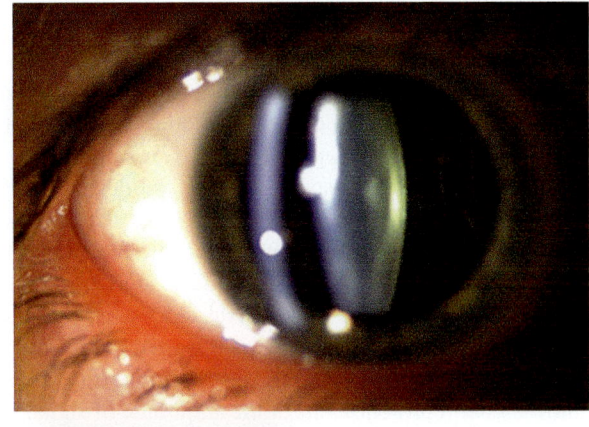

a

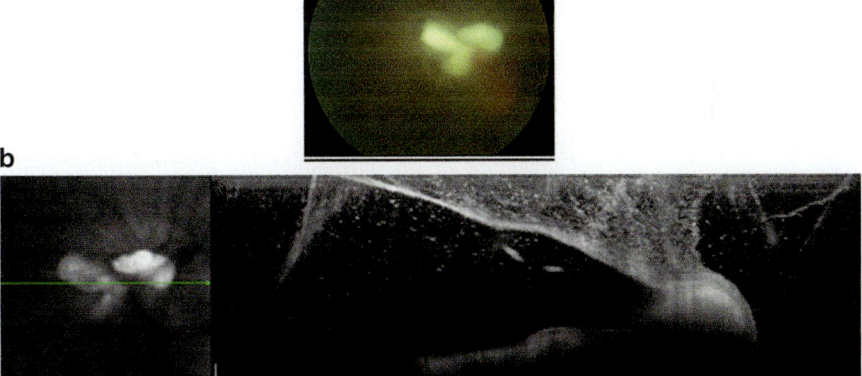

b

**Fig. 3** (**a**) Fundus photo photograph (Topcon, NJ) focused the posterior vitreous of a 27-year-old female with endogenous candida endophthalmitis. There are fluffy white opacities with surrounding vitreous membranes. Vitreous haze and condensed vitreous almost completely obscure the posterior view to the macula, retinal vessels, and optic nerve. (**b**) Enface scanning laser ophthalmoscopy (SLO) and SD-OCT (Spectralis® HRA+OCT, Heidelberg Engineering, Germany) line scan through posterior vitreous opacities in Candida endophthalmitis. Vitreous membranes extend from a central spherical vitreous opacity. The scattered hyperreflective spots represent vitritis

macula, or a ring of glial tissue detached from the peripapillary retina known as a Weiss ring. Posterior vitreous pathology such as inflammatory opacities and vitreous membranes may also be visible (Fig. 3a, b). Widefield contact lenses can be placed directly on the cornea allowing for excellent stereopsis and peripheral vitreous and retinal examination.

# 4   Ultrasonography

Ultrasonography is a technique that uses pulses of high frequency acoustic waves to produce images based on differential acoustic impedance of components in a structure. In typical ophthalmic B-scan ultrasound, frequencies range from 8 to 10 MHz [10]. This sound wave is produced by piezoelectric crystals in the transducer of the ultrasound probe. After leaving the transducer, sound waves propagate through a medium until encountering an interface with a different density, or acoustic impedance. Sound can be scattered, refracted, or reflected back to the transducer as an echo. These reflected waves vibrate the piezoelectric element which in turn generates an electrical signal. More echogenic structures reflect more signal and are thus displaced as a higher intensity signal. A-scan refers to the one-dimensional representations of the amplitude of signal over the axial length. A-scan ultrasound is useful to identify intraocular calcium or foreign bodies, to characterize intraocular tumors, and to measure axial length preoperatively for cataract surgery [11, 12]. The more familiar B-scan provides a two-dimensional cross-sectional image of the tissue.

Widely used as a noninvasive imaging technique across disciplines including cardiology, obstetrics/gynecology, and vascular medicine, ultrasound has been used to study the vitreous since the late 1950s. In 1958, Oksala and Lehtinen published their *Investigations on the Structure of the Vitreous Body by Ultrasound* using a 4-MHz transducer to produce one-dimensional A-scans of the vitreous [13]. Despite advances in acquisition and processing, the resolution of current B-scan ultrasound still remains on the order of 200 $\mu$m [14]. By comparison, the posterior cortex is just over 100 $\mu$m at its thickest point [15]. However, ultrasound is still useful for evaluation of the vitreous and retina in eyes with vitreous hemorrhage, floaters, foreign bodies, vitritis, membranes, vitreoschisis, retinal detachment, and asteroid hyalosis (Fig. 4a–c) [16–18]. Posterior vitreous detachment can be assessed with ultrasound with 80.9% sensitivity and 33.3% specificity compared to exam [19]. Visualization of the posterior cortex is more easily identified in proliferative diabetic vitreoretinopathy related to thickening of the vitreous–retinal interface in diabetic eyes. Beyond imaging in aging and pathologic changes in the vitreous, ultrasound has been used to investigate the location and stability of pharmacologic drug implants injected into the vitreous [20].

Even higher frequency acoustic waves are used for ultrasound biomicroscopy (UBM). With these higher frequencies, the resolution improves but penetration into tissue is reduced. Utilizing frequencies of up to 100 MHz, UBM axial resolution approaches 25 $\mu$m but can only penetrate 4 mm into the eye, thus limiting its applications in vitreous imaging [21]. The utility of UBM has been primarily confined to the anterior segment, although UBM has been utilized to demonstrate vitreous incarceration at post-vitrectomy sclerotomy sites and at intravitreal drug injection sites [22–24].

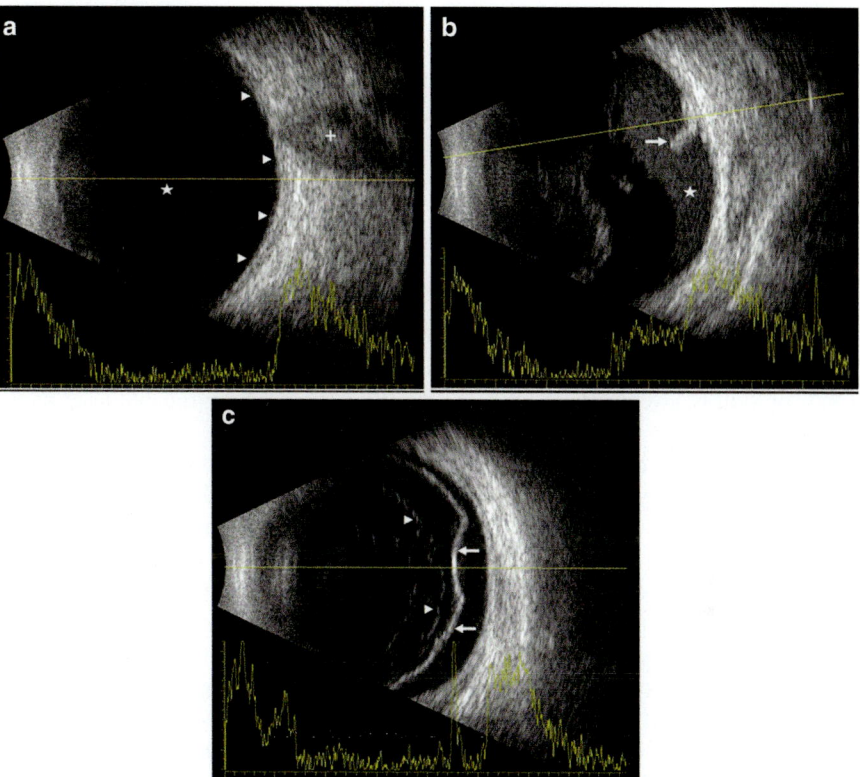

**Fig. 4** (**a**) Ultrasound B-scan (Aviso™, Quantel Medical, France) of normal vitreous, retina, and posterior sclera right eye in a 25-year-old male. Ultrasound waves propagate through ocular tissue from left to right until the sound wave is scattered or reflected back by an interface. Normal vitreous appears hypoechoic (asterisk). The retina and sclera (arrowheads) are often difficult to resolve separately in a normal eye. The optic nerve (plus) appears hypoechoic compared to the surrounding tissue. (**b**) Ultrasound B-scan (Aviso™, Quantel Medical, France) right eye of a 66-year-old man demonstrates traction and a retinal tear masked by vitreous hemorrhage. Hyperechoic material in the vitreous cavity correlates to vitreous hemorrhage (star). A horseshoe tear (arrow) and traction are noted superiorly at the hyperechoic tuft arising from the retina. (**c**) Ultrasound B-scan (Aviso™, Quantel Medical, France) right eye of a 68-year-old female reveals a retinal detachment (arrows) visible as a hyperechoic band with a corresponding high A-scan spike separated from the RPE/choroid. Anterior to the retinal detachment, there are strands of condensed vitreous fibers (arrowheads)

# 5    Optical Coherence Tomography

The ability of optical coherence tomography (OCT) to produce in vivo, noninvasive, depth-resolved images of ocular structures has led to its widespread use in ophthalmology since its introduction in the early 1990s. In a manner analogous to ultrasound, OCT relies on the analysis of backscattering from tissue interfaces but uses properties of reflected light rather than sound [25]. Light emitted from the instrument's light source is split into two beams with one aimed at a mirror to create a reference and the second beam directed at the tissue of interest. The backscattered light from the tissue and the reference beam recombine at the photodetector to produce a pattern of coherence that is used to measure the tissue properties [26, 27]. In the first iteration of the technology, time-domain (TD) OCT, the reference mirror moved (over time) to scan over the range of depths being imaged. The mechanical movement of the mirror limits acquisition speed. Current OCT technology uses spectral-domain (SD) OCT, where the reference mirror is stationary, and the spectrum of the interference pattern produced by the backscattered light is measured. The light source of spectrometer-based SD-OCT is typically a diode laser that produces a broadband light. Swept-source (SS) SD-OCT, or simply swept-source OCT (SS-OCT), uses a rapidly tunable laser as the light source [26].

The introduction of SD-OCT in the mid-2000s opened the door to high-resolution imaging of the vitreous and vitreoretinal interface. While TD-OCT delivers axial resolution of up to 10 $\mu$m, spectral domain achieves resolution of 3–7 $\mu$m [28]. Improved resolution enables identification of the normal anatomical structures of the vitreous and describes the changes that produce pathologic conditions. While post-mortem studies had previously revealed the presence of Cloquet's canal, Kagemann et al. used a prototype ultra-high-resolution SD-OCT to demonstrate its presence in vivo and show that it persists in adults [29]. Spectral-domain imaging by Uchino and Johnson showed that detachment of the posterior vitreous is typically a chronic, slow process that begins with the separation of the vitreous from the perifoveal macula rather than an acute process [30, 31]. Further SD-OCT studies produced three-dimensional representations of the vitreoretinal interface in VMT, ERM, lamellar holes, pseudoholes, and full thickness macular holes [28, 32, 33]. OCT showed that VMT typically coexists with ERM, which likely enhances adherence between the ILM and posterior vitreous cortex. Despite improved visualization of the vitreous interface, authors found that SD-OCT had relatively less success in definitively identifying PVD status [34]. Cases of complete PVD could be missed because the posterior cortex may have moved anteriorly and be outside of the imaging range of the OCT scans. More recently, several authors have used SD-OCT to quantify the vitreous inflammation in uveitis and evaluate its response to treatment [35–37].

Since the mid-2010s, swept-source OCT (SS-OCT) has ushered in a new era of studying the vitreous and vitreoretinal interface. Prior to SS-OCT, spectrometer-based spectral-domain OCT (SD-OCT) was limited to scanning rates of 70,000. In clinical practice, low scanning rates correspond to more movement artifacts such as doubling artifacts caused by saccades [38, 39]. Current SS-OCT systems are able to acquire up to 100,000 A-scans/s, and operate at wavelengths over 1000 nm to achieve axial resolution of 3–5 $\mu$m. At such high scanning rates, movement artifacts are decreased. Further, the longer SS-OCT wavelength provides better resolution, longer imaging range, and less light scattering off dense media [40]. These correspond to less motion artifact and improved resolution [41]. SS-OCT also demonstrates less sensitivity roll-off for better imaging of deeper structures

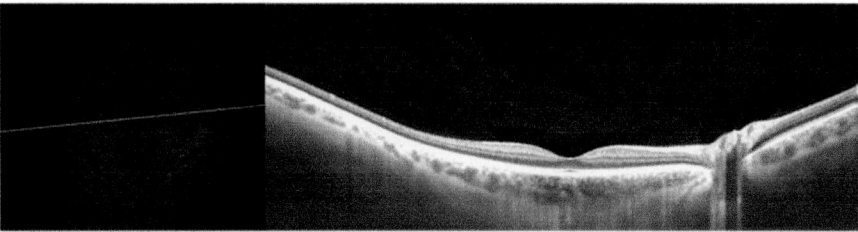

**Fig. 5** PLEX Elite 9000 SS-OCT (Carl-Zeiss Meditech, Dublin, CA) line scan through the fovea and optic nerve of a 30-year-old man. SS-OCT provides longer imaging range and less sensitivity roll-off for better imaging of deeper structures such as the choroid. In the scan above, high resolution is maintained from the posterior vitreous to the choroid

such as the choroid (Fig. 5). In addition to improved resolution and speed, SS-OCT imaging can be processed to improve visualization of the vitreous. Enhanced vitreous imaging or EVI, analogous to windowing in radiology, optimizes the visualization of vitreous structures by adjusting the logarithmic gray scale presented in the image [41].

SS-OCT is sensitive in the identification of posterior vitreous features including the posterior precortical vitreous pocket, Cloquet's canal, Bergmeister papillae, posterior hyaloid detachment, and papillomacular detachment, which has broadened our understanding of these structures [41]. The posterior precortical vitreous pocket (PPVP), or bursa premacularis, which was first identified via anatomical dye studies by Worst, appears as a boat-shaped hypointense pocket on the SS-OCT B-scan anterior to the posterior pole [41, 42]. The PPVP, present in young children, enlarges with time, and connections form between the PPVP and Cloquet's canal [43, 44]. Further, SS-OCT imaging of the PPVP shows that the anterior border of the PPVP shifts anteriorly in supine position compared to upright position [45]. Alasil and colleagues compared features of SD-OCT to SS-OCT in an eye with asteroid hyaloid (AH) [39]. Affecting 1 in 200 persons, this relatively common vitreous disorder refers to calcium–lipid refractile deposits that develop in the vitreous cavity (Fig. 6a). Imaging is particularly important to differentiate AH from other causes of vitreous opacities and to characterize the retina when the refractile asteroid bodies obscure direct visualization. Cirrus HD-OCT 3-D imaging (Carl-Zeiss Meditech, Dublin, CA) reveals multiple hyperreflective spots which are axially stretched as well as artifacts related to positioning of the asteroid hyalosis deposits near the zero-delay line (Fig. 6b). The SD-OCT one-line raster (Cirrus) failed to show the asteroid deposits (Fig. 6c). However, widefield SS-OCT was able to more appropriately identify the positioning of the asteroid deposits in the vitreous, and imaging could be adjusted to remove reflection artifacts (Fig. 6d).

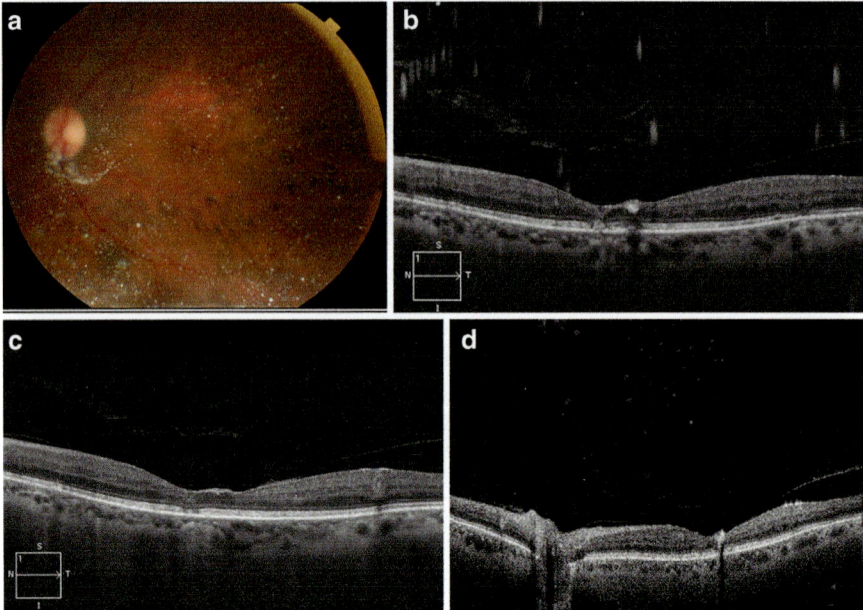

**Fig. 6** (**a**) Color fundus photography (Topcon, NJ) shows asteroid hyalosis in a 65-year-old diabetic male. The refractile asteroid deposits obscure the finer retina details, including prior panretinal laser photocoagulation for proliferative diabetic retinopathy. (**b**) Cirrus HD-OCT 3-D (Carl-Zeiss Meditech, CA) imaging of the vitreous in asteroid hyalosis reveals multiple hyperreflective spots which are axially stretched as well as artifacts related to positioning of the asteroid hyalosis deposits near the zero-delay line. (**c**) Cirrus SD-OCT scan (Carl-Zeiss Meditech, CA) single line of a patient with asteroid hyalosis. The single-line scan failed to show the asteroid deposits. (**d**) Widefield SS-OCT prototype (Massachusetts Institute of Technology, Cambridge, MA) scan. The scan appropriately demonstrates the position of the asteroid deposits (arrow) in the vitreous without introducing reflection artifacts as seen in the HD-OCT 3-D imaging

# 6    Magnetic Resonance Imaging

Magnetic resonance imaging (MRI) is a noninvasive imaging technique that ana-lyzes changes in the protons induced by a magnetic field to produce an image. MRI instruments consist of three basic components: a strong electromagnet, a radiofrequency coil that transmits and receives radio signals, and a computer to reconstruct the signal. When protons of water molecules in a tissue are placed in the strong magnetic field of the MRI, they act as dipoles and orient themselves either parallel or antiparallel direction to the vector of the field. The radiofrequency coil then delivers a pulse of radiofrequency which magnetizes the proton longitudinally along the axis. At the same time, the magnetic field induces transverse rotation of the protons off of the magnetic axis. As the protons return to their resting state, they release a radiofrequency corresponding to the energy absorbed. The energy release during dephasing along the magnetic axis is called T1 relaxation, and T2 relaxation is due to dephasing off the magnetic axis. T1 and T2 relaxation depend on the concentration and mobility of water within the tissue [15, 46]. Coils receive the weak radiofrequency emitted by the protons returning to their resting state and convert it to a signal that can be interpreted by the computer. Certain tissue characteristics can be optimized by delivering radiofrequency pulses of variable number, strength, and pattern.

MRI was first used to image the vitreous in 1984 [47]. Since then, magnetic resonance imaging has been used to detect aging changes of the vitreous [48], intraocular masses [49, 50], and persistent hypertrophic primary vitreous [51]. In addition to structural data, MRI can be utilized to noninvasively measure physiologic parameters including the partial pressure of vitreous oxygen [52, 53], clearance of drugs from the vitreous [54, 55], diffusion of water in the vitreous [56], and elasticity of the vitreous [57]. In a study of the effect of vitrectomy on vitreous oxygen, Simpson et al. showed that the partial pressure of oxygen in the vitreous cavity increases from 13.2 mmHg to 34.5 mmHg after pars plana vitrectomy [58]; the increased oxygen and subsequent reactive oxygen species may be linked to the pathophysiology of post-vitrectomy cataract formation.

# 7   In Vitro Imaging Techniques

The structure of the collagen fibers is best observed with dark-field microscopy. Until the twentieth century, histologic approaches for preparing the vitreous were limited by the introduction of artifacts in the fixation process. As is seen in the aging and diabetic vitreous, collagen fibrils tend to cross-link and condense when the microenvironment established by hyaluronan and other molecules is disrupted. Pioneered by Sebag and Balazs, the technique of dark-field microscopy on minimally preserved vitreous limits artifacts introduced by early processes. In their technique, Sebag and Balazs dissect fresh, unfixed eyes by removing the sclera and choroid before hydrodissection of the retina from the vitreous [59, 60]. The intact vitreous—still adherent to the anterior structures—is then fixed in position via sutures through the limbus before being placed in balanced salt solution. The overall effect is an undisturbed vitreous that can be imaged with orthogonal light to maximize the visualization of its internal anatomy. Through this technique, vitreous fibrils can be seen running anteroposterior through the center of the vitreous—especially in the peripapillary region—and inserting into the vitreous base anteriorly. Dark-field microscopy reveals that collagen fibers become more prominent and less organized with increasing age [7].

To visualize the internal bursas and canals of the vitreous, Worst developed a technique of injecting dye into the vitreous to delineate the structures. In a method similar to that of Sebag and Balazs, the sclera, choroid, and retina are removed leaving the intact vitreous body. With the vitreous suspended in an aqueous solution, dyes are injected into the bursas and spaces. The technique reveals the bursa macularis surrounded by smaller cisterns, which Worst describes as the corona petaliformis, as well as Cloquet's canal and the area of Martegiani [42, 61]. By observing the flow of these colored dyes, Worst demonstrated that fluid could move between structures such as Berger's space into the bursa premacularis.

# 8  Conclusion

The vitreous plays a key role in embryologic development of the eye, pathogenesis of retinal disorders, inflammatory processes, and intravitreal drug delivery. However, the vitreous is not necessary to maintain ocular health after birth as demonstrated by the excellent lifelong vision quality maintained in eyes after vitrectomy surgery. Advancements in vitreous imaging have led to current theories about the role of the vitreous in the pathogenesis of various retinal disorders. Future therapies targeting vitreous drug release may benefit from high-resolution in vivo imaging.

# References

1. *Diseases of the vitreo-macular interface.* New York: Springer; 2013.
2. Read SP, Fortun JA. Visualization of the retina and vitreous during vitreoretinal surgery: new technologies. *Curr Opin Ophthalmol.* 2017;28(3):238-241.
3. Yanoff M, Duker JS, Augsburger JJ. *Ophthalmology.* 2nd ed. St. Louis, MO: Mosby; 2004.
4. Sebag J. Vitreous: the resplendent enigma. *Br J Ophthalmol.* 2009;93(8):989-991.
5. Sebag J. *The vitreous: structure, function, and pathobiology.* New York: Springer-Verlag; 1989.
6. Foos RY, Wheeler NC. Vitreoretinal juncture. Synchysis senilis and posterior vitreous detachment. *Ophthalmology.* 1982;89(12):1502-1512.
7. *Vitreous: in health and disease.* New York: Springer; 2014.
8. Haddock LJ, Kim DY, Mukai S. Simple, inexpensive technique for high-quality smartphone fundus photography in human and animal eyes. *J Ophthalmol.* 2013;2013:518479.
9. Sharma S, Walker R, Brown GC, Cruess AF. The importance of qualitative vitreous examination in patients with acute posterior vitreous detachment. *Arch Ophthalmol.* 1999;117(3):343-346.
10. Silverman RH. Focused ultrasound in ophthalmology. *Clin Ophthalmol.* 2016;10:1865-1875.
11. Ossoinig KC. Echographic detection and classification of posterior hyphemas. *Ophthalmologica.* 1984;189(1-2):2-11.
12. Ossoinig KC. Modern examination methods of orbital disease. A-scan echography of the orbit. *Trans Am Acad Ophthalmol Otolaryngol.* 1974;78(4):OP581-586.
13. Oksala A, Lehtinen A. Investigations on the structure of the vitreous body by ultrasound. *Am J Ophthalmol.* 1958;46(3 Pt 1):361-366.
14. Webb S. *The Physics of medical imaging.* Bristol; Philadelphia: Hilger; 1988.
15. Sebag J. To see the invisible: the quest of imaging vitreous. *Dev Ophthalmol.* 2008;42:5-28.
16. Chu TG, Lopez PF, Cano MR, et al. Posterior vitreoschisis. An echographic finding in proliferative diabetic retinopathy. *Ophthalmology.* 1996;103(2):315-322.
17. Mamou J, Wa CA, Yee KM, et al. Ultrasound-based quantification of vitreous floaters correlates with contrast sensitivity and quality of life. *Invest Ophthalmol Vis Sci.* 2015;56(3):1611-1617.

18. Stringer CEA, Ahn JS, Kim DJ. Asteroid Hyalosis: A Mimic of Vitreous Hemorrhage on Point of Care Ultrasound. *CJEM.* 2017;19(4):317-320.
19. Woo MY, Hecht N, Hurley B, Stitt D, Thiruganasambandamoorthy V. Test characteristics of point-of-care ultrasonography for the diagnosis of acute posterior ocular pathology. *Can J Ophthalmol.* 2016;51(5):336-341.
20. Manna S, Banerjee RK, Augsburger JJ, Al-Rjoub MF, Correa ZM. Ultrasonographical assessment of implanted biodegradable device for long-term slow release of methotrexate into the vitreous. *Exp Eye Res.* 2016;148:30-32.
21. Pavlin CJ, Foster FS. Ultrasound biomicroscopy in glaucoma. *Acta Ophthalmol Suppl.* 1992(204):7-9.
22. Bhende M, Agraharam SG, Gopal L, et al. Ultrasound biomicroscopy of sclerotomy sites after pars plana vitrectomy for diabetic vitreous hemorrhage. *Ophthalmology.* 2000;107(9):1729-1736.
23. Hodjatjalali K, Riazi M, Faghihi H, Khorami A. Ultrasound biomicroscopy study of vitreous incarceration subsequent to intravitreal injections. *Can J Ophthalmol.* 2012;47(1):24-27.
24. Helvaci S, Sahinoglu-Keskek N, Kiziloglu M, Oksuz H, Cevher S. Vitreous incarceration after ranibizumab injection: an ultrasound biomicroscopy study. *Ophthalmic Surg Lasers Imaging Retina.* 2015;46(4):471-474.
25. Huang D, Swanson EA, Lin CP, et al. Optical coherence tomography. *Science.* 1991;254(5035):1178-1181.
26. Yaqoob Z, Wu J, Yang C. Spectral domain optical coherence tomography: a better OCT imaging strategy. *Biotechniques.* 2005;39(6 Suppl):S6-13.
27. Lavinsky F, Lavinsky D. Novel perspectives on swept-source optical coherence tomography. *Int J Retina Vitreous.* 2016;2:25.
28. Barak Y, Ihnen MA, Schaal S. Spectral domain optical coherence tomography in the diagnosis and management of vitreoretinal interface pathologies. *J Ophthalmol.* 2012;2012:876472.
29. Kagemann L, Wollstein G, Ishikawa H, et al. Persistence of Cloquet's canal in normal healthy eyes. *Am J Ophthalmol.* 2006;142(5):862-864.
30. Uchino E, Uemura A, Ohba N. Initial stages of posterior vitreous detachment in healthy eyes of older persons evaluated by optical coherence tomography. *Arch Ophthalmol.* 2001;119(10):1475-1479.
31. Johnson MW. Posterior vitreous detachment: evolution and complications of its early stages. *Am J Ophthalmol.* 2010;149(3):371-382 e371.
32. Chang LK, Fine HF, Spaide RF, Koizumi H, Grossniklaus HE. Ultrastructural correlation of spectral-domain optical coherence tomographic findings in vitreomacular traction syndrome. *Am J Ophthalmol.* 2008;146(1):121-127.
33. Koizumi H, Spaide RF, Fisher YL, Freund KB, Klancnik JM, Jr., Yannuzzi LA. Three-dimensional evaluation of vitreomacular traction and epiretinal membrane using spectral-domain optical coherence tomography. *Am J Ophthalmol.* 2008;145(3):509-517.
34. Kicova N, Bertelmann T, Irle S, Sekundo W, Mennel S. Evaluation of a posterior vitreous detachment: a comparison of biomicroscopy, B-scan ultrasonography and optical coherence tomography to surgical findings with chromodissection. *Acta Ophthalmol.* 2012;90(4):e264-268.
35. Sreekantam S, Macdonald T, Keane PA, Sim DA, Murray PI, Denniston AK. Quantitative analysis of vitreous inflammation using optical coherence tomography in patients receiving sub-Tenon's triamcinolone acetonide for uveitic cystoid macular oedema. *Br J Ophthalmol.* 2017;101(2):175-179.
36. Keane PA, Karampelas M, Sim DA, et al. Objective measurement of vitreous inflammation using optical coherence tomography. *Ophthalmology.* 2014;121(9):1706-1714.
37. Keane PA, Balaskas K, Sim DA, et al. Automated Analysis of Vitreous Inflammation Using Spectral-Domain Optical Coherence Tomography. *Transl Vis Sci Technol.* 2015;4(5):4.
38. Pang CE, Freund KB, Engelbert M. Enhanced vitreous imaging technique with spectral-domain optical coherence tomography for evaluation of posterior vitreous detachment. *JAMA Ophthalmol.* 2014;132(9):1148-1150.

With ageing, the vitreous typically experiences a degradation process that leads to gel liquefaction (synchysis) and to the formation of liquid pockets in the vitreous chamber. It can also undergo a shrinking process (syneresis) that can cause the vitreous cortex (the outermost layer of the vitreous body) to detach from the retina. This typically happens at the posterior part of the eye, and the process is called posterior vitreous detachment (PVD).

The vitreous body has the optical role of allowing the light that enters the eye through the pupil to reach the retina, which is why it is an avascular and acellular tissue. It also has important mechanical roles. It fills the vitreous chamber, and since the fluid there is under pressure, it contributes to maintaining the almost spherical shape of the eye globe. As discussed in Chap. 9 of this volume, this pressure is determined by a balance between aqueous humour production rate and resistance to outflow. The vitreous also acts as a diffusion barrier that protects the posterior segment of the eye from heat and molecule (particularly oxygen) transport from the front of the eye globe. In fact, it has been demonstrated that when the vitreous body is replaced after vitrectomy with aqueous solution, the intraocular oxygen tension increases, which can lead to the formation of nuclear cataracts and primary open-angle glaucoma [53]. Finally, the vitreous has the important role of holding the retina in contact with the retinal pigment epithelium (RPE). The possible generation of vitreoretinal tractions can lead to retinal tears and, potentially, retinal detachment (RD). Understanding the underlying mechanics is thus relevant for both the physiology and the pathophysiology of the vitreous.

The difficulty in visualising the vitreous body, which is by design invisible [47], makes it extremely challenging to directly measure its dynamics and the role this has on the occurrence of vitreoretinal pathological conditions. Mathematical modelling has significantly contributed to overcoming these difficulties by providing insight in these areas. Section 3 of this chapter revises mathematical models that have been proposed to explain and quantify the generation of vitreoretinal tractions. We consider first the case of the vitreous modelled as a homogeneous viscoelastic fluid (Sects. 3.2, 3.3, and 3.4), which applies to eyes in healthy conditions. We then address the case, that is clinically more relevant, in which vitreous degradation has led to discontinuities in the material properties of the vitreous, such as PVD (Sect. 3.5).

Vitreoretinal tractions are the main factors responsible for the generation of retinal tears, which can, in some circumstances, evolve into a RD. The fluid mechanics of RDs is poorly understood, however, it is generally accepted that it plays a key role in the detachment process. This is a field in which mathematical modelling can provide a significant contribution. In Sect. 4 of this chapter, we describe models of fluid motion in the presence of a RD and discuss how they have contributed to improving our understanding of this pathology. In particular, we consider both rhegmatogenous RD (RRD), which occurs in the presence of a retinal tear (Sect. 4.2) and exudative RD (Sect. 4.3), which occurs when fluid accumulates in the subretinal space.

Two chapters on the biomechanics of the vitreous humour have been recently written by the senior author of this work [35, 41], to which we refer the reader for completeness. We notice that the present chapter focuses on the fluid mechanics of the vitreous and does not address problems associated with the biomechanics of the vitreous that involve solid mechanics. Transport processes in the vitreous chamber are also omitted as they are discussed in detail in [35]. Finally, we focus on mathematical modelling, and only briefly mention experimental works related to vitreous mechanics.

## 2 Rheology of the Vitreous Humour

In a linear elastic solid, the stress depends linearly on the strain, which is a dimensionless measure of the local deformation of the material, with respect to a reference configuration (typically the relaxed one). On the other hand, the concept of reference configuration (and consequently the concept of strain) has little meaning in a fluid, since fluid particles have no preferential position relative to one another. In fact, in a viscous Newtonian fluid the stress depends on the rate of strain. This means that an elastic solid is a material with memory (it has a permanent notion of its relaxed configuration), whereas a Newtonian fluid has no memory at all, since the stress depends only on the actual kinematics. Viscoelastic materials exhibit both solid and fluid behaviour, since in such materials the state of stress depends both on the strain and the rate of strain. In other words, stress is affected by the history of motion through a "fading memory", meaning that the strength of memory decreases for states further back in the past.

The vitreous body is a viscoelastic material and owes its properties to the combined action of hyaluronic acid and collagen fibres. The vitreous body of various animal species has been subjected to various studies, aimed at characterising its rheological behaviour. Most authors focused their measurements on the vitreous complex modulus, $\tilde{G}$, which is a measure of the fast response of a viscoelastic material. In order to measure it, the material sample is subjected to harmonic periodic strains of small amplitude and the corresponding stress is measured, or vice versa [57]. The real part of the complex modulus, denoted as $G' = \Re(\tilde{G})$, is a measure of the elasticity of the fluid and is named storage modulus. The imaginary part, $G'' = \Im(\tilde{G})$, on the other hand, is a measure of the viscosity and is referred to as loss modulus.

In Fig. 1, we show the dependency of the storage and loss moduli on the testing frequency as measured by various authors on different animal species and on humans. In spite of the relatively large variability in the data, there is a general trend in the measurements. In particular, it is invariably found that the storage modulus is larger than the loss modulus ($G' > G''$).

Owing to its fragile structure, the vitreous gel easily tends to separate into two phases: a solid gel and a liquid. Silva et al. [50] and Tram and Swindle-Reilly [58] managed to test the rheological properties of these two phases separately and, interestingly, found that also the liquid phase has viscoelastic properties, with a value of the storage modulus $G'$ comparable to the loss modulus $G''$ (see the open circles in Fig. 1).

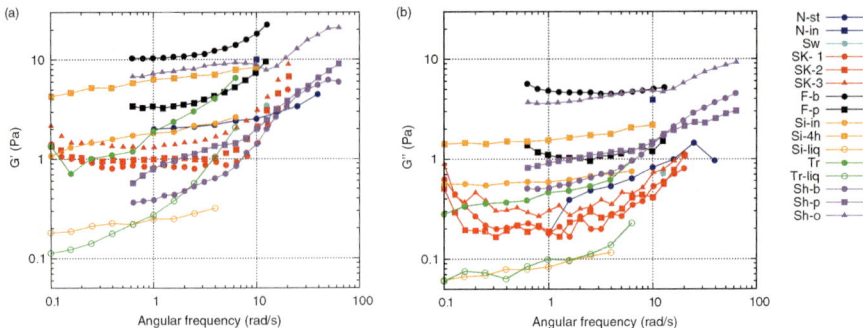

**Fig. 1** Storage (**a**) and loss (**b**) moduli as a function of the testing frequency. N-st: Nickerson et al. [31] steady state values, porcine vitreous; N-in: Nickerson et al. [31] initial values, porcine vitreous; Sw: Swindle et al. [55], porcine vitreous; SK-$i$: Sharif-Kashani et al. [49] ($i = 1, 2, 3$ denotes different eyes), porcine vitreous; F-b: Filas et al. [14], bovine vitreous; F-p: Filas et al. [14], porcine vitreous; Si-in: Silva et al. [50] initial values, rabbit vitreous; Si-4h: Silva et al. [50] 4 h after dissection, rabbit vitreous; Si-liq: Silva et al. [50] liquid phase, rabbit vitreous; Tr: Tram and Swindle-Reilly [58], human vitreous; Tr-liq: Tram and Swindle-Reilly [58], liquid phase, human vitreous; Sh-b: Shafaie et al. [48], bovine vitreous; Sh-p: Shafaie et al. [48], porcine vitreous; Sh-o: Shafaie et al. [48], ovine vitreous

The fact that the measurements shown in Fig. 1 are quite sparse can in part be attributed to the fact that different species might have different vitreous properties. However, another reason is that rheological measurements on the vitreous are very challenging. This is because the vitreous body has a very delicate structure that tends to degrade as soon as it is extracted from the vitreous chamber. Nickerson et al. [31] found that the moduli measured just after excision were significantly higher than those measures after some time (see Fig. 1). Silva et al. [50] also found that vitreous properties change significantly after dissection; however, differently from Nickerson et al. [31], they observed that the moduli immediately after dissection are lower than 4 h later (Fig. 1). Vitreous properties are also variable in space within the vitreous chamber, because the fibre networks density increases radially and peaks at the most external layers of the vitreous, which makes the vitreous there stiffer [23, 24]. Finally, vitreous properties significantly change with age [8, 58].

Sharif-Kashani et al. [49] performed creep tests on vitreous samples. In such tests, the sample is subjected to a stress kept constant over time and the corresponding strain is measured. The authors identified three different regions of response: an elastic region (lasting $\approx 1$ s), a retardation region ($\approx 80$ s) and, finally, a viscous region.

Silva et al. [50] showed that, in steady shear flow experiments, the liquid phase of the vitreous exhibits a strong shear thinning behaviour, with the viscosity decreasing more than four orders of magnitude (from $\approx 10$ to $\approx 10^{-3}$ Pa · s) as the shear rate is varied from $10^{-2}$ to $10^{3}$ 1/s.

# 3 Dynamics of the Vitreous Humour Induced by Eye Rotations

## 3.1 Saccadic Rotations

Motion of the vitreous body induces stresses on the retina that can have an important role in the generation of retinal tears. Moreover, flow of the vitreous is also relevant for transport processes in the vitreous chamber.

Dyson et al. [13] showed by a simple order of magnitude argument that the most important mechanism for generating motion in the vitreous body is eye rotations. In the following of this section we, therefore, focus on the dynamics of the vitreous induced by rotations of the eye bulb and, in particular, by saccadic movements. These are rapid eye rotations that are performed when redirecting the sight from one target to another. The characteristics of saccadic rotations (in terms of amplitude, duration, angular velocity and acceleration) are described in detail in [3].

Based on these metrics, Repetto et al. [37] proposed to model mathematically the rotation of the eyeball during a saccadic rotation using the following fifth order polynomial function of time $\beta(t)$, where $\beta$ represents the angle of the eye with respect to a reference position:

$$\beta(t) = c_0 + c_1 t + c_2 t^2 + c_3 t^3 + c_4 t^4 + c_5 t^5. \tag{1}$$

The six coefficients in this expression are computed imposing the following constraints: $\beta(0) = 0$, $\beta(D) = A$, $\dot{\beta}(0) = 0$, $\dot{\beta}(D) = 0$, $\dot{\beta}(t_p) = \Omega_p$ and $\ddot{\beta}(t_p) = 0$. In the above expressions, $A$ represents the saccade amplitude, $D$ its duration, $\Omega_p$ the peak angular velocity and $t_p$ the time at which the angular velocity reaches its maximum. All the above parameters are taken from expressions provided by Becker [3]. In Fig. 2, we report the angular velocity of the eye bulb as a function

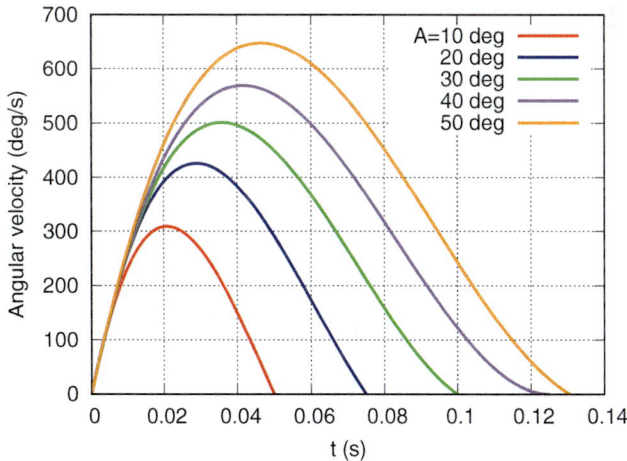

**Fig. 2** Angular velocity of saccades vs time, reproduced using the polynomial law proposed by Repetto et al. [37]

of time, as predicted by the model of Repetto et al. [37]. The figure shows that saccadic eye rotations are short-lasting and swift movements, during which the eye globe reaches fairly large angular velocities. The duration of the motion increases with the amplitude and so does the angular velocity, up to a saturation value of about 650 deg/s.

In many theoretical, numerical and experimental works, a sequence of saccadic rotations has been modelled as a harmonic function of time.

## 3.2   An Analytical Model of Vitreous Dynamics: Motion of a Viscoelastic Fluid in a Sphere

Due to its transparency, visualising vitreous motion induced by eye rotations is extremely challenging. Some authors have attempted to obtain quantitative information about vitreous dynamics in vivo, analysing ultrasound scan films, either by tracking speckles [60] or using particle image velocimetry [43]. The latter can provide very interesting information about the dynamics of the vitreous, but has the drawback that the vitreous itself, especially in healthy conditions, is poorly echogenic (i.e. it does not effectively return ultrasound signals). An alternate, very promising, technique is based on the use of films generated through magnetic resonance imaging and was adopted by Piccirelli et al. [33] to estimate vitreous deformation during saccades.

In vitro experiments on physical models of the eye represent a useful complementary tool to in vivo measurements and some contributions are available in the literature (e.g. [4, 37, 54]).

Notwithstanding the importance of these exploratory works, our knowledge on vitreous dynamics remains patchy, and mathematical modelling has proven to be a very useful tool to improve our understanding of it, especially in relation to the generation of vitreoretinal stresses. The first mathematical model of vitreous motion induced by eye rotations was proposed by David et al. [10]. The authors modelled the vitreous chamber as a rigid sphere and the vitreous body as a viscoelastic fluid. The authors' analysis was reconsidered and extended by Meskauskas et al. [25] and we describe this last model in the following.

We consider a hollow rigid spherical domain of radius $R$, which is assumed to be completely filled with a homogeneous viscoelastic and incompressible fluid. Following David et al. [10] and Meskauskas et al. [25], we assume to have a "slow flow" and linearise the equations that govern fluid motion, which can be written as

$$\rho \frac{\partial \mathbf{u}}{\partial t} + \nabla p - \nabla \cdot \mathbf{d} = 0, \tag{2a}$$

$$\nabla \cdot \mathbf{u} = 0. \tag{2b}$$

where $\rho$ is fluid density, $\mathbf{u}$ is fluid velocity, $p$ denotes pressure and, finally, $\mathbf{d}$ denotes the deviatoric part of the Cauchy stress tensor $\boldsymbol{\sigma}$, so that $\mathbf{d} = \boldsymbol{\sigma} - \frac{1}{3}(\text{tr}\boldsymbol{\sigma})\mathbf{I} = \boldsymbol{\sigma} + p\mathbf{I}$, with $\mathbf{I}$ the unit second order tensor. The no slip boundary condition has to be imposed at the wall.

We model the linear viscoelastic behaviour of the fluid with the Boltzmann–Volterra constitutive equation

$$\mathbf{d} = 2 \int_{-\infty}^{t} G(t - \tau)\mathbf{D}(\tau)d\tau, \tag{3}$$

where $\mathbf{D}$ represents the rate of deformation tensor, $\mathbf{D} = \text{sym}(\nabla\mathbf{u})$, and $G$ is the fluid relaxation function. The above expression can be substituted into Eq. (2a) to obtain

$$\rho\frac{\partial\mathbf{u}}{\partial t} + \nabla p - \int_{-\infty}^{t} G(t - \tau)\nabla^2\mathbf{u}\, d\tau = 0. \tag{4}$$

Following Meskauskas et al. [25], we first consider the relaxation behaviour of the system. In other words, we study how an initial state of motion decays in time within the sphere. Mathematically, this implies solving a nonlinear eigenvalue problem, by seeking a solution in the form

$$\mathbf{u}(\mathbf{x}, t) = \mathbf{u}_\lambda(\mathbf{x})e^{\lambda t} + \text{c.c.}, \qquad p(\mathbf{x}, t) = p_\lambda(\mathbf{x})e^{\lambda t} + \text{c.c.} \tag{5a, b}$$

In the above expressions, $\lambda$ represent the eigenvalues of the system ($\lambda \in \mathbb{C}$) and $\mathbf{u}_\lambda$ and $p_\lambda$ the corresponding eigenfunctions. In particular, we note that the real part of $\lambda$, $\Re(\lambda)$, is expected to be negative for all eigenvalues, since the system is dissipative and any initial velocity within the sphere will decay in time. We can interpret $1/\Re(\lambda)$ as a relaxation time. Moreover, if $\lambda$ is complex, its imaginary part, $\Im(\lambda)$, represents the natural frequency of oscillation of the system.

Substituting (5a, b) into (4), we obtain

$$\rho\lambda\mathbf{u}_\lambda + \nabla p_\lambda - \frac{\tilde{G}(\lambda)}{\lambda}\nabla^2\mathbf{u}_\lambda = 0, \tag{6}$$

which has to be solved together with (2b) and subjected to a zero velocity condition at the wall, $\mathbf{u}_\lambda = 0$ at $r = R$. In the expression (6), we have introduced the complex modulus of the fluid, which was mentioned in Sect. 2, and is formally defined as

$$\tilde{G}(\lambda) = \lambda \int_0^\infty G(s)e^{-\lambda s}ds. \tag{7}$$

Following Meskauskas et al. [25], the above problem can be solved by expanding the velocity field in terms of vector spherical harmonics and the pressure field in terms of scalar spherical harmonics. The mathematical details are quite technical and the interested reader should refer to the original paper. Explicit expressions for

**Fig. 3** Sketch of (**a**) the two-parameter Kelvin–Voigt model and (**b**) the four-parameter Burgers–Kelvin model

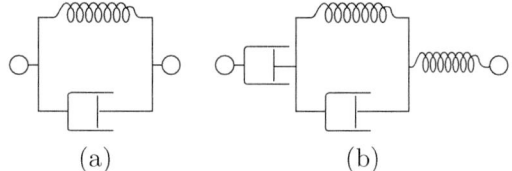

(a)                               (b)

the eigenvalues of the system can be found, once the dependency of the complex modulus on $\lambda$ is prescribed. The authors provide expressions for $\tilde{G}(\lambda)$ based on simple rheological models of the vitreous. In particular, they use the two-parameter Kelvin–Voigt model and the four-parameter Burgers–Kelvin model (see [57]). The former consists of spring and dashpot elements arranged in parallel; the second is a Kelvin–Voigt element in series with a dashpot and a spring (see Fig. 3). The parameters of the rheological models can be obtained using measurements of the complex modulus of the vitreous humour available in the literature. Details about the procedure are reported in Meskauskas et al. [25].

The interesting result of the analysis is that, in all cases corresponding to experimentally measured parameters, complex eigenvalues exist. This implies that the system admits natural frequencies of oscillation. Moreover, it is found that the natural frequency of the less damped mode, i.e. with the smallest $|\Re(\lambda)|$, is always approximately in the range of typical frequencies of eye rotations, i.e. 10–25 rad/s. This means that eye rotations can possibly resonantly excite vitreous motion.

To appreciate the consequences of the possible occurrence of resonance, it is useful to study the motion induced in the sphere by harmonic torsional oscillations of the domain, which was the problem originally considered by David et al. [10]. We assume that the rotations of the domain have frequency $\omega$ and amplitude $\epsilon$, so that the angular displacement of the domain $\beta$ with respect to a reference position reads

$$\beta = -\epsilon \cos(\omega t). \tag{8}$$

The amplitude of eye rotations is assumed small ($\epsilon \ll 1$), which allows us to use the linearised equation of motion (4). We note that the assumption of small amplitude rotations applies well to many real eye movements.

Working in terms of spherical polar coordinates $(r, \theta, \phi)$, the no slip condition at the wall reads

$$u_r = 0, \qquad u_\theta = 0, \qquad u_\phi = \epsilon \omega R \sin\theta \sin(\omega t) \qquad (r = R), \tag{9a–c}$$

where $u_r$, $u_\theta$ and $u_\phi$ are the radial, zenithal and azimuthal components of the velocity, respectively.

Since the problem is linear and motion is forced by the boundary condition (9) at the wall, which imposes that the wall velocity depends harmonically on time, we can now seek a solution in the form

$$\mathbf{u}(\mathbf{x}, t) = \mathbf{U}(\mathbf{x})e^{i\omega t} + c.c., \qquad p(\mathbf{x}, t) = P(\mathbf{x})e^{i\omega t} + c.c., \qquad (10a, b)$$

where $\mathbf{U} = (U_r, U_\theta, U_\phi)$ and $P$ are complex functions of space. Substituting (10a, b) into (4), we obtain an equation analogous to (6), in which $\mathbf{u}_\lambda$ has to be replaced with $\mathbf{U}$, $p_\lambda$ with $P$ and $\lambda$ with $i\omega$.

It is easy to show that the equation for $U_\phi$ decouples from the others and the solution reads

$$U_r = 0, \qquad U_\theta = 0, \qquad U_\phi = -\frac{i\epsilon R\omega (\sin(ar) - ar\cos(ar))}{2r^2(\sin a - a\cos a)}\sin\theta, \qquad P = \text{const.},$$
$$(11a\text{–}d)$$

where

$$a = \sqrt{\frac{i\rho\omega^2 R^2}{\tilde{G}(i\omega)}}e^{-i\pi/4}. \qquad (12)$$

In Fig. 4, we show different profiles of the azimuthal velocity component ($u_\phi$) on the equatorial plane orthogonal to the axis of rotation ($\theta = \pi/2$), at different times during a period of oscillation. The complex modulus of the fluid has been computed using the two-parameter Kelvin–Voigt model and the rheological properties of the vitreous measured by Swindle et al. [55], following the approach proposed by Meskauskas et al. [25]. The two plots correspond to two different oscillation frequencies. In the first case (Fig. 4a), $\omega = 10$ rad/s and the maximum velocity in the domain is attained at the wall ($r = R$), as it would happen in the case of a purely viscous fluid. On the contrary, in Fig. 4b the frequency has been chosen equal to the first natural frequency of oscillation of the fluid. In this case, resonant excitation occurs and the velocity profiles are remarkably different from those of the previous

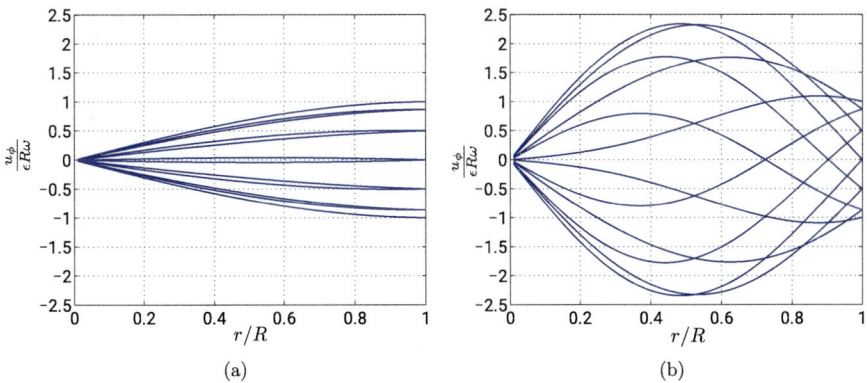

**Fig. 4** Azimuthal velocity profiles at different times during a period of oscillation. The rheological parameters are based on the measurements of Swindle et al. [55] (for details see Meskauskas et al. [25]). (**a**) $\omega = 10$ rad/s, (**b**) $\omega = 21.61$ rad/s

case. In particular, the velocity $u_\phi$ peaks in the core of the domain and the velocity there is more than twice as large as at the boundary. This in turn produces large stresses on the wall (the retina). These theoretical results are in good agreement with experimental observations of Bonfiglio et al. [4]. The authors performed experiments on a real scale eye model, adopting viscoelastic fluids with properties similar to those of the real vitreous. They also found the occurrence of resonance, at frequencies very close to those predicted theoretically. Good agreement between theoretical results and measurements was also found in terms of velocity profiles (see Fig. 7 of Bonfiglio et al. [4]).

The possible occurrence of resonant excitation of fluid motion in the vitreous chamber also has important implications for the choice of fluids to be used as vitreous substitutes. Soman and Banerjee [51] and Swindle and Ravi [56] discuss the characteristics of an ideal vitreous humour replacement. Fluid elasticity is considered beneficial to avoid excessive flow within the vitreous chamber [9]. However, if the elastic modulus is large and the fluid not dissipative enough, resonant excitation of fluid motion can be induced by eye rotations. This, in turn, would generate large stresses on the retina, which can have harmful consequences.

Repetto et al. [40] also considered the problem of the motion of a viscoelastic fluid in a sphere, rotating periodically with frequency $\omega$, by expanding of all variables in terms of the small amplitude of the oscillations of the domain $\epsilon$. At leading order, $\mathcal{O}(\epsilon)$, they recover the solution of Meskauskas et al. [25] described above. At order $\epsilon^2$, nonlinear interactions produces a component of the flow that oscillates at frequency $2\omega$ and also a steady component, which is normally referred to as steady streaming component in the fluid mechanics literature. Steady streaming originating in a flow forced by a zero average periodic forcing is known to be very relevant for mixing processes [42]. Repetto et al. [40] found that the steady streaming flow has several, complicated and highly three-dimensional structures, depending on the values of the controlling parameters.

## 3.3   The Effect of a Realistic Geometry

The real shape of the vitreous chamber is not exactly spherical, owing to the presence of the crystalline lens, which induces a change in curvature in its anterior part. The importance of the real geometry on the characteristics of the flow field was first highlighted by Stocchino et al. [54]. The authors studied the motion of a viscous fluid in a magnified scale physical model of the eye and observed the generation of fairly complicated three-dimensional structures, in particular in the anterior part of the domain, close to the indentation produced by the crystalline lens.

Repetto [34] and Repetto et al. [38] proposed mathematical models of vitreous motion that accounted for the actual shape of the domain. The vitreous was modelled as an inviscid fluid by Repetto [34] and as a purely viscous fluid by Repetto et al. [38]. Both works are based on the assumption that the vitreous cavity can be thought of as a "weakly deformed" sphere. In other words, the authors model the shape of the domain according to the following equation (in spherical coordinates):

$$r = R(1 + \delta \mathcal{R}(\theta, \phi)), \tag{13}$$

where $\mathcal{R}$ is an order one function of space and $\delta \ll 1$. In both works, the authors find an analytical solution by performing an asymptotic expansion in terms of the small parameter $\delta$. The authors managed to reproduce a three-dimensional structure of the flow field in agreement with the experimental observations of Stocchino et al. [54]. In particular, the theoretical works reproduce flow circulations that are observed in the experiments in the anterior part of the vitreous chamber, owing to the indentation produced by the lens (Figs. 2 and 3 in [54] and Fig. 4 in [38]).

The above works have been complemented to the case of a viscoelastic fluid by Meskauskas et al. [26], who considered how vitreoretinal stresses change from normal to myopic eyes, which have elongated shape in the antero-posterior direction. Their main finding is that vitreoretinal stresses, as well as the state of stress within the vitreous, significantly increase with increasing degree of myopia. The authors' work shows that during each eye rotation the vitreous body of a highly myopic subject experiences shear stresses 1.5-fold higher than that of an emmetropic eye. It is thus speculated that this can provide a mechanical explanation for the observation that vitreous liquefaction occurs, on average, at a younger age in myopic subjects.

## 3.4 Numerical Simulations of Vitreous Motion

The models described in the previous section are based on analytical approaches and, therefore, rely on simplifying assumptions regarding the geometry of the vitreous chamber and the description of eye movements. Numerical simulations of the flow in the vitreous chamber have also been proposed, which can help to overcome some of the limitations of analytical models.

Abouali et al. [1] studied the flow of a Newtonian fluid in a model of the vitreous cavity with a shape similar to that considered in the experiments of Stocchino et al. [54], consisting of a sphere indented on one side, to simulate the effect of the natural lens. The assumption of Newtonian fluid applies to the case of completely liquefied vitreous, which approximately has the same mechanical properties of water, or to tamponade fluids used after vitrectomy, such as silicone oils. The authors considered both harmonic eye rotations and single saccades. In the work, the authors propose relationships between the wall shear stress and the characteristics of the fluid and the eye rotations, which they establish numerically.

Modarreszadeh and Abouali [28] extended the above work to the case of a viscoelastic fluid but only considered harmonic rotations of the domain. They showed that in the case of a normal vitreous the shear stress on the retina is around ten times higher than when the vitreous is liquefied.

## 3.5   Generation of Vitreoretinal Tractions

In the previous sections, we discussed that motion of the vitreous humour induced by eye rotations generates stresses on the retina. In some circumstances, such stresses can be localised on certain regions of the retina and, when this happens, they can produce particularly large vitreoretinal tractions.

The most notable example are vitreoretinal tractions induced by an anomalous posterior vitreous detachment (APVD). PVD is the separation between the vitreous body from the internal limiting lamina of the retina. This is a pathophysiological process that, typically, occurs in elderly subjects. For PVD to develop harmlessly, two processes must take place simultaneously degradation of the macromolecular structure of vitreous, which results in gel liquefaction, and alterations in the extracellular matrix at the vitreoretinal interface, which decreases the adhesive force between the posterior vitreous cortex and the internal limiting membrane. When these two processes are not concurrent and, in particular, when vitreoretinal adhesion remains large, APVD occurs [46].

In the presence of an APVD, tractions on the retina can be generated during eye rotations as a consequence of the motion of the detached vitreous. This problem was studied by Repetto et al. [39], who proposed a simple mathematical model of the dynamics in the vitreous cavity in the presence of APVD. The authors considered the case in which the vitreous is detached from the back pole of the eye but remains attached to the retina, approximately at the equator (the equatorial plane is orthogonal to the antero-posterior axis of the eye). In Fig. 5a we show a sketch of a PVD and in Fig. 5b the simplified geometry adopted by Repetto et al. [39]. The authors considered a two-dimensional geometry and modelled the vitreous chamber as a rigid circle. The circle is occupied by a gel on its right side, which represents the detached vitreous, and by a viscous fluid on its left side, which models the liquefied vitreous that occupies the space between the vitreous cortex and the retina, where detachment has occurred. The authors modelled the gel phase as a hyperelastic, viscous solid and the liquid phase as a Newtonian liquid with the same properties as

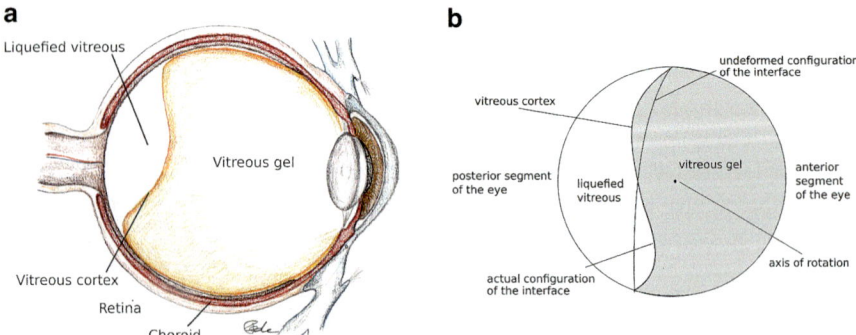

**Fig. 5** (a) Sketch of a PVD. (b) Simplified model adopted by Repetto et al. [39]

water. Both phases were assumed incompressible. The gel and the liquid phases are separated by a membrane, which models the vitreous cortex, and is attached at the rigid wall at two points. Several shapes of the detachment were considered by the authors, with different attachment angles and points of attachment of the membrane to the wall.

The domain performs rotations about its centre and the authors adopted the model of saccades, described by Eq. (1). The motion of the solid and fluid phases were solved using the arbitrary Lagrangian–Eulerian method, and implemented in a finite element method solver.

Results provide a feasible, even if complicated, picture of the dynamics of detached vitreous produced by eye rotations. For all considered cases, the authors found that during the movement large values of the traction force are attained at the attachment points of the membrane on the wall. This provides a physically based confirmation that localised tractions are exerted by the detached vitreous on the retina during eye rotations.

Repetto et al. [39] found that vitreoretinal tractions are larger when the equilibrium curvature of the interface between the two phases has the same sign as that of the vitreous chamber wall, which is the situation that is more often observed clinically.

Adopting reasonable values for the mechanical properties of the vitreous, the authors showed that the maximum vitreoretinal traction attains values of the same order of magnitude as the retinal adhesive force, as measured by Kita and Marmor [20]. Therefore, the model suggests that dynamic vitreoretinal tractions in the presence of an APVD are likely to play a significant role in generating retinal tears.

Vroon et al. [59] considered a similar problem, with a more realistic three-dimensional geometry. They performed numerical simulations and considered both head and eye rotations. Their results suggest that head movements are the major factor in the generation of vitreoretinal tractions.

Repetto et al. [36] studied with an analytical model the dynamics of vitreous membranes surrounded by liquefied vitreous on both sides and showed that their oscillations, secondary to eye rotations, may induce the development of large vitreoretinal tractions.

Recently, Rossi et al. [44] studied the effect on vitreoretinal tractions of a pre-macular bursa (PMB). This is a small reservoir of liquid vitreous, located anteriorly to the fovea and present in virtually all eyes of young subjects [52]. The work is motivated by the hypothesis that PMB might play the role of protecting the fovea from mechanical insult by reducing the shear stress on it. The authors performed three-dimensional numerical simulations of the flow induced in the vitreous cavity by saccadic eye rotations. The vitreous cavity was described as a domain with a realistic geometry, similar to that adopted by Modarreszadeh and Abouali [28] and the vitreous was modelled as an Oldroyd-B viscoelastic fluid [32]. The presence of the PMB, owing to the small thickness of this region, was simulated adopting a slip boundary condition, by assuming the existence of a linear law of transmission of the shear stress through the depth of the bursa.

The authors computed the wall shear stress on the retinal surface during saccadic movements, in the presence and absence of the PMB and compared the results. The numerical simulations show that the shear stress on the macula is significantly reduced, owing to the presence of the PMB, which confirms the speculation that PMB might have the mechanical role of protecting the most delicate part of the retina.

## 3.6 Non-technical Summary on the Dynamics of Vitreous Humour

Vitreous humour motion induced by eye rotations generates mechanical stresses on the retina that are known to be related to the possible occurrence of RD. For this reason, it is of great clinical relevance to understand vitreous dynamics. Visualising and measuring vitreous motion in vivo is, however, extremely challenging, owing to its transparency. As an alternative investigation tool, mathematical modelling has proven to be very useful in improving our understanding of vitreous dynamics.

Mathematical models have informed us about how the vitreous humour responds to eye rotations and how this response depends on the vitreous mechanical properties, which are known to change with ageing. It has been shown that vitreous motion, owing to the viscoelastic properties of the material, can be resonantly excited by eye rotations and, therefore, vitreous velocities within the vitreous cavity might achieve values significantly larger than one would expect in the case of a viscous fluid. This has important implications both for the generation of high vitreoretinal stresses and for the choice of tamponade fluids.

Theoretical models have also improved our understanding of the mechanics of vitreoretinal tractions. It has been proven that, in the case of PVD, significant tractions are generated in correspondence of adhesion points between the vitreous and the retina, which can be large enough to produce a retinal tear. Moreover, shapes of the detached vitreous to which large tractions on the retina are likely to be associated have been identified.

Finally, a numerical model has recently been used to show that the existence of a PMB is likely to have the mechanical role of protecting macula against mechanical insult.

## 4    Mathematical Models of Retinal Detachment

### 4.1  Introduction to Retinal Detachment

RD occurs when the neurosensory retina separates from the retinal pigment epithelium (RPE) and the space created between these two layers is occupied by fluid. RD can be classified into three different categories: (1) rhegmatogenous (RRD), (2) exudative (ERD) and (3) tractional [6].

RRD is the most common type of RD and it occurs when liquefied vitreous flows through a retinal tear from the vitreous cavity into the subretinal space, causing separation between the neurosensory retina and the RPE. Thus, a prerequisite for RRD is the existence of a break in the retina, which is typically caused by vitreoretinal tractions [18]. We note, however, that not all retinal tears result in a RD. In the general population, RRD occurs in approximately 10 out of 100,000 patients [27].

ERD occurs when fluid accumulates in the subretinal space, without the presence of a retinal break. Under normal conditions, there is a net water flux from the vitreous chamber to the choroid. This flow is regulated by a hydraulic pressure jump and also active water pumping by the RPE from the subretinal space to the choroid. Mismatch between inflow and outflow in the subretinal space may lead to fluid accumulation there, and the formation of a blister that detaches the sensory retina from the RPE.

Finally, tractional RD occurs when pathological vitreoretinal adhesions or membranes pull the retina away from the RPE. This process happens without a retinal break, which distinguishes tractional from rhegmatogenous RD.

RD heavily involves mechanics, and so mathematical modelling can provide a significant contribution in understanding the underlying physics. However, the existing literature on mathematical models of RDs is very limited. Bottega et al. [5] proposed a mechanics-based model of RD, which was further improved by Lakawicz et al. [21]. The authors considered both retinas with and without tears. In these models, vitreoretinal tractions are produced by vitreous contraction that induces extension of its fibrils. The models predict that, once detachment ensues, it propagates catastrophically. Results also show that the presence of a retinal tear may have a stabilising effect on detachment propagation.

These models of RD are purely based on the effect of vitreoretinal tractions and do not account for the role of fluid flow. However, rhegmatogenous and exudative RDs are inherently associated with the motion of fluids and we will focus on mathematical models of these types of RD in the following.

## 4.2 Models of Rhegmatogenous Retinal Detachment

As discussed above, RRD requires the existence of a retinal break, through which fluid enters the subretinal space from the vitreous cavity. From the mechanical point of view, the problem is extremely complicated and, therefore, the authors who attempted to study it made use of several simplifying assumptions.

Natali et al. [30] recently proposed a simple model of the flow in the presence of a retinal flap. They considered the two-dimensional geometries sketched in Fig. 6. The surface of the retina is modelled as a flat surface and the detached retinal flap as a slender elastic beam, clamped at a given angle on the wall. The assumption that the

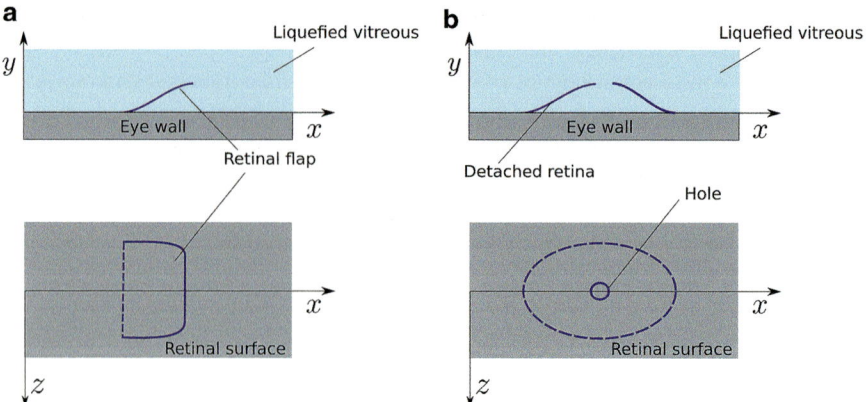

**Fig. 6** Sketches of the domains considered by Natali et al. [29]. (**a**) giant retinal tear; (**b**) retinal hole. The top panels show lateral cross-sections and the bottom panels plan views

retinal surface is flat is valid in the limit in which the detached flap is short compared to the radius of the eye, which is often the case. Similarly, the domain is assumed to have infinite extent in the $y$-direction, orthogonal to the wall. Eye rotations are modelled by letting the retinal surface move along its own plane ($x$-direction in Fig. 6), following the saccadic law described by the Eq. (1).

Natali et al. [30] considered two cases: a giant retinal tear (GRT), Fig. 6a, and a retinal hole (RH), Fig. 6b. A giant retinal tear consists of a single detached flap with 90 deg or more of circumferential extent. This means that a realistic aspect ratio of the flap (length to circumferential extent) is at least 1/40, which makes the two-dimensional approximation acceptable. A retinal hole, on the other hand, is a localised hole in the retina and its cross-section consists of two retinal flaps (left in Fig. 6). In order to mimic the three-dimensionality of the hole configuration in a (highly) simplified way, the authors assumed that the tips of the two flaps are connected to each other by a virtual linear elastic spring with an asymmetric behaviour, such that it reacts when the two tips move apart but allows them to get close to each other. The half plane $y \geq 0$ is occupied by a Newtonian fluid with the same properties as water that models the liquefied vitreous.

During eye rotations, i.e. translations of the wall, the fluid moves and interacts with the detached flap, which undergoes very complicated dynamics. The Reynolds number of the flow, computed with the length of the flap, is of order $10^2$, which means that the flow is laminar but the dynamics of the system is dominated by inertial effects. During its motion, the flap exerts both forces and moments at the attaching point. Forces and moments are combined using Winkler's theory of a beam over an elastic substrate [62] to determine a "tendency of the retina to detach" further.

The governing equations were solved numerically by the authors and an immersed boundary method was used to model the interaction between the elastic retinal flap and the moving fluid.

The authors found that the tendency to detach increases with the retinal flap length, both for the RH and the GRT. Moreover, the model predicts that detachment is more likely to progress in the case of a RH than GRT. This agrees with the clinical observation that the fovea off rate for rhegmatogenous retinal detachment excluding GRTs is 56% [61], while it is 45.2% for GRTs [2], and provides some physical basis to interpret this finding.

## 4.3   Models of Exudative Retinal Detachment

The basic processes that contribute to generation of forces associated with ERD are pressure driven flows, RPE active pumping, cellular adhesion between the sensory retina and the RPE and, possibly, vitreoretinal tractions. Chou and Siegel [7] proposed a mathematical model of the above processes that allows one to establish when ERD can occur.

The authors first proposed a one-dimensional model based on the geometry sketched in Fig. 7, consisting of several tissue layers. They modelled the adhesive stress $\sigma$ between the retina and the RPE as

$$\sigma(z) = \frac{dU(z)}{dz}, \qquad (z < z_{\max}), \tag{14}$$

where $U$ is the cellular adhesion potential, $z$ denotes the separation between the two layers and $z_{\max}$ is the maximum value of $z$ that the two layers can withstand. The corresponding yield stress is $\sigma_{\max}$.

Water fluxes between the choroid and the subretinal space, $J_c$, and between the subretinal space and the vitreous cavity, $J_r$, are modelled according to the following expressions:

$$J_c = L_c(p_c - p_b), \qquad J_r = L_r(p_b - p_{IOP}), \tag{15a, b}$$

**Fig. 7** Schematic of the layers considered by Chou and Siegel [7] and associated water fluxes

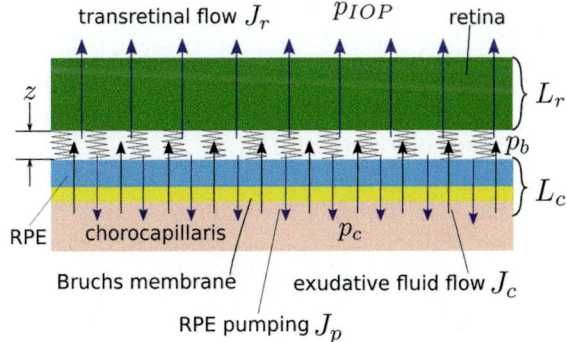

where $L_i$ ($i = c, r$) are the corresponding hydraulic conductivities, $p_c$ and $p_b$ are the values of pressure in the choriocapillaris and in the space between the retina and the RPE, respectively, and $p_{IOP}$ is the pressure in the vitreous chamber (the intraocular pressure). The authors estimate that $p_c$ is typically larger than $p_{IOP}$, which implies that, in the absence of active pumping by the RPE, the passive contribution due to mechanical pressure differences would drive fluid flow from the choroid into the vitreous chamber and would tend to separate the retina from the RPE.

Active pumping by the RPE is related to RPE metabolism. Chou and Siegel [7] assumed that it does not depend on pressure and modelled it as a constant value. The corresponding flux is denoted by $J_p$, taken positive if directed towards the choroid.

Conservation of mass across the tissue layers and use of (15a, b) leads to the following relationship for the mechanical pressure difference $p_b - p_{IOP}$, which tends to separate the retina from the RPE,

$$p_b - p_{IOP} = \frac{L_c(p_c - p_{IOP}) - J_p}{L_r + L_c} = \sigma(z). \tag{16}$$

In order for the forces to be in equilibrium, $p_b - p_{IOP}$ needs to be equal to the adhesive stress $\sigma(z)$, which provides an equation for the equilibrium separation distance $z$.

When $p_c - p_{IOP} < 0$, Eq. (16) implies $\sigma(z) < 0$, meaning that the retina presses onto the RPE and detachment cannot occur. On the other hand, for positive values of $L_c(p_c - p_{IOP}) - J_p$, the cellular adhesive bounds between the retina and the RPE are stretched. In particular, when

$$\frac{L_c(p_c - p_{IOP}) - J_p}{L_r + L_c} > \sigma_{max}, \tag{17}$$

retinal detachment would occur along the whole retinal surface.

Failure of the RPE pumping ($J_p \to 0$) is considered one of the possible causes of ERD. In this case (17) shows that detachment would occur if $L_c(p_c - p_{IOP})/(L_r + L_c) > \sigma_{max}$. Using experimentally measured values for all parameters, the authors speculate that loss of the RPE pumping is insufficient to produce a retinal detachment.

Chou and Siegel [7] further developed their model by accounting for the presence of a retinal blister, i.e. a localised region of detachment between the retina and the RPE. The authors showed that RPE pumping failure can very significantly affect the size and stability of a retinal blister.

In Chou and Siegel's [7] model, water flux pumped by the RPE was set to a constant value. Recently, Dvoriashyna et al. [12] proposed a mathematical model of water and ion transport across the RPE that helps to elucidate the physical mechanisms at the basis of this active water pumping. In particular, the authors focused on two possible mechanisms: osmosis and electroosmosis. Osmosis is the spontaneous water flux through a semipermeable membrane from a region of low to a region of high osmolarity, i.e. concentration of molecules dissolved in the fluid and

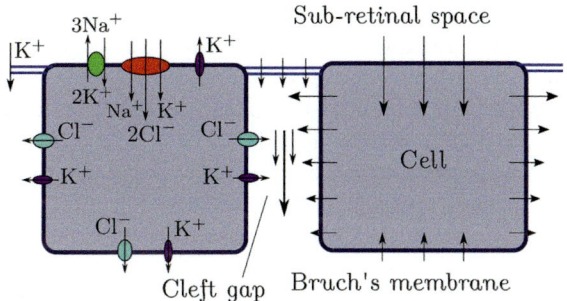

**Fig. 8** Sketch of the idealised geometry of the RPE considered by Dvoriashyna et al. [12]. The authors considered that the apical membrane has $Na^+/K^+$-ATPases, $K^+$ channels and $Na^+/K^+/2$ $Cl^-$ co-transporters. $Cl^-$ and $K^+$ channels are present in the basolateral membrane. Arrows represent schematically the directions and magnitude of the fluid flow across the membranes. In the figure, the thickness of the cleft gap between the cells is magnified

that cannot freely pass through the cell membrane. The osmolarity on the retinal and choroidal sides of the RPE is approximately the same. Therefore, osmosis cannot be responsible for water flux across the RPE, unless the presence of the cleft gap between adjacent cells, where a build-up of ion concentration can occur, plays a role. This phenomenon is known as local osmosis.

Electroosmosis is the flux of a fluid with dissolved ions in the presence of charged walls (that attract ions of opposite charge) and an external electrical field, which results in a Coulomb body force acting on the fluid. Electroosmosis could occur along the cleft gap since the cell membrane is negatively charged and a transepithelial electrical potential exists.

Dvoriashyna et al. [12] developed a spatially resolved mathematical model aimed at studying both ion and water transport across the RPE. They considered three ions species: $Na^+$, $K^+$ and $Cl^-$. The authors' model predicts that local osmosis is the main mechanism for water transport across the RPE and that electroosmosis is largely subdominant. The predicted directions of water fluxes across the cell membranes and the cleft gap are shown in Fig. 8 with black arrows. The overall water flux is directed from the subretinal space to the choroid and has an order of magnitude compatible with experimental observations.

## 4.4 Non-technical Summary on Retinal Detachment

RD is a very complicated phenomenon that heavily involves mechanics. Owing to its complexity, mathematical modelling has the potential of being an extremely useful tool as it can help isolating the role of single factors. In spite of this, very little modelling work has been carried out so far.

The tendency of a retinal flap to further detach has been studied with a simple and highly idealised mathematical model that accounts for liquefied vitreous flow and flap elasticity and deformability. The model predicts that in the case of a GRT detachment is less likely to progress than in the case of a RH. This counter-intuitive result qualitatively agrees with clinical observations.

Exudative RD has also been modelled under idealised conditions. This model suggests that failure of RPE pumping is insufficient to produce a retinal detachment and other factors are involved. Moreover, the model predicts that failure of the RPE pumping can significantly affect the size and stability of a retinal blister.

Various mathematical models have also been proposed aimed at understanding the physical problems associated with vitreoretinal surgeries, such as gas and perfluoron retinopexy [16], scleral buckling [15, 17, 63], removal of epiretinal membrane [11] and transpupillary thermotherapy [19]. We do not cover these topics here for the sake of space.

**Acknowledgements** The authors thank Prof. Federica Grillo, University of Genoa (Italy), for drawing Fig. 5a. Mariia Dvoriashyna acknowledges the Department of Civil, Chemical and Environmental Engineering of the University of Genoa (Italy), where she worked as a PhD student when the original version of this chapter was written.

# References

1. O. Abouali, A. Modareszadeh, A. Ghaffariyeh, and J. Tu. Numerical simulation of the fluid dynamics in vitreous cavity due to saccadic eye movement. *Medical Engineering & Physics*, 34(6):681–692, July 2012. ISSN 1350-4533. http://dx.doi.org/10.1016/j.medengphy.2011.09. 011.
2. G. S. Ang, J. Townend, and N. Lois. Epidemiology of giant retinal tears in the united kingdom: the british giant retinal tear epidemiology eye study (bgees). *Investigative Ophthalmology & Visual science*, 51(9):4781–4787, 2010.
3. W. Becker. Metrics In R. Wurtz and M. Goldberg, editors, *The neurobiology of saccadic eye movements*. Elsevier Science Publisher BV (Biomedical Division), 1989.
4. A. Bonfiglio, A. Lagazzo, R. Repetto, and A. Stocchino. An experimental model of vitreous motion induced by eye rotations. *Eye and Vision*, 2(1):10, 2015.
5. W. J. Bottega, P. L. Bishay, J. L. Prenner, and H. F. Fine. On the mechanics of a detaching retina. *Math. Med. Biol.*, 30:287–310, 2013. http://dx.doi.org/10.1093/imammb/dqs024.
6. D. A. Brinton and C. P. Wilkinson. *Retinal Detachment: Priniciples and Practice*, volume 1. Oxford University Press, 2009.
7. T. Chou and M. Siegel. A mechanical model of retinal detachment. *Physical Biology*, 9(4): 046001, 2012.
8. J. Colter, A. Williams, P. Moran, and B. Coats. Age-related changes in dynamic moduli of ovine vitreous. *Journal of the Mechanical Behavior of Biomedical Materials*, 41:315–324, 2015.
9. P. D. Dalton, T. V. Chirila, Y. Hong, and A. Jefferson. Oscillatory shear experiments as criteria for potential vitreous substitutes. *Polymer Gels and Networks*, 3(4):429–444, 1995. ISSN 0966-7822. http://dx.doi.org/10.1016/0966-7822(94)00011-5.
10. T. David, S. Smye, T. Dabbs, and T. James. A model for the fluid motion of vitreous humour of the human eye during saccadic movement. *Phys. Med. Biol.*, 43:1385–1399, 1998.
11. M. Dogramaci and T. H. Williamson. Dynamics of epiretinal membrane removal off the retinal surface: a computer simulation project. *British Journal of Ophthalmology*, 97(9):1202–1207, 2013.

12. M. Dvoriashyna, A. J. Foss, E. A. Gaffney, O. E. Jensen, and R. Repetto. Osmotic and electroosmotic fluid transport across the retinal pigment epithelium: A mathematical model. *Journal of Theoretical Biology*, 456:233–248, 2018.

13. R. Dyson, A. J. Fitt, O. E. Jensen, N. Mottram, D. Miroshnychenko, S. Naire, R. Ocone, J. H. Siggers, and A. Smithbecker. Post re-attachment retinal re-detachment. In *Proceedings of the Fourth Medical Study Group, University of Strathclyde, Glasgow*, 2004.

14. B. A. Filas, Q. Zhang, R. J. Okamoto, Y.-B. Shui, and D. C. Beebe. Enzymatic degradation identifies components responsible for the structural properties of the vitreous body. *Investigative Ophthalmology & Visual Science*, 55(1):55–63, 2014.

15. W. J. Foster. Bilateral patching in retinal detachment: fluid mechanics and retinal settling. *Investigative Ophthalmology & Visual Science*, 52(8):5437–5440, 2011.

16. W. J. Foster and T. Chou. Physical mechanisms of gas and perfluoron retinopexy and sub-retinal fluid displacement. *Physics in Medicine & Biology*, 49(13):2989, 2004.

17. W. J. Foster, N. Dowla, S. Y. Joshi, and M. Nikolaou. The fluid mechanics of scleral buckling surgery for the repair of retinal detachment. *Graefe's Archive for Clinical and Experimental Ophthalmology*, 248(1):31, 2010.

18. W. S. Foulds. Role of vitreous in the pathogenesis of retinal detachment. In S. J., editor, *Vitreous in Health and Disease*, pages 375–393. Springer-Verlag, 2014.

19. O. P. Garcia, P. R. M. Lyra, A. Fernandes, and R. d. C. F. de Lima. The influence of the vitreous humor viscosity during laser-induced thermal damage in choroidal melanomas. *International Journal of Thermal Sciences*, 136:444–456, 2019.

20. M. Kita and M. Marmor. Retinal adhesive force in living rabbit, cat, and monkey eyes. normative data and enhancement by mannitol and acetazolamide. *Invest. Ophthalmol. Vis. Sci.*, 33(6):1879–1882, May 1992.

21. J. M. Lakawicz, W. J. Bottega, J. L. Prenner, and H. F. Fine. An analysis of the mechanical behaviour of a detached retina. *Math. Med. Biol.*, 32:137–161, 2015. http://dx.doi.org/10.1093/imammb/dqt023.

22. M. M. Le Goff and P. N. Bishop. Adult vitreous structure and postnatal changes. *Eye (London, England)*, 22(10):1214–1222, Oct. 2008. ISSN 1476-5454. http://dx.doi.org/10.1038/eye.2008.21. PMID: 18309340.

23. B. Lee, M. Litt, and G. Buchsbaum. Rheology of the vitreous body. Part I: viscoelasticity of human vitreous. *Biorheology*, 29:521–533, 1992.

24. B. Lee, M. Litt, and G. Buchsbaum. Rheology of the vitreous body: Part 2. viscoelasticity of bovine and porcine vitreous. *Biorheology*, 31(4):327–338, Aug. 1994. ISSN 0006-355X. PMID: 7981433.

25. J. Meskauskas, R. Repetto, and J. H. Siggers. Oscillatory motion of a viscoelastic fluid within a spherical cavity. *Journal of Fluid Mechanics*, 685:1–22, 2011. http://dx.doi.org/10.1017/jfm.2011.263.

26. J. Meskauskas, R. Repetto, and J. H. Siggers. Shape change of the vitreous chamber influences retinal detachment and reattachment processes: Is mechanical stress during eye rotations a factor? *Investigative Ophthalmology & Visual Science*, 53(10):6271–6281, Oct. 2012. ISSN 1552-5783. http://dx.doi.org/10.1167/iovs.11-9390. PMID: 22899755.

27. D. Mitry, J. Chalmers, K. Anderson, L. Williams, B. W. Fleck, A. Wright, and H. Campbell. Temporal trends in retinal detachment incidence in scotland between 1987 and 2006. *Br. J. Ophthalmol.*, 95(3):365–369, 2009.

28. A. Modarreszadeh and O. Abouali. Numerical simulation for unsteady motions of the human vitreous humor as a viscoelastic substance in linear and non-linear regimes. *Journal of Non-Newtonian Fluid Mechanics*, 204:22–31, 2014.

29. D. Natali, R. Repetto, J. H. Tweedy, T. H. Williamson, and J. O. Pralits. A simple mathematical model of rhegmatogenous retinal detachment. *Journal of Fluids and Structures*, 82:245–257, 2018. https://doi.org/10.1016/j.jfluidstructs.2018.06.020

30. D. Natali, R. Repetto, J. H. Tweedy, T. H. Williamson, and J. O. Pralits. A simple mathematical model of rhegmatogenous retinal detachment. *Journal of Fluids and Structures*, 82:245–257, 2018.

31. C. S. Nickerson, J. Park, J. A. Kornfield, and H. Karageozian. Rheological properties of the vitreous and the role of hyaluronic acid. *Journal of Biomechanics*, 41(9):1840–1846, 2008.
32. J. Oldroyd. On the formulation of rheological equations of state. *Proc. R. Soc. Lond. A*, 200 (1063):523–541, 1950.
33. M. Piccirelli, O. Bergamin, K. Landau, P. Boesiger, and R. Luechinger. Vitreous deformation during eye movement. *NMR in Biomedicine*, 25(1):59–66, 2012.
34. R. Repetto. An analytical model of the dynamics of the liquefied vitreous induced by saccadic eye movements. *Meccanica*, 41:101–117, 2006. http://dx.doi.org/10.1007/s11012-005-0782-5.
35. R. Repetto and J. H. Siggers. Biomechanics of the vitreous humor. In C. J. Roberts, W. J. Dupps Jr., and J. C. Downs, editors, *Biomechanics of the eye*. Kugler Publications, 2018.
36. R. Repetto, I. Ghigo, G. Seminara, and C. Ciurlo. A simple hydro-elastic model of the dynamics of a vitreous membrane. *J. Fluid Mech.*, 503:1–14, 2004. http://dx.doi.org/10.1017/S0022112003007389.
37. R. Repetto, A. Stocchino, and C. Cafferata. Experimental investigation of vitreous humour motion within a human eye model. *Phys. Med. Biol.*, 50:4729–4743, 2005. http://dx.doi.org/10.1088/0031-9155/50/19/021.
38. R. Repetto, J. H. Siggers, and A. Stocchino. Mathematical model of flow in the vitreous humor induced by saccadic eye rotations: effect of geometry. *Biomechanics and Modeling in Mechanobiology*, 9(1):65–76, 2010. ISSN 1617-7959. http://dx.doi.org/10.1007/s10237-009-0159-0.
39. R. Repetto, A. Tatone, A. Testa, and E. Colangeli. Traction on the retina induced by saccadic eye movements in the presence of posterior vitreous detachment. *Biomechanics and Modeling in Mechanobiology*, 10:191–202, 2010b. http://dx.doi.org/10.1007/s10237-010-0226-6.
40. R. Repetto, J. H. Siggers, and J. Meskauskas. Steady streaming of a viscoelastic fluid within a periodically rotating sphere. *Journal of Fluid Mechanics*, 761:329–347, 2014. http://dx.doi.org/10.1017/jfm.2014.546.
41. R. Repetto, J. H. Siggers, and J. Meskauskas. Fluid mechanics of the vitreous chamber. In P. Causin, G. Guidoboni, R. Sacco, and A. Harris, editors, *Integrated multidisciplinary approaches in the study and care of the human eye*. Kugler Publications, 2014.
42. N. Riley. Steady streaming. *Ann. Rev. Fluid Mech.*, 33:43–65, 2001.
43. T. Rossi, G. Querzoli, G. Pasqualitto, M. Iossa, L. Placentino, R. Repetto, A. Stocchino, and G. Ripandelli. Ultrasound imaging velocimetry of the human vitreous. *Experimental eye Research*, 99(1):98–104, June 2012. ISSN 1096-0007. http://dx.doi.org/10.1016/j.exer.2012.03.014. PMID: 22516112.
44. T. Rossi, M. G. Badas, G. Querzoli, C. Trillo, S. Telani, L. Landi, R. Gattegna, and G. Ripandelli. Does the bursa pre-macularis protect the fovea from shear stress? A possible mechanical role. *Experimental Eye Research*, 175:159–165 2018.
45. J. Sebag. *The vitreous: Structure, Function and Pathobiology*. Springer and Verlag, 1989.
46. J. Sebag. Anomalous posterior vitreous detachment: a unifying concept in vitreo-retinal disease. *Graefe's Archive for Clinical and Experimental Ophthalmology*, 242(8):690–698, Aug. 2004. ISSN 0721-832X, 1435-702X. http://dx.doi.org/10.1007/s00417-004-0980-1.
47. J. Sebag. Seeing the invisible: the challenge of imaging vitreous. *Journal of Biomedical Optics*, 9(1):38–46, 2004.
48. S. Shafaie, V. Hutter, M. B. Brown, M. T. Cook, and D. Y. Chau. Diffusion through the ex vivo vitreal body–bovine, porcine, and ovine models are poor surrogates for the human vitreous. *International Journal of Pharmaceutics*, 550(1–2):207–215, 2018.
49. P. Sharif-Kashani, J. Hubschman, D. Sassoon, and H. P. Kavehpourlee. Rheology of the vitreous gel: effects of macromolecule organization on the viscoelastic properties. *Journal of Biomechanics*, 44(3):419–423, Feb. 2011. ISSN 1873-2380. http://dx.doi.org/10.1016/j.jbiomech.2010.10.002. PMID: 21040921.
50. A. F. Silva, M. A. Alves, and M. S. Oliveira. Rheological behaviour of vitreous humour. *Rheologica Acta*, 56(4):377–386, 2017.

51. N. Soman and R. Banerjee. Artificial vitreous replacements. *Bio-Medical Materials and Engineering*, 13(1):59–74, 2003. ISSN 0959-2989.
52. R. F. Spaide. Visualization of the posterior vitreous with dynamic focusing and windowed averaging swept source optical coherence tomography. *American Journal of Ophthalmology*, 158(6):1267–1274, 2014.
53. E. Stefánsson. Physiology of vitreous surgery. *Graefes Arch. Clin. Exp. Ophthalmol.*, 247:147–163, 2009. http://dx.doi.org/10.1007/s00417-008-0980-7.
54. A. Stocchino, R. Repetto, and C. Cafferata. Eye rotation induced dynamics of a newtonian fluid within the vitreous cavity: the effect of the chamber shape. *Phys. Med. Biol.*, 52:2021–2034, 2007. http://dx.doi.org/10.1088/0031-9155/52/7/016.
55. K. Swindle, P. Hamilton, and N. Ravi. In situ formation of hydrogels as vitreous substitutes: Viscoelastic comparison to porcine vitreous. *Journal of Biomedical Materials Research - Part A*, 87A(3):656–665, Dec. 2008. ISSN 1549-3296.
56. K. E. Swindle and N. Ravi. Recent advances in polymeric vitreous substitutes. *Expert Rev Ophthalmol.*, 2(2):255–265, 2007.
57. R. I. Tanner. *Engineering Rheology*. Oxford University Press, USA, 2 edition, May 2000. ISBN 0198564732.
58. N. K. Tram and K. E. Swindle-Reilly. Rheological properties and age-related changes of the human vitreous humor. *Frontiers in Bioengineering and Biotechnology*, 6:199, 2018.
59. J. Vroon, J. de Jong, A. Aboulatta, A. Eliasy, F. van der Helm, J. van Meurs, D. Wong, and A. Elsheikh. Numerical study of the effect of head and eye movement on progression of retinal detachment. *Biomechanics and Modeling in Mechanobiology*, pages 1–9, 2018.
60. K. A. Walton, C. H. Meyer, C. J. Harkrider, T. A. Cox, and C. A. Toth. Age-related changes in vitreous mobility as measured by video b scan ultrasound. *Experimental Eye Research*, 74(2):173–80, Feb. 2002. ISSN 0014-4835. http://dx.doi.org/10.1006/exer.2001.1136. PMID: 11950227.
61. T. H. Williamson, E. J. Lee, and M. Shunmugam. Characteristics of rhegmatogenous retinal detachment and their relationship to success rates of surgery. *Retina*, 34(7):1421–1427, 2014.
62. E. Winkler. *Die Lehre von der Elasticitaet und Festigkeit: mit besonderer Rücksicht auf ihre Anwendung in der Technik für polytechnische Schulen, Bauakademien, Ingenieue, Maschinenbauer, Architecten, etc*, volume 1. Dominicus, 1867.
63. D. Wong, Y. Chan, T. Bek, I. Wilson, and E. Stefansson. Intraocular currents, Bernoulli's principle and non-drainage scleral buckling for rhegmatogenous retinal detachment. *Eye*, 32(2):213, 2018.

# The Tear Film: Anatomy and Physiology

**Vikram Paranjpe, Lam Phung, and Anat Galor**

**Abstract** This chapter explores the normal structure and function of the tear film. Normal properties and composition of tears as well as the structure of the aqueous, mucin, and meibomian layers are discussed. The role of supporting structures such as the eyelids, meibomian glands, and goblet cells in the lacrimal functional unit (LFU) is also reviewed. Finally, the physiologic roles of the tear film, including lubrication and nutrition of the ocular surface, visual function, anti-inflammatory and healing properties, and its role in ocular surface nerve modulation are all examined.

## 1 Introduction

The tear film, corneal and conjunctival epithelia, lacrimal and meibomian glands, goblet cells, eyelids, sensory and motor nerves together form the lacrimal functional unit [1]. This review will cover the anatomy and physiology of the tear film and its supporting structures (lacrimal and meibomian glands and goblet cells) in humans, compare this to animal data, and summarize questions and controversies.

V. Paranjpe · L. Phung · A. Galor (✉)
Miami Veterans Administration Medical Center, Miami, FL, USA

Bascom Palmer Eye Institute, University of Miami, Miami, FL, USA
e-mail: agalor@med.miami.edu

© Springer Nature Switzerland AG 2019
G. Guidoboni et al. (eds.), *Ocular Fluid Dynamics*, Modeling and Simulation in Science, Engineering and Technology, https://doi.org/10.1007/978-3-030-25886-3_14

## 2  Anatomy

## 2.1  Tear Film Properties and Composition

### 2.1.1  Structure

The precorneal tear film architecture has been previously proposed to comprise of three distinct layers: an innermost mucin layer (provided by goblet cells), intermediate aqueous layer (provided by aqueous layer), and outermost lipid layer (provided by meibomian glands). However, newer models studied in rats and mice advocate for a two-layered structure where there is a more homogenous aqueous–mucin gel under the lipid layer, in which mucins are distributed in a concentration gradient (Fig. 1) [2–4]. It is not known however, if this model holds in the human tear film.

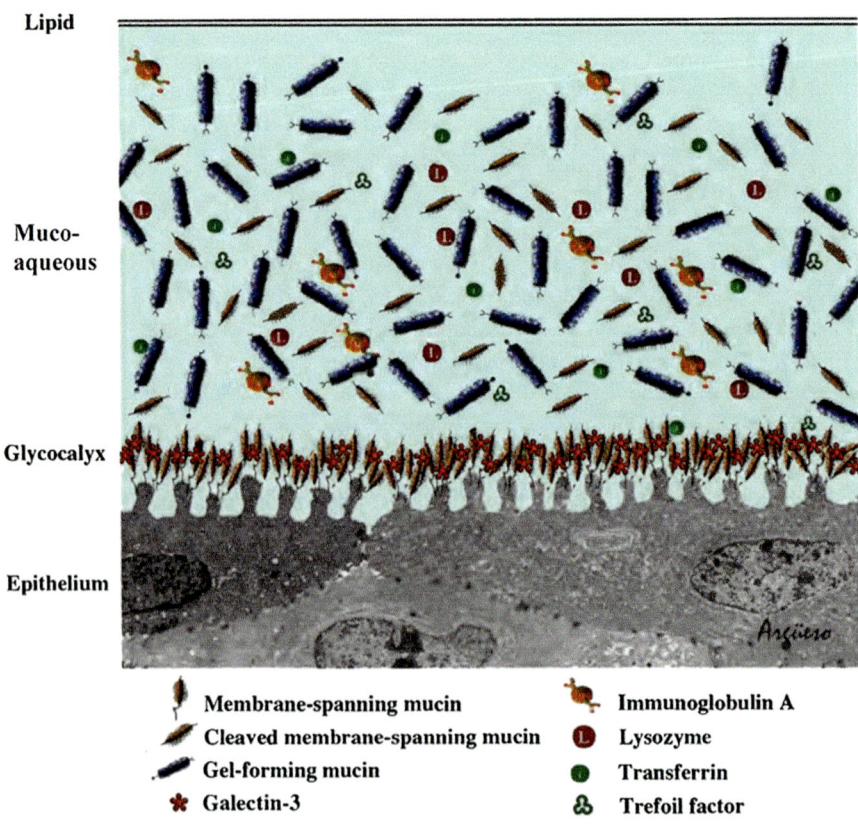

| | |
|---|---|
| ⬇ Membrane-spanning mucin | 🦐 Immunoglobulin A |
| ◣ Cleaved membrane-spanning mucin | Ⓛ Lysozyme |
| ◣ Gel-forming mucin | ● Transferrin |
| ✱ Galectin-3 | ♌ Trefoil factor |

**Fig. 1** Proposed model of the precorneal tear film demonstrating the two-layered architecture of lipid and aqueous–mucin layer. Reproduced with permission [5]

## 2.2 Mucin Layer Composition

The mucin component of the tear film is produced by conjunctival goblet cells and by corneal and conjunctival stratified squamous cells. Mucins are complex glycoproteins with polypeptide backbone, tandem repeats of amino acids rich in serine, threonine, and proline, and oligosaccharide side chains [2]. There are two categories of mucins: secreted and membrane associated. Secreted mucins have cysteine-rich domains that allow them to form large polymers, giving viscous characteristic to these gel-forming mucins [6]. Epithelial (membrane associated) mucins (e.g. MUC1, MUC4, MUC16) have transmembrane domains that traverse the plasma membrane of corneal and conjunctival epithelial cells, forming a glycocalyx on the ocular surface that interacts with gel-forming mucins, primarily MUC5AC, secreted by conjunctival goblet cells [2, 3, 7]. Gel-forming mucin MUC5AC is a large molecule that has four cysteine-rich domains that form disulfide bridges with other MUC5AC molecules. These interactions between secreted and membrane-associated mucins anchor the mucin layer to the corneal and conjunctival epithelium, providing a continuous and smooth layer of tear film over the ocular surface. It was widely believed that mucins played a major role in maintaining tear film viscosity and surface tension based on observations of this behavior in mucin solutions of 5 mg/mL [8]. However, it has recently been reported [9] that the concentration of mucins in normal human tears is much lower and highly variable, with a range of 0–100 μg/mL, making it highly unlikely that mucins play a major role in the maintenance of tear surface tension or viscosity [8].

## 2.3 Aqueous Layer Composition

The aqueous tear layer constitutes the main portion of the tear film and is produced by the lacrimal and accessory lacrimal glands in response to ocular surface and nasal mucosa stimulation. It is composed of 98.2% water and 1.8% other constituents including electrolytes, proteins, growth factors, and vitamin (Table 1) [10, 11].

In humans, secreted electrolytes include sodium, potassium, calcium, magnesium, bicarbonate and chloride ions. The predominant extracellular and intracellular cations are sodium and potassium, respectively, but while concentration of sodium is similar in tears and serum, potassium concentration is three to five times higher in tears than in serum (Table 1) [10]. Chloride is an essential anion, and together with sodium and potassium plays an important role in osmotic regulation, maintaining tear osmolarity at approximately 302 ± 6 mosm/kg (Table 1) [2, 10, 11]. Bicarbonate ions constitute a buffer system that maintains a near neutral tear pH of 7.2–7.6 [2]. Calcium and magnesium are cations found at lower concentrations that associate with cofactors and enzyme molecules to control membrane permeability (Table 1) [2, 12].

332

**Table 1** Electrolytes and organic solutes in human and rabbit tears, serum and aqueous humor.[a] table and legend reproduced with permission from Berman et al. [11] with additional data from Iwata et al. [10]

| Component | Human tears [11] (mM) | Human serum [11] (mM) | Human aqueous humor [10] ($\mu$Eq/mL) | Rabbit tears [10] ($\mu$Eq/mL) | Rabbit serum [10] ($\mu$Eq/mL) | Rabbit aqueous humor [10] ($\mu$Eq/mL) |
|---|---|---|---|---|---|---|
| Electrolytes | | | | | | |
| $Na^+$ | 120–165 | 130–145 | – | 137.5–151 | 146–151.1[b] | 138–153.1 |
| $Cl^-$ | 118–135 | 95–125 | 131 | 131–138 | 106.2[b]–111 | 101–105.7 |
| $HCO_3^-$ | 20–26 | 24–30 | 20.15 | 20 | 22–27.4[b] | 30.2–33.6 |
| $Mg^{2+}$ | 0.5–0.9 | 0.7–1.1 | – | – | – | – |
| $K^+$ | 20–42 | 3.5–5 | – | 13.8–29 | 4–5.50[b] | 5–5.25 |
| $Ca^{2+}$ | 0.4–1.1 | 2.0–2.6 | – | – | – | – |
| Organic solutes | | | | | | |
| Glucose | 0.1–0.60 | 4–6 | – | – | – | – |
| Urea | 3.0–6.0 | 3.3–6.5 | – | – | – | – |
| Lactate | 2–5 | 0.5–0.8 | – | – | – | – |
| Pyruvate | 0.05–0.35 | 0.1–0.2 | – | – | – | – |
| Ascorbate | 0.008–0.04 | 0.04–0.06 | – | – | – | – |
| All-*trans* retinoic acid ($\mu$g/dL) | 0.04–1.06[c] | 30–60 | – | – | – | – |
| | Trace-3.3[d] | | | | | |

[a] Adapted from van Haeringen (1981) and Stanifer et al. (1985)
[b] In plasma
[c] Range of retinol values in nine adult volunteers determined by HPLC (Speek et al. 1986)
[d] The mean concentration of retinol in 11 human tear samples determined by HPLC is $11.6 \pm 0.2$ $\mu$g/dL (Ubels and MacRae 1984)

**Table 2** Concentrations of major proteins in aqueous component of normal human tears

| Aqueous tear composition | Human tears |
|---|---|
| Lysozyme (mg/mL) | 0.6–2.6 [13] |
| | 2.07 [2] |
| Lactoferrin (mg/mL) | 2.05 ± 1.12 [14] |
| | 1.65 [2] |
| EGF (ng/mL) | 5.09 ± 3.74 [14] |
| AQP5 (pg/μL) | 31.1 ± 23.9 [14] |
| Lipocalin (mg/mL) | 1.55 [2] |
| sIgA (mg/mL) | 0.1–0.6 [11] |
| | 1.93 [2] |

Major proteins found in aqueous tears include lysozyme, lactoferrin, lipocalin, secretory IgA, aquaporin 5 (AQP5), and epidermal growth factor (EGF) (Table 2). Lysozyme is a long-chain high molecular weight, antibacterial enzyme produced by lysosomes which has a normal concentration of 0.6–2.6 mg/mL in human tears [13]. Lactoferrin is another antibacterial enzyme secreted from acini of lacrimal gland [13]. EGF produced by lacrimal gland helps to maintain ocular surface by controlling corneal wound healing [13]. AQP5 is a selective water channel protein found on apical membrane of lacrimal acinar and ductal cells, which regulates water transport [13]. The use of direct and indirect ELISA allowed for the quantification of these proteins in 16 healthy controls (75% female) with an average age of 30 ± 4 years, compared to 103 patients with dry eye (71 with non-Sjogren syndrome, 23 with Sjogren syndrome, and 9 with Stevens-Johnson syndrome) [14]. This prospective case–control study revealed the normal concentration of lactoferrin, EGF, and AQP5 to be 2.05 ± 1.12 mg/mL, 5.09 ± 3.74 ng/mL, and 31.1 ± 23.9 pg/μL, respectively, in the healthy controls [14]. Immunoglobulin A (IgA), secreted by plasma cells of lacrimal glands and conjunctiva, is the predominant antibody found in aqueous tears. It is composed of two monomeric IgA molecules held together by a J piece, which is transported through the acini and the secretory component is added to form secretory IgA (sIgA).

## 2.4 Meibomian Layer Composition

The lipid component of the tear film is produced primarily by the meibomian glands located in the tarsal plates of the eyelids. Due to its low melting point of 35 °C, the lipid layer is always fluid regardless of ambient temperature to which the ocular surface is exposed [11]. The majority of meibomian lipid consists of nonpolar lipids, whereas the remainder is composed of polar lipids, free fatty acids, alcohols, and neutral fats [7]. The polar and nonpolar lipids organize into two layers with the polar lipids forming the lower sublayer and nonpolar lipids forming the upper layer that is in contact with the air. The lipid products orient perpendicularly with their

**Fig. 3** Electron micrograph of human conjunctival goblet cell, demonstrating a polarized cell with a thin layer of cytoplasm at the periphery, nucleus and organelles at the base, and mucin granules toward the apical surface. Reproduced with permission from Gipson et al. [27]

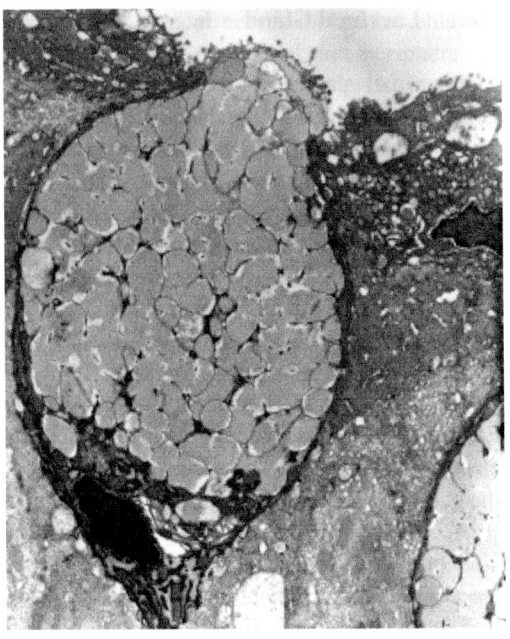

epithelium. Fluid from conjunctiva can hydrate excreted mucins, explaining the known phenomenon of mucin expansion upon secretion from goblet cell (Fig. 3).

### 2.6.4   Eyelids

The eyelid structure is maintained by a collagen plate called the tarsus, which contains a row of meibomian glands that secrete the lipid component of the tear film. The levator palpebral muscle tendon inserts on the tarsus. The anterior surface of the eyelid is covered by a thin, stratified, keratinized squamous epithelium, while its posterior surface makes up the palpebral conjunctiva, which is composed of stratified, nonkeratinized squamous epithelium with small goblet cells [26]. The orbicularis oculi muscle is a striated muscle located in the middle layer of the eyelid (Fig. 4).

## 3   Physiology

The tear film is essential in maintaining the integrity of the ocular surface by providing lubrication and nutrients, forming a smooth optical surface for light refraction, removing and protecting against foreign material, and in promoting wound healing.

**Fig. 4** Histology of the eyelid in sagittal plane: (1) superior tarsal muscle (smooth muscle), (2) accessory lacrimal gland (Krause gland), (3) fornical conjunctiva, (4) tarsus, (5) posterior surface or palpebral conjunctiva, (6) meibomian gland, (7) ciliary (Moll's) gland, (8) eyelash, (9) orbicularis oculi muscle, and (10) anterior surface. Reproduced with permission from Paulsen et al. [26]

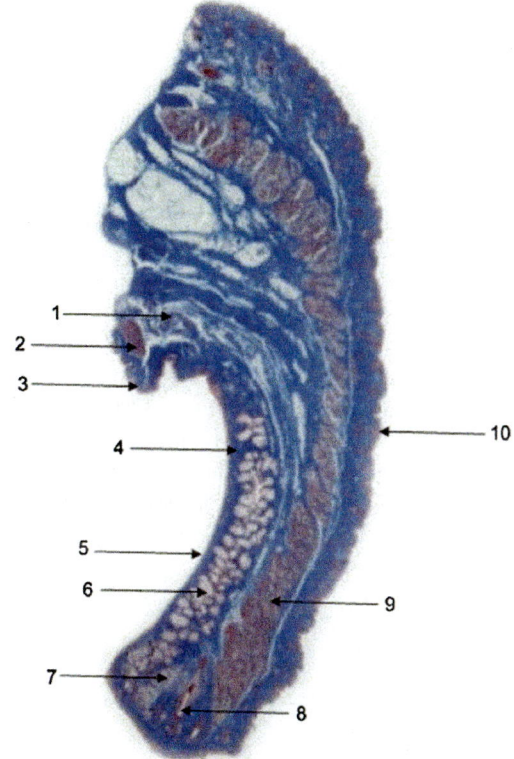

## 3.1 Lubrication and Nutrition

Transmembrane mucins of the glycocalyx interact with secreted mucins of the tear film to anchor it onto the corneal and conjunctival epithelium, to provide lubrication and to prevent shear damage to the ocular surface from the high speed of blinking (~20 cm/s) [2]. The hydrophilic nature of the glycocalyx also allows for even spreading of the overlying aqueous layer on the ocular surface [8].

To maintain the transparency of the cornea for visual function, it is avascular. Gases and nutrients are delivered to the cornea by the tear film anteriorly, with some contribution by the aqueous humor on the endothelial surface. When the eyes are open, the tear film is thought to be saturated with oxygen (~155 mmHg) and is sufficient to adequately provide oxygen to the cornea; however, when the eyes are closed, oxygen must be provided through diffusion of oxygen from the conjunctival vasculature (55 mmHg) [2]. There are few other nutrients found in significant quantities in the cornea, and it is thought that glucose, which has a low concentration in tears (Table 1), is delivered to the cornea entirely from the endothelial side [2].

## 3.2 Visual Function

When comparing the precorneal tear film, cornea, and lens, the tear film is the eye's first and most important refractive interface due to the greatest change in refractive index (RI) that occurs between the air (1.00) and tear film (1.34) interface. Therefore, aberrations in tear film will result in visual quality changes, such that a 0.2-mm variation in the tear film radius can cause a 1.3-diopter change in power [28]. To test the importance of tear film on visual quality, optical aberrations were measured in 15 dry eye volunteers (40% female with an average age of $27.5 \pm 3.4$ years), which revealed a statistically significant decrease in aberrations after artificial tear instillation, on average by a factor of 2–3 [29]. The tear film improves corneal regularity and symmetry, thus enhancing quality of the retinal image. A closer look at the tear film structure demonstrates how the lipid layer contributes to the tear film stability for optimal light refraction. The organization of the polar lipid layer below the nonpolar layer allows it to serve as a surfactant at the aqueous and lipid interface, which helps to stabilize the nonpolar sublayer above against rupture and prevent evaporation of aqueous tear. The long-chain (C20-31) fatty acids, fatty alcohols, and hydrocarbons increase cohesiveness of the nonpolar layer to decrease vapor transmission rate of water, carbon dioxide, oxygen, and ions. Furthermore, lipocalin protein found in the aqueous layer binds excess lipid from the ocular surface or mucous layer to avoid development of hydrophobic patches that would lead to tear film breakup [2].

## 3.3 Defense Mechanisms

The ocular surface is constantly exposed to environmental influences and therefore relies heavily on an integrated defense system to maintain its integrity. These defense mechanisms include intrinsic anatomic barriers, mucous and antimicrobial proteins, local humoral (sIgA) antibody, and local T-lymphocyte cellular responses, which work closely together to mount a highly efficient system of host defense. The mucin layer provides physical protection to the ocular surface by forming a continuous barrier that prevents pathogen penetration. It also has structures mimicking receptor sites for microorganisms that facilitate trapping of bacteria and viruses, which are removed as the mucus is swept toward the lower fornix and eventually to the skin of the inner canthus [2, 26]. Lighter materials including dust, hair, and bacteria, especially hydrophilic particles, are also reflected by the lipid layer of the tear film [2]. The aqueous component of the tear film contains various antibacterial proteins. Lysozyme has muramidase activity to break down the outer cell wall of Gram-positive bacteria, while both lactoferrin and lipocalin can sequester iron to prevent the growth of siderophilic bacteria [2]. Secretory IgA provides immunological protection through the priming of plasma cells against specific microorganisms. This occurs at mucosa-associated lymphoid tissues either in the conjunctiva or elsewhere [2].

## 3.4 Anti-inflammatory

Studies have shown an increase in tear film evaporation rate by fourfold with removal of lipid layer as well as a negative correlation between tear film lipid layer thickness and rate of tear film thinning, indicating the importance of lipid layer in maintaining a stable tear film [18]. Tear film stability is an important aspect in maintaining a healthy ocular surface because it prevents tear film evaporation, hyperosmolarity, and in turn, ocular inflammation.

A new concept in inflammation is that resolution of inflammation is an active process mediated by *pro-resolving lipid mediators*, such as resolvins, protectins, and maresins. In experimental models of acute, self-resolving inflammation, early oxygenation of arachadonic acid (AA) into pro-inflammatory eicosanoids was followed by a resolving phase, in which pro-resolving lipid mediators were prevalent [30]. Most pro-resolving lipid mediators are derived from the oxygenation of omega-3 (ω-3) PUFA precursors, such as docosahexaenoic acid (DHA) and eicosapentaenoic acid (EPA). These pro-resolving mediators are part of a lipid circuit that is important in the resolution of acute inflammatory processes [21, 31, 32]. Gronert et al., have identified 15-lipoxygenase (LOX) and 5-LOX as resident enzymes in the cornea and retina of humans and mice. 15-LOX expression is upregulated during acute inflammation and is a key enzyme in the production of EPA and DHA-derived resolvins and protectins [21]. Pro-resolving molecules function to limit polymorphonuclear leukocyte (PMN) and lymphocyte activation, dendritic cell function, and the production of pro-inflammatory cytokines. They also promote the clearance of apoptotic PMN and regulate wound healing [21]. As previously mentioned, the oxygenation of AA, an omega-6 (ω-6) PUFA leads to the formation of prostaglandins (PG), which are part of a wide range of pro-inflammatory lipid mediators [33]. Recent research suggests that an imbalance between ω-6 PUFA and ω-3 PUFA may contribute to nonresolving inflammation on the ocular surface leading to pathologies like dry eye syndrome [20]. In a study of 41 predominantly male subjects ($62 \pm 11$ years old), AA and $PGE_2$, markers of inflammation, as well as DHA and EPA, markers of anti-inflammatory pathways were detected in over 90% of samples. Furthermore, the ratio of ω-6 PUFA (AA) to ω-3 PUFA (EPA + DHA) was correlated with multiple clinical parameters of tear film dysfunction including tear breakup time, Schirmer's test scores, and corneal staining [20].

## 3.5 Nerve Modulation

Serotonin (5-HT) was presented as a potential marker of corneal nociceptor sensitization [24] since 5-HT is a known sensitizer of peripheral nerves [34, 35] and its levels are known to increase in response to inflammation and peripheral nerve abnormalities [36–38]. 5-HT is thought to exert its sensitizing effect both directly and indirectly. Directly by amplifying conductance through tetrodotoxin-resistant sodium channels in primary afferents, thus shifting the conductance–voltage curve toward a hyperpolarized state. This has the effect of changing the activation threshold and sensitivity of peripheral nerves and ultimately resulting in 5-HT enhancing pain perception [24]. 5-HT is thought to indirectly act by binding to G-protein-coupled receptors causing downstream modulation via second messengers, ultimately leading to activation of ligand-gated ion channels and rapid depolarization of nerves [24].

## 3.6 Healing

Several molecules have been theorized to be involved in healing of the ocular surface. Endogenous peptides such as epidermal growth factor (EGF), transforming growth factor alpha (TGF-$\alpha$), transforming growth factor beta (TGF-$\beta$), and $\alpha$ and $\beta$ defensins are all thought to play a vital role in the wound healing process at the ocular surface [39–41]. Defensins are highly conserved peptides and are one of the earliest effectors of innate immunity. They are thought to accelerate wound healing through mitogenic effects on epithelial cells and fibroblasts [39]. In humans, $\alpha$1-4 defensins are produced by neutrophils, $\alpha$5-6 defensins are produced by the Paneth cells of the small intestine, $\beta$1 defensins are produced by the pancreas, kidney, and respiratory eplthelium, and $\beta$2 defensins are produced by the skin and bronchial mucosa [41]. A few defensins were detected in the eye, including $\alpha$1-3 defensins, which are thought to be released into the fluid on the ocular surface by passing neutrophils or lacrimal gland epithelial cells and $\beta$1 and 2 defensins which are produced by the ocular surface. Finally, peptide hormones including insulin-like factor 3 [42] and relaxin 2 [43] were recently also shown to promote wound healing at the ocular surface.

## 3.7 Comparison to Animal Tear Film

Multiple previous studies have compared the composition of the human tear film to the composition of the tear film of many animal species including the gerbil, the rat, the rabbit, the dog, the hamster, and the mouse, with a focus on comparing the lipid layer of the tear film [44–52]. Butovich et al. [53] carried out a comprehensive study comparing the lipidomes of meibum collected from humans, canines, mice, and rabbits using high pressure liquid chromatography (HPLC) and gas–liquid chromatography (GC) in combination with mass spectrometry. They found that the lipid composition of the tear film of mice and canines is more similar to the human tear film than the tear film of rabbits. Major lipid classes found in human, mouse, and canine tears included wax esters, cholesterol esters, and o-acyl-ω-hydroxy fatty acids (OAHFA). However, in rabbit tears, the major lipid classes were found to be DiHL esters (24,25-dihydro-Δ-lanosterol esters of C $_{26:2}$, $C_{28:2}$, and $C_{30:2}$), di-acylated diols, and OAHFA, with low or trace amounts of cholesterol and wax esters [53]. Another physiologically significant parameter is the tear film breakup time (TBUT), which was shown to be greater than 30 s in rabbits, less than 20 s in canines, and $5 \pm 1$ s in mice, compared to 5–30 s in humans [53, 54]. An additional advantage of using mice is that there are anatomical similarities between mouse and human meibomian glands. Both secrete meibum in holocrine fashion via small ductules connected to a central duct with an orifice that opens onto the ocular surface [53]. Given these findings, Butovich et al. [53] argue that the rabbit is too different to serve as a valid animal model for human meibomian gland secretions, and that the mouse and the canine serve as better animal models, from a biochemical standpoint. The advantage of using mice over canines is that the mouse genome, physiology, and biochemistry is well studied, that there are many gene knock-in and knock-out mice available, that the mouse life cycle is short, and, finally, that the maintenance costs for mice are relatively low. Canine models may be useful in less intensive experiments where larger volumes of tears are needed, or in animal models of dry eye syndrome, given that several canine species including spaniels, terries, Lhasa Apso, and the Shih-Tzu are prone to developing dry eye syndrome [53].

**Funding** Supported by the Department of Veterans Affairs, Veterans Health Administration, Office of Research and Development, Clinical Sciences Research EPID-006-15S (Dr. Galor), NIH Center Core Grant P30EY014801, R01EY026174 (Dr. Galor), Research to Prevent Blindness Unrestricted Grant.

# References

1. The definition and classification of dry eye disease: report of the Definition and Classification Subcommittee of the International Dry Eye WorkShop (2007). *The ocular surface.* 2007;5(2):75–92.
2. Tiffany JM. The normal tear film. *Dev Ophthalmol.* 2008;41:1–20.
3. Rolando M, Zierhut M. The ocular surface and tear film and their dysfunction in dry eye disease. *Surv Ophthalmol.* 2001;45 Suppl 2:S203–210.
4. Willcox MDP, Argueso P, Georgiev GA, et al. TFOS DEWS II Tear Film Report. *The ocular surface.* 2017;15(3):366–403.
5. Methodologies to diagnose and monitor dry eye disease: report of the Diagnostic Methodology Subcommittee of the International Dry Eye WorkShop (2007). *The ocular surface.* 2007;5(2):108–152.
6. Gipson IK. The ocular surface: the challenge to enable and protect vision: the Friedenwald lecture. *Invest Ophthalmol Vis Sci.* 2007;48(10):4390; 4391–4398.

7. Bron AJ, Tiffany JM, Gouveia SM, Yokoi N, Voon LW. Functional aspects of the tear film lipid layer. *Exp Eye Res.* 2004;78(3):347–360.
8. Tiffany JM. Tears in health and disease. *Eye (London, England).* 2003;17(8):923–926.
9. Zhao H, Jumblatt JE, Wood TO, Jumblatt MM. Quantification of MUC5AC protein in human tears. *Cornea.* 2001;20(8):873–877.
10. Iwata S. Chemical composition of the aqueous phase. *International ophthalmology clinics.* 1973;13(1):29–46.
11. Berman ER. Tears. In: Blakemore C, ed. *Biochemistry of the Eye* New York Springer Science+Business Media; 1991:63–84.
12. Ubels JL, Williams KK, Lopez Bernal D, Edelhauser HF. Evaluation of effects of a physiologic artificial tear on the corneal epithelial barrier: electrical resistance and carboxyfluorescein permeability. *Advances in experimental medicine and biology.* 1994;350:441–452.
13. Ohashi Y, Dogru M, Tsubota K. Laboratory findings in tear fluid analysis. *Clin Chim Acta.* 2006;369(1):17–28.
14. Ohashi Y, Ishida R, Kojima T, et al. Abnormal protein profiles in tears with dry eye syndrome. *Am J Ophthalmol.* 2003;136(2):291–299.
15. Green-Church KB, Butovich I, Willcox M, et al. The international workshop on meibomian gland dysfunction: report of the subcommittee on tear film lipids and lipid-protein interactions in health and disease. *Invest Ophthalmol Vis Sci.* 2011;52(4):1979–1993.
16. McCulley JP, Shine WE. Meibomian gland function and the tear lipid layer. *The ocular surface.* 2003;1(3):97–106.
17. Butovich IA, Millar TJ, Ham BM. Understanding and Analyzing Meibomian Lipids-A Review. *Current eye research.* 2008;33(5):405–420.
18. Pucker AD, Nichols JJ. Analysis of meibum and tear lipids. *Ocul Surf.* 2012;10(4):230–250.
19. Nakamura Y, Sotozono C, Kinoshita S. Inflammatory cytokines in normal human tears. *Current eye research.* 1998;17(6):673–676.
20. Walter SD, Gronert K, McClellan AL, Levitt RC, Sarantopoulos KD, Galor A. omega-3 Tear Film Lipids Correlate With Clinical Measures of Dry Eye. *Investigative ophthalmology & visual science.* 2016;57(6):2472–2478.
21. Gronert K. Resolution, the grail for healthy ocular inflammation. *Experimental eye research.* 2010;91(4):478–485.
22. Lambiase A, Micera A, Sacchetti M, Cortes M, Mantelli F, Bonini S. Alterations of tear neuromediators in dry eye disease. *Archives of ophthalmology (Chicago, Ill: 1960).* 2011;129(8):981–986.
23. Martin XD, Brennan MC. Serotonin in human tears. *Eur J Ophthalmol.* 1994;4(3):159–165.
24. Chhadva P, Lee T, Sarantopoulos CD, et al. Human Tear Serotonin Levels Correlate with Symptoms and Signs of Dry Eye. *Ophthalmology.* 2015;122(8):1675–1680.
25. Stern ME, Gao J, Siemasko KF, Beuerman RW, Pflugfelder SC. The role of the lacrimal functional unit in the pathophysiology of dry eye. *Experimental eye research.* 2004;78(3):409–416.
26. Paulsen FP, Berry MS. Mucins and TFF peptides of the tear film and lacrimal apparatus. *Prog Histochem Cytochem.* 2006;41(1):1–53.
27. Gipson IK. Goblet cells of the conjunctiva: A review of recent findings. *Prog Retin Eye Res.* 2016;54:49–63.
28. Montes-Mico R, Alio JL, Charman WN. Dynamic changes in the tear film in dry eyes. *Invest Ophthalmol Vis Sci.* 2005;46(5):1615–1619.
29. Montes-Mico R, Caliz A, Alio JL. Changes in ocular aberrations after instillation of artificial tears in dry-eye patients. *J Cataract Refract Surg.* 2004;30(8):1649–1652.
30. Serhan CN. Resolution phase of inflammation: novel endogenous anti-inflammatory and proresolving lipid mediators and pathways. *Annual review of immunology.* 2007;25:101–137.
31. Bazan NG, Molina MF, Gordon WC. Docosahexaenoic acid signalolipidomics in nutrition: significance in aging, neuroinflammation, macular degeneration, Alzheimer's, and other neurodegenerative diseases. *Annual review of nutrition.* 2011;31:321–351.

32. Kenchegowda S, Bazan HE. Significance of lipid mediators in corneal injury and repair. *Journal of lipid research.* 2010;51(5):879–891.
33. Srinivasan BD, Kulkarni PS. The role of arachidonic acid metabolites in the mediation of the polymorphonuclear leukocyte response following corneal injury. *Investigative ophthalmology & visual science.* 1980;19(9):1087–1093.
34. Bardin L. The complex role of serotonin and 5-HT receptors in chronic pain. *Behav Pharmacol.* 2011;22(5–6):390–404.
35. Sommer C. Serotonin in pain and analgesia: actions in the periphery. *Mol Neurobiol.* 2004;30(2):117–125.
36. Hong Y, Abbott FV. Behavioural effects of intraplantar injection of inflammatory mediators in the rat. *Neuroscience.* 1994;63(3):827–836.
37. Anden NE, Olsson Y. 5-hydroxytryptamine in normal and sectioned rat sciatic nerve. *Acta Pathol Microbiol Scand.* 1967;70(4):537–540.
38. Vogel C, Mossner R, Gerlach M, et al. Absence of thermal hyperalgesia in serotonin transporter-deficient mice. *J Neurosci.* 2003;23(2):708–715.
39. Murphy CJ, Foster BA, Mannis MJ, Selsted ME, Reid TW. Defensins are mitogenic for epithelial cells and fibroblasts. *J Cell Physiol.* 1993;155(2):408–413.
40. Schultz G, Chegini N, Grant M, Khaw P, MacKay S. Effects of growth factors on corneal wound healing. *Acta Ophthalmol Suppl.* 1992(202):60–66.
41. Haynes RJ, Tighe PJ, Dua HS. Antimicrobial defensin peptides of the human ocular surface. *The British journal of ophthalmology.* 1999;83(6):737–741.
42. Hampel U, Klonisch T, Sel S, et al. Insulin-like factor 3 promotes wound healing at the ocular surface. *Endocrinology.* 2013;154(6):2034–2045.
43. Hampel U, Klonisch T, Makrantonaki E, et al. Relaxin 2 is functional at the ocular surface and promotes corneal wound healing. *Investigative ophthalmology & visual science.* 2012;53(12):7780–7790.
44. Harvey DJ. Identification by gas chromatography/mass spectrometry of long-chain fatty acids and alcohols from hamster meibomian glands using picolinyl and nicotinate derivatives. *Biomed Chromatogr.* 1989;3(6):251–254.
45. Harvey DJ. Long-chain fatty acids and alcohols from gerbil meibomian lipids. *J Chromatogr.* 1989;494:23–30.
46. Harvey DJ, Tiffany JM. Identification of meibomian gland lipids by gas chromatography-mass spectrometry: application to the meibomian lipids of the mouse. *J Chromatogr.* 1984;301(1):173–187.
47. Harvey DJ, Tiffany JM, Duerden JM, Pandher KS, Mengher LS. Identification by combined gas chromatography-mass spectrometry of constituent long-chain fatty acids and alcohols from the meibomian glands of the rat and a comparison with human meibomian lipids. *J Chromatogr.* 1987;414(2):253–263.
48. Miyazaki M, Man WC, Ntambi JM. Targeted disruption of stearoyl-CoA desaturase1 gene in mice causes atrophy of sebaceous and meibomian glands and depletion of wax esters in the eyelid. *J Nutr.* 2001;131(9):2260–2268.
49. Butovich IA, Borowiak AM, Eule JC. Comparative HPLC-MS analysis of canine and human meibomian lipidomes: many similarities, a few differences. *Sci Rep.* 2011;1:24.
50. Greiner JV, Glonek T, Korb DR, Booth R, Leahy CD. Phospholipids in meibomian gland secretion. *Ophthalmic Res.* 1996;28(1):44–49.
51. Tiffany JM. The meibomian lipids of the rabbit. I. Overall composition. *Experimental eye research.* 1979;29(2):195–202.
52. Tiffany JM, Marsden RG. The meibomian lipids of the rabbit. II. Detailed composition of the principal esters. *Experimental eye research.* 1982;34(4):601–608.
53. Butovich IA, Lu H, McMahon A, Eule JC. Toward an animal model of the human tear film: biochemical comparison of the mouse, canine, rabbit, and human meibomian lipidomes. *Investigative ophthalmology & visual science.* 2012;53(11):6881–6896.
54. Lee JH, Kee CW. The significance of tear film break-up time in the diagnosis of dry eye syndrome. *Korean J Ophthalmol.* 1988;2(2):69–71.

# The Tear Film: Pathological Conditions

**Vikram Paranjpe and Anat Galor**

**Abstract** This chapter first reviews the pathogenesis of dry eye (DE) and the pathological effects DE has on the properties and composition of the tear film, lacrimal and meibomian glands, and goblet cells. Next, the pathologic changes that occur with two DE related risk factors: aging and the use of glaucoma eye drops are discussed. Finally, the clinical consequences of tear film, lacrimal gland, meibomian gland, and goblet cell alterations, focusing on visual function, nerve modulation, quality of life, and economic burden, are explored.

## 1 Introduction

It is well known that alteration of the properties and composition of the human tear film leads to an unstable and dysfunctional tear film as well as changes to the ocular surface—resulting in a condition otherwise known as dry eye (DE). This chapter will discuss mechanisms underlying DE and examine the changes in tear film properties and composition as well as changes in lacrimal glands, meibomian glands, and goblet cells. There are numerous risk factors leading to a condition of DE, and a discussion of all of them is beyond the scope of this chapter, but specific changes seen with aging and long-term use of glaucoma eye drops will be reviewed. Finally, the clinical consequences of tear film alteration associated with DE will be examined.

V. Paranjpe · A. Galor (✉)
Miami Veterans Administration Medical Center, Miami, FL, USA

Bascom Palmer Eye Institute, University of Miami, Miami, FL, USA
e-mail: agalor@med.miami.edu

© Springer Nature Switzerland AG 2019         347
G. Guidoboni et al. (eds.), *Ocular Fluid Dynamics*, Modeling and Simulation in
Science, Engineering and Technology, https://doi.org/10.1007/978-3-030-25886-3_15

# 2 Dry Eye

## 2.1 Defining Dry Eye

Per the Definition and Classification Subcommittee of the International Dry Eye Workshop (DEWS), dry eye is "a multifactorial disease of the tears and ocular surface that results in symptoms of discomfort, visual disturbance, and tear film instability with potential damage to the ocular surface. It is accompanied by increased osmolarity of the tear film and inflammation of the ocular surface" [1]. The disorder, however, also involves other components of the lacrimal functional unit (LFU) [2], including the ocular surface epithelium and the interconnecting innervation. DE is classified into two subgroups—aqueous deficiency dry eye (ADDE) and evaporative dry eye (EDE) with EDE found to be the more prevalent type in both population and clinic based studies [3].

## 2.2 Inflammation in the Pathogenesis of Dry Eye

Chronic inflammation at the ocular surface is understood to be the hallmark in the pathophysiology of both ADDE and EDE. Inflammation in DE is mediated both by antigen-presenting cells (APCs) such as lymphocytes, dendritic cells, Langerhans cells, macrophages, and epithelial cells of the cornea and conjunctiva, and by inflammatory mediators including cytokines [4–6], chemokines [6], matrix-metalloproteinases (MMPs) [7], and secretory phospholipid A2 (sPLA2) [8, 9]. Evidence for inflammation as the underlying process of DE has been well established in cell culture and animal models of DE as well as human DE studies. Cell culture models demonstrated that short-term desiccation (0–30 min) induced human corneal epithelial cells to express proinflammatory cytokines including IL-6, IL-8, and TNF-$\alpha$, and longer periods of desiccation (>8 h) also lead to cell death [10]. Various animals including rabbits [11–14], dogs [13], and most commonly, mice [15–22] have been used to model DE. Through mouse models, it has been elucidated that the progression of DE is fundamentally driven by autoimmune inflammatory cycles, primarily mediated by Th-17 cytokines [8]. Other important proinflammatory molecules include toll-like receptors (TLRs) and sPLA2, which specifically plays a role in eliminating bacteria from the tear film, but is also the rate-limiting enzyme in the production of prostaglandin E2 and other inflammatory mediators that drive inflammation at the ocular surface [8, 19, 21]. Finally, human studies have shown that both innate and adaptive immune mechanisms are critical to the pathogenesis of DE [8]. CD3 and CD4 T cells, and HLA-DR positive APCs have been identified in greater quantities in the conjunctival epithelium of both non-Sjögren syndrome (defined as patients with Schirmer's wetting of <8 mm at 5 min and no symptoms of Sjögren syndrome) and primary Sjögren syndrome DE patients as compared to normal controls [23]. Furthermore, interferon gamma has been found to upregulate HLA-DR in conjunctival cells, and in another study, HLA-DR positive cell populations were found to positively correlate with the severity of DE [8, 24]. These data strongly support the idea that chronic inflammation underlies the pathogenesis of DE and contributes to the disruption of the tear film, which in turn allows for further inflammation at the ocular surface. The specific role of inflammation in the lacrimal glands in ADDE and inflammation in the meibomian glands in EDE will be discussed below.

## 2.3 Changes in Tear Film Properties and Composition

Several properties of the tear film are altered in dry eye (DE). One of the hallmarks of DE is hyperosmolar tears, which occur due to either increased evaporative loss of tears or decreased secretion of tears [1, 25]. The normal osmolarity of tears is about 300 mOsm/L, whereas in DE the osmolarity of tears has shown to be between 330 and 400 mOsm/L [26, 27]. A human model of aqueous deficiency DE found a significant increase ($p < 0.001$) in osmolarity between 36 samples from 31 normal eyes ($302 \pm 6.3$ mOsm/L) and 38 samples from 30 eyes with keratoconjunctivitis sicca (defined as one or more of the following: tear film debris, increased tear film viscosity, decreased marginal tear strip, or superficial punctate staining with fluorescein) ($343 \pm 32.3$ mOsm/L) [28]. Tear film instability is another parameter that is classically associated with DE. Tear break-up time (TBUT) is thought to be a direct measure of tear film stability [25]. Tear break-up is initiated in areas where epithelial cells have recently sloughed off, with new epithelial cells having slightly less wettability. Thus, in ocular surface diseases like DE, which feature increased cell apoptosis and turnover, TBUT is reduced [25]. In normal subjects, TBUT can range from 3 to 132 s, with an average of 27 s [29]. Times of less than 10 s are considered to represent an abnormal tear film and times of less than 5 s are indicative of DE [30, 31].

Tears contain many constituent molecules including electrolytes and proteins, and the composition of tears has also been shown to be altered in DE. In a prospective study, Ohashi et al. [32] compared concentrations of three proteins: lactoferrin, epidermal growth factor (EGF), and aquaporin 5 (AQP5). These proteins are produced by the lacrimal gland. Lactoferrin is secreted from the acini of lacrimal glands and has antibacterial properties and scavenges free ions. EGF acts in a regulatory role on the ocular surface and aids in wound healing. AQP5 is a selective $H_2O$ channel and is expressed in increasing concentrations in lacrimal and salivary glands [32]. A study analyzed tears of 16 normal healthy volunteers, 23 patients with Sjögren syndrome (SS), 9 patients with Stevens–Johnson syndrome (SJS), and 71 patients with non-SS dry eye (defined as patients not meeting criteria for SS with Schirmer's wetting of <5 mm after 5 min and ocular surface abnormalities visualized with Rose Bengal or fluorescein greater than 3+). Lactoferrin concentrations (mg/mL) were found to be $2.05 \pm 1.12$ in controls, $0.69 \pm 0.55$ in non-SS, $0.13 \pm 0.22$ in SS, and $0.26 \pm 0.33$ in SJS. EGF concentrations (ng/mL) were

$5.09 \pm 3.74$ in controls, $2.30 \pm 3.04$ in non-SS, $0.58 \pm 0.60$ in SS, and $0.32 \pm 0.16$ in SJS. AQP5 concentrations (pg/$\mu$L) were $31.1 \pm 23.9$ in controls, $60.5 \pm 105.5$ in non-SS, $124.1 \pm 137.9$ in SS, and $27.7 \pm 25.7$ in SJS [32]. Lactoferrin and EGF were significantly decreased in SS ($p = 0.00002$), non-SS ($p = 0.0005$), and SJS ($p = 0.0001$) as compared to controls, whereas the concentration of AQP5 was significantly increased only in SS patients compared to controls ($p = 0.01$) and SS patients compared to non-SS patients ($p = 0.007$) [32]. These results indicate that tear components are altered in DE, and that in SS, AQP5 is released from the acini into tears when the lacrimal ducts are infiltrated by inflammatory cells [32]. Other biologically relevant components of tears that have been implicated in DE are $\omega$-3 polyunsaturated fatty acids (PUFAs) [33]. These lipids have demonstrated anti-inflammatory properties in several animal and cell culture models of DE [34–38]. Conversely, $\omega$-6 PUFAs, which are oxygenated to form arachidonic acid and later downstream prostaglandins are shown to be proinflammatory molecules [33]. The ratio of $\omega$-6 PUFA to $\omega$-3 PUFA is associated with parameters of DE including tear volume (Schirmer's test), TBUT, and corneal staining [33]. Furthermore, both mouse and human models have shown that dietary supplements of $\omega$-3 PUFA improve signs and symptoms of DE [38–40].

## 2.4   Lacrimal Glands

Inflammation of the lacrimal glands is the hallmark finding in aqueous deficiency dry eye (ADDE). There are a number of mechanisms that are thought to play a role in the dysfunction of the lacrimal glands including apoptosis, hormone changes, neural dysfunction, auto-antibodies, and proinflammatory cytokines [41]. Apoptosis is activated through pathways involving Fas, Fas ligand, Bax, caspases, perforin, and granzyme B, and can be perpetuated by proinflammatory cytokines including IL-1 and TNFα and T-cells, which express Fas [42–44]. The importance of apoptosis in lacrimal gland dysfunction has been debated and it is thought that perhaps the resistance of epithelial cells to apoptosis rather than the death of cells may alter the physiology of the lacrimal gland [41]. Hormones also play an important role in the maintenance of lacrimal gland function. Androgens tend to exert anti-inflammatory effects whereas estrogens and prolactin are known to be more proinflammatory [41]. Androgen deficiency may allow proinflammatory cytokines to mediate inflammation of the lacrimal glands as decreased levels of androgens are seen in SS patients [45–47]. The effects of androgens are exerted through androgen receptors on the epithelial cell surface, which affect the expression of proinflammatory cytokines [41]. The role of neural dysfunction has been extensively studied as part of the mechanism of lacrimal gland dysfunction, but it is clear that lacrimal gland acinar cells retain the ability to respond to exogenous agonist stimuli, so perhaps the role of neural dysfunction is not as significant in lacrimal gland dysfunction [41]. The same can be said for auto-antibodies. There is no denying that anti-muscarinic antibodies are present in SS, but their role in damaging the lacrimal gland is controversial [41]. Finally, proinflammatory cytokines IL-1 and TNFα are important in the pathogenesis of all inflammatory lacrimal gland diseases. They perpetuate inflammation by attracting immune cells and also interfere with the normal function of the lacrimal gland [41]. In a mouse model of SS, Xiao et al. [48] found increased levels of IL-1, IL-6, IL-8, and TNFα, as well as increased expression of CD4 (helper T cells), CD8 (cytotoxic T cells), CD103 (intraepithelial lymphocytes), CD11b (monocyte/macrophages), and CD45 (leukocytes).

Evaporative dry eye (EDE) does not seem to cause inflammation of the lacrimal glands so much as changes to the ocular surface. In a mouse model of EDE, Xiao et al. [48] found no signs of lacrimal gland inflammation, although they noted some structural changes including larger lacrimal gland acini as compared to normal mice.

## 2.5 Meibomian Glands

Meibomian gland dysfunction (MGD) is the major driving force in the development of evaporative dry eye (EDE). Meibomian glands secrete meibum, which forms the outermost layer of the tear film and protects it from evaporation. Meibum is made of a mixture of synthesized lipids, membrane derived lipids, and bacterial degradation products [49]. Non-polar components (wax and sterol esters) function as a barrier to water and as a lubricant, whereas polar components (phospholipids) act as a surfactant to ensure proper spreading of the tear film lipid layer across the ocular surface [49]. The prevailing mechanism leading to MGD is thought to involve inflammation and hyperkeratinization of the meibomian glands. Animal models of MGD showed that hyperkeratinization of the acini of meibomian gland ducts lead to plugging and eventually dilation of the orifices, thus leading to impaired meibum secretion into the tear film [14, 16, 50]. Animal and human models have also shown that the composition of meibum is altered in MGD, which destabilizes the tear film lipid layer. In the animal models, meibum contained an increase in lipids such as sterols and ceramides, which are more consistently found on epidermal surfaces [50]. Human models found that meibum contained increased concentrations of unsaturated triacylglycerols, phosphatidylcholine, and branched-chain fatty acids (wax, cholesterol, and triglycerides), all of which disrupt the stability of the tear film lipid layer [51, 52]. Alterations of the tear film lipid layer are correlated with the severity of DE symptoms, and increased evaporation of tears due to a compromised tear film lipid layer is one of the most common causes of hyperosmolar tears, a hallmark of DE [53]. Other mechanisms thought to be involved in MGD include microbial changes on the ocular surface leading to increased bacterial lipases, which in turn degrade meibum lipids into free-fatty acids and disrupt the lipid layer [54].

More recently, a depletion of meibocyte stem cells and meibomian gland atrophy via progenitor cell senescence pathways in the absence of hyperkeratinization have been proposed as major drivers of MGD [15, 18, 55, 56]. Expression of lipid-sensitive nuclear receptor, PPARγ, has been of interest. PPARγ is found to be highly expressed in meibomian gland acinar cells and meibocytes, both in the cytoplasm and nucleus and is involved in regulating the expression of genes involved in

**Table 1** Mean normal values of tear protein concentrations for each decade in mg/dL

| Age | 20–29 years 18 eyes | 30–39 years 18 eyes | 40–49 years 18 eyes | 50–59 years 18 eyes | 60–69 years 17 eyes | 70–79 years 19 eyes |
|---|---|---|---|---|---|---|
| Lactoferrin | 136 (26) | 139 (18) | 118 (24) | 90 (22) | 114 (32) | 87 (22) |
| Lysozyme | 110 (22) | 105 (18) | 99 (20) | 79 (16) | 73 (24) | 72 (28) |
| Ceruloplasmin | 12 (2) | 13 (2) | 10 (0.8) | 15 (4) | 15 (2) | 30 (14) |
| IgA | 52 (14) | 63 (14) | 41 (12) | 41 (12) | 52 (24) | 45 (14) |
| IgG | 1.7 (1.0) | 0.7 (0.2) | 0.7 (0.4) | 0.7 (0.2) | 2.8 (2.2) | 6.5 (22) |

Reproduced with permission from McGill et al. [74]
Numbers in parenthesis are standard deviation

## 3.3   Lacrimal Gland

The lacrimal gland is responsible for producing the aqueous component of the tear film. Increasing age has shown to cause morphological changes in the lacrimal glands. Obata et al. [76] reported increases in periductal fibrosis, diffuse fibrosis, and diffuse atrophy in the orbital lobes in older women, and increased periductal fibrosis and interlobular duct dilation in the palpebral lobes in older men. They hypothesize that periductal fibrosis leads to decreased tear fluid outflow, and stenosis of excretory ducts of the conjunctival fornix may lead to interlobular duct dilation [76]. Ueno et al. [77] studied the lacrimal gland thickness and area in 104 normal subjects aged 2–79 years old and found that lacrimal gland thickness and area significantly decreased ($P < 0.01$) in women, but there were no significant differences in men, thus indicating that gender has a significant influence on lacrimal gland morphology during aging [78].

Additionally, the prevalence of aqueous deficiency dry eye increases with age and animal studies have shown a decreased response of the lacrimal gland to neural agonists [20, 41, 79–81]. Dramatic structural changes have also been observed and are thought to be driven by infiltration of inflammatory cells including T and B cells, mast cells, and accumulation of lipofuscin. Finally, in mouse models, lacrimal glands of old but not young mice were found to have increased concentrations of proinflammatory cytokines IL-1β and TNFα [41].

## 3.4   Meibomian Gland

Mouse models have shown loss of meibomian gland acini in older mice in addition to a decrease in total gland volume, primarily associated with loss of lipid volume [82, 83]. Nien et al. [83] stained eyelids of mice at 2, 6, 12, and 24 months of age with specific antibodies against peroxisome proliferator activated receptor (PPAR) gamma (to identify differentiating meibocytes), Oil Red O (ORO, to identify lipid), and Ki67 nuclear antigen (to identify cycling cells). They found that eyelids of young mice (2 and 6 months) showed cytoplasmic and perinuclear

PPARgamma staining and abundant intracellular ORO staining. Older mice (12 and 24 months) only showed perinuclear PPARgamma staining and less ORO staining, with significantly decreased Ki67 staining as well [83]. These data support the hypothesis that PPARgamma signaling is altered in older mice that may contribute to changes in cell cycle entry, lipid formation, and meibomian gland drop out associated with aging [83]. Mathers et al. [75] demonstrated a similar pattern of age-related meibomian gland drop out in humans, with a significant linear regression ($p < 0.05$) between drop out and age. Finally, changes in the tear film lipid layer have been associated with a decrease in tear film stability seen in aging. Borchman et al. [84] used nuclear magnetic resonance to analyze meibum from 43 normal donors aged 1–88 years old. They found that meibum from infants and children contained fewer $CH_3$ and $C=C$ groups as compared to meibum from adolescents and adults. Fewer $CH_3$ and $C=C$ groups as well as increased protein are thought to allow for tighter lipid–lipid interactions, resulting in a more ordered tear film lipid layer that is less susceptible to disruption and subsequent evaporation of tears [84].

## 3.5 Goblet Cells

Goblet cells produce the mucous component of the tear film and allow tears to adhere to the ocular surface [67]. The literature on changes in the morphology and number of goblet cells with increasing age contains discrepancies. Marquardt [85] found no significant changes in the number of goblet cells with age, although reporting a slight decrease in number after age 61. Kessing [86] found increased goblet cell occlusion and altered morphology of goblet cells in up to 50% of patients, with older patients having goblet cells that contained neutral mucopolysaccharides rather than the acidic mucopolysaccharides seen in goblet cells of younger patients. Abdel-Kahlek et al. [87] collected bulbar conjunctival biopsies from 49 patients between 50 and 89 years old and found no significant morphological changes in goblet cells of patients below the age of 80. In patients 80 and older, 25% were found to have goblet cells containing periodic acid-Schiff (PAS) positive "hyaline bodies" which may represent broken down secretory granules [87]. They also found that the number of goblet cells decreased with age, with patients 50–59 years old having $10.2 \pm 3.5$ cells and patients 80–89 years old having $6.4 \pm 3$ cells [87].

## 4 Long-Term Use of Glaucoma Eye Drops

Glaucoma eye drops are the mainstays in the management of glaucoma and include beta adrenergic antagonists, alpha adrenergic agonists, prostaglandin analogues, and carbonic anhydrase inhibitors. As glaucoma is a chronic condition, these medications are used for long periods, and adverse effects are a major concern. Along with adverse effects caused by the drugs themselves, benzalkonium chloride

356 V. Paranjpe and A. Galor

(BAK), a common preservative in these eye drops, has been consistently implicated in inducing a number of changes to the ocular surface and other structures in laboratory, experimental, and clinical trials [88]. Changes include tear film instability, loss of goblet cells, conjunctival squamous metaplasia and apoptosis, oxidative stress, and direct interactions with the lipid layer of the tear film [88]. The effects of glaucoma eye drops on the tear film, lacrimal and meibomian glands, and goblet cells will be discussed below.

## 4.1 Tear Film Properties and Composition

The chronic use of glaucoma eye drops has significant effects on tear film stability. In a rabbit model using drops containing 0.01% BAK, Wilson et al. [89] found a fourfold increase in drying of the precorneal tear film, whereas a human model demonstrated a twofold increase in drying of the tear film. In a more recent model of albino rabbits receiving eye drops for 60 days, the TBUT was significantly decreased in the group receiving timolol maleate drops with 0.01% BAK preservative compared to the group receiving timolol maleate drops with no preservative [90]. A human trial with three groups: group I (20 individuals with no anterior segment pathology and receiving no medication), group II (20 individuals with primary open-angle glaucoma receiving 0.5% timolol maleate with 0.01% BAK), and group III (20 individuals with primary open-angle glaucoma receiving 0.5% timolol maleate and 0.1% dipivefrin HCL with 0.04% BAK) showed significant decreases in both Schirmer's test and TBUT [91]. Schirmer's test values for groups I, II, and III were $12.70 \pm 2.21$ mm, $10.40 \pm 1.58$ mm, and $8.20 \pm 1.55$ mm, respectively ($p < 0.001$). TBUT times for groups I, II, and III were $14.40 \pm 2.67$ s, $8.00 \pm 1.89$ s, and $6.90 \pm 1.97$ s, respectively ($p < 0.001$) [91]. These studies demonstrate that long-term use of BAK significantly affects tear film stability in a dose-dependent manner. The disruption of the tear film can lead to dry eye and corneal surface damage [92].

## 4.2 Lacrimal Glands

Effects of glaucoma eye drops have been reported on the lacrimal draining system as well as on tear secretion. Beta-adrenergic antagonists such as timolol have been associated with lacrimal drainage system obstruction (LDSO), a decrease in the width of the nasolacrimal drainage system, and a decrease in tear volume on the ocular surface [93]. Kashkouli et al. [94] compared 128 glaucomatous eyes being treated with eye drops to 277 healthy control eyes and demonstrated a significant increase in LDSO ($p = 0.008$), upper LDSO in 76.92% of glaucomatous eyes compared to 37.5% of control eyes ($p = 0.01$), and nasolacrimal drainage obstruction in 19.2% of glaucomatous eyes. In another study, Kaskhouli et al. [95]

found LDSO in 20% of the glaucomatous eyes compared to 8.57% of the control eyes ($p = 0.02$). Finally, Nuzzi et al. [96] found that both beta-adrenergic antagonist timolol and the cholinergic agent pilocarpine were associated with decreased tear secretion.

## 4.3   Meibomian Glands

Long-term use of glaucoma eye drops has been associated with changes in the morphology and function of meibomian glands in several studies [97–99]. A cross-sectional analysis of 70 patients with glaucoma being treated with prostaglandin analogue (PGA) or non-PGA drops found that 92% of those treated with PGA compared with 58.3% of those treated with non-PGA drops ($p = 0.02$) had mei-bomian gland dysfunction (MGD), defined as terminal duct obstruction determined by slit-lamp examination [97]. The PGA-treated patients also had worse ocular surface disease index (OSDI) scores and ocular surface test scores (Schirmer's test with anesthesia, TBUT, and lissamine green staining) as compared to the non-PGA treated patients ($p < 0.001$) [97]. Another cross-sectional analysis compared 70 patients on either 1, 2, or 3 glaucoma eye drops for at least 1 year with 45 healthy controls [98]. They again defined MGD as showing signs of terminal duct obstruction and performed ocular surface tests (Schirmer's test with anesthesia, TBUT, and lissamine green staining), along with the OSDI questionnaire. MGD was found in 80% of glaucoma patients and ocular surface test scores for patients with or without MGD were significantly worse for all parameters when compared with the healthy controls ($p < 0.01$). However, there were no significant differences observed between the OSDI scores or any ocular surface test parameters of glaucoma patients with or without MGD, indicating that while mild to moderate MGD is seen frequently in glaucoma patients, the presence of MGD does not cause significantly greater detrimental effects to the ocular surface than that already induced by chronic glaucoma eye drops [98].

## 4.4   Goblet Cells

The use of glaucoma eye drops has been shown to have a detrimental effect on goblet cell density and as a result, the mucus layer of the tear film in a number of studies [91, 100, 101]. The human model described above, consisting of three groups: group I (no treatment), group II (monotherapy with timolol maleate + 0.01% BAK), and group III (bi-therapy, 0.5% timolol maleate and 0.1% dipivefrin HCL with 0.04% BAK) found detrimental effects on goblet cells. Goblet cell density was found to be $43 \pm 10.59/1000$ cells, $17 \pm 8.23/1000$ cells, and $15 \pm 4.08/1000$ cells in groups I, II, and III, respectively ($p < 0.0001$). In normal patients, no changes in goblet cell density are observed until after the age of 80, so it is significant that decreases were seen in goblet cell density in this study, with all patients under the age of 80 [87, 91]. The exact mechanisms by which goblet cell damage occurs are unclear, but proposed mechanisms include beta-blockers causing an interference with the blood flow to the conjunctiva, and induction of chronic inflammation or fibroblast proliferation, the latter changes likely due to cytotoxic effects of BAK and other preservatives used in commercial preparations of these medications [12, 102–104].

# 5  Clinical Consequences of Tear Film Alteration

It is important to understand the mechanisms and risk factors that lead DE and the subsequent changes in the tear film, lacrimal glands, meibomian glands, and goblet cells; however, ultimately, it is perhaps more important to realize how these changes manifest clinically and the effects they have on individual's daily lives. The remainder of this chapter will explore the effects of the alterations discussed so far on visual function, ocular morbidity, ocular surface nerve function, economic considerations, and overall quality of life.

## 5.1  Visual Function

The tear film is the outermost optical surface and has the largest refractive index step from the air to tears, giving it the most optical power [105]. It is commonly believed that the tear film does not majorly contribute to optical power given that the refractive index of the tear film/cornea combination and the cornea alone is almost identical, if the tear film is uniformly thick. This condition is often not met, as there are local disruptions in the tear film between blinks known clinically as tear break-up, which create aberrations in the optical system [105]. Tear break-up is thought to initiate in locations where epithelial corneal cells have recently sloughed off, with reductions in tear break-up time (TBUT), seen in pathological conditions including DE, representing increased rates of cell sloughing [25]. Complete breaks in the tear film expose the irregular corneal surface to the air, creating optical scatter and resulting in a poor quality retinal images [105]. Reductions in visual quality are common complaints of patients with DE. Lee et al. [106] reported that 8% of DE patients complained of "fluctuating blurry vision" that improved with blinking. Other studies found that between 42 and 80% of patients with primary Sjögren syndrome reported disturbed vision [107, 108].

Tutt et al. [105] assessed optical quality of the eye during periods of non-blinking by quantifying retinal vessel contrast and psychophysical contrast sensitivity in patients with and without soft contact lenses. They found similar patterns of image degradation with and without soft contact lenses with optical aberrations caused by tear break-up producing both objective (retinal vessel contrast) and subjective (contrast sensitivity) declines in image quality. This decline in image quality could be the cause of blurry vision in DE patients [105].


## 5.2   Ocular Morbidity: Increased Risk of Infection?

DE is characterized by a chronic state of inflammation that results in alteration of the tear film and disruption of the ocular surface [109]. It is commonly thought that disruptions in the ocular surface and tear film in DE then lead to an increased risk of infection at the ocular surface (keratitis). The rationale for this argument is supported by findings such as those of Kwong et al. [110] who demonstrated in an in vivo model that human tear fluid conferred protection to corneas against infiltration by *P. aeruginosa*. However, overall the literature supporting an association between DE and microbial keratitis is not very strong. A study of patients in a nursing home found that 26% of staphylococcal keratitis cases were associated with DE, but most of the patients in the study also had rheumatoid arthritis and were using topical and/or systemic corticosteroids, which on their own, increase the risk of infection [111]. Boiko et al. [112] found increased rates of DE in patients with chlamydial conjunctivitis, but another study [113] found that treating chlamydial conjunctivitis also improved the DE condition, suggesting that the relationship between DE and keratitis might actually be reversed. Other studies [114, 115] often cited as evidence of an association between DE and keratitis have demonstrated a relationship between "ocular surface disease" and infection; however, they lumped together herpetic corneal infection, bullous keratopathy, DE, blepharitis, and other eyelid diseases, making it difficult to assess the prevalence of DE in their study populations, and truly determine the strength of the association [109]. Interestingly, a number of studies [116–124] have found that microbial loads of many species including *Corynebacterium* species, *Propionibacterium* species, coagulase negative *Staphylococcus, S. aureus, B. subtilis, Rhodococcus* species, *P. aeruginosa,* and *H. influenza* are increased at the ocular surface are increased in DE. However, the literature as a whole does not support an increase in the rate of ocular surface infection in DE [109].

   To understand why there does not seem to be an increased risk of keratitis in DE, it is important to examine the innate and adaptive (T and B cell mediated) immune mechanisms at play. Innate immunity at the ocular surface includes pattern recognition receptors, mucins, antimicrobial peptides, lactoferrin, lysozyme, lipocalin, secretory IgA (sIgA), and secretory phospholipase A2 (sPLA2) [109]. DE seems to alter these innate immune mechanisms and confer dry eyes the ability to avoid increased rates of microbial invasion, as would be expected with the disruption of the tear film and ocular surface. These changes will be discussed below, and are summarized in Table 2.

   Pattern recognition peptides (PRRs) serve as the primary mechanism for detecting invades at the ocular surface and include the Toll-like receptors (TLRs) as well as the NOD-like receptors (NLRs), which recognize specific pathogen-

**Table 2** Modulation of innate immune molecules in dry eye

|  | Classification of dry eye | Fluid/tissue tested | Change (see original paper for references) |
|---|---|---|---|
| *Molecule* | | | |
| Lysozyme | Non-SS[a] and SS | Tears | Decreased |
|  | Non-SS[a] and SS |  | Unchanged |
| Lactoferrin | Non-SS[a] and SS | Tears | Decreased |
| Lipocalin | Non-SS[a] and SS | Tears | Decreased |
| sIgA | Non-SS[a] and SS | Tears | Decreased |
|  |  |  | Mixed results++ |
|  | SS |  | Unchanged |
| sPLA2 | Non-SS[a] and SS | Tears | Increased |
|  | Non-SS[a]* | Conjunctival epithelium* | Increased |
| *Mucins* | | | |
| MUC1 | Non-SS[a] and SS | Tears, CIC | Increased |
| MUC5AC | SS | CIC, conjunctival biopsy, tears | Decreased |
| MUC16 | Non-SS[a] | CIC | Altered distribution/glycosylation |
| MUC19 | SS | CIC, conjunctival biopsy | Decreased |
| *AMPs* | | | |
| hBD-1 | Non-SS[a] | CIC | Unchanged |
| hBD-2 | Non-SS[a] and SS | CIC | Increased |
| hBD-3 | Non-SS[a] | CIC | Unchanged |
| hBD-9 | Unspecified | Corneal and conjunctival impression cytology | Decreased |
| LL-37 | Non-SS[a] | CIC | Unchanged |
| *TLRs* | | | |
| TLR2 | Non-SS[a] | CIC | Increased (mRNA but not protein) |
| TLR4 | Non-SS[a]* | Corneal epithelium and stroma* | Increased |
| TLR9 | Non-SS[a] | CIC | Decreased |

All data pertain to human studies, except where * denotes a murine model of dry eye. Tears were collected using Whatman filter paper, microcapillary tube, or micropipette, Schirmer strip, or surgical sponge extraction, or by a tear wash

*SS* Sjögren syndrome, *CIC* conjunctival impression cytology, *MUC* mucin, *AMPs* antimicrobial peptides, *TLRs* Toll-like receptors; ++IgA was decreased in 3 of 12 eyes, and was in the low normal range for the others

Reproduced with permission from Narayan et al. [109]

[a]Non-SS DE defined as Schirmer's wetting <5 mm after 5 min or TBUT<5 s or ocular surface abnormalities graded >3+ as visualized with Rose Bengal or fluorescein dye

associated molecular patterns [109]. Ocular surface epithelial cells and corneal epithelial-associated Langerhans cells express a wide range of TLRs and NLRs, which recognize pathogens and respond by secreting chemokines, cytokines, and antimicrobial peptides. Enhanced expression of PRRs might make the dry eye better equipped to evade microbial attack, but alternatively, they may also serve as a source of the inflammation seen in DE, because many of the proinflammatory cytokines secreted by TLRs are the same molecules that are known to be found at the ocular surface in DE [109, 125, 126]. In a mouse model of DE, Lee et al. [127] found increased expression of TLR2–4 and TLR9 and that inhibiting TLR4 decreased the severity corneal staining and cytokine expression, providing further evidence that TLR expression is modulated at the ocular surface in DE, however more research is required to fully understand the relationship between PRRs, DE, and ocular surface infections [109].

Lactoferrin, lysozyme, lipocalin, sIgA, and sPLA2 are also all important components of the innate immune system at the ocular surface and are found to be modulated in DE. Lactoferrin binds iron in the tear film and prevents bacterial colonization, whereas lysozyme attacks the cell wall of bacteria [119]. Lipocalin prevents siderophore-mediated iron uptake by bacteria and sIgA provides antimicrobial protection of mucosal tissue, and may play a role in preventing bacterial adhesion to contact lenses, which are associated with increased risk of infection [128, 129]. All four of these molecules are found to be decreased in DE [109, 119, 130]. Secretory PLA2 catalyzes the initial step of the arachidonic pathway, and binds to the surface of bacteria to kill them via phospholipolytic enzyme activity [131]. Secretory PLA2 has been found to be increased in the tear film of DE patients [132].

Mucins maintain a wet ocular surface and prevent adverse environmental conditions, with ocular surface expression of 9 out of the 18 known mucins being reported in the literature [109, 133]. MUC1 and MUC16 are unique in that they exist both as soluble mucins in the tear film and in the membrane-bound form on the ocular surface epithelia, where they function to bind bacteria and aid in bacterial clearance [134]. Loss of mucins results in the loss of a physical barrier against pathogens, thus alteration of mucins increases the risk of infection in DE. MUC1 expression was found to be increased in DE [130], whereas the expression of other mucins was either decreased or dysregulated [109].

Finally, antimicrobial peptides include defensins and cathelicidins. Human beta-defensin-1 (hBD1) and hbD3 are constitutively expressed at the ocular surface, whereas hBD2 is only expressed in response to inflammation, infection, or injury [109, 135]. hBD2 is found to have strong antimicrobial action, and is also found to be increased in DE [135]. Similarly, enhanced expression of cathelicidin LL-37 is seen in response to inflammation, however, its expression is found to be unchanged in DE. Overall, a recent study showed that defensins and cathelicidins were essential in preventing bacteria from traversing the epithelial barrier [136].

While DE is a state of chronic inflammation characterized by disruption of the tear film as well as the ocular surface, there does not seem to be an increased prevalence of ocular surface infections, likely due to the actions of the innate immune defenses, which are modulated in DE (as seen in Table 2) [109].

## 5.3   Nerve Modulation

The ocular surface is extensively innervated by both sensory and autonomic nerve fibers. Lacrimal gland function is regulated by these autonomic and sensory nerves, which locally release neuropeptides, primarily substance P (SP), calcitonin gene-related peptide (CGRP), and neuropeptide Y (NPY) [137]. These neuropeptides are released during inflammatory processes and each play a unique role in the modulation of neurogenic inflammation [137]. SP induces inflammation through the release of cytokines and chemokines to activate and recruit cells of the innate immune system [138]. CGRP plays a role in inflammation by facilitating dilation of blood vessels, thus allowing extravasation of leukocytes [137]. NPY, which is released from sympathetic nerves, modulates the immune response and its effects include allowing for increased adhesion and migration of macrophages, decreasing the activity of natural killer cells, and inhibiting T cell proliferation [139]. It is thought that neuropeptides act as the first line of defense for the ocular surface, because they are released in response to ocular surface irritants or pathogens and lead to the release of leukocytes into the tears, which then triggers a systemic immune response [140, 141]. Conditions that alter the tear film and ocular surface such as aging, DE, and the use of glaucoma medications can lead to inflammatory states, which then alter nerve modulation on the corneal surface [142]. Alteration of these nerve fibers causes decreased corneal sensitivity, reduced reflex tear secretion, and neurogenic inflammation [141, 142]. SP and CGRP have both been found in higher concentrations in individuals with DE, and conversely, NPY is found to be decreased in autoimmune diseases including Sjögren syndrome, and may in fact also be decreased in the lacrimal glands of non-Sjögren DE as well [140, 143]. Ultimately, nerve modulation in DE leads to the symptoms that individuals often experience—dryness, burning, stinging, foreign-body sensation, and pain.

## 5.4   Quality of Life and Economic Considerations in Dry Eye

DE like other chronic conditions has a significant impact on patients' daily activity, workplace productivity, and overall quality of life (QoL). In a cross-sectional study of 190 individuals with DE and 399 individuals without DE, individuals with DE were found to be significantly more likely to face issues performing professional work (OR 3.49, 95% CI 1.72–7.09), computer use (OR 3.37, 95% CI 2.11–5.38), television watching (OR 2.84, 95% CI 1.05–7.74), daytime driving (OR 2.80, 95% CI 1.58–4.96), and nighttime driving (OR 2.20, 95% CI 1.48–3.28) [144]. Another study of 210 individuals, 130 with non-Sjögren DE, 32 with primary Sjögren Syndrome, and 48 controls all answering the short-form 36 (SF-36) health survey similarly found that individuals with DE reported vision-related impacts on their QoL [145]. Nelson et al. [146] in a study of 73 non-Sjögren DE individuals with an average time from diagnosis of $9.23 \pm 7.19$ years found that 73% of individuals reported that DE interfered with activities of daily life, and on average, in a year, lost 2 days of work, and went to work on 191 days despite DE symptoms.

DE also has a significant economic impact on both the patient and health system levels. The major cost drivers are direct medical costs (office visits, prescription and over-the-counter medications, specialized eye wear, and surgical procedures), direct non-medical costs (transportation), indirect costs (lost work time, lost productivity, and change in the type of work due to DE), and finally intangible costs (impaired social, emotional, and physical functioning), which primarily are felt on the patient level [147, 148]. Kozma et al. [149] found that on average, individuals with DE lost 184 h of productivity at an annual expense of $5362 per individual. From a health system perspective, Lee et al. [150] developed a Markov economic model and estimated the cost of managing a population with DE for 1 year with palliative medications, punctal plugs, and/or surgery to be $357,000 for an organization covering 500,000 lives. In a case review of a managed care database, Smeeding et al. [151] found that DE patients had greater total healthcare costs after diagnosis as compared to matched control patients. They discussed how this increase could be attributed to the propensity of DE patients to have comorbid ophthalmic conditions, but regardless, found a 22% increase ($p < 0.001$) in total costs with DE patients, which were attributed primarily to more frequent physician visits, but also to artificial tears, ophthalmic antibiotics, and to corticosteroids, which are expensive and have a wide range of possible side-effects potentially requiring further clinical evaluation [151].

The mainstays in DE treatment—artificial tears, ophthalmic antibiotics, corticosteroids, and punctual plugs—mostly manage the symptoms of DE and do not alter the progression of the disease. As a result, patients for the most part do not see significant improvements in their QoL, nor do they tend to see any decrease in disease-related costs [152]. The emergence and increased use of topical cyclosporin A (CsA) for the treatment of DE has yielded promising outcomes in subjective patient measures of DE as well as in the economic impact of DE. A double-blinded phase 3 study of CsA found a 55% decrease in total medication orders, including those for non-steroidal anti-inflammatory drugs, anti-histamines, artificial tears, ophthalmic antibiotics, and ophthalmic corticosteroids post-CsA treatment compared to pre-CsA treatment at 1 year of follow-up [153]. A review of medical records of 181 patients using topical CsA twice daily for the treatment of DE found a decrease in average discomfort scores, an improvement in patient satisfaction scores, and a decrease in prescriptions for ancillary medications and DE-related physician visits [154]. These studies demonstrate potential for topical CsA to improve patient QoL measures as well as their potential to relieve some of the economic burden at both a patient and healthcare system level as compared to more traditional measures such as artificial tears, ophthalmic antibiotics, corticosteroids, and punctal plugs, which serve as more palliative measures that do not alter the progression of the disease.

# 6   Conclusion

This chapter reviewed the pathogenesis of DE as well as DE-associated changes seen in the tear film, lacrimal and meibomian glands, and goblet cells. Specific changes associated with aging and long-term use of glaucoma eye drops, both risk factors for the development of DE were also discussed. The clinical implications of tear film alterations associated with DE include a decrease in visual quality and function, no significant increase in risk for ocular surface infections, and nerve modulation leading to many of the symptoms of DE such as pain and burning. Finally, it is important to recognize that tear film alterations and DE negatively impact quality of life; leading to issues with computer use, daytime and nighttime driving, watching television, and other activities of daily living. Treatment of these conditions also carries a significant economic burden to patients and healthcare system, and currently, treatment strategies mainly target symptomology and not the underlying processes causing the alterations of the tear film. Further research should aim to fully understand the various processes that can alter the stability of the tear film, how these changes manifest clinically, and devise treatment modalities that target the underlying mechanisms to truly improve the quality of life of patients.

**Funding** This work was supported by the Department of Veterans Affairs, Veterans Health Administration, Office of Research and Development, Clinical Sciences Research EPID-006-15S (Dr. Galor), NIH Center Core Grant P30EY014801, R01EY026174 (Dr. Galor), Research to Prevent Blindness Unrestricted Grant.

# References

1. The definition and classification of dry eye disease: report of the Definition and Classifi cation Subcommittee of the International Dry Eye WorkShop (2007). *The ocular surface.* 2007;5(2):75-92.
2. Stern ME, Gao J, Siemasko KF, Beuerman RW, Pflugfelder SC. The role of the lacrimal functional unit in the pathophysiology of dry eye. *Exp Eye Res.* 2004;78(3):409-416.
3. Stapleton F, Alves M, Bunya VY, et al. TFOS DEWS II Epidemiology Report. *The ocular surface.* 2017;15(3):334-365.
4. Boehm N, Riechardt AI, Wiegand M, Pfeiffer N, Grus FH. Proinflammatory cytokine profiling of tears from dry eye patients by means of antibody microarrays. *Investigative ophthalmology & visual science.* 2011;52(10):7725-7730.
5. Corrales RM, Villarreal A, Farley W, Stern ME, Li DQ, Pflugfelder SC. Strain-related cytokine profiles on the murine ocular surface in response to desiccating stress. *Cornea.* 2007;26(5):579-584.
6. Lam H, Bleiden L, de Paiva CS, Farley W, Stern ME, Pflugfelder SC. Tear cytokine profiles in dysfunctional tear syndrome. *Am J Ophthalmol.* 2009;147(2):198-205 e191.
7. Corrales RM, Stern ME, De Paiva CS, Welch J, Li DQ, Pflugfelder SC. Desiccating stress stimulates expression of matrix metalloproteinases by the corneal epithelium. *Investigative ophthalmology & visual science.* 2006;47(8):3293-3302.
8. Wei Y, Asbell PA. The core mechanism of dry eye disease is inflammation. *Eye Contact Lens.* 2014;40(4):248-256.

9. Wei Y, Pinhas A, Liu Y, Epstein S, Wang J, Asbell P. Isoforms of secretory group two phospholipase A (sPLA2) in mouse ocular surface epithelia and lacrimal glands. *Investigative ophthalmology & visual science.* 2012;53(6):2845-2855.

10. Higuchi A, Kawakita T, Tsubota K. IL-6 induction in desiccated corneal epithelium in vitro and in vivo. *Molecular vision.* 2011;17:2400-2406.

11. Bandamwar KL, Papas EB, Garrett Q. Fluorescein staining and physiological state of corneal epithelial cells. *Contact lens & anterior eye: the journal of the British Contact Lens Association.* 2014;37(3):213-223.

12. Burstein NL. The effects of topical drugs and preservatives on the tears and corneal epithelium in dry eye. *Trans Ophthalmol Soc U K.* 1985;104 (Pt 4):402-409.

13. Butovich IA, Lu H, McMahon A, Eule JC. Toward an animal model of the human tear film: biochemical comparison of the mouse, canine, rabbit, and human meibomian lipidomes. *Investigative ophthalmology & visual science.* 2012;53(11):6881-6896.

14. Jester JV, Nicolaides N, Kiss-Palvolgyi I, Smith RE. Meibomian gland dysfunction. II. The role of keratinization in a rabbit model of MGD. *Investigative ophthalmology & visual science.* 1989;30(5):936-945.

15. Jester JV, Potma E, Brown DJ. PPARgamma Regulates Mouse Meibocyte Differentiation and Lipid Synthesis. *The ocular surface.* 2016;14(4):484-494.

16. Jester JV, Rajagopalan S, Rodrigues M. Meibomian gland changes in the rhino (hrrhhrrh) mouse. *Investigative ophthalmology & visual science.* 1988;29(7):1190-1194.

17. McMahon A, Lu H, Butovich IA. A Role for ELOVL4 in the Mouse Meibomian Gland and Sebocyte Cell Biology. *Investigative ophthalmology & visual science.* 2014;55(5):2832-2840.

18. Parfitt GJ, Brown DJ, Jester JV. Transcriptome analysis of aging mouse meibomian glands. *Molecular vision.* 2016;22:518-527.

19. Redfern RL, Patel N, Hanlon S, et al. Toll-like receptor expression and activation in mice with experimental dry eye. *Investigative ophthalmology & visual science.* 2013;54(2):1554-1563.

20. Rios JD, Horikawa Y, Chen LL, et al. Age-dependent alterations in mouse exorbital lacrimal gland structure, innervation and secretory response. *Experimental eye research.* 2005;80(4):477-491.

21. Zheng X, Bian F, Ma P, et al. Induction of Th17 differentiation by corneal epithelial-derived cytokines. *J Cell Physiol.* 2010;222(1):95-102.

22. Stern ME, Schaumburg CS, Pflugfelder SC. Dry eye as a mucosal autoimmune disease. *International reviews of immunology.* 2013;32(1):19-41.

23. Stern ME, Gao J, Schwalb TA, et al. Conjunctival T-cell subpopulations in Sjogren's and non-Sjogren's patients with dry eye. *Investigative ophthalmology & visual science.* 2002;43(8):2609-2614.

24. De Saint Jean M, Brignole F, Feldmann G, Goguel A, Baudouin C. Interferon-gamma induces apoptosis and expression of inflammation-related proteins in Chang conjunctival cells. *Investigative ophthalmology & visual science.* 1999;40(10):2199-2212.

25. Tiffany JM. Tears in health and disease. *Eye (London, England).* 2003;17(8):923-926.

26. Tiffany JM. The normal tear film. *Developments in ophthalmology.* 2008;41:1-20.

27. Farris RL. Tear osmolarity–a new gold standard? *Advances in experimental medicine and biology.* 1994;350:495-503.

28. Gilbard JP, Farris RL, Santamaria J, 2nd. Osmolarity of tear microvolumes in keratoconjunctivitis sicca. *Archives of ophthalmology (Chicago, Ill: 1960).* 1978;96(4):677-681.

29. Norn MS. Dead, degenerate, and living cells in conjunctival fluid and mucous thread. *Acta Ophthalmol (Copenh).* 1969;47(5):1102-1115.

30. Mengher LS, Bron AJ, Tonge SR, Gilbert DJ. Effect of fluorescein instillation on the pre-corneal tear film stability. *Current eye research.* 1985;4(1):9-12.

31. Pflugfelder SC, Tseng SC, Sanabria O, et al. Evaluation of subjective assessments and objective diagnostic tests for diagnosing tear-film disorders known to cause ocular irritation. *Cornea.* 1998;17(1):38-56.

32. Ohashi Y, Dogru M, Tsubota K. Laboratory findings in tear fluid analysis. *Clin Chim Acta.* 2006;369(1):17-28.
33. Walter SD, Gronert K, McClellan AL, Levitt RC, Sarantopoulos KD, Galor A. omega-3 Tear Film Lipids Correlate With Clinical Measures of Dry Eye. *Investigative ophthalmology & visual science.* 2016;57(6):2472-2478.
34. Cortina MS, He J, Li N, Bazan NG, Bazan HE. Neuroprotectin D1 synthesis and corneal nerve regeneration after experimental surgery and treatment with PEDF plus DHA. *Investigative ophthalmology & visual science.* 2010;51(2):804-810.
35. Cortina MS, He J, Russ T, Bazan NG, Bazan HE. Neuroprotectin D1 restores corneal nerve integrity and function after damage from experimental surgery. *Investigative ophthalmology & visual science.* 2013;54(6):4109-4116.
36. Dartt DA, Hodges RR, Li D, Shatos MA, Lashkari K, Serhan CN. Conjunctival goblet cell secretion stimulated by leukotrienes is reduced by resolvins D1 and E1 to promote resolution of inflammation. *J Immunol.* 2011;186(7):4455-4466.
37. Esquenazi S, Bazan HE, Bui V, He J, Kim DB, Bazan NG. Topical combination of NGF and DHA increases rabbit corneal nerve regeneration after photorefractive keratectomy. *Investigative ophthalmology & visual science.* 2005;46(9):3121-3127.
38. Harauma A, Saito J, Watanabe Y, Moriguchi T. Potential for daily supplementation of n-3 fatty acids to reverse symptoms of dry eye in mice. *Prostaglandins Leukot Essent Fatty Acids.* 2014;90(6):207-213.
39. Kangari H, Eftekhari MH, Sardari S, et al. Short-term consumption of oral omega-3 and dry eye syndrome. *Ophthalmology.* 2013;120(11):2191-2196.
40. Kawakita T, Kawabata F, Tsuji T, Kawashima M, Shimmura S, Tsubota K. Effects of dietary supplementation with fish oil on dry eye syndrome subjects: randomized controlled trial. *Biomed Res.* 2013;34(5):215-220.
41. Zoukhri D. Effect of inflammation on lacrimal gland function. *Experimental eye research.* 2006;82(5):885-898.
42. Manganelli P, Fietta P. Apoptosis and Sjogren syndrome. *Semin Arthritis Rheum.* 2003;33(1):49-65.
43. Mariette X. [Pathophysiology of Sjogren's syndrome]. *Ann Med Interne (Paris).* 2003;154(3):157-168.
44. Tapinos NI, Polihronis M, Tzioufas AG, Skopouli FN. Immunopathology of Sjogren's syndrome. *Ann Med Interne (Paris).* 1998;149(1):17-24.
45. Sullivan DA. Tearful relationships? Sex, hormones, the lacrimal gland, and aqueous-deficient dry eye. *The ocular surface.* 2004;2(2):92-123.
46. Sullivan DA, Belanger A, Cermak JM, et al. Are women with Sjogren's syndrome androgen-deficient? *The Journal of rheumatology.* 2003;30(11):2413-2419.
47. Sullivan DA, Krenzer KL, Sullivan BD, Tolls DB, Toda I, Dana MR. Does androgen insufficiency cause lacrimal gland inflammation and aqueous tear deficiency? *Investigative ophthalmology & visual science.* 1999;40(6):1261-1265.
48. Xiao B, Wang Y, Reinach PS, et al. Dynamic ocular surface and lacrimal gland changes induced in experimental murine dry eye. *PloS one.* 2015;10(1):e0115333.
49. McCulley JP, Shine WE. Meibomian gland function and the tear lipid layer. *The ocular surface.* 2003;1(3):97-106.
50. Nicolaides N, Santos EC, Smith RE, Jester JV. Meibomian gland dysfunction. III. Meibomian gland lipids. *Investigative ophthalmology & visual science.* 1989;30(5):946-951.
51. Lam SM, Tong L, Yong SS, et al. Meibum lipid composition in Asians with dry eye disease. *PloS one.* 2011;6(10):e24339.
52. Joffre C, Souchier M, Gregoire S, et al. Differences in meibomian fatty acid composition in patients with meibomian gland dysfunction and aqueous-deficient dry eye. *The British journal of ophthalmology.* 2008;92(1):116-119.
53. Foulks GN. The correlation between the tear film lipid layer and dry eye disease. *Surv Ophthalmol.* 2007;52(4):369-374.

54. Baudouin C, Messmer EM, Aragona P, et al. Revisiting the vicious circle of dry eye disease: a focus on the pathophysiology of meibomian gland dysfunction. *The British journal of ophthalmology.* 2016;100(3):300-306.
55. Hwang HS, Parfitt GJ, Brown DJ, Jester JV. Meibocyte differentiation and renewal: Insights into novel mechanisms of meibomian gland dysfunction (MGD). *Experimental eye research.* 2017.
56. Jester JV, Parfitt GJ, Brown DJ. Meibomian gland dysfunction: hyperkeratinization or atrophy? *BMC ophthalmology.* 2015;15 Suppl 1:156.
57. Rosen ED, Sarraf P, Troy AE, et al. PPAR gamma is required for the differentiation of adipose tissue in vivo and in vitro. *Mol Cell.* 1999;4(4):611-617.
58. Rosen ED, Spiegelman BM. PPARgamma: a nuclear regulator of metabolism, differentiation, and cell growth. *The Journal of biological chemistry.* 2001;276(41):37731-37734.
59. Petroll WM, Jester JV, Bean JJ, Cavanagh HD. Myofibroblast transformation of cat corneal endothelium by transforming growth factor-beta1, -beta2, and -beta3. *Investigative ophthalmology & visual science.* 1998;39(11):2018-2032.
60. Argueso P, Gipson IK. Epithelial mucins of the ocular surface: structure, biosynthesis and function. *Experimental eye research.* 2001;73(3):281-289.
61. Gipson IK. Distribution of mucins at the ocular surface. *Experimental eye research.* 2004;78(3):379-388.
62. Mantelli F, Argueso P. Functions of ocular surface mucins in health and disease. *Curr Opin Allergy Clin Immunol.* 2008;8(5):477-483.
63. Stephens DN, McNamara NA. Altered Mucin and Glycoprotein Expression in Dry Eye Disease. *Optom Vis Sci.* 2015;92(9):931-938.
64. Sweeney DF, Millar TJ, Raju SR. Tear film stability: a review. *Experimental eye research.* 2013;117:28-38.
65. Shimazaki-Den S, Dogru M, Higa K, Shimazaki J. Symptoms, visual function, and mucin expression of eyes with tear film instability. *Cornea.* 2013;32(9):1211-1218.
66. Tiffany JM, Winter N, Bliss G. Tear film stability and tear surface tension. *Current eye research.* 1989;8(5):507-515.
67. Van Haeringen NJ. Aging and the lacrimal system. *The British journal of ophthalmology.* 1997;81(10):824-826.
68. Xu KP, Tsubota K. Correlation of tear clearance rate and fluorophotometric assessment of tear turnover. *The British journal of ophthalmology.* 1995;79(11):1042-1045.
69. Nava A, Barton K, Monroy DC, Pflugfelder SC. The effects of age, gender, and fluid dynamics on the concentration of tear film epidermal growth factor. *Cornea.* 1997;16(4):430-438.
70. van Best JA, Benitez del Castillo JM, Coulangeon LM. Measurement of basal tear turnover using a standardized protocol. European concerted action on ocular fluorometry. *Graefes Arch Clin Exp Ophthalmol.* 1995;233(1):1-7.
71. Furukawa RE, Polse KA. Changes in tear flow accompanying aging. *Am J Optom Physiol Opt.* 1978;55(2):69-74.
72. Hirase K, Shimizu A, Yokoi N, Nishida K, Kinoshita S. [Age-related alteration of tear dynamics in normal volunteers]. *Nippon Ganka Gakkai zasshi.* 1994;98(6):575-578.
73. Hagele JE, Guzek JP, Shavlik GW. Lacrimal testing. Age as a factor in Jones testing. *Ophthalmology.* 1994;101(3):612-617.
74. McGill JI, Liakos GM, Goulding N, Seal DV. Normal tear protein profiles and age-related changes. *The British journal of ophthalmology.* 1984;68(5):316-320.
75. Mathers WD, Lane JA, Zimmerman MB. Tear film changes associated with normal aging. *Cornea.* 1996;15(3):229-234.
76. Obata H, Yamamoto S, Horiuchi H, Machinami R. Histopathologic study of human lacrimal gland. Statistical analysis with special reference to aging. *Ophthalmology.* 1995;102(4):678-686.
77. Ueno H, Ariji E, Izumi M, Uetani M, Hayashi K, Nakamura T. MR imaging of the lacrimal gland. Age-related and gender-dependent changes in size and structure. *Acta Radiol.* 1996;37(5):714-719.

78. Methodologies to diagnose and monitor dry eye disease: report of the Diagnostic Methodology Subcommittee of the International Dry Eye WorkShop (2007). *The ocular surface.* 2007;5(2):108-152.
79. Bromberg BB, Cripps MM, Welch MH. Sympathomimetic protein secretion by young and aged lacrimal gland. *Current eye research.* 1986;5(3):217-223.
80. Bromberg BB, Welch MH. Lacrimal protein secretion: comparison of young and old rats. *Experimental eye research.* 1985;40(2):313-320.
81. Draper CE, Adeghate E, Lawrence PA, Pallot DJ, Garner A, Singh J. Age-related changes in morphology and secretory responses of male rat lacrimal gland. *J Auton Nerv Syst.* 1998;69(2-3):173-183.
82. Jester BE, Nien CJ, Winkler M, Brown DJ, Jester JV. Volumetric reconstruction of the mouse meibomian gland using high-resolution nonlinear optical imaging. *Anat Rec (Hoboken).* 2011;294(2):185-192.
83. Nien CJ, Paugh JR, Massei S, Wahlert AJ, Kao WW, Jester JV. Age-related changes in the meibomian gland. *Experimental eye research.* 2009;89(6):1021-1027.
84. Borchman D, Foulks GN, Yappert MC, Milliner SE. Changes in human meibum lipid composition with age using nuclear magnetic resonance spectroscopy. *Investigative ophthalmology & visual science.* 2012;53(1):475-482.
85. Marquardt R, Wenz FH. [Histological studies of goblet cell counts in human conjunctiva (author's transl)]. *Klin Monbl Augenheilkd.* 1979;175(5):692-696.
86. Kessing SV. Mucous gland system of the conjunctiva. A quantitative normal anatomical study. *Acta Ophthalmol (Copenh).* 1968:Suppl 95:91+.
87. Abdel-Khalek LM, Williamson J, Lee WR. Morphological changes in the human conjunctival epithelium. I. In the normal elderly population. *The British journal of ophthalmology.* 1978;62(11):792-799.
88. Baudouin C, Labbe A, Liang H, Pauly A, Brignole-Baudouin F. Preservatives in eyedrops: the good, the bad and the ugly. *Prog Retin Eye Res.* 2010;29(4):312-334.
89. Wilson WS, Duncan AJ, Jay JL. Effect of benzalkonium chloride on the stability of the precorneal tear film in rabbit and man. *The British journal of ophthalmology.* 1975;59(11):667-669.
90. Pisella PJ, Fillacier K, Elena PP, Debbasch C, Baudouin C. Comparison of the effects of preserved and unpreserved formulations of timolol on the ocular surface of albino rabbits. *Ophthalmic Res.* 2000;32(1):3-8.
91. Yalvac IS, Gedikoglu G, Karagoz Y, et al. Effects of antiglaucoma drugs on ocular surface. *Acta Ophthalmol Scand.* 1995;73(3):246-248.
92. Baudouin C. Detrimental effect of preservatives in eyedrops: implications for the treatment of glaucoma. *Acta Ophthalmol.* 2008;86(7):716-726.
93. Servat JJ, Bernardino CR. Effects of common topical antiglaucoma medications on the ocular surface, eyelids and periorbital tissue. *Drugs Aging.* 2011;28(4):267-282.
94. Kashkouli MB, Pakdel F, Hashemi M, et al. Comparing anatomical pattern of topical anti-glaucoma medications associated lacrimal obstruction with a control group. *Orbit.* 2010;29(2):65-69.
95. Kashkouli MB, Rezaee R, Nilforoushan N, Salimi S, Foroutan A, Naseripour M. Topical antiglaucoma medications and lacrimal drainage system obstruction. *Ophthal Plast Reconstr Surg.* 2008;24(3):172-175.
96. Nuzzi R, Finazzo C, Cerruti A. Adverse effects of topical antiglaucomatous medications on the conjunctiva and the lachrymal (Brit. Engl) response. *Int Ophthalmol.* 1998;22(1):31-35.
97. Mocan MC, Uzunosmanoglu E, Kocabeyoglu S, Karakaya J, Irkec M. The Association of Chronic Topical Prostaglandin Analog Use With Meibomian Gland Dysfunction. *J Glaucoma.* 2016;25(9):770-774.
98. Uzunosmanoglu E, Mocan MC, Kocabeyoglu S, Karakaya J, Irkec M. Meibomian Gland Dysfunction in Patients Receiving Long-Term Glaucoma Medications. *Cornea.* 2016;35(8):1112-1116.

99. Arita R, Itoh K, Maeda S, et al. Comparison of the long-term effects of various topical antiglaucoma medications on meibomian glands. *Cornea.* 2012;31(11):1229-1234.
100. Derous D, de Keizer RJ, de Wolff-Rouendaal D, Soudijn W. Conjunctival keratinisation, an abnormal reaction to an ocular beta-blocker. *Acta Ophthalmol (Copenh).* 1989;67(3):333-338.
101. Herreras JM, Pastor JC, Calonge M, Asensio VM. Ocular surface alteration after long-term treatment with an antiglaucomatous drug. *Ophthalmology.* 1992;99(7):1082-1088.
102. Arici MK, Arici DS, Topalkara A, Guler C. Adverse effects of topical antiglaucoma drugs on the ocular surface. *Clin Exp Ophthalmol.* 2000;28(2):113-117.
103. Tseng SC, Hirst LW, Maumenee AE, Kenyon KR, Sun TT, Green WR. Possible mechanisms for the loss of goblet cells in mucin-deficient disorders. *Ophthalmology.* 1984;91(6):545-552.
104. Takahashi N. A new method evaluating quantitative time-dependent cytotoxicity of ophthalmic solutions in cell culture. Beta-adrenergic blocking agents. *Graefes Arch Clin Exp Ophthalmol.* 1983;220(6):264-267.
105. Tutt R, Bradley A, Begley C, Thibos LN. Optical and visual impact of tear break-up in human eyes. *Investigative ophthalmology & visual science.* 2000;41(13):4117-4123.
106. Lee SH, Tseng SC. Rose bengal staining and cytologic characteristics associated with lipid tear deficiency. *Am J Ophthalmol.* 1997;124(6):736-750.
107. Bjerrum KB. Test and symptoms in keratoconjunctivitis sicca and their correlation. *Acta Ophthalmol Scand.* 1996;74(5):436-441.
108. Vitali C, Moutsopoulos HM, Bombardieri S. The European Community Study Group on diagnostic criteria for Sjogren's syndrome. Sensitivity and specificity of tests for ocular and oral involvement in Sjogren's syndrome. *Ann Rheum Dis.* 1994;53(10):637-647.
109. Narayanan S, Redfern RL, Miller WL, Nichols KK, McDermott AM. Dry eye disease and microbial keratitis: is there a connection? *The ocular surface.* 2013;11(2):75-92.
110. Kwong MS, Evans DJ, Ni M, Cowell BA, Fleisig SM. Human tear fluid protects against Pseudomonas aeruginosa keratitis in a murine experimental model. *Infect Immun.* 2007;75(5):2325-2332.
111. Jhanji V, Constantinou M, Taylor HR, Vajpayee RB. Microbiological and clinical profiles of patients with microbial keratitis residing in nursing homes. *The British journal of ophthalmology.* 2009;93(12):1639-1642.
112. Boiko EV, Chernysh VF, Pozniak AL, Ageev VS. [To the role of Chlamydia infection in the development of dry eye]. *Vestn Oftalmol.* 2008;124(4):16-19.
113. Krasny J, Hruba D, Netukova M, Kodat V, Tomasova BJ. [Chlamydia pneumoniae in the etiology of the keratoconjunctivitis sicca in adult patients (a pilot study)]. *Cesk Slov Oftalmol.* 2009;65(3):102-106.
114. Bourcier T, Thomas F, Borderie V, Chaumeil C, Laroche L. Bacterial keratitis: predisposing factors, clinical and microbiological review of 300 cases. *The British journal of ophthalmology.* 2003;87(7):834-838.
115. Keay L, Edwards K, Naduvilath T, et al. Microbial keratitis predisposing factors and morbidity. *Ophthalmology.* 2006;113(1):109-116.
116. Albietz JM, Lenton LM. Effect of antibacterial honey on the ocular flora in tear deficiency and meibomian gland disease. *Cornea.* 2006;25(9):1012-1019.
117. Graham JE, Moore JE, Jiru X, et al. Ocular pathogen or commensal: a PCR-based study of surface bacterial flora in normal and dry eyes. *Investigative ophthalmology & visual science.* 2007;48(12):5616-5623.
118. Dougherty JM, McCulley JP. Comparative bacteriology of chronic blepharitis. *The British journal of ophthalmology.* 1984;68(8):524-528.
119. Seal DV, McGill JI, Mackie IA, Liakos GM, Jacobs P, Goulding NJ. Bacteriology and tear protein profiles of the dry eye. *The British journal of ophthalmology.* 1986;70(2):122-125.
120. Sharma S. Diagnosis of external ocular infections: microbiological processing and interpretation. *The British journal of ophthalmology.* 2000;84(2):229.

121. Ta CN, Chang RT, Singh K, et al. Antibiotic resistance patterns of ocular bacterial flora: a prospective study of patients undergoing anterior segment surgery. *Ophthalmology.* 2003;110(10):1946-1951.
122. Cuello OH, Caorlin MJ, Reviglio VE, et al. Rhodococcus globerulus keratitis after laser in situ keratomileusis. *J Cataract Refract Surg.* 2002;28(12):2235-2237.
123. Fleiszig SM, Evans DJ. Contact lens infections: can they ever be eradicated? *Eye Contact Lens.* 2003;29(1 Suppl):S67-71; discussion S83-64, S192-194.
124. Shine WE, Silvany R, McCulley JP. Relation of cholesterol-stimulated Staphylococcus aureus growth to chronic blepharitis. *Investigative ophthalmology & visual science.* 1993;34(7):2291-2296.
125. Redfern RL, McDermott AM. Toll-like receptors in ocular surface disease. *Experimental eye research.* 2010;90(6):679-687.
126. Pearlman E, Johnson A, Adhikary G, et al. Toll-like receptors at the ocular surface. *The ocular surface.* 2008;6(3):108-116.
127. Lee HS, Hattori T, Park EY, Stevenson W, Chauhan SK, Dana R. Expression of toll-like receptor 4 contributes to corneal inflammation in experimental dry eye disease. *Investigative ophthalmology & visual science.* 2012;53(9):5632-5640.
128. Lan J, Willcox MD, Jackson GD. Effect of tear-specific immunoglobulin A on the adhesion of Pseudomonas aeruginosa I to contact lenses. *Aust N Z J Ophthalmol.* 1999;27(3-4):218-220.
129. Mantis NJ, Rol N, Corthesy B. Secretory IgA's complex roles in immunity and mucosal homeostasis in the gut. *Mucosal Immunol.* 2011;4(6):603-611.
130. Caffery B, Joyce E, Boone A, et al. Tear lipocalin and lysozyme in Sjogren and non-Sjogren dry eye. *Optom Vis Sci.* 2008;85(8):661-667.
131. Buckland AG, Wilton DC. The antibacterial properties of secreted phospholipases A(2). *Biochim Biophys Acta.* 2000;1488(1-2):71-82.
132. Aho VV, Nevalainen TJ, Saari KM. Group IIA phospholipase A2 content of tears in patients with keratoconjunctivitis sicca. *Graefes Arch Clin Exp Ophthalmol.* 2002;240(7):521-523.
133. Argueso P, Balaram M, Spurr-Michaud S, Keutmann HT, Dana MR, Gipson IK. Decreased levels of the goblet cell mucin MUC5AC in tears of patients with Sjogren syndrome. *Investigative ophthalmology & visual science.* 2002;43(4):1004-1011.
134. Govindarajan B, Gipson IK. Membrane-tethered mucins have multiple functions on the ocular surface. *Experimental eye research.* 2010;90(6):655-663.
135. McDermott AM. The role of antimicrobial peptides at the ocular surface. *Ophthalmic Res.* 2009;41(2):60-75.
136. Huang LC, Reins RY, Gallo RL, McDermott AM. Cathelicidin-deficient (Cnlp -/-) mice show increased susceptibility to Pseudomonas aeruginosa keratitis. *Investigative ophthalmology & visual science.* 2007;48(10):4498-4508.
137. Mantelli F, Massaro-Giordano M, Macchi I, Lambiase A, Bonini S. The cellular mechanisms of dry eye: from pathogenesis to treatment. *J Cell Physiol.* 2013;228(12):2253-2256.
138. Mosimann BL, White MV, Hohman RJ, Goldrich MS, Kaulbach HC, Kaliner MA. Substance P, calcitonin gene-related peptide, and vasoactive intestinal peptide increase in nasal secretions after allergen challenge in atopic patients. *J Allergy Clin Immunol.* 1993;92(1 Pt 1):95-104.
139. Kovacs I, Ludany A, Koszegi T, et al. Substance P released from sensory nerve endings influences tear secretion and goblet cell function in the rat. *Neuropeptides.* 2005;39(4):395-402.
140. Mantelli F, Micera A, Sacchetti M, Bonini S. Neurogenic inflammation of the ocular surface. *Curr Opin Allergy Clin Immunol.* 2010;10(5):498-504.
141. Beuerman RW, Stern ME. Neurogenic inflammation: a first line of defense for the ocular surface. *The ocular surface.* 2005;3(4 Suppl):S203-206.
142. Kulka M, Sheen CH, Tancowny BP, Grammer LC, Schleimer RP. Neuropeptides activate human mast cell degranulation and chemokine production. *Immunology.* 2008;123(3):398-410.

143. Tuisku IS, Konttinen YT, Konttinen LM, Tervo TM. Alterations in corneal sensitivity and nerve morphology in patients with primary Sjogren's syndrome. *Experimental eye research.* 2008;86(6):879-885.
144. Miljanovic B, Dana R, Sullivan DA, Schaumberg DA. Impact of dry eye syndrome on vision-related quality of life. *Am J Ophthalmol.* 2007;143(3):409-415.
145. Mertzanis P, Abetz L, Rajagopalan K, et al. The relative burden of dry eye in patients' lives: comparisons to a U.S. normative sample. *Investigative ophthalmology & visual science.* 2005;46(1):46-50.
146. Nelson JD, Helms H, Fiscella R, Southwell Y, Hirsch JD. A new look at dry eye disease and its treatment. *Advances in therapy.* 2000;17(2):84-93.
147. Hirsch JD. Considerations in the pharmacoeconomics of glaucoma. *Manag Care.* 2002;11(11 Suppl):32-37.
148. Reddy P, Grad O, Rajagopalan K. The economic burden of dry eye: a conceptual framework and preliminary assessment. *Cornea.* 2004;23(8):751-761.
149. Kozma CMH, J.D.; Wojcik, A.R.;. Economic and quality of life impact of dry eye symptoms. Poster at the Annual Meeting of the Association for Research in Vision and Ophthalmology; April 30-May 5, 2000, 2000; Fort Lauderdale, Florida.
150. Lee JTT, C.W. Development of an economic model to assess costs and outcomes associated with dry eye disease. Poster at the 2000 Spring Practice and Research Forum of the American College of Clinical Pharmacy; April 2-5, 2000, 2000; Monterey, California.
151. Smeeding JEM, C.; Walt, J.G. Dry-eye increases in health care and utilization and expenditures. Poster at the Sixth Annual International Meeting of the International Society for Pharmacoeconomics and Outcomes Research; May 20-23, 2001, 2001; Arlington, Virginia.
152. Pflugfelder SC. Prevalence, burden, and pharmacoeconomics of dry eye disease. *Am J Manag Care.* 2008;14(3 Suppl):S102-106.
153. Sall K, Stevenson OD, Mundorf TK, Reis BL. Two multicenter, randomized studies of the efficacy and safety of cyclosporine ophthalmic emulsion in moderate to severe dry eye disease. CsA Phase 3 Study Group. *Ophthalmology.* 2000;107(4):631-639.
154. Cross WD, Lay LF, Jr., Walt JG, Kozma CM. Clinical and economic implications of topical cyclosporin A for the treatment of dry eye. *Manag Care Interface.* 2002;15(9):44-49.

# 1    Fluorescent Imaging

Fluorescein is one of the most widely used fluorescent agents for the investigation of the physiology of the tear film and the ocular surface and for clinical evaluation of DED. This agent is commonly used in tear film breakup time (TBUT) measurement, which has been used as a diagnostic test to evaluate the tear film stability [4, 5]. The TBUT test requires instilling fluorescein sodium into the tear film with a strip or pipette; the tear film is then illuminated with a cobalt blue light and observed with a microscope through a yellow filter [6]. TBUT is defined as the interval between the last complete blink and the first appearance of a dry spot, or disruption in the tear film [7]. The normal eyes have an average TBUT value of 27 s (ranging from 3 s to 132 s) [8]; TBUT that is less than 5 s indicates DED [9].

Besides the TBUT measurement, fluorescent imaging provides a tool to visualize and study the overall patterns of the tear film. Begley et al. videotaped the fluorescent images after the first break in the tear film and quantified the total area of tear breakup (AB) of the exposed cornea with image processing algorithms [10]. It was found that there is a significant difference in tear film breakup pattern between normal controls and dry eye patients, as shown in Fig. 1 [11, 12]. The rate of tear breakup or dry area growth rate (DAGR) was observed to be four times greater in DED subjects, who also demonstrated a greater breakup in the central cornea than controls [11].

Fluorescent images have been also used to investigate the mechanism associated with tear film thinning, based on the principle of fluorescein self-quenching. Self-quenching is a phenomenon that plays an important role in determining the fluorescent intensity in which high fluorescein concentration greatly reduces the efficiency of fluorescence. This phenomenon was applied to investigate the relative contributions of two major tear film thinning mechanisms—evaporation

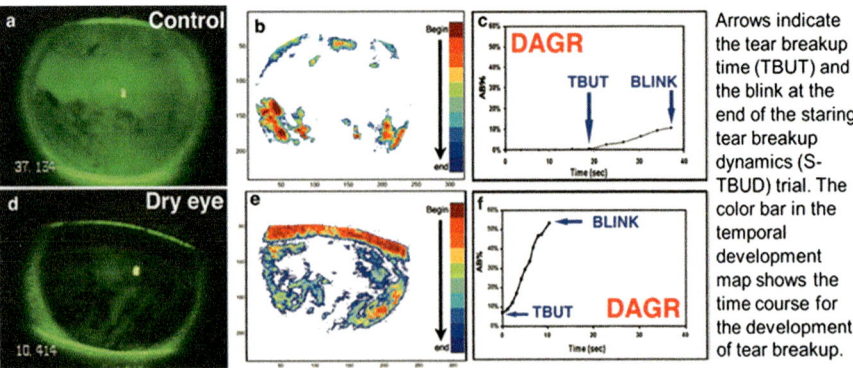

**Fig. 1** Fluorescein image and analysis before a blink for normal (**a–c**) and DED patients (**d–f**); (**b, e**) temporal development of the area of tear breakup (AB) plotted in (**c, f**) showing the dry area growth rate (DAGR) for breakup. (Adapted from [11])

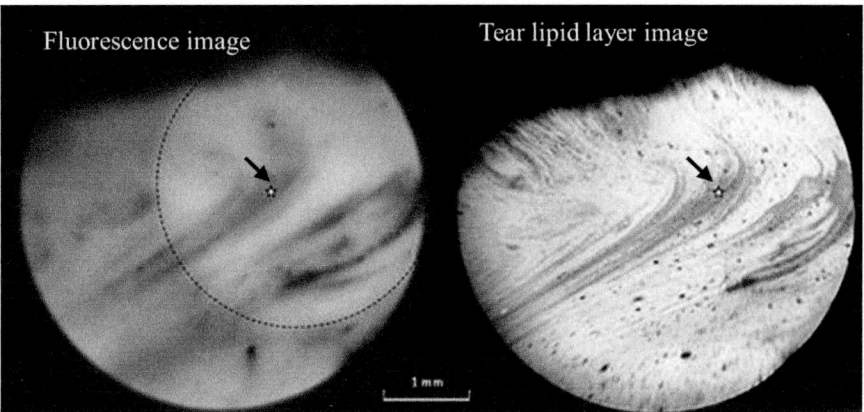

**Fig. 2** Tear film thinning is associated with thin lipid (e.g., in region indicated by the arrows). (Adapted from [16])

and tangential flow; tear evaporation should increase fluorescein concentration, and thus is expected to cause fluorescence dimming for high concentration condition but not for low concentration [13, 14]. King-Smith et al. recorded the tear film fluorescent images for both high and low fluorescein concentration conditions [15]. The fluorescent images showed that conditions with high fluorescein concentration have a greater fluorescence decay rate, which is expected if tear film thinning is mainly due to evaporation [15].

Tear film fluorescent imaging is also often combined with other imaging approaches to study tear film dynamics. For example, tear film fluorescent images have been simultaneously recorded with the lipid layer images to study the role of the lipid layer in tear film thinning [16]. It was observed that areas of tear thinning and breakup can generally be matched to the corresponding regions of the lipid layer image, as illustrated in Fig. 2. If evaporation has the largest contribution to tear film thinning, this observation is consistent with the expectation [16]. Fluorescent images have also been combined with thermal images to capture the ocular surface temperature and tear film fluorescence simultaneously, which showed that tear film evaporation accelerates on the fluorescein tear film breakup area, resulting in a lower temperature area on the tear film [17].

## 2 Interferometric Approaches

Based on the phenomenon of light waves superposition, various interferometric approaches have been developed to evaluate the tear film dynamics [18–20]. The non-invasive nature of interferometric approaches is one great advantage compared to invasive measurements, since the invasive methods may disturb the tear film dynamics.

To explain the fundamental principle of the interferometric approaches, the light interaction with the tear film (assuming a one layer structure) is schematically shown in Fig. 3. The reflected light from different interfaces of the tear film interfere with each other and the interference signal is recorded; constructive or destructive interference occurs when the reflected light are in phase or out of phase, respectively. The maximal interference for a tear film occurs when

$$m\lambda = 2nd \cos\phi, \tag{1}$$

where $m$ represents the order of the constructive interference, $\lambda$ is the wavelength of light in a vacuum, $n$ is the refractive index of the tear film, and $\phi$ is the angle of refraction in the tear film [19].

Depending on the implementation of how the interference fringes are recorded, there are three general interferometric approaches to evaluate the tear film; the three approaches correspond to varying one of the three parameters in Eq. (1), and as such include the thickness-dependent fringes method, the wavelength-dependent fringes method, and the angle-dependent fringes method.

In the thickness-dependent fringes method, the contour of the tear film thickness is represented by the cycles of bright and dark fringes of light. Doane developed an interferometer that uses thickness-dependent fringes method with nearly monochromatic illumination (full width half maximum of spectrum <8 nm; centered at 546 nm), to evaluate the tear film on contact lens surface. It was demonstrated that interferometry is feasible to provide a means for examining the dynamic changes in tear film thickness and its distribution. One limitation of this method is that it needs a zero thickness as a starting point for fringe counting, which means one needs to wait until the tear film breaks up in order to evaluate its thickness [18].

The thickness-dependent fringes method has also been demonstrated with broadband light sources, which can be used to image the interference fringes of the lipid layer, since the contrast of the fringes from the whole tear film is greatly attenuated due to the short coherence length of the light source. The color of the interference fringes depends on the lipid layer thickness, thus color clues of the fringe patterns have been used to quantify the lipid layer thickness. Goto et al. developed a color chart to map the color information of interference patterns to thicknesses distributions of the lipid layer [21]. One example of quantifying the lipid layer thickness by using the color chart is shown in Fig. 4.

**Fig. 3** Schematic illustration of light interaction within the tear film

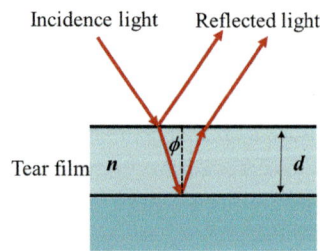

**Fig. 4** Quantification of the lipid layer thickness from the interference pattern color (the numbers indicate the thickness in nanometers) (Adapted from [21])

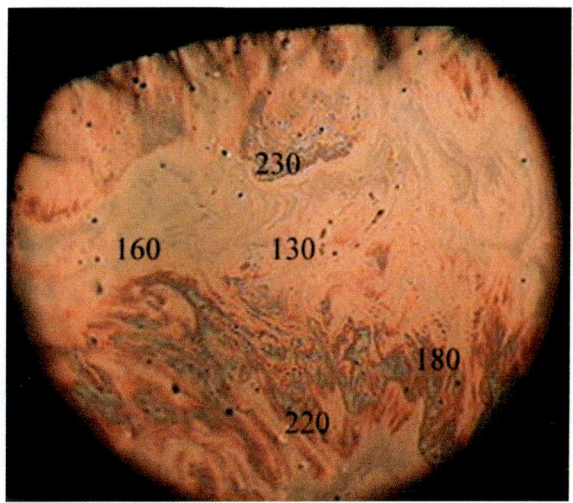

In the wavelength-dependent fringes method, the reflectance of the tear film with a given thickness is a sinusoidal function of the wavenumber. Thus, the wavelength-dependent interferometry system records the reflectance of the tear film for different wavenumbers. One implementation of the optical systems is shown in Fig. 5, which consists of a broadband light source as the illumination and a spectrophotometer to detect the interference fringes by wavenumbers [19]. The data processing step involves performing a Fourier transform to the spectral reflectance, in which the frequency of the spectral oscillation is mapped into the tear film thickness. With the system shown in Fig. 5, the authors found that there is a relatively strong interference effect for the surface at a depth of approximately 3 μm beneath the air surface and indicates that the human precorneal tear thickness is about 3 μm. To capture the tear film dynamics, measurements were taken every 27.5 ms over a 20-s time frame, with a blink approximately 1 s after the start of recording. The captured thinning process of the tear film is shown in Fig. 6; it was observed that the tear film gets thinning rapidly right after the blink to approximately 2.7 μm, followed by a slow thinning to approximately 2.3 μm over the remaining 18 s [22]. Compared to the thickness-dependent fringes method, one disadvantage of the wavelength-dependent fringes method lies in the fact that it only evaluates the tear film dynamics at one spot, rather than a distribution pattern on the ocular surface.

For the angle-dependent fringes method, the fringes are observed as a function of the incidence angle on the tear film. Prydal et al. developed an instrument to evaluate the tear film, using the angle-dependent fringes [23]. Results showed that the tear film has a thickness of about 40 μm, which is considerably thicker than the results from other approaches [19].

**Fig. 5** Optical system implementation of the wavelength-dependent fringes method (L indicates Lens) (implemented in [19])

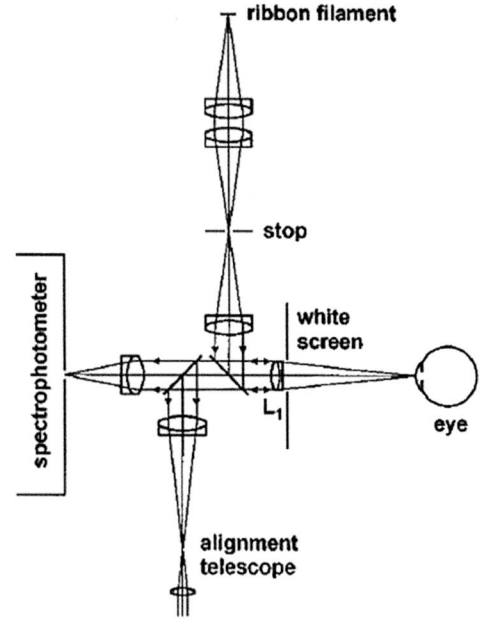

**Fig. 6** (a) Depth of the 3-μm layer and its thinning over a 20-s period. (b) Reflectance from the eye [22]

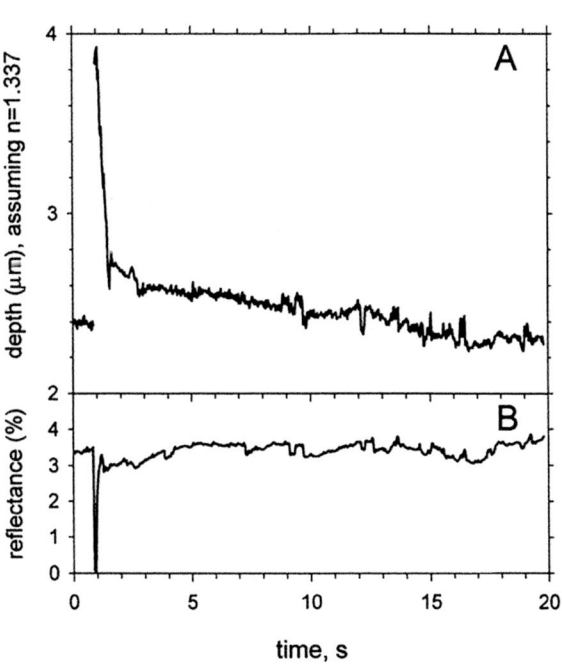

# 3   Optical Coherence Tomography

Optical coherence tomography (OCT) has developed rapidly over the past two decades. As a non-invasive 3D volumetric imaging technique with micrometer-scale resolution, OCT has been widely used in various biomedical fields, especially for ophthalmology applications [24].

The physics principle of OCT is based on low coherence interferometry, which is achieved by using a broadband light source [25]. In an OCT system, the broadband light is focused on the sample surface, and OCT collects the backscattered or back-reflected light from the internal microstructures in materials and biological systems. The time of flight of the collected light is measured through interference, which can be mapped into the depth information of the sample at one lateral point (known as one A-scan). When the light beam scans across the sample or if the sample is translated with stages, a 3D volumetric image is acquired.

The early OCT systems were based on the time domain configuration, in which a photodiode was adopted as the detector and a moving mirror was placed in the reference arm [25]. By moving the reference mirror axially, different interfaces of the sample along the depth could be observed, which was then used to reconstruct the image profile for one A-scan. The Fourier domain OCT (FD-OCT) system was first described by Fercher et al. in 1995 [26]. In a FD-OCT system, the coherence gating involves acquiring the interferometric signal as a function of optical wavenumber. The depth profile of the sample is directly encoded in the interferometric signal in the wavenumber domain, which eliminates the need of a moving reference mirror. Studies have shown that FD-OCT has a superior sensitivity advantage over TD-OCT, especially for high-speed imaging [27]. The development of high-speed FD-OCT systems enables a comprehensive imaging over larger fields of view and undoubtedly changes the clinical implementation and adoption of OCT [28]. Most of the state-of-the-art OCT systems nowadays are built in the Fourier domain.

A FD-OCT system has two distinct implementations. The first approach, spectral domain OCT (SD-OCT), employs a spectrometer as the detector, which disperses the light spatially, and the wavenumber is encoded by the pixel number of the camera in the spectrometer. A schematic layout of the SD-OCT is demonstrated in Fig. 7, which builds upon a Michelson interferometer structure. The other approach, swept-source OCT (SS-OCT), uses a wavelength-swept optical source, in which the wavenumber is encoded in the time stamps of the wavelength sweeps.

The non-invasive nature, high-speed and high-resolution capabilities make OCT a great imaging modality for tear film evaluation. In SD-OCT, a fast Fourier transform is typically performed to the acquired spectrum and is followed by a peak detection technique to extract the tear film thickness information. Wang et al. demonstrated the feasibility of indirectly measuring the precorneal tear film

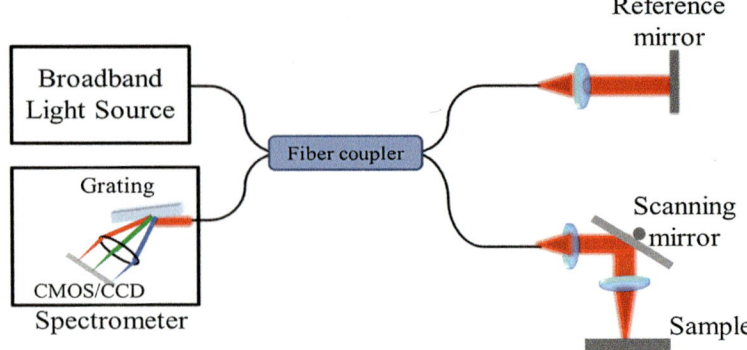

**Fig. 7** Schematic layout of a spectrometer-based OCT (CMOS/CCD are arrays of pixel sensors)

thickness, by using a contact lens to better define the interface of the corneal surface [29]. Direct thickness measurements of the tear film were also reported with OCT peak detection techniques and tear film thickness estimates range from 2 to 5.5 $\mu$m [30, 31]. Figure 8 shows the cross-sectional OCT image of the cornea and tear film, as well as the peak detection technique for tear film thickness estimation, through searching peaks along the A-scan intensity profile.

The axial resolution of the peak detection method is fundamentally limited by the width of the axial point spread function (PSF), which is in the order of 1 $\mu$m in state-of-the-art systems, thus OCT has been mainly applied to measure the total thickness of the lipid and aqueous layers combined. To overcome this limitation, an approach that combines the axial selectivity capability of OCT with statistical decision theory was introduced [32, 33]. In this approach, a maximum-likelihood (ML) estimator was implemented through a comprehensive mathematical modeling [34]. The ML estimator is applied directly to each raw spectrum acquired by the OCT system to estimate the thickness configuration that has most likely generated the given spectrum.

To mathematically illustrate the principle of the ML estimation, a spectrum collected at the OCT detector is denoted as $N_g(x, \Delta t)$, where $x$ represents the pixel index number at the line-scan camera and $\Delta t$ is the integration time. For a given tear film with lipid layer thickness $d_l$ and aqueous layer thickness $d_a$, the expected mean spectrum is denoted as $<<< N_{g|(d_l, d_a)} (x, \Delta t)>>>$, where the brackets represent the average over the Poisson noise of the photon counting process, source intensity noise of the broadband source, and the dark noise of the detector. The conditional likelihood that the measured spectrum is generated by different lipid and aqueous thickness pairs is given by

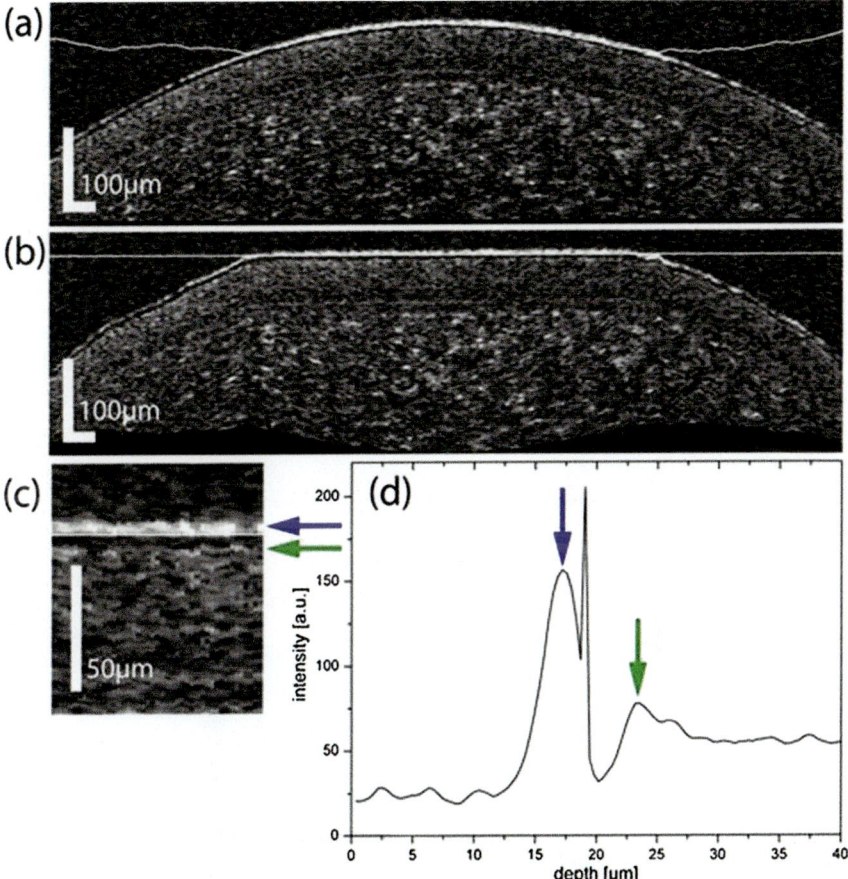

**Fig. 8** Detection scheme for tear film thickness: (**a**) Line-wise contrast enhancement and automatic detection of air–tear film interface; (**b**) Flattening of the tear film; (**c**) Definition of the tear film in the linearized image; (**d**) Intensity profile of an A-scan and definition of the tear film thickness as the distance between the tear film front surface (blue arrow) and the cornea front surface (green arrow) [30]

$$
P\left(N_g | d_l, d_a\right) = \frac{1}{(2\pi)^{\frac{M}{2}} \prod\limits_{x} \left[ K_{N_{g|(d_l, d_a)}}(x, \Delta t) \right]^{\frac{1}{2}}}
$$

$$
\times \exp\left[ -\frac{1}{2} \sum_{x} \frac{\left( N_g(x, \Delta t) - \left\langle\!\left\langle\!\left\langle N_{g|(d_l, d_a)}(x, \Delta t) \right\rangle\!\right\rangle\!\right\rangle \right)^2}{K_{N_{g|(d_l, d_a)}}(x, \Delta t)} \right], \tag{2}
$$

where $K_{Ng|(dl, da)}(x, \Delta t)$ is the variance of the corresponding expected spectrum and is a second-order polynomial of the ensembles mean, and $M$ is the number of pixels in the line-scan camera [35]. The estimator makes an estimate by maximizing the conditional likelihood in Eq. (2). As an example to illustrate the principle, one estimation process of a spectrum acquired from a human subject is visualized in Fig. 9 [36].

This method was further applied to simultaneously estimate the thickness dynamics of both the lipid and aqueous layers. As shown in Fig. 10, the dynamics shows that the lipid layer gets thicker rapidly right after a complete blink, with a thickening rate ranging from 5 to 18 nm/s with an average of 10 nm/s, and the lipid layer stabilizes after an average of 2.5 s. The aqueous layer gets thinner gradually with an average thinning rate of 0.29 μm/s. These findings are consistent with the prior measurements with interferometry and other OCT techniques [30, 37–39].

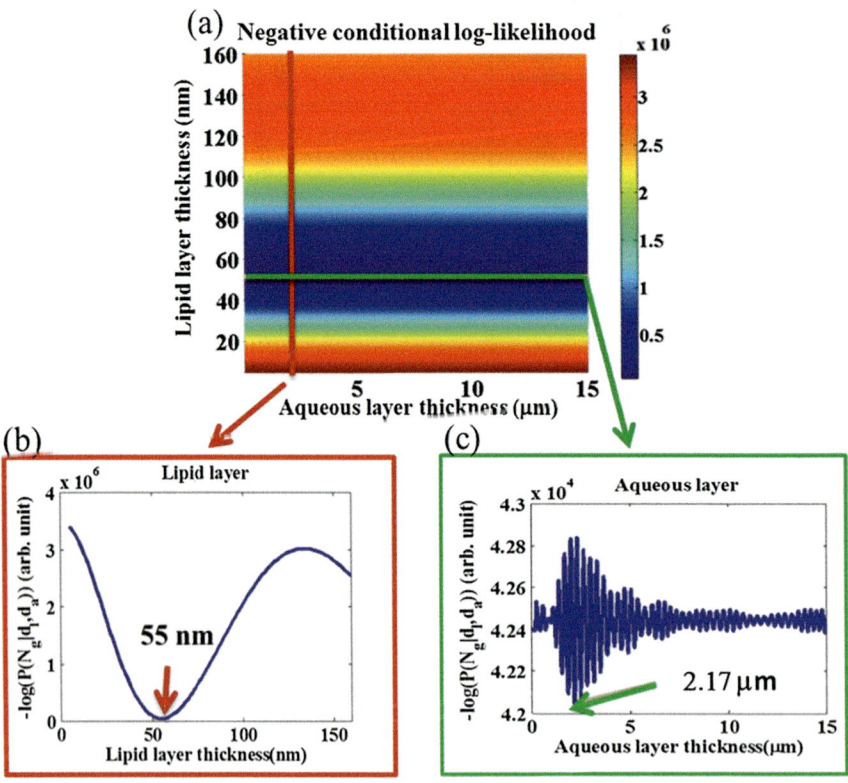

**Fig. 9** (a) Negative conditional log-likelihood of one measured spectrum is generated by different thickness pairs of lipid and aqueous layer thickness; (b) conditional log-likelihood along the red line in (a) and the lipid layer thickness estimate; (c) conditional log-likelihood along the green line in (a) and the aqueous layer thickness estimate [36]

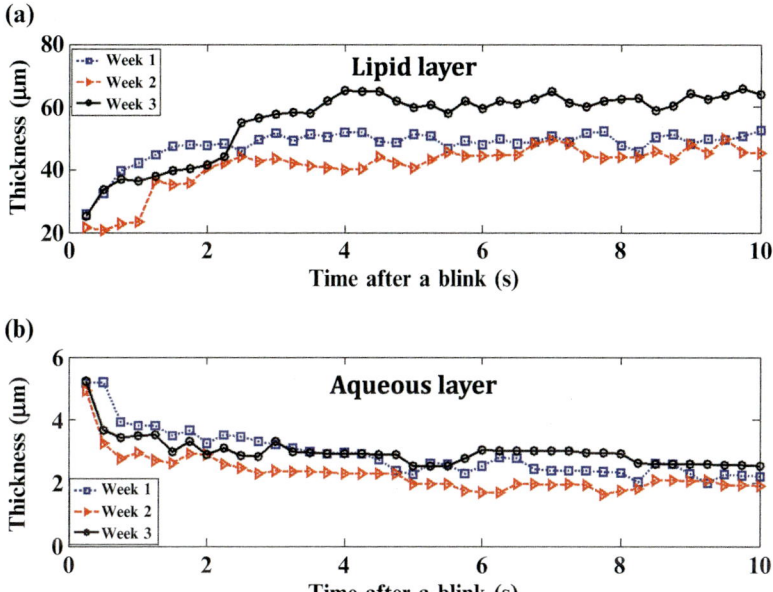

**Fig. 10** Repeated measurements of the tear film dynamics of both (**a**) the lipid and (**b**) the aqueous layers over the corneal apex of one subject over a 3-week period [36]

## 4   Summary

Tear film fluid dynamics plays an important role in maintaining the health of the ocular system. To visualize and understand the tear film dynamics, great efforts have been made with the development of various imaging techniques. This chapter discussed three major imaging techniques that have been used in tear film dynamics studies: fluorescent imaging, interferometry, and OCT. Among these three major techniques, interferometry and OCT are non-invasive by nature and thus are preferred modalities for tear film dynamics evaluation. With the rapid advancement of broadband light sources and the development of advanced algorithms, the potential of OCT in the field of tear film measurement is still increasing.

## References

1. Rieger, G., 1992. The importance of the precorneal tear film for the quality of optical imaging. British journal of ophthalmology, 76(3), pp.157-158.
2. Tomlinson, A. and Khanal, S., 2005. Assessment of tear film dynamics: quantification approach. The ocular surface, 3(2), pp.81-95.
3. Lemp, M.A., Baudouin, C., Baum, J., Dogru, M., Foulks, G.N., Kinoshita, S., Laibson, P., McCulley, J., Murube, J., Pflugfelder, S.C. and Rolando, M., 2007. The definition and

classification of dry eye disease: report of the Definition and Classification Subcommittee of the International Dry Eye WorkShop (2007). Ocular Surface, 5(2), pp.75-92.

4. Norn, M.S., 1969. Desiccation of the precorneal film. Acta ophthalmologica, 47(4), pp.865-880.

5. Korb, D.R., 2000. Survey of preferred tests for diagnosis of the tear film and dry eye. Cornea, 19(4), pp.483-486.

6. Cho, P. and Douthwaite, W., 1995. The relation between invasive and noninvasive tear break-up time. Optometry & vision science, 72(1), pp.17-22.

7. Bron, A.J., Abelson, M.B., Ousler, G., Pearce, E., Tomlinson, A., Yokoi, N., Smith, J.A., Begley, C., Caffery, B., Nichols, K. and Schaumberg, D., 2007. Methodologies to diagnose and monitor dry eye disease: report of the Diagnostic Methodology Subcommittee of the International Dry Eye WorkShop (2007). Ocular surface, 5(2), pp.108-152.

8. Norn, M.S., 1969. Dead, degenerate, and living cells in conjunctival fluid and mucous thread. Acta ophthalmologica, 47(5-6), pp.1102-1115.

9. Pflugfelder, S.C., Tseng, S.C., Sanabria, O., Kell, H., Garcia, C.G., Felix, C., Feuer, W. and Reis, B.L., 1998. Evaluation of subjective assessments and objective diagnostic tests for diagnosing tear-film disorders known to cause ocular irritation. Cornea, 17(1), p.38.

10. Begley, C.G., Liu, H., Chalmers, R.L., Renner, D. and Wilkinson, J., 2005. The forced staring tear breakup dynamics model: a quantitative method to measure tear film stability in dry eye. The ocular surface, 3, p.S47.

11. Liu, H., Begley, C.G., Chalmers, R., Wilson, G., Srinivas, S.P. and Wilkinson, J.A., 2006. Temporal progression and spatial repeatability of tear breakup. Optometry & vision science, 83(10), pp.723-730.

12. Begley, C., Simpson, T., Liu, H., Salvo, E.,Wu, Z., Bradley, A. and Situ, P., 2013. Quantitative analysis of tear film fluorescence and discomfort during tear film instability and thinning. Investigative ophthalmology & visual science, 54(4), pp.2645-2653.

13. Webber, W.R.S. and Jones, D.P., 1986. Continuous fluorophotometric method of measuring tear turnover rate in humans and analysis of factors affecting accuracy. Medical and biological engineering and computing, 24(4), p.386.

14. Joshi, A., Maurice, D. and Paugh, J.R., 1996. A new method for determining corneal epithelial barrier to fluorescein in humans. Investigative ophthalmology & visual science, 37(6), pp.1008-1016.

15. King-Smith, P.E., Ramamoorthy, P., Braun, R.J. and Nichols, J.J., 2013. Tear film images and breakup analyzed using fluorescent quenching. Investigative ophthalmology & visual science, 54(9), p.6003.

16. King-Smith, P.E., Reuter, K.S., Braun, R.J., Nichols, J.J. and Nichols, K.K., 2013. Tear film breakup and structure studied by simultaneous video recording of fluorescence and tear film lipid layer images. Investigative ophthalmology & visual science, 54(7), pp.4900-4909.

17. Su, T.Y., Chang, S.W., Yang, C.J. and Chiang, H.K., 2014. Direct observation and validation of fluorescein tear film break-up patterns by using a dual thermal-fluorescent imaging system. Biomedical optics express, 5(8), pp.2614-2619.

18. Doane, M.G., 1989. An instrument for in vivo tear film interferometry. Optometry and Vision Science, 66(6), pp.383-388.

19. Fogt, N., King-Smith, P.E. and Tuell, G., 1998. Interferometric measurement of tear film thickness by use of spectral oscillations. JOSA A, 15(1), pp.268-275.

20. King-Smith, P.E., Fink, B.A. and Fogt, N., 1999. Three interferometric methods for measuring the thickness of layers of the tear film. Optometry & vision science, 76(1), pp.19-32.

21. Goto, E., Dogru, M., Kojima, T. and Tsubota, K., 2003. Computer-synthesis of an interference color chart of human tear lipid layer, by a colorimetric approach. Investigative ophthalmology & visual science, 44(11), pp.4693-4697.

22. King-Smith, P.E., Fink, B.A., Fogt, N., Nichols, K.K., Hill, R.M. and Wilson, G.S., 2000. The thickness of the human precorneal tear film: evidence from reflection spectra. Investigative ophthalmology & visual science, 41(11), pp.3348-3359.

23. Prydal, J.I. and Campbell, F.W., 1992. Study of precorneal tear film thickness and structure by interferometry and confocal microscopy. Investigative ophthalmology & visual science, 33(6), pp.1996-2005.
24. Wojtkowski, M., Leitgeb, R., Kowalczyk, A., Bajraszewski, T. and Fercher, A.F., 2002. In vivo human retinal imaging by Fourier domain optical coherence tomography. Journal of biomedical optics, 7(3), pp.457-463.
25. Huang, D., Swanson, E.A., Lin, C.P., Schuman, J.S., Stinson, W.G., Chang, W., Hee, M.R., Flotte, T., Gregory, K., Puliafito, C.A. and Fujimoto, J.G., 1991. Optical coherence tomography. Science (New York, NY), 254(5035), p.1178.
26. Fercher, A.F., Hitzenberger, C.K., Kamp, G. and El-Zaiat, S.Y., 1995. Measurement of intraocular distances by backscattering spectral interferometry. Optics communications, 117(1-2), pp.43-48.
27. De Boer, J.F., Cense, B., Park, B.H., Pierce, M.C., Tearney, G.J. and Bouma, B.E., 2003. Improved signal-to-noise ratio in spectral-domain compared with time-domain optical coherence tomography. Optics letters, 28(21), pp.2067-2069.
28. Bouma, B.E., Yun, S.H., Vakoc, B.J., Suter, M.J. and Tearney, G.J., 2009. Fourier-domain optical coherence tomography: recent advances toward clinical utility. Current opinion in biotechnology, 20(1), pp.111-118.
29. Wang, J., Fonn, D., Simpson, T.L. and Jones, L., 2003. Precorneal and pre-and postlens tear film thickness measured indirectly with optical coherence tomography. Investigative ophthalmology & visual science, 44(6), pp.2524-2528.
30. Werkmeister, R.M., Alex, A., Kaya, S., Unterhuber, A., Hofer, B., Riedl, J., Bronhagl, M., Vietauer, M., Schmidl, D., Schmoll, T. and Garhöfer, G., 2013. Measurement of tear film thickness using ultrahigh-resolution optical coherence tomography. Investigative ophthalmology & visual science, 54(8), pp.5578-5583.
31. Yadav, R., Lee, K.S., Rolland, J.P., Zavislan, J.M., Aquavella, J.V. and Yoon, G., 2011. Micrometer axial resolution OCT for corneal imaging. Biomedical optics express, 2(11), pp.3037-3046.
32. Huang, J., Clarkson, E., Kupinski, M., Lee, K.S., Maki, K.L., Ross, D.S., Aquavella, J.V. and Rolland, J.P., 2013. Maximum-likelihood estimation in optical coherence tomography in the context of the tear film dynamics. Biomedical optics express, 4(10), pp.1806-1816.
33. Huang, J., Lee, K.S., Clarkson, E., Kupinski, M., Maki, K.L., Ross, D.S., Aquavella, J.V. and Rolland, J.P., 2013. Phantom study of tear film dynamics with optical coherence tomography and maximum-likelihood estimation. Optics letters, 38(10), pp.1721-1723.
34. Huang, J., Yuan, Q., Zhang, B., Xu, K., Tankam, P., Clarkson, E., Kupinski, M.A., Hindman, H.B., Aquavella, J.V., Suleski, T.J. and Rolland, J.P., 2014. Measurement of a multi-layered tear film phantom using optical coherence tomography and statistical decision theory. Biomedical optics express, 5(12), pp.4374-4386.
35. Huang, J., Yao, J., Cirucci, N., Ivanov, T. and Rolland, J.P., 2015. Performance analysis of optical coherence tomography in the context of a thickness estimation task. Journal of biomedical optics, 20(12), p.121306.
36. Huang, J., Hindman, H.B. and Rolland, J.P., 2016. In vivo thickness dynamics measurement of tear film lipid and aqueous layers with optical coherence tomography and maximum-likelihood estimation. Optics letters, 41(9), pp.1981-1984.
37. King-Smith, P.E., Fink, B.A., Nichols, J.J., Nichols, K.K., Braun, R.J. and McFadden, G.B., 2009. The contribution of lipid layer movement to tear film thinning and breakup. Investigative ophthalmology & visual science, 50(6), pp.2747-2756.
38. King-Smith, P.E., Hinel, E.A. and Nichols, J.J., 2010. Application of a novel interferometric method to investigate the relation between lipid layer thickness and tear film thinning. Investigative ophthalmology & visual science, 51(5), pp.2418-2423.
39. dos Santos, V.A., Schmetterer, L., Gröschl, M., Garhofer, G., Schmidl, D., Kucera, M., Unterhuber, A., Hermand, J.P. and Werkmeister, R.M., 2015. In vivo tear film thickness measurement and tear film dynamics visualization using spectral domain optical coherence tomography. Optics express, 23(16), pp.21043-21063.

# Mathematical Models of the Tear Film

Richard J. Braun, Tobin A. Driscoll, and Carolyn G. Begley

**Abstract** The complex dynamics of the tear film are affected by its many processes and components. Mathematical models of the tear film allow the selective elimination or inclusion of various effects that are not otherwise possible in human subjects. Such models have been able to provide local estimates of osmolarity in tear break up (TBU), for example, which to our knowledge cannot be measured directly. Models also suggest that different modes of TBU must be considered to make sense of in vivo data regarding response of the ocular epithelia and causes of dry eye. More complex models that include tear film formation via blinking are within our grasp, which will no doubt extend our knowledge of the tear film, its dynamics, and its role in ocular surface health.

## 1 Overview

In this chapter, we introduce and develop mathematical models for a few of the many dynamic phenomena associated with the tear film. The tear film is a very thin multicomponent fluid that coats the surface of the eye after each blink. It undergoes rapid processes such as blinks, during which the upper lid descends and then returns upward in about 0.28 s (on average for unconscious blinks) [32]. Interblink interval can range from just a few seconds [27] to minutes [32, 50] for concentration-intensive tasks. During the interblink period, the tear film may remain stable and relatively uniform over the surface, as evaporation occurs. It may also become highly unstable and be disrupted, a phenomenon known as tear break up (TBU). (TBU is often called film rupture in the fluid mechanics literature.) The time from a blink to when TBU first occurs, the TBU time, can range from about a tenth of

R. J. Braun (✉) · T. A. Driscoll
Department of Mathematical Sciences, University of Delaware, Newark, DE, USA
e-mail: rjbraun@udel.edu; driscoll@udel.edu

C. G. Begley
School of Optometry, Indiana University, Bloomington, IN, USA
e-mail: cbegley@indiana.edu

© Springer Nature Switzerland AG 2019
G. Guidoboni et al. (eds.), *Ocular Fluid Dynamics*, Modeling and Simulation in
Science, Engineering and Technology, https://doi.org/10.1007/978-3-030-25886-3_17

Here $c'$ is the osmolarity, $D_o$ is its diffusivity, and $f'$ is the fluorescein concentration with diffusivity $D_f$.

At the film/substrate (cornea) interface located at $z' = 0$, we need to satisfy no slip,

$$u' = 0, \tag{5}$$

and osmosis through a perfect semipermeable membrane

$$w' = P_o V_w (c' - c_0). \tag{6}$$

Here $P_o$ is the permeability of the membrane, $V_w$ is the molar volume, $c_0$ is the isotonic (serum) concentration. Any contribution of the fluorescein is neglected according to the analysis of [69].

For the solutes, we also have the flux boundary conditions

$$D_o \partial'_z c' + w' c' = 0, \text{ and} \tag{7}$$

$$D_f \partial'_z f' + w' f' = 0. \tag{8}$$

At the film/air interface, $z' = h'(r', t)$, we need to satisfy the kinematic condition

$$\rho \left( \partial_{t'} h' + u' \nabla'_{II} h' - w' \right) / \sqrt{1 + |\nabla_{II} h'|^2} = -J' \tag{9}$$

where the mass flux of evaporation is given by the constitutive relation

$$J' = \rho w_0 J'_f (r') + \alpha_0 (p' - p'_v), \tag{10}$$

$$J'_f(r') = w_1/w_0 + \left[ 1 - (w_1/w_0) \right] \exp\left( -(r'/r'_w)^2/2 \right), \tag{11}$$

and $\nabla'_{II}$ is in the plane of the substrate. We assume that there is an isothermal film; $\alpha_0$ is effectively $\alpha/K$ from [1]. We assume that the pressure causes deviation from a uniform rate of evaporation of $\rho w_0$ where $w_0$ is a measured thinning rate of the tear film. We will use the $\rho w_0$ to nondimensionalize the evaporative mass flux, where $J_f(r) = 1$ is a uniform evaporation rate. Note that different functions $J_f(r)$ will specify Gaussian evaporation profiles of different hole size and evaporation rate.

The normal stress condition is given by

$$- p_v - \mathbf{n}' \cdot \mathbf{T}' \cdot \mathbf{n}' = \gamma \nabla_s \cdot \mathbf{n}' - \phi, \tag{12}$$

where the $\mathbf{T}' = -p' \mathbf{I} + \mu (\nabla \mathbf{u}' + \nabla \mathbf{u}'^T)$ is the Newtonian stress tensor. The terms from the rate of strain do not survive the thin film (lubrication) approximation to follow, and the terms we need include

$$p' - p'_v = -\left(\sigma \nabla_s \cdot \mathbf{n}' + \frac{A^*}{h'^3}\right), \tag{13}$$

where $\nabla'_s = (I - \mathbf{nn}) \cdot \nabla'$ [108] and $\mathbf{n} = (-\nabla'_{II}h', 1)/(1 + \nabla'_{II}h'^2)^{1/2}$ is the outward normal to the free surface. The surface tension $\sigma$ may depend on the surface concentration $\Gamma'$ of a surfactant representing polar lipids. The tangential stress is given by

$$\mathbf{t}' \cdot \mathbf{T} \cdot \mathbf{n}' = \mathbf{t}' \cdot \nabla'_s \sigma. \tag{14}$$

For this chapter, we assume the simplest possible case of a linearized equation of state for $\sigma$, namely

$$\sigma = \sigma_0 + (\partial_{\Gamma'}\sigma)\big|_{\Gamma'_0} (\Gamma' - \Gamma'_0). \tag{15}$$

In the limiting case of a strong surfactant, this equation will result in a simplified condition of tangential immobility [84, 107], namely $\mathbf{t}' \cdot \mathbf{u}' = 0$. We use that result in the derivation presented in this section.

The no flux condition for solutes across the film surface is given by

$$D_o \mathbf{n} \cdot \nabla' c' + \mathbf{n} \cdot \mathbf{u} c' = 0, \text{ and } D_f \mathbf{n} \cdot \nabla' f' + \mathbf{n} \cdot \mathbf{u} f' = 0. \tag{16}$$

For boundary conditions at $r' = 0$, we apply symmetry, and no flux conditions at $r = R'_L$ at the outside edge of the domain. For initial conditions, we shall apply constant thickness and concentration, and let evaporation (or other driving forces) modify the dependent variables; more details are below.

The thin film equations may be derived by adapting previous approaches to the axisymmetric case [13, 45, 124]. The following relations are used to nondimensionalize the governing set of equations:

$$r' = \ell r, \quad z' = d'z, \quad \epsilon = d'/\ell, \quad t' = \frac{d}{w_0}t, \quad h' = dh, \quad u' = (v_0/\epsilon)u, \tag{17}$$

$$w' = w_0 w, \quad p' = \frac{\mu w_0}{\ell \epsilon^3}p, \quad J' = \rho w_0 J, \quad c' = c_0 c, \text{ and } f' = f_{cr}f. \tag{18}$$

Here $\epsilon \ll 1$ represents the ratio of typical vertical to horizontal length scales.

The resulting set of axisymmetric system of equations in the liquid region $0 < z < h(r, t)$ is

$$\frac{1}{r}\partial_r(ru) + \partial_z w = 0 \tag{19}$$

$$\epsilon \text{Re}\left(\partial_t u + u \partial_r u + w \partial_z u\right) = \epsilon^2 \left[\frac{1}{r}\partial_r(r\partial_r u) - \frac{u}{r^2}\right] + \partial_z^2 u - \partial_r(p + Ah^{-3}) \tag{20}$$

$$\epsilon^3 \mathrm{Re} \left( \partial_t w + u \partial_r w + w \partial_z w \right) = \epsilon^4 \left[ \frac{1}{r} \partial_r (r \partial_r w) - \frac{w}{r^2} \right] + \epsilon^2 \partial_z^2 w - \partial_z (p + A h^{-3})$$
$$\tag{21}$$

$$\partial_t c + u \partial_r c + w \partial_z c = \mathrm{Pe}_c^{-1} \left[ \frac{1}{r} \partial_r (r \partial_r c) - \frac{c}{r^2} + \epsilon^{-2} \partial_z^2 c \right] \tag{22}$$

$$\partial_t f + u \partial_r f + w \partial_z f = \mathrm{Pe}_f^{-1} \left[ \frac{1}{r} \partial_r (r \partial_r f) - \frac{f}{r^2} + \epsilon^{-2} \partial_z^2 f \right] \tag{23}$$

where $\mathrm{Re} = \rho w_0 \ell / \mu$.

The leading order boundary conditions at $z = 0$ are

$$u = 0, \tag{24}$$

$$w = P_c(c - 1), \text{ and} \tag{25}$$

$$\epsilon^{-2} \mathrm{Pe}_c^{-1} \partial_z c = wc. \tag{26}$$

The boundary conditions at $z = h(r, t)$ are

$$\frac{h_t + u \partial_r h - w}{\sqrt{1 + (\epsilon \partial_r h)^2}} = -J, \tag{27}$$

$$p = -S \frac{1}{r} \partial_r (r \partial_r h) - A h^{-3} + o(1) \tag{28}$$

$$u - (\epsilon \partial_r h) w = 0, \tag{29}$$

$$(\mathrm{Pe}_c \epsilon)^{-1} \left( -\epsilon^2 \partial_r h \partial_r c + \partial_z c \right) = u c \partial_r h - wc, \text{ and} \tag{30}$$

$$(\mathrm{Pe}_f \epsilon)^{-1} \left( -\epsilon^2 \partial_r h \partial_r f + \partial_z f \right) = u f \partial_r h - wf, \tag{31}$$

where

$$J = J_f(r) - \alpha \left[ \frac{1}{r} (r h_r)_r + \frac{A}{h^3} \right], \tag{32}$$

and

$$J_f(r) = w_b - (1 - w_b) \exp \left( -(r/r_w)^2 / 2 \right). \tag{33}$$

Here $w_b = w_1 / w_0$. In the stress conditions (28) and (29), we have only shown simplified versions [70, 75].

The dimensional parameters we use are shown in Table 1; the nondimensional parameters that arise for this evaporative case (as well as for subsequent models) are in Table 2.

**Table 1** Dimensional parameters

| Parameter | Description | Value | Reference |
|---|---|---|---|
| $\mu$ | Viscosity | $1.3 \times 10^{-3}$ Pa·s | [109] |
| $\sigma_0$ | Surface tension | $0.045$ N·m$^{-1}$ | [83] |
| $k$ | Tear film thermal conductivity | $0.68$ W·m$^{-1}$·K$^{-1}$ | Water |
| $\rho$ | Density | $10^3$ kg·m$^{-3}$ | Water |
| $g$ | Gravitational acceleration | $9.81$ m·s$^{-2}$ | Estimated |
| $A^*$ | Hamaker constant | $3.5 \times 10^{-19}$ Pa·m$^3$ | [112] |
| $d'$ | Characteristic thickness | 3.5 or $5 \times 10^{-6}$ m | [52] |
| $\ell_0$ | Capillary/viscous length (evap, 1D) | $\left[\sigma_0/(\mu w_0)\right]^{1/4} d'$ | [14] |
| $\ell_1$ | Rapid TBU length scale (glob, 1D) | $\left[\sigma_0/|\Delta\sigma|_0\right]^{1/2} d'$ | [123] |
| $\ell_2$ | Half-width of palpebral fissure (2D) | $5 \times 10^{-3}$ m | Estimated |
| $w_0$ | Peak thinning rate | $1--25\,\mu$m/min | [86] |
| $w_1$ | Background thinning rate | $1\,\mu$m/min | [86] |
| $U_0$ | Characteristic speed (evap, 1D) | $w_0/\epsilon$ | [14] |
| $U_1$ | Characteristic speed (glob, 1D) | $\sigma_0 d\mu/[|\Delta\sigma|_0]^2$ | [123] |
| $U_2$ | Characteristic speed (2D) | $5 \times 10^{-3}$ m/s | [53, 70] |
| $\tau_0$ | Characteristic time scale | $d'/w_0,\ (i=1)$ | [14] |
| $\tau_i$ | Characteristic time scale | $\ell_i/U_i,\ i=2,3$ | Calculated |
| $|\partial_\Gamma \sigma|_0$ | Composition dependence | $0.01$ N/m | [5] |
| $|\Delta\sigma|_0$ | Change in surface tension | $\Gamma_0|\partial_\Gamma\sigma|_0 = 0.1$ mN/m | |
| $P^{\text{tiss}}_{\text{corn}}$ | Tissue permeability of cornea | $12.0\,\mu$m/s | [17] |
| $P^{\text{tiss}}_{\text{conj}}$ | Tissue permeability of conjunctiva | $55.4\,\mu$m/s | [17] |
| $V_w$ | Molar volume of water | $1.8 \times 10^{-5}$ m$^3$·mol$^{-1}$ | Water |
| $D_c$ | Diffusivity of osmolarity in water | $1.6 \times 10^{-9}$ m$^2$/s | [93] |
| $D_f$ | Diffusivity of fluorescein in water | $0.39 \times 10^{-9}$ m$^2$/s | [23] |
| $\kappa$ | Napierian extinction coefficient | $1.75 \times 10^7$ m$^{-1}$M$^{-1}$ | [82] |
| $c'_0$ | Isotonic osmolarity | $302$ mOsM$^3$ | [36] |
| $f_{cr}$ | Critical fluorescein concentration | $0.2\%$ (by mass) | [85] |
| $D_s$ | Surface diffusion coefficient | $3 \times 10^{-8}$ m$^2$/s | [97] |
| $\mathcal{L}_m$ | Latent heat of vaporization | $2.3 \times 10^6$ J·kg$^{-1}$ | Water |
| $T'_s$ | Saturation temperature | $27\,^\circ$C | Estimated |
| $T'_B$ | Body temperature | $37\,^\circ$C | Estimated |
| $\alpha$ | Pressure coefficient for evaporation | $3.6 \times 10^{-2}$ K·Pa$^{-1}$ | [112] |
| $K$ | Non-equilibrium coefficient | $1.5 \times 10^5$ K·m$^2$·s·kg$^{-1}$ | Estimated |

Here $K$ corresponds to a nominal thinning rate of $4\,\mu$m/min; to obtain a $20\,\mu$m/min thinning rate, this quantity is reduced by a factor of 5. Additional sources for the thinning rates include [53] and [90]. The molar extinction coefficient given in [82] has been multiplied by $\ln(10)$ to convert it to the Napierian form; this corrects our previous use of the extinction coefficient [16, 17, 85]

to the air, and osmolarity $c(t)$ responds inside the tear film causing osmosis from the tear/cornea interface. In this case, the tear film equations become [13, 16]

$$\dot{h} = -J + P_c(c - 1), \quad h(0) = 1, \tag{53}$$

$$h\dot{c} = \left[J - P_c(c - 1)\right]c, \quad c(0) = 1. \tag{54}$$

Here the dot denotes ordinary differentiation with respect to time. The fluid is stagnant inside the film in this case ($\bar{u} = 0$), and simply leaves the free surface via evaporation. The osmolarity is constant throughout the depth, but it increases in response to the lost water from the tear film. These two equations can be combined to find $d(ch)/dt = 0$, so that integration gives the constant value of the product $hc = 1$ using our scalings.

The two equations can be integrated with an ODE solver as in [13], or the osmolarity can be eliminated to obtain a single equation for $h$ as in [17]. The single equation is

$$\dot{h} = -1 + P_c\left(h^{-1} - 1\right). \tag{55}$$

An analytical solution is the Lambert-W function, but for our purposes numerical solution is more convenient. The equilibrium solution from $h' = 0$ is

$$h_{eq} = \frac{P_c}{P_c + 1}. \tag{56}$$

The larger the permeability, the closer the equilibrium is to the initial value of unity. From solute conservation, we also have that $c_{eq} = (P_c + 1)/P_c$.

If a van der Waals wetting term and a pressure correction is included ($A$, $\alpha \neq 0$), then $J = 1 - \alpha A h^{-3}$, and the equilibrium location comes from solving a cubic polynomial for the thickness $h_{eq}$. This cubic was solved in [105] for a different scaling, and sigmoidal dependence on $P_c$ was found in that case as well. This case is not considered for uniform films here.

When fluorescein is also present, then it is a passive scalar in this problem. The additional equation is

$$h\dot{f} = \left[J - P_c(c - 1)\right]f, \quad f(0) = f_0. \tag{57}$$

This equation can also be combined with the equation for the thickness and integrated to obtain $hf = f_0$. Dividing by $f_0$, we find that $(f/f_0)h = 1$, and we can solve the problem for a unit initial condition and simply scale the result for different concentrations. This will be true for all of our fluorescein transport equations below as well.

Results for the evaporation-only problem are shown as solid curves in Fig. 4. The thinning for evaporation only is roughly linear until the thickness nears the equilibrium where osmosis and evaporation balance; this is the equilibrium solution given in Eq. (56). If evaporation occurs, the osmolarity rises smoothly during the thinning, satisfying $ch = 1$ in the scalings used here.

### 2.2.2 Divergent Flow and Evaporation

In this case, we first hypothesize a divergent flow given by $u = ax$; it is sketched in the left column of Fig. 3. This imposed flow could be viewed as pure extension resulting from relaxing the no slip conditions on the film. Despite the crude approximation to the flow, it will capture some essential elements of problems discussed below.

From mass conservation, $\partial_z w = -\partial_x u = a$, so that $w = -az$ is consistent with the velocity distribution along the film in the absence of osmosis. Then, the integrating conservation of mass $\int_0^h dz$, and using the kinematic condition at $z = h$ and the boundary condition at $z = 0$, requires that

$$\dot{h} + ah = -J + P_c(c - 1), \quad h(0) = 1; \quad h\dot{c} = [J - P_c(c - 1)]c, \quad c(0) = 1. \quad (58)$$

In the absence of evaporation, $J = 0$; then $c(t) = 1$ and

$$h(t) = e^{-at}. \quad (59)$$

For divergent flow without evaporation, the thinning is exponential as given in Eq. (59). For our parameters, with $a = 2.97$, the divergent flow thins the film faster than for evaporation alone. But, in this case, the osmolarity remains unchanged at $c = 1$ for all time. See the dashed curves in Fig. 4.

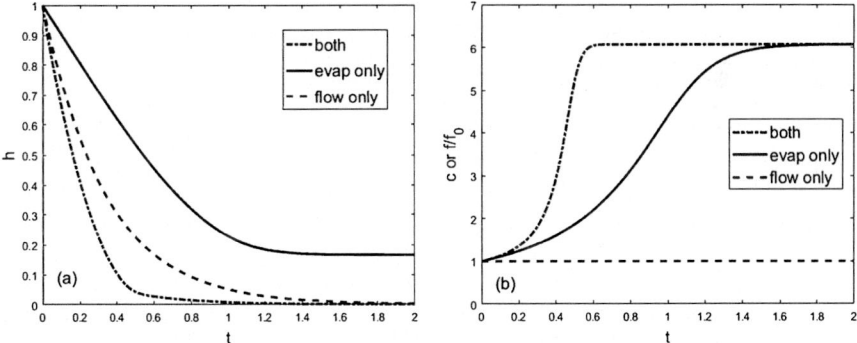

**Fig. 4** When active, the thinning rate is a high value of $w_0 = 20 \, \mu\text{m/s}$ ($P_c = 0.196$), and if divergent flow is present, $a' = (10^{-4} \text{m/s})/\ell = 0.283 \, \text{s}^{-1}$; nondimensionally, $a = a'd'/w_0 = 2.97$. Here $d = 3.5 \, \mu\text{m}$. (a) The thickness $h$ is shown for three cases: evaporation only ($a' = 0$), divergent flow only ($w_0 = 0$), and for a combination of the two ($a'$, $w_0$ both nonzero). (b) The normalized concentrations $c$ or $f/f_0$ are shown for three cases: evaporation only ($a' = 0$), divergent flow only ($w_0 = 0$), and for combination of the two ($a'$, $w_0$ both nonzero)

When both evaporation and divergent flow are present, the two equations can be combined as before to obtain a single equation for $h$. We find $d(ch)/dt + a(ch) = 0$, so that $ch = e^{-at}$. We can again eliminate $c$ and obtain the single equation

$$h' = -ah - J + P_c \left( e^{-at} h^{-1} - 1 \right). \tag{60}$$

For the computed solutions shown in Fig. 4 (dash-dot curves), we see that $h(t)$ may approach zero from above for large times. This equation also has a non-physical equilibrium with $h_{eq} = -(1 + P_c)/a < 0$; at a sufficiently long time, a numerically computed $h$ may become negative due to roundoff error, and then rapidly decrease to an equilibrium value that balances the term $-ah$ with the constant terms (not shown). This equilibrium is clearly not relevant for eyes, but prior to crossing $h = 0$, we believe that there is relevant information from the model.

Up to about $t = 0.5$, there is rapid thinning from the cooperation of both effects. After that, osmosis has slowed, but not stopped, the thinning; a slow thinning continues from the exponential time decay in the right-hand side of Eq. (60). In previous work [16], we have stopped the thinning at about 0.07 (0.25 μm dimensionally), which approximates the thickness of the glycocalyx; with that assumption, the slow thinning regime for $t > 0.5$ would not be observed. It may be possible to see it with other parameter choices, however. We also see that the osmolarity rises to its equilibrium value fastest in this case. When evaporation and flow are present, we find $c_{eq} = (1 + P_c)/P_c \approx 6.1$ for $P_c = 0.196$ as in Fig. 4.

### 2.2.3 Fluorescent Imaging

When the spatially uniform tear film thins by evaporation, mass conservation requires that $hf = f_0$ where the subscript zero indicates initial value. Once $h$ is known, we can compute $f$, and then evaluate the fluorescent intensity from Eq. (52). This sequence of solution will apply to all of our problems in this chapter.

The results in Fig. 5 utilize the $h$ and $f$ from the previous two figures and evaluate Eq. (52) using different initial $f_0$. The resulting intensity depends on all three of $h$, $f$, and $f_0$, and we evaluate the intensity for each of the three different flow conditions. Starting with $f_0 = 2$, we obtain the right-most set of curves. The intensity here is in the self-quenching regime. The solid curve with evaporation only has been computed in [16, 17]; to a good approximation, $I \propto f^{-2}$ in this regime. For the case with flow only, one obtains the dashed curve oriented vertically; this case is where $f$ does not change, and $h$ is reduced due to flow only. This case is a limiting form of the results from Zhong et al. [121]. For the case with both evaporation and flow, the dashed-dotted curve shows an increase faster than for evaporation alone, then bends to a vertical orientation such as that in the flow-only case. The bend occurs at about $t = 0.5$ in the thickness plot. The bend in the curve will not be seen experimentally for these parameters, but could be for slower flow contributions. The (middle) set of curves with $f_0 = 0.7$ follow a similar pattern for these three cases.

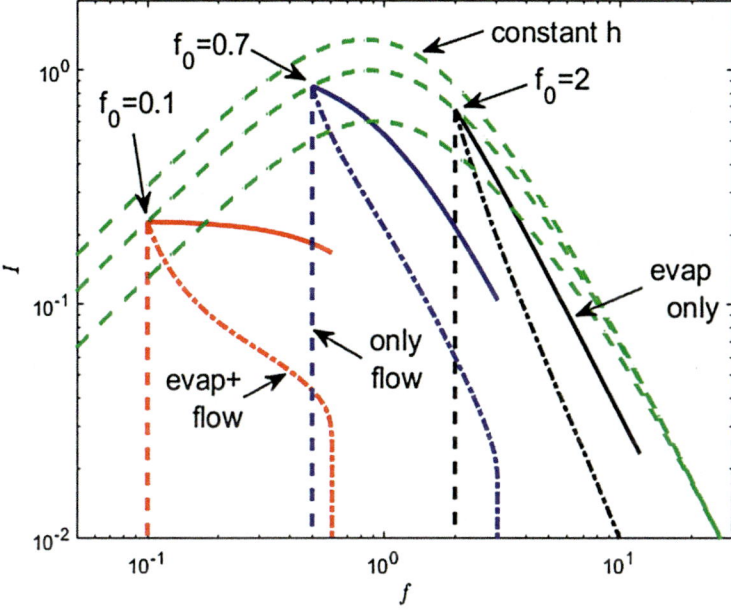

**Fig. 5** The relative intensity $I$ as a function of $f$; here $t$ is a parameter. The three sets of curves correspond to $f_0 = 0.1, 0.7$, and 2, respectively. The required thickness and concentration results for all of those curves come from the simulations shown in the previous two figures. The initial values are at the common points labeled with values of $f_0$. The dashed, roughly parabolic, curves are constant thickness plots of Eq. (52) shown for reference; from top to bottom, the corresponding thicknesses are $d' = 5, 3.5$, and $2\,\mu m$

For $f_0 = 0.1$ (leftmost set of curves), the intensity is in the dilute regime. The intensity decreases at constant $f$ for divergent flow only. As computed previously, the evaporation-only case shows little change in intensity [16]. With both evaporation and flow present, then the decrease in $I$ is intermediate to the two individual effects, and a bend is seen corresponding to about $f = 0.6$ (at about $t = 0.5$) for these parameters. For sufficiently fast divergent flow, the decrease in intensity is rapid, and due solely to the decrease in thickness [121]. When the flow is slower than the evaporation, then the dynamics roughly follow the evaporation-only case.

### 2.2.4 Key Findings

- It is possible to devise simple models that usefully caricature the complex dynamics of the tear film. New results of this type are presented here.
- For evaporative TBU, it is best use a fluorescein concentration in the self-quenching regime [14].

- For rapid TBU, it is better to use a little lower value than the critical fluorescein concentration [121].
- Flow and evaporation can combine to cause intermediate dynamics for the resulting images. It seems likely that most instances of TBU involve more than one mechanism.
- It is difficult to know in advance which will occur in a given subject. To quantify thinning and TBU with fluorescence imaging, it is best to know the starting fluorescein concentration and thickness as well as possible.

## 2.3   TBU Models

We now turn to solving the problem for the same dependent variables that vary with both time and one space dimension. Such models can approximate and produce comparable data for simple TBU regions as described in Sect. 1.1.3. There are at least two types of TBU: evaporative and rapid glob-driven. Figure 6 shows sketches of the different effects that are present and their relative importance in these types of TBU.

We first consider evaporative TBU, then proceed to rapid TBU, and finally consider multilayer models.

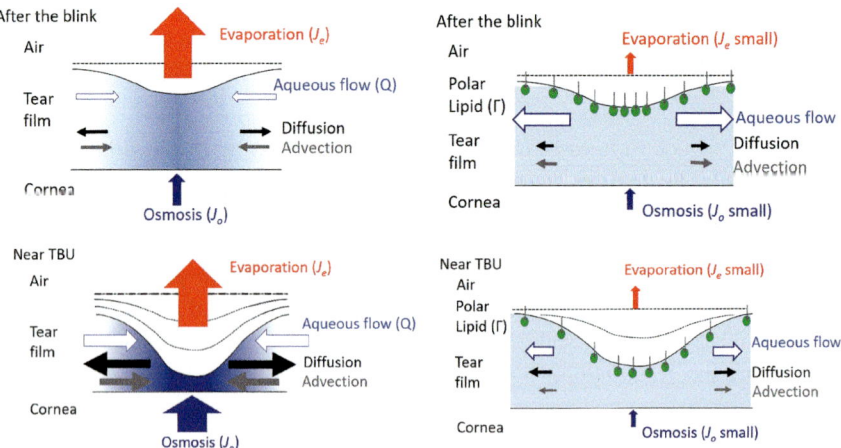

**Fig. 6** Left column: TBU driven by evaporation. Increased evaporation causes thinning and hyperosmolarity. TBU occurs if the evaporation is faster than capillarity can push fluid into the thin region. The darker shade indicates higher osmolarity. Right column: TBU driven by lipid globs. The Marangoni drives rapid divergent flow away from center, causing thinning but without hyperosmolarity. These are limiting cases that can be observed in vivo [51]; many cases may be a combination of the two mechanisms

### 2.3.1 Evaporative TBU: Fixed Distributions

In this case, we show results from [14] for the problem with PDEs given by Eqs. (36), (39), (40), (47), and (48) subject to the boundary conditions Eq. (49) and the initial conditions Eq. (50). For illustration, the Gaussian distribution has peak thinning rate $w_0 = 20\,\mu$m/min and background rate $w_1 = 1\,\mu$m/min, and width $r_w = 1$; this width is the same as the length scale along the film (0.35 mm in this case). The time scale is $d'/w_0 = 10.5$ s. This approximates a hole in the lipid layer allowing localized fast evaporation compared to a slow background rate. In this example, wetting forces are active ($A \neq 0$) and there is a pressure correction to the evaporation rate ($\alpha \neq 0$).

Figure 7 shows tear film thickness $h$, evaporative flux $J$, osmolarity $c$, and FL concentration $f$ as a function of space for different times. We see that the tear film thickness decreases to a low level that we interpret as TBU and that this takes slightly longer than one unit of time; this confirms the suitability of our choice

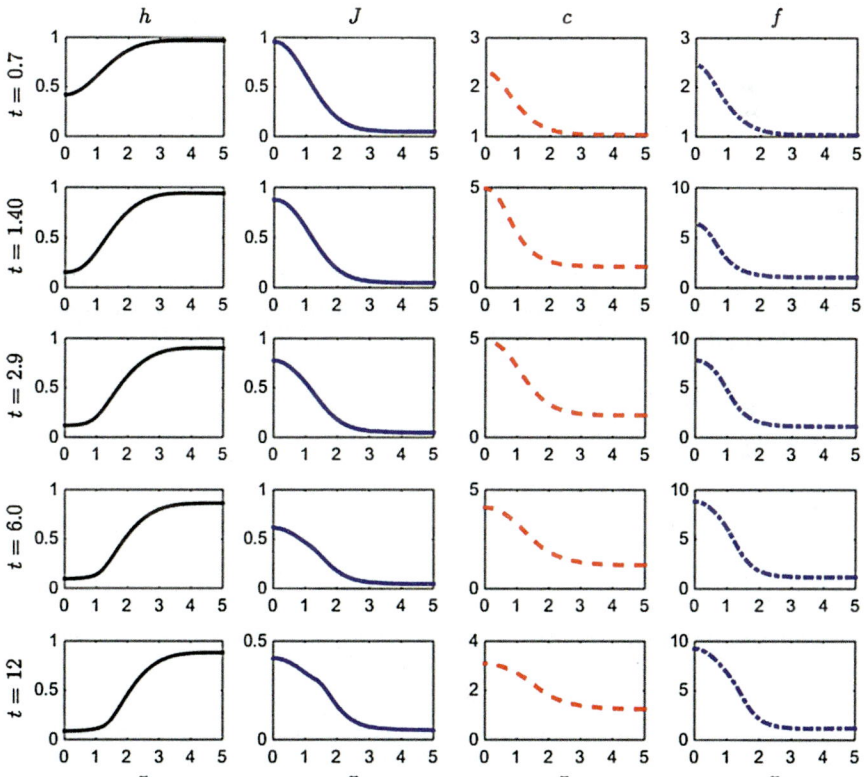

**Fig. 7** From left to right in each column: thickness $h$, evaporation rate $J$, osmolarity $c$, and FL concentration $f$ at several times for a Gaussian $J_f$ with $r_w = 1$, $w_0 = 20\,\mu$m/min and $d = 3.5\,\mu$m. Reprinted from [14] by permission of Oxford University Press

of time scale for a thinning rate distribution of this width. When the thickness of the tear film gets small enough, the evaporative flux is reduced according to the model. The osmolarity and FL concentrations develop elevated values in the TBU region as has been found elsewhere, with the peak in $f$ several times larger than the peak in $c$ due to its smaller diffusivity. However, these peak values of osmolarity are not sufficient to stop the thinning, which progresses to TBU. The evaporative flux $J$ decreases at the thinnest $h$ values due to the presence of the wetting term $Ah^{-3}$. This prevents $h$ from decreasing below about $0.25\,\mu$m to mimic the thickness of the glycocalyx [112]. If desired, we could compute solutions beyond the initial appearance of TBU, but we do not treat the spreading of TBU in this model because it is limited by the assumed evaporation distribution. While the TBU time is often used as a measure of tear film instability, the progression of TBU as measured by increasing area of TBU during the interblink is an important clinical aspect [71]. We note that when different evaporation models are used, the thinning may stop at much smaller values than we use here [90, 106]. For a more complete discussion, see [14].

When different initial fluorescein concentrations are used to compute the intensity for this kind of result, it turns out that the self-quenching regime, $f_0 > 1$ (relative to the critical concentration $f_{cr}$) works best to approximate the thickness [14, 16, 85]. In particular, the approximation $h \propto \sqrt{I}$ holds, and when the tear film thins due to evaporation the intensity decreases markedly. The regime corresponds to the solid curves for $f_0 \geq 1$ in Fig. 5. If a small $f_0$ is used, then the situation is much like the solid curve for $f_0 = 0.1$ in Fig. 5 where little happens during thinning. The dilute regime does not work well for evaporative thinning and TBU. Detailed parametric study and discussion can be found in [14].

In Fig. 8, we show TBU time as a function of spot size $r_w$ for different peak thinning rates $w_0$. The TBU time is defined as when the tear film thickness reaches twice the wetting equilibrium thickness $h_{eq} = (A\alpha)^{1/3}$ where evaporation stops due to the balance with the wetting term. In Fig. 8, we use $h_{eq} = 0.25\,\mu$m, so the TBU time was taken to be when the tear film reached $0.5\,\mu$m. This size was chosen because the mathematical equilibrium is only slowly approached, which artificially extends the TBU time. In these results, osmosis is active with $P_{corn}^{(tiss)} = 12.1\,\mu$m/s permeability [17]. The results show that for relatively large spots with $r_w > 0.8$ and $w_0 \geq 10\,\mu$m/min, the TBU time is about one unit of time. This indicates that the time to get to TBU is about $d/w_0$ which is the time for a flat film to evaporate away. If the peak thinning rate is small, such as $w_0 = 5\,\mu$m/min, then the thinning is slow and the TBU time increases because there is sufficient time for capillary flow and osmosis to supply fluid to the thinning region [14].

In the range of moderate spot sizes $0.4 \leq r_w \leq 0.8$ and with $w_0 \geq 10\,\mu$m/min, the TBU time increases as $r_w$ decreases. In this regime, capillary flow is increasing which brings more fluid into the TBU region. However, osmolarity diffuses out rapidly enough that the peak value of $c$ does not approach the flat film case, and thinning can proceed to TBU in similar times to the large spots.

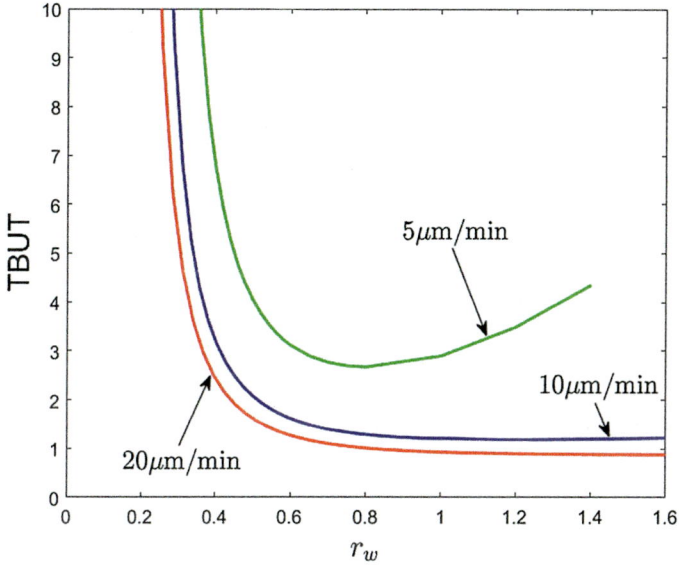

**Fig. 8** TBU time (TBUT) for $d = 3.5\,\mu$m for three different maximum thinning rates $w_0$ (indicated) as function of the size of the evaporation distribution $r_w$. The spot size is relative to the length scale $\ell$

Finally, in the range of small spots with $r_w \leq 0.4$, the TBU time is very long compared to typical in vivo results. For example, with $r_w = 0.25$, it takes more than six time units (more than a minute) to reach TBU. This happens because capillary flow into the TBU region (healing flow due to capillarity) is fast relative to the diffusion of solutes out of the TBU region. Increased advection causes osmolarity to reach values near the flat film case, and then osmosis and capillarity cooperate to slow thinning dramatically. This suggests that a different mechanism is present in small TBU spots.

Besides affecting TBU time via flow and osmolarity distributions, these dynamics have important consequences for imaging the tear film from the distributions of $f$ in TBU. For details, the interested reader is referred to [14].

### 2.3.2 Rapid Glob-Driven TBU

Zhong et al. [123] developed a model that hypothesized that a localized region of excess lipid could supply polar lipid (surfactant) for a period of time long enough to drive rapid TBU. Experimental evidence suggests that this occurs in some kinds of rapid TBU [56]. Their model has a region under the glob that is deformable, is tangentially immobile, and has constant high concentration of surfactant. Outside the glob, the surface is mobile, and a standard surfactant transport equation conserves surfactant. No flux conditions apply at the center of the glob and at the edge of the film away from the glob.

These different conditions are combined by weighting them with a smooth blend function that has a narrow transition at the edge of the glob. For the streak geometry, the cartesian model applies, and the glob edge is at the fixed location $r = R_i$. The nondimensional blend function is given by

$$B(r; R_i, R_w) = 0.5 + 0.5 \tanh \left( \frac{r - R_i}{R_w} \right). \tag{61}$$

We choose $R_w$ to be small so that the transition width of $B(R; R_i, R_w)$ is very narrow compared to the domain size. $B(r; R_i, R_w)$ then approximates a step function which is zero to the left of $R_i$ but one to the right; $1 - B(r; R_i, R_w)$ behaves the opposite way. The equations on the two subdomains are then combined into one continuous equation; schematically, we have,

$$\left[ \text{BCs on } (0, R_I) \right] \left[ 1 - B(r; R_I, R_w) \right] + \left[ \text{BCs on } (R_I, R_L) \right] B(r; R_I, R_w). \tag{62}$$

Once the boundary conditions are blended using this function, then lubrication theory can be applied to the thin film throughout the domain to obtain the following equations for the tear film thickness $h$, the pressure $p$, polar lipid (insoluble surfactant) concentration $\Gamma$. The reduced equations in this case govern the tear film thickness $h(r, t)$, pressure $p(r, t)$, the solutes $c(r, t)$ and $f(r, t)$, and surfactant surface concentration $\Gamma(r, t)$:

$$\partial_t h + \frac{1}{r} \partial_r (rh\bar{u}_r) = -J + P_c(c - 1), \tag{63}$$

$$p = -\frac{1}{r} \partial_r (r \partial_r h) - Ah^{-3}, \tag{64}$$

and

$$h \left( \partial_t m + \bar{u} \partial_r m \right) = \mathrm{Pe}_s^{-1} \frac{1}{r} \partial_r (rh \partial_r m) + \left[ J - P_c(c - 1) \right] m, \quad m = c, f, \tag{65}$$

where

$$\bar{u}_r = \frac{-\frac{1}{3} \partial_r ph^2 [B + (1 - B)(\beta + \frac{1}{4}h)] - \frac{1}{2} hB \partial_r \Gamma}{B + (1 - B)h}, \tag{66}$$

The solute transport equations remain very similar with only osmolarity contributing to osmotic supply. We now have a surfactant transport equation governing the surface concentration of $\Gamma$, namely

$$\partial_t \Gamma = \left[ Pe_s^{-1} \left( \frac{1}{r} \partial_r (r \partial_r \Gamma) \right) - \frac{1}{r} \partial_r (r \Gamma u_r) \right] B. \tag{67}$$

where

$$u_r(r, h, t) = \frac{1}{2} h^2 \partial_r p + \frac{-h(\partial_r p)[(\frac{1}{3}h + \beta)(1 - B) + B] - (\partial_r \Gamma) B}{B + (1 - B)h} h \tag{68}$$

Here $\bar{u}_r$ is the depth-averaged flow, $u_r$ is the surface velocity at the tear film free surface, and $B$ is the blend function in cylindrical coordinates.

At the glob center, $r = 0$, we use symmetry to arrive at the following equations for numerical solution:

$$\partial_t h = -h \partial_r \bar{u}_r - J, \quad 0 = p + \partial_r^2 h + Ah^{-3}, \text{ and } \partial_r c = \partial_r f = \partial_r \Gamma = 0. \tag{69}$$

The first two equations are obtained from (63) and (64).

We again make the domain size large enough so that the no flux conditions have no effect on the overall dynamics; at $r = R_L$, we have

$$\partial_r h = \partial_r c = \partial_r f = 0, \quad \partial_r p = 0, \quad \text{and} \quad \partial_r \Gamma = 0. \tag{70}$$

We choose the initial conditions

$$h(r, 0) = c(r, 0) = 1, \quad f(r, 0) = f_0, \quad \text{and} \quad \Gamma(r, 0) = (1 - B) \cdot 1 + 0.1B, \tag{71}$$

along with the consistent pressure initial condition $p = -A$. The initial condition for $\Gamma$ is such that $\Gamma = 1$ in the glob and 0.1 outside the glob with the smooth, narrow transition between them.

Figure 9 portrays a spot TBU result for our default parameters from Tables 1 and 2 and with no evaporation ($J = 0$). In this case, TBU occurs and is induced only by the different composition of lipid inside the glob. The locally elevated surfactant concentration there spreads out immediately over the tear/air interface, and the surfactant concentration variation generates a significant shear stress, which in turn drives a strong divergent tangential flow. A significant depression of the tear/air interface forms at 0.15 s. In 0.59 s, the aqueous layer thins from 3.5 to 0.25 μm, the thickness which we defined as TBU. This TBU time is near unity, which indicates that the time scales we chose work well. The Marangoni flow dominates the thinning process in this case, which leads to TBU in less than a second. In some in vivo observations with simultaneous fluorescence and lipid layer interferometry [56], the dark spot has already appeared by 0.14 s (practically instantaneously); the bright circular glob in the lipid layer spreads rapidly but its center remains relatively bright and thus relatively thick. This experimental observation corresponds well with the model.

The floating lipid layer is dominated by extension and lacks a time derivative in the equation for the velocity along the film in that layer, which remains as a dependent variable in the problem. A more typical equation results for the thickness of the lipid layer. The aqueous layer is dominated by shear and its equation resembles a typical thin film equation, being first order in time and fourth order in space. The insoluble surfactant transport equation had a large Péclet number, and as a result they neglected diffusion of the surfactant. The resulting complex problem was solved on a moving domain that included the opening and interblink phases. They also found a number of useful simplifying limits to their model. For example, they found a limit where the surfactant concentration on the lipid/aqueous interface exactly follows the lipid layer thickness.

Zubkov and coworkers [124, 125] investigated several other aspects of the tear film related to this approach. In [124], they included evaporation and osmolarity transport with tear film formation and a standard interblink of 5 s. Their aim was to evaluate the normal tear film with a typical low-concentration task. The previous film formation results were effectively unchanged, and the osmolarity rose very little in the cases they studied. Zubkov et al. [125] studied the flow near the Marx line, which is on eyelid margin, in contrast to most other tear film studies. They found that the lubrication approximation did not provide the same level of detail as direct numerical solution of the Navier–Stokes equations in that region.

### 2.5.2   Key Findings

- Including a moving end can correctly capture trends seen in vivo with tear film formation, drainage, and blink cycles.
- Such features as black line formation and the thin region left from lid turnaround are recovered.
- The Marangoni effect from an insoluble surfactant helps make the deposited tear film more uniform, and better matches observed tear film profiles. This includes the uniform stretching limit.
- Osmolarity increases need sufficient time between blinks to become, given most physiological evaporation rates.
- To correctly capture flows very close to the lid, it may be necessary to abandon lubrication theory.

## 2.6   Overall Flows

The flow of the tear film over the exposed ocular surface, and other elements of the lacrimal system, may be of interest. Besides models that compute the flow over the exposed ocular surface, compartment models have been used for this purpose. Gaffney et al. [35] developed a compartment model that is a mass and solute balance of the tear film that focuses on osmolarity. The model accounts for osmolarity in discrete regions of the tear film, such as the broad middle, the menisci, and the fornices. Input parameters include the tear supply from the lacrimal gland, the tear evaporation rate, and the blink rate. Osmosis from the ocular surface was neglected. The model indicates that the osmolarity should increase over most of the eye from evaporation, possibly to high enough levels to cause noticeable sensation [72]. Parameters relevant for dry eye predict more elevated concentration than for normals. Results from the model also suggest that increased blink rate, in an effort to supply more tears, may have a limited benefit for aqueous-deficient dry eye. Osmosis from the ocular surface was added by Cerretani et al. [24]; they reached similar conclusions.

We now turn to models where the detailed flow and solute distributions over the ocular surface are obtained.

### 2.6.1 Fluid Motion in the Interblink

We first discuss the problem in an open eye-shaped domain without solutes [70, 75]. Then we add osmolarity, followed by fluorescein. After nondimensionalization and simplification using lubrication theory (e.g., [13, 45]), we arrive at a system of PDEs for the dimensionless variables $h(x, y, t)$ and $c(x, y, t)$:

$$\partial_t h + EJ + \nabla \cdot \mathbf{Q} - P_c(c - 1) = 0, \tag{72}$$

$$h\partial_t c + \nabla c \cdot \mathbf{Q} = EcJ + \frac{1}{Pe_c}\nabla \cdot (h\nabla c) - P_c(c - 1)c. \tag{73}$$

The evaporative mass flux $J$ is given by

$$J = \frac{1 - \delta\left(S\Delta h + Ah^{-3}\right)}{\bar{K} + h},$$

and the fluid flux $\mathbf{Q}$ across any cross-section of the film is given by

$$\mathbf{Q} = \frac{h^3}{12}\nabla\left(S\Delta h + Ah^{-3}\right).$$

The conjoining pressure in modeling evaporation plays an important role in the tear film model. It is meant to mimic the effect of the glycocalyx, whose transmembrane mucins are strongly wet by water and we assume that they arrest the thinning of the tear film. A secondary benefit is that the model allows solutions to be computed past the initial tear film breakup because the tear film thickness never reaches zero in this model. These aspects of evaporation competing with conjoining pressure are discussed by Winter et al. [112] in the context of eyes, but the idea

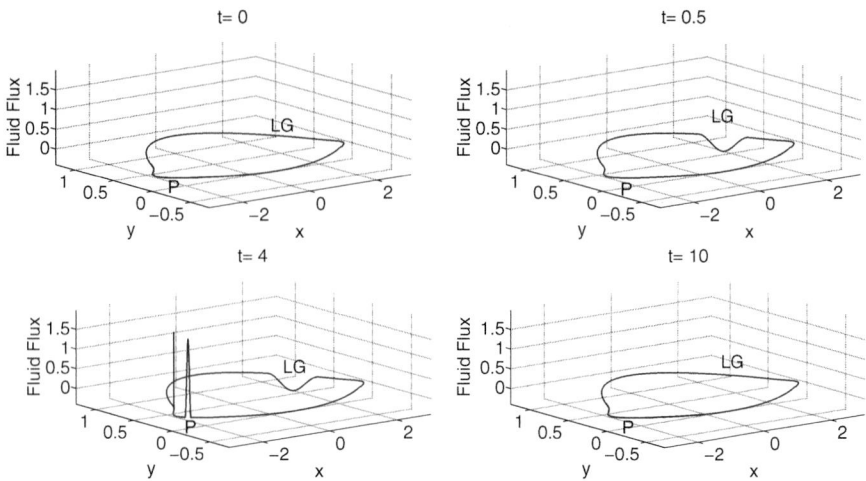

**Fig. 10** Time sequences of fluid flux boundary condition during one flux cycle. LG labels the location of influx from the lacrimal gland near the temporal canthus; P labels the puncta on the nasal side

was developed by Potash and Wayner [91] and Moosman and Homsy [80]. More recent versions of the approach may be found in [81] and [1]. The nondimensional parameters that arise are defined and given values in the following section and in Table 2. The dimensional parameters used in those expressions are given in Table 1. The boundary conditions are the constant Dirichlet condition $h_0 = 13$ together with time-dependent fluxes into the domain from the lacrimal gland and out from the domain at the puncta [70]. These boundary conditions mimic the supply and draining throughout the blink cycle, but the lids do not move in this model. Several time values of the flux boundary conditions are plotted in Fig. 10.

We now go on to state the equations for solute transport and discuss results for those cases.

### 2.6.2 Osmolarity and Fluorescence

Li et al. extended the 2D overall flow models to include transport of osmolarity [68] and fluorescein [69]. Transport of solutes satisfies

$$h\partial_t m + \nabla m \cdot \mathbf{Q} = \frac{1}{\text{Pe}_s}\nabla \cdot (h\nabla m) + \left[EJ - P_c(c-1)\right]m. \tag{74}$$

where $m = c, f$ for the two different solutes. The boundary conditions we use for solutes are $c = 1$ and $\mathbf{n}_b \cdot \nabla_{II} f = 0$ on the boundary [68, 69]. The initial conditions are still uniform in space, with $c(\mathbf{x}, 0) = 1$ and $f(\mathbf{x}, 0) = f_0$. The concentration $f$ is still a passive scalar; the solution can be computed for a single value of $f_0$, then rescaled for a different value; this reduces computation for evaluating the fluorescence results for comparison with experiment [69].

In the 1D studies of TBU above, the pre-corneal tear film was studied. When considering osmosis from the exposed ocular surface, one should allow for different permeabilities of the cornea ($P_{corn}^{tiss}$) and conjunctiva ($P_{conj}^{tiss}$) which are roughly a factor of 4 different (Table 1, and [17]). This was accomplished in the model using a tanh-like function to smoothly and narrowly change between the two values in the different areas of the model substrate of the film. The cornea was represented as a unit circle with low permeability and the surrounding conjunctiva had a higher permeability.

We first consider results for the osmolarity [68], shown in Fig. 11. This case corresponds to $i = 2$ in the parameter tables, with $w_0 = 4\,\mu\text{m/min}$ uniformly over

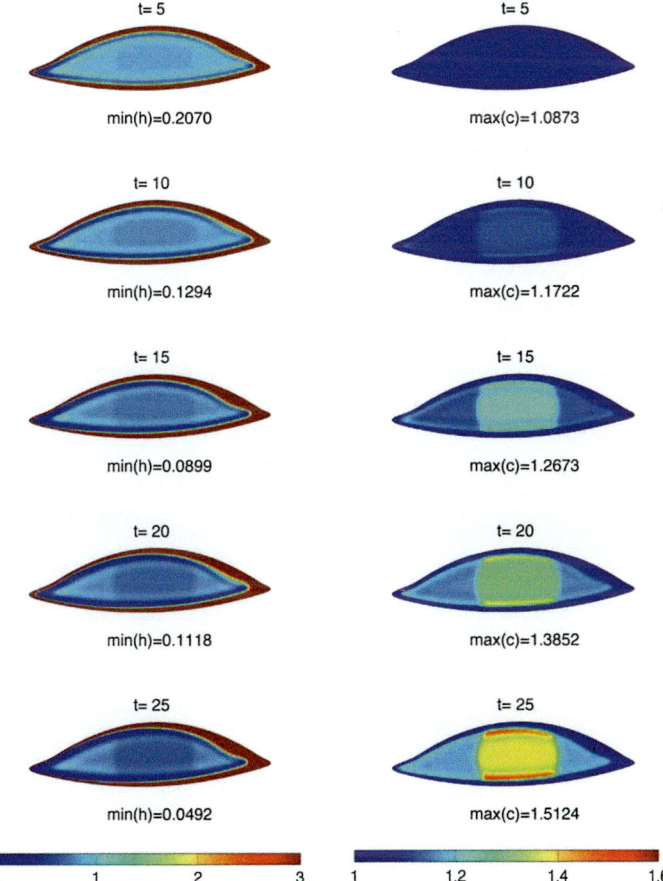

**Fig. 11** Contours of tear film thickness (left column) and osmolarity (right column). The thinning rate is $4\,\mu\text{m/min}$. Reprinted from [68] by permission of Oxford University Press

the domain. The left column gives $h$ and the right column $c$ at different times. There is more decrease in thickness and increase in osmolarity over the cornea due to its lower permeability. The thinning rate is at the upper end of what could be considered normal [86], and the rise in osmolarity is not too large, to about 450 mOsM; this is just at the edge of what can be felt by a subject in a controlled experiment [72]. The location of thinnest tear film and highest osmolarity occurs in the black lines over the cornea.

When the thinning rate is increased to 20 μm/min, there is a dramatic increase in the maximum value of the osmolarity to roughly six times the isotonic value, or 1800 mOsM. This value is well above the pain threshold, and would most likely only be achieved in a controlled study where the subject is asked to hold the eye open as long as possible. It is also well above the maximum value estimated in vivo, from subject feedback after known solution instillation via turning a pain knob, of 800–900 mOsM [72]. This value is similar to that observed for TBU in models with fixed evaporation profiles [14, 90]. These estimates are important because, at this time, it is not possible to measure osmolarity in regions of TBU or the black line. The predominant measurement used in the clinic, using calibrated resistance measurements in the inferior meniscus [62], is from a region that did not have much osmolarity variation. While there may still be utility in that measurement, the actual osmolarity experienced by the ocular surface is apparently much larger in regions of evaporative tear thinning (Fig. 12).

Transport of fluorescein was also included in [69]. In that paper, any contribution of fluorescein to osmosis was found to be small, and so the fluorescein concentration $f$ was still a passive scalar transported by the flow which was determined by the equations for $h$ and $c$. Computed results for $h$ and $f$ could then be used in Eq. (52) to generate images analogous to those from fluorescein imaging experiments in vivo. Features such as the black line were readily observed, and dynamics of the different

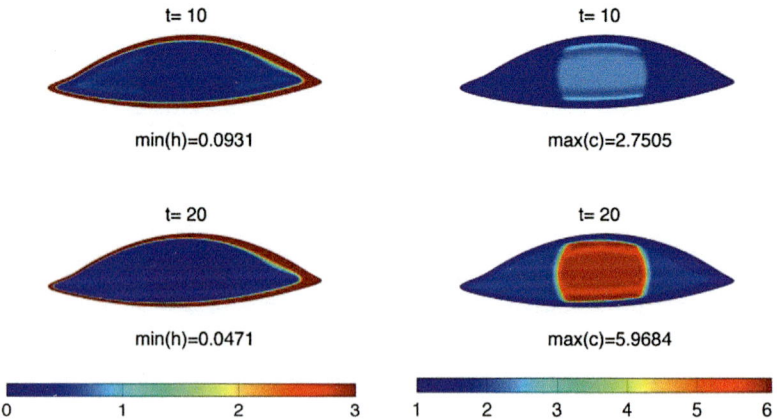

**Fig. 12** Contours of tear film thickness (left column) and osmolarity (right column). The thinning rate is 20 μm/min. Reprinted from [68] by permission of Oxford University Press

$f$ regimes (dilute vs. self-quenching) were also explored. These results provide a useful basis for interpretation of fluorescein images, and they take the approach a step closer to quantitative, dynamic measurements of tear film thickness.

### 2.6.3  Blinking in 2D

We first explore the geometric complication from the eye shape and the moving boundary introduces more mathematical challenges for modeling the tear film, and it is appropriate to start with models that account for relatively few physical effects. We reduce the model to one that is effectively pure water on a blinking eye-shaped domain. The eyelids are arcs of circles, which only approximate actual eye shapes. However, these simplified lid margins may be mapped to convenient shapes for numerical solution via conformal mapping [33]. We solve the model PDE

$$h_t + \nabla \cdot \mathbf{q} = 0, \qquad (x, y) \in \mathcal{E}(t),$$

$$\mathbf{q} = \frac{h^3}{3} \nabla p, \qquad (75)$$

$$p = -S\nabla^2 h - Ah^{-3},$$

subject to boundary conditions given below. Here $h(x, y, t)$ is the thickness of the fluid layer, $p(x, y, t)$ is the fluid pressure, and $\mathbf{q}$ is a known, smooth flux function. The constants $S$ and $A$, respectively, determine the strengths of surface tension and van der Waals forces, as described in [70] and Table 2 above for $i = 2$.

The domain $\mathcal{E}(t)$ is an idealized eye shape bounded at all times by four circular arcs. As explained in detail in [33], this region is the transformation of a box $\mathcal{R}(t) = \{|\tilde{x}| \le L, -1 \le \tilde{y} \le Y(t)\}$ under a hyperbolic tangent in complex coordinates. The resulting region is then scaled so that $|x| \le 3$ and the aspect ratio matches that of the smallest box containing the realistic eye shape in [70].

The motion of $Y(t)$ is that prescribed as realistic lid motion in [41]. Specifically, given a value $\lambda \in (0, 1]$ for the fraction of the maximum opening width that is left open at the end of the upper lid downstroke, we define

$$Y(t) = \begin{cases} -1 + 2\lambda + 2(1 - \lambda)t_1^2 e^{1-t_1^2}, & 0 \le t \le \Delta t_{co}, \\ 1, & \Delta t_{co} \le t \le \Delta t_{co} + \Delta t_o, \\ 1 - 2(1 - \lambda)t_2^2 e^{1-t_2^2}, & \Delta t_{co} + \Delta t_o \le t \le \Delta t_{bc}. \end{cases} \qquad (76)$$

Here $t_1 = t/\Delta t_{co}$, $t_2 = (t - \Delta t_{co} - \Delta t_o)/\Delta t_{oc}$, $\Delta t_{co}$ is the time interval of the lid upstroke, $\Delta t_o$ is the length of the interblink interval, $\Delta t_{oc}$ is the lid downstroke interval, and $\Delta t_{bc} = \Delta t_{co} + \Delta t_o + \Delta t_{oc}$ is the period of one complete blink cycle (all given in nondimensional time units). The motion satisfies $2\lambda - 1 = Y(0) \le Y(t) \le 1$ for all times. The values of the time intervals given in [41] are nondimensionalized assuming a characteristic closing speed $U_m = 10\,\text{cm/s}$. In order to better compare with [70], we instead use the latter's value of $U_2 = 5$ mm/s and accordingly take $\Delta t_{co} = 0.176$, $\Delta t_o = 5$, and $\Delta t_{oc} = 0.0820$, for a cycle duration of $\Delta t_{bc} = 5.258$.

Boundary conditions are imposed on both $h$ and $p$. As in [70], we choose $h = 13$ uniformly at the boundary. For pressure we use the two-dimensional equivalent of FPLM (flux proportional to lid motion) as described in [41]:

$$\mathbf{n} \cdot \mathbf{q} = (\mathbf{v} \cdot \mathbf{n})(h - h_e/2) \quad \text{on} \partial \mathcal{E}(t), \tag{77}$$

where $\mathbf{n}$ is the unit outward normal, $\mathbf{v}$ is the velocity of a point on the boundary, and $h_e$ represents an amount of fluid remaining underneath the upper eyelid as it moves.

Because the model includes neither evaporative nor osmotic effects, if $h_e = 0$ the total volume of exposed fluid, given by

$$V(t) = \int_{\mathcal{E}(t)} h(x, y, t) \, dA, \tag{78}$$

remains constant. We use this fact below to assess the accuracy of our computational results. Even when $h_e > 0$, we should expect $V(t)$ to be periodic if the motion of $\mathcal{E}(t)$ is periodic. Equation (77) can be generalized to prescribe any flux that varies in time and space along the boundary, but we have not yet explored this freedom.

For the numerical solution of this problem, we closely follow the procedure described in [33]. The model domain $\mathcal{E}(t)$ is mapped to the box $\mathcal{R}(t)$ by a constant conformal map. This map preserves angles, making normals and the FPLM boundary condition (77) straightforward to transplant to $\mathcal{R}(t)$. This domain in turn is mapped by a time-varying transformation to the square $[-1, 1]^2$, which is easy to discretize on an $N_x \times N_y$ grid by a Chebyshev spectral collocation method.

The dynamic equation for $h$, the algebraic equation for $p$, and the boundary conditions form a differential-algebraic equation (DAE) system in $2N_x N_y$ unknowns. This system is solved by `ode15s` in MATLAB. The chief step consuming computational time is the factorization of the Jacobian in the nonlinear iteration for finding implicitly defined time steps.

Results are shown in Fig. 13. The initial state ($t = 0$) is the uniform thickness $h = 13$, and the lids are 60% closed. The image at $t = 0.079$ is about halfway through the opening phase. The upper lid has moved upward, but the bulk of the film is left behind. This is apparent at the next time, $t = 0.182$, which is shortly after the opening phase. The meniscus along the upper lid is quite narrow, and upper half of the tear film is thin. In the next image ($t = 5.182$), near the end of the interblink phase, the tear film has spread slowly toward the upper lid, but there is still a thin region separating most of the tear film and the upper meniscus. The black line, a thin region adjacent to the meniscus around the edge of the domain, has also developed noticeably by this time. In the next image, $t = 5.205$, the closing phase has begun, and it can be seen that the thickness is increasing near the canthi and along the upper lid. In the last image, at the end of the closing phase (and the blink cycle), the tear film has exceeded the initial thickness in the canthi because of the proximity of the two lids, and in the center, because that is where the fastest lid motion occurs. On the time scale of the blink, the tear film cannot redistribute the tear film laterally

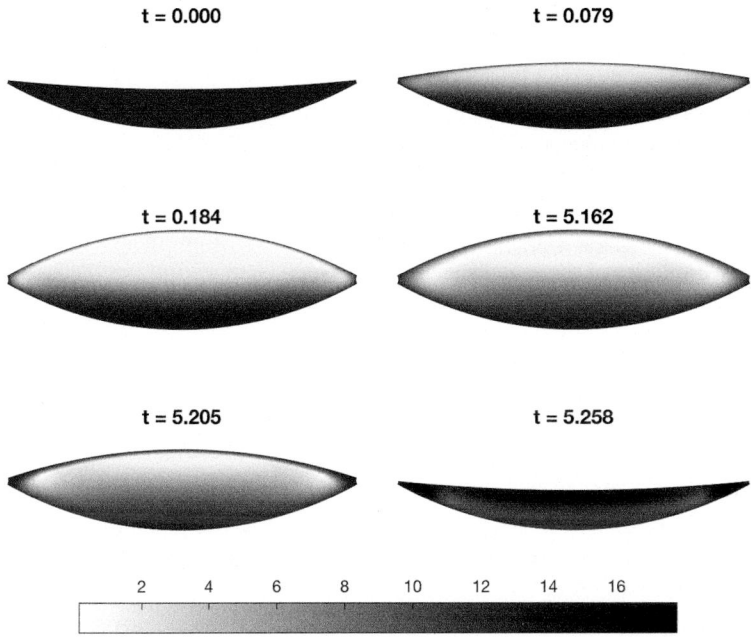

**Fig. 13** Dynamics of the tear film thickness $h(x, y, t)$ during one blink cycle. Here $h_e = 2$, implying that the layer under the lids is $10\,\mu$m thick

from canthus to canthus. However, it is practically erased any thin regions across the gap between the lids vertically.

Further work is needed to include Marangoni effects due to the polar lipids (and perhaps other surface active contents) of the tear film. The polar lipids help the tear film to keep up with the moving upper lid and cause it to keep moving upward in the first part of the interblink [5]. These effects help to make the tear film more uniform after the blink.

An overset grid approach is also viable for modeling the blink computationally. The domain boundary can be captured from digital video frames and an interpolation method used to generate the boundary at desired times. An overset grid method based on finite differencing [40] previously used for tear film models [68, 75] may be adapted to treat the moving domain problem. Additionally, other methods such as finite elements could be used [38].

### 2.6.4 Key Findings

- The 2D models on eye-shaped domains correctly capture a number of aspects of tear film flows during the interblink [70, 74, 75]. These include fluorescence images [69] and the results improve our understanding of tear film dynamics.

- The model with dynamic changes in osmolarity provides a detailed estimate of osmolarity over the corneal surface [68]. To the best of our knowledge, this is the only model of this type to date.
- These estimates are consistent with TBU results [14, 90], and experimental estimates from subject feedback via a pain knob [72].
- The osmolarity model has a high potential to affect clinical care of patients because hyperosmolarity of the tear film is considered a major etiological factor in dry eye disease [60]. Current techniques only allow measurement of tear film osmolarity in the inferior meniscus, which changes little in the simulations compared to the pre-corneal tear film.

## 3  Summary

Mathematical models for the tear film have achieved some successes. Some standard observations such as the black line near the menisci and a thin region at the turnaround point of the partial blink have been explained. More in line with the interests of optometrists and ophthalmologists, these models have been able to estimate the high levels of osmolarity that occur in evaporative water loss in localized areas of TBU or over the open eye. These estimates are important given that they cannot be measured in vivo for humans at this time. The models have also provided justification for the notion that there different mechanisms of TBU, and that these different mechanisms have different effects on the ocular surface. Perhaps most importantly, rapid TBU is not accompanied by hyperosmolarity, while the slower evaporative mechanism may have severe hyperosmolarity [14, 51, 122]. Recognizing these different consequences may be of great help for interpreting seemingly contradictory data and symptoms.

Recent models have also aided understanding of imaging methods, particularly for fluorescein. Fluorescein images result from a solute distribution and the thickness of the film. The intensity is affected by the flow inside the film because fluorescein diffuses less readily than other ions in the tear film, causing it to be more susceptible to advection. The advection may narrow the apparent width of TBU or may lead to higher than expected concentrations of fluorescein compared to a uniform concentration assumption [17].

Work on the pre-corneal tear film is far ahead of that for a contact lens with respect to the fluid dynamics and interaction with blinking. On the other hand, interaction of oxygen transport [63] and drug delivery [49] with contact lenses may be ahead of modeling for interaction of the pre-corneal tear film with the cornea.

Much remains to be done. The successes listed above relied primarily on physical-science and mechanics style models to make good progress. In the same vein, more sophisticated blinking simulations and other effects will soon come. It is now a good time to use both old and new results to probe biological consequences on the ocular surface. This could include dynamics of the epithelial cells, including nerves (e.g., [43]), both water and ion transport (e.g., [64, 65]) and inflammatory and other biochemical pathways (e.g., [67]). A number of biochemical pathways were discussed in previous tear film chapters in this book, and starting to address those together with tear film modeling is highly likely to be of interest.

Incorporating recent measurements of properties is also needed, as found in, e.g., [9, 22, 94]. The properties, structure, and composition are all important aspects of the lipid layer that have profound effects on tear film dynamics [21]. Building in this relatively new information regarding the tear film will likely lead to new advances.

**Acknowledgements** This work is funded by NSF grant 1412085 (Braun, Driscoll) and NIH grant 1R01EY021794 (Begley). The opinions expressed here are those of the authors and do not represent the official position of either the NSF or the NIH.

# References

1. Ajaev, V.S., Homsy, G.: Steady vapor bubbles in rectangular microchannels. J. Colloid Interface Sci. **240**, 259–271 (2001)
2. Allouche, M., Abderrahmane, H.A., Djouadi, S.M., Mansouri, K.: Influence of curvature on tear film dynamics. Euro. J. Mech. B Fluids **66**, 81–91 (2017)
3. Arnold, S., Bruenner, H., Langenbucher, A.: Simultaneous examination of tear film break-up and the lipid layer of the human eye: A novel model eye for time course simulation of physiologic tear film behavior (part 2). Z. Med. Phys. **20**, 316–319 (2010)
4. Arnold, S., Walter, A., Eppig, T., Bruenner, H., Langenbucher, A.: Simultaneous examination of tear film break-up and the lipid layer of the human eye: A novel sensor design (part 1). Z. Med. Phys. **20**, 309–315 (2010)
5. Aydemir, E., Breward, C.J.W., Witelski, T.P.: The effect of polar lipids on tear film dynamics. Bull. Math. Biol. **73**, 1171–1201 (2010)
6. Begley, C.G., Himebaugh, N., Renner, D., Liu, H., Chalmers, R., Simpson, T., Varikooty, J.: Tear breakup dynamics: a technique for quantifying tear film instability. Optom. Vis. Sci. **83**(1), 15–21 (2006)
7. Begley, C.G., Simpson, T., Liu, H., Salvo, E., Wu, Z., Bradley, A., Situ, P.: Quantative analysis of tear film fluorescence and discomfort during tear film instability and thinning. Invest. Ophthalmol. Vis. Sci. **54**, 2645–2653 (2013)
8. Berger, R.E., Corrsin, S.: A surface tension gradient mechanism for driving the pre-corneal tear film after a blink. J. Biomech. **7**, 225–238 (1974)
9. Bhamla, M.S., Chai, C., Alvarez-Valenzuela, M.A., Tajuelo, J., Fuller, G.G.: Interfacial mechanisms for stability of surfactant-laden films. PLOS One **12**, e0175,753 (2017)
10. Bhamla, M.S., Chai, C., Rabiah, N.I., Frostad, J.M., Fuller, G.G.: Instability and breakup of model tear films. Invest. Ophthalmol. Vis. Sci. **57**, 949–958 (2016)
11. Bhamla, M.S., Giacomin, C.E., Balemansc, C., Fuller, G.G.: Influence of interfacial rheology on drainage from curved surfaces. Soft Matter **10**, 6917–6925 (2014)
12. Blackie, C.A., Solomon, J.D., Scaffidi, R.C., Greiner, J.V., Lemp, M.A., Korb, D.R.: The relationship between dry eye symptoms and lipid layer thickness. Cornea **28**, 789–794 (2009)
13. Braun, R.J.: Dynamics of the tear film. Annu. Rev. Fluid Mech. **44**, 267–297 (2012)
14. Braun, R.J., Driscoll, T.A., Begley, C.G., King-Smith, P.E., Siddique, J.I.: On tear film breakup (TBU): Dynamics and imaging. Math. Med. Biol. **45**, 145–180 (2018)
15. Braun, R.J., Fitt, A.D.: Modeling the drainage of the precorneal tear film after a blink. Math. Med. Biol. **20**, 1–28 (2003)
16. Braun, R.J., Gewecke, N., Begley, C.G., King-Smith, P.E., Siddique, J.I.: A model for tear film thinning with osmolarity and fluorescein. Invest. Ophthalmol. Vis. Sci. **55**, 1133–1142 (2014)

17. Braun, R.J., King-Smith, P.E., Begley, C.G., Li, L., Gewecke, N.R.: Dynamics and function of the tear film in relation to the blink cycle. Prog. Retin. Eye Res. **45**, 132–164 (2015)

18. Braun, R.J., Usha, R., McFadden, G.B., Driscoll, T.A., Cook, L.P., King-Smith, P.E.: Thin film dynamics on a prolate spheroid with application to the cornea. J. Eng. Math. **73**, 121–138 (2012)

19. Bron, A., Tiffany, J., Gouveia, S., Yokoi, N., Voon, L.: Functional aspects of the tear film lipid layer. Exp. Eye Res. **78**, 347–360 (2004)

20. Bruna, M., Breward, C.J.W.: The influence of nonpolar lipids on tear film dynamics. J. Fluid Mech. **746**, 565–605 (2014)

21. Butovich, I.A.: Tear film lipids. Exp. Eye Res. **117**, 4–27 (2013)

22. Butovich, I.A., Lu, H., McMahon, A., Kelelson, H., Senchyna, M., Meadows, D., Campbell, E., Molai, M., Linsenhardt, E.: Biophysical and morphological evaluation of human normal and dry eye meibum using hot stage polarized light microscopy. Invest. Ophthalmol. Vis. Sci. **55**, 87–101 (2014)

23. Casalini, T., Salvalaglio, M., Perale, G., Masi, M., Cavallotti, C.: Diffusion and aggregation of sodium fluorescein in aqueous solutions. J. Phys. Chem. B **115**, 12,896–12,904 (2011)

24. Cerretani, C.F., Radke, C.J.: Tear dynamics in healthy and dry eyes. Curr. Eye Res. **39**, 580–595 (2014)

25. Craig, J.P., Nichols, K.K., Akpek, E.K., Caffery, B., Dua, H.S., Joo, C.K., Liu, Z., Nelson, J.D., Nichols, J.J., Tsubota, K., et al.: TFOS DEWS-II definition and classification report. Ocul. Surf. **15**(3), 276–283 (2017)

26. Craig, J.P., Tomlinson, A.: Importance of the lipid layer in human tear film stability and evaporation. Optom. Vis. Sci. **74**, 8–13 (1997)

27. Cruz, A.A.V., Garcia, D.M., Pinto, C.T., Cechetti, S.P.: Spontaneous eyeblink activity. Ocul. Surf. **9**, 29–41 (2011)

28. Dartt, D.A.: Neural regulation of lacrimal gland secretory processes: Relevance in dry eye diseases. Prog. Ret. Eye Res. **28**(3), 155–177 (2009)

29. Dartt, D.A., Hodges, R.R., Zoukhri, D.: Tears and their secretion. In: J. Fischbarg (ed.) The Biology of the Eye, *Advances in Organ Biology*, vol. 10, pp. 21–82. Elsevier (2005)

30. Deng, Q., Braun, R.J., Driscoll, T.A.: Heat transfer and tear film dynamics over multiple blink cycles. Phys. Fluids **26**, 071,901 (2014)

31. Deng, Q., Driscoll, T.A., Braun, R.J., King-Smith, P.E.: A model for the tear film and ocular surface temperature for partial blinks. Interfacial Phen. Ht. Trans. **1**, 357–381 (2013)

32. Doane, M.G.: Interaction of eyelids and tears in corneal wetting and the dynamics of the normal human eyeblink. Am. J. Ophthalmol. **89**, 507–516 (1980)

33. Driscoll, T.A., Braun, R.J., Brosch, J.K.: Simulation of parabolic flow on an eye-shaped domain with moving boundary. J. Eng. Math. **111**, 111–126 (2018)

34. Efron, N., Young, G., Brennan, N.A.: Ocular surface temperature. Curr. Eye. Res. **8**, 901–906 (1989)

35. Gaffney, E.A., Tiffany, J.M., Yokoi, N., Bron, A.J.: A mass and solute balance model for tear volume and osmolarity in the normal and the dry eye. Prog. Retinal Eye Res. **29**, 59–78 (2010)

36. Gilbard, J.P., Farris, R.L., Santamaria, J.: Osmolarity of tear microvolumes in keratoconjunctivitis sicca. Arch. Ophthalmol. **96**, 677–681 (1978)

37. Goto, E., Tseng, S.C.G.: Kinetic analysis of tear interference images in aqueous tear deficiency dry eye before and after punctal occlusion. Invest. Ophthalmol. Vis. Sci. **44**, 1897–1905 (2003)

38. Greer, J.B., Bertozzi, A.L., Sapiro, G.: Fourth order partial differential equations on general geometries. J. Comput. Phys. **216**(1), 216–246 (2006)

39. Harrison, W.W., Begley, C.G., Liu, H., Chen, M., Garcia, M., Smith, J.A.: Menisci and fullness of the blink in dry eye. Optom. Vis. Sci. **85**, 706–714 (2008)

40. Henshaw, W.D.: Ogen: the overture overlapping grid generator. Tech. Rep. UCRL-MA-132237, Lawrence Livermore National Laboratory (2002)

41. Heryudono, A., Braun, R.J., Driscoll, T.A., Cook, L.P., Maki, K.L., King-Smith, P.E.: Single-equation models for the tear film in a blink cycle: Realistic lid motion. Math. Med. Biol. **24**, 347–377 (2007)

42. Himebaugh, N., Wright, A.R., Bradley, A., Begley, C.G., Thibos, L.N.: Use of retroillumination to visualize optical aberrations caused by tear film break-up. Optom. Vis. Sci. **80**, 69–78 (2003)
43. Hirata, H., Mizerska, K., Dallacasagrande, V., Rosenblatt, M.I.: Estimating the osmolarities of tears during evaporation through the "eyes of the corneal nerves. Invest. Ophthalmol. Vis. Sci. **58**, 168–178 (2017)
44. Holly, F.: Formation and rupture of the tear film. Exp. Eye Res. **15**, 515–525 (1973)
45. Jensen, O.E., Grotberg, J.B.: The spreading of heat or soluble surfactant along a thin liquid film. Phys. Fluids A **75**, 58–68 (1993)
46. Jones, M.B., McElwain, D.L.S., Fulford, G.R., Collins, M.J., Roberts, A.P.: The effect of the lipid layer on tear film behavior. Bull. Math. Biol. **68**, 1355–81 (2006)
47. Jones, M.B., Please, C.P., McElwain, D.L.S., Fulford, G.R., Roberts, A.P., Collins, M.J.: Dynamics of tear film deposition and draining. Math. Med. Biol. **22**, 265–88 (2005)
48. Jossic, L., Lefevre, P., de Loubens, C., Magnin, A., Corre, C.: The fluid mechanics of shear-thinning tear substitutes. J. Non-Newtonian Fluid Mech. **161**, 1–9 (2009)
49. Kapoor, Y., Thomas, J.C., Tan, G., John, V.T., Chauhan, A.: Surfactant-laden soft contact lenses for extended delivery of ophthalmic drugs. Invest. Ophthalmol. Vis. Sci. **30**, 867–78 (2009)
50. King-Smith, P., Fink, B., Fogt, N., Nichols, K.K., Hill, R., Wilson, G.S.: The thickness of the human precorneal tear film: Evidence from reflection spectra. Invest. Ophthalmol. Vis. Sci. **41**, 3348–3359 (2000)
51. King-Smith, P.E., Begley, C.G., Braun, R.J.: Mechanisms, imaging and structure of tear film breakup. Ocul. Surf. **16**, 4–30 (2018)
52. King-Smith, P.E., Fink, B.A., Hill, R.M., Koelling, K.W., Tiffany, J.M.: The thickness of the tear film. Curr. Eye. Res. **29**, 357–368 (2004)
53. King-Smith, P.E., Fink, B.A., Nichols, J.J., Nichols, K.K., Braun, R.J., McFadden, G.B.: The contribution of lipid layer movement to tear film thinning and breakup. Invest. Ophthalmol. Visual Sci. **50**, 2747–2756 (2009)
54. King-Smith, P.E., Hinel, E.A., Nichols, J.J.: Application of a novel interferometric method to investigate the relation between lipid layer thickness and tear film thinning. Invest. Ophthalmol. Vis. Sci. **51**, 2418–2423 (2010)
55. King-Smith, P.E., Ramamoorthy, P., Braun, R.J., Nichols, J.J.: Tear film images and breakup analyzed using fluorescent quenching. Invest. Ophthalmol. Vis. Sci. **54**, 6003—6011 (2013)
56. King-Smith, P.E., Reuter, K.S., Braun, R.J., Nichols, J.J., Nichols, K.K.: Tear film breakup and structure studied by simultaneous video recording of fluorescence and tear film lipid layer, TFLL, images. Invest. Ophthalmol. Vis. Sci. **54**, 4900–4909 (2013)
57. Leiske, D.L., Leiske, C.I., Leiske, D.R., Toney, M.F., Senchyna, M., Ketelson, H.A., Meadows, D.L., Fuller, G.G.: Temperature-induced transitions in the structure and interfacial rheology of human meibum. Biophys. J. **102**, 369–376 (2011)
58. Leiske, D.L., Miller, C.E., Rosenfeld, L., Cerretani, C., Ayzner, A., Lin, B., Meron, M., Senchyna, M., Ketelson, H.A., Meadows, D., Srinivasan, S., Jones, L., Radke, C.J., Toney, M.F., Fuller, G.G.: Molecular structure of interfacial human meibum films. Langmuir **28**, 11,858—11,865 (2012)
59. Leiske, D.L., Raju, S.R., Ketelson, H.A., Millar, T.J., Fuller, G.G.: The interfacial viscoelastic properties and structures of human and animal Meibomian lipids. Exp. Eye Res. **90**, 598–604 (2010)
60. Lemp, M.A.: The definition and classification of dry eye disease: report of the definition and classification subcommittee of the international dry eye workshop. Ocul. Surf. **5**, 75–92 (2007)
61. Lemp, M.A., Baudouin, C., Baum, J., Dogru, M., Foulks, G.N., Kinoshita, S., Laibson, P., McCulley, J., Murube, J., Pflugfelder, S.C., et al.: The definition and classification of dry eye disease: report of the definition and classification subcommittee of the international dry eye workshop (2007). Ocul. Surf. **5**(2), 75–92 (2007)

62. Lemp, M.A., Bron, A.J., Baudoin, C., Benitez Del Castillo, J.M., Geffen, D., Tauber, J., Foulks, G.N., Pepose, J.S., Sullivan, B.D.: Tear osmolarity in the diagnosis and management of dry eye disease. Am. J. Ophthalmol. **151**, 792–798 (2011)

63. Leung, B., Bonanno, J., Radke, C.: Oxygen-deficient metabolism and corneal edema. Prog. Ret. Eye Res. **30**, 471–492 (2011)

64. Levin, M.H., Kim, J.K., Hu, J., Verkman, A.S.: Potential difference measurements of ocular surface na$^+$ absorption analyzed using an electrokinetic model. Invest. Ophthalmol. Vis. Sci. **47**, 312–316 (2006)

65. Levin, M.H., Verkman, A.S.: Aquaporin-dependent water permeation at the mouse ocular surface: In vivo microfluorometric measurements in cornea and conjunctiva. Invest. Ophthalmol. Vis. Sci. **45**, 4423–4432 (2004)

66. Li, C.C., Chauhan, A.: Modeling ophthalmic drug delivery by soaked contact lenses. Ind. Eng. Chem. Res. **45**, 3718–3734 (2006)

67. Li, D.Q., Luo, L., Chen, Z., Kim, H.S., Song, X.J., Pflugfelder, S.C.: JNK and ERK MAP kinases mediate induction of IL-1$\beta$, TNF-$\alpha$ and IL-8 following hyperosmolar stress in human limbal epithelial cells. Exp. Eye Res. **82**(4), 588–596 (2006)

68. Li, L., Braun, R.J., Henshaw, W.D., King-Smith, P.E.: Computed tear film and osmolarity dynamics on an eye-shaped domain. Math. Med. Biol. **33**, 123–157 (2016)

69. Li, L., Braun, R.J., Henshaw, W.D., King-Smith, P.E.: Computed flow and fluorescence over the ocular surface. Math. Med. Biol. **35**, 51–85 (2018)

70. Li, L., Braun, R.J., Maki, K.L., Henshaw, W.D., King-Smith, P.E.: Tear film dynamics with evaporation, wetting and time-dependent flux boundary condition on an eye-shaped domain. Phys. Fluids **26**, 052,101 (2014)

71. Liu, H., Begley, C.G., Chalmers, R., Wilson, G., Srinivas, S.P., Wilkinson, J.A.: Temporal progression and spatial repeatability of tear breakup. Optom. Vis. Sci. **83**, 723–730 (2006)

72. Liu, H., Begley, C.G., Chen, M., Bradley, A., Bonanno, J., McNamara, N.A., Nelson, J.D., Simpson, T.: A link between tear instability and hyperosmolarity in dry eye. Invest. Ophthalmol. Vis. Sci. **50**, 3671–3679 (2009)

73. Maki, K.L., Braun, R.J., Driscoll, T.A., King-Smith, P.E.: An overset grid method for the study of reflex tearing. Math. Med. Biol. **25**, 187–214 (2008)

74. Maki, K.L., Braun, R.J., Henshaw, W.D., King-Smith, P.E.: Tear film dynamics on an eye-shaped domain I. pressure boundary conditions. Math. Med. Biol. **27**, 227–254 (2010a)

75. Maki, K.L., Braun, R.J., Ucciferro, P., Henshaw, W.D., King-Smith, P.E.: Tear film dynamics on an eye shaped domain. Part 2. Flux boundary conditions. J. Fluid Mech. **647**, 361–390 (2010b)

76. Maki, K.L., Ross, D.S.: A new model for the suction pressure under a contact lens. J. Biol. Sys. **22**, 235–248 (2014)

77. Matar, O.K., Craster, R.V., Warner, M.R.E.: Surfactant transport on highly viscous surface films. J. Fluid Mech. **466**, 85–111 (2002)

78. Maurice, D.M.: The dynamics and drainage of tears. Int. Ophthalmol. Clin. **13**, 103–116 (1973)

79. Miller, K.L., Polse, K.A., Radke, C.J.: Black line formation and the "perched" human tear film. Curr. Eye Res. **25**, 155–62 (2002)

80. Moosman, S., Homsy, G.: Evaporating menisci of wetting fluids. J. Colloid Interface Sci. **73**, 212–223 (1980)

81. Morris, S.J.S.: Contact angles for evaporating liquids predicted and compared with existing experiments. J. Fluid Mech. **432**, 1–30 (2001)

82. Mota, M.C., Carvalho, P., Ramalho, J., Leite, E.: Spectrophotometric analysis of sodium fluorescein aqueous solutions. determination of molar absorption coefficient. Int. Ophthalmol. **15**, 321–326 (1991)

83. Nagyová, B., Tiffany, J.M.: Components of tears responsible for surface tension. Curr. Eye Res. **19**, 4–11 (1999)

84. Naire, S., Braun, R.J., Snow, S.A.: Limiting cases of gravitational drainage of a vertical free film for evaluating surfactants. SIAM J. Appl. Math. **61**, 889–913 (2000)

85. Nichols, J.J., King-Smith, P.E., Hinel, E.A., Thangavelu, M., Nichols, K.K.: The use of fluorescent quenching in studying the contribution of evaporation to tear thinning. Invest. Ophthalmol. Visual Sci. **53**, 5426–5432 (2012)
86. Nichols, J.J., Mitchell, G.L., King-Smith, P.E.: Thinning rate of the precorneal and prelens tear films. Invest. Ophthalmol. Vis. Sci. **46**, 2353–2361 (2005)
87. Nong, K., Anderson, D.M.: Thin tilm evolution over a thin porous layer: Modeling a tear film over a contact lens. SIAM J. Appl. Math. **70**, 2771–95 (2010)
88. Owens, H., Phillips, J.: Spreading of the tears after a blink: Velocity and stabilization time in healthy eyes. Cornea **20**, 484–487 (2001)
89. Pandit, J.C., Nagyová, B., Bron, A.J., Tiffany, J.M.: Physical properties of stimulated and unstimulated tears. Exp. Eye Res. **68**, 247–53 (1999)
90. Peng, C.C., Cerretani, C., Braun, R.J., Radke, C.J.: Evaporation-driven instability of the precorneal tear film. Adv. Coll. Interface Sci. **206**, 250–264 (2014)
91. Potash, M., Wayner, P.: Evaporation from a two-dimensional extended meniscus. Intl J. Heat Mass Transfer **15**, 1851–1863 (1972)
92. Purslow, C., Wolffsohn, J.S.: Ocular surface temperature: A review. Eye Contact Lens **31**, 117–123 (2005)
93. Riquelme, R., Lira, I., Pérez-López, C., Rayas, J.A., Rodríguez-Vera, R.: Interferometric measurement of a diffusion coefficient: comparison of two methods and uncertainty analysis. J. Phys. D. Appl. Phys. **40**, 2769–2776 (2007)
94. Rosenfeld, L., Cerretani, C., Leiske, D.L., Toney, M.F., Radke, C.J., Fuller, G.G.: Structural and rheological properties of meibomian lipid. Invest. Ophthalmol. Vis. Sci. **54**, 2720–2732 (2013)
95. Rosenfeld, L., Fuller, G.G.: Consequences of interfacial viscoelasticity on thin film stability. Langmuir **28**(40), 14,238–14,244 (2012)
96. Ross, D.S., Maki, K.L., Holz, E.K.: Existence theory for the radially symmetric contact lens equation. SIAM J. Appl. Math. **76**, 827–844 (2016)
97. Sakata, E., Berg, J.: Surface diffusion in monolayers. Ind. Eng. Chem. Fundam. **8**(3), 570–573 (1969)
98. Sharma, A.: Surface properties of normal and damaged corneal epithelia. J. Dispersion Sci. Technol. **13**, 459–478 (1992)
99. Sharma, A.: Acid-base interactions in the cornea-tear film system: surface chemistry of corneal wetting, cleaning, lubrication, hydration and defense. J. Dispersion Sci. Technol. **19**(6-7), 1031–1068 (1998)
100. Sharma, A.: Surface-chemical pathways of the tear film breakup. In: D.A. Sullivan, D.A. Dartt, M.A. Meneray (eds.) Lacrimal Gland, Tear Film, and Dry Eye Syndromes 2, *Lacrimal Gland, Tear Film, and Dry Eye Syndromes 2*, vol. 438, pp. 361–370. Springer, Berlin (1998)
101. Sharma, A., Ruckenstein, E.: Mechanism of tear film rupture and formation of dry spots on cornea. J. Colloid Interface Sci. **106**, 12–27 (1985)
102. Sharma, A., Ruckenstein, E.: An analytical nonlinear theory of thin film rupture and its application to wetting films. J. Coll. Interface Sci. **113**, 8–34 (1986)
103. Sharma, A., Ruckenstein, E.: The role of lipid abnormalities, aqueous and mucus deficiencies in the tear film breakup, and implications for tear substitutes and contact lens tolerance. J. Colloid Interface Sci. **111**, 456–479 (1986)
104. Sharma, A., Tiwari, S., Khanna, R., Tiffany, J.: Hydrodynamics of meniscus-induced thinning of the tear film. In: D.A. Sullivan, D.A. Dartt, M.A. Meneray (eds.) Lacrimal Gland, Tear Film, and Dry Eye Syndromes 2, pp. 425–31. New York: Plenum Press (1998)
105. Siddique, J.I., Braun, R.J.: Tear film dynamics with evaporation, osmolarity and surfactant transport. Appl. Math. Model. **39**, 255–269 (2015)
106. Stapf, M.R., Braun, R.J., King-Smith, P.E.: Duplex tear film evaporation analysis. Bull. Math. Biol. **79**(12), 2814–2846 (2017)
107. Stebe, K.J., Maldarelli, C.: Remobilizing surfactant retarded fluid particle interfaces: Ii. controlling the surface mobility at interface of solutions containing surface active components. J. Colloid Interface Sci. **163**, 177–189 (1994)

108. Stone, H.A.: A simple derivation of the time-dependent convective-diffusion equation for surfactant transport along a deforming interface. Phys. Fluids A **2**, 111–112 (1990)
109. Tiffany, J.M.: The viscosity of human tears. Int. Ophthalmol. **15**, 371–376 (1991)
110. Usha, R., Anjalaiah, Sanyasiraju, Y.V.S.S.: Dynamics of a pre-lens tear film after a blink: Model, evolution, and rupture. Phys. Fluids **25**, 112,111 (2013)
111. Webber, W.R.S., Jones, D.P.: Continuous fluorophotometric method measuring tear turnover rate in humans and analysis of factors affecting accuracy. Med. Biol. Eng. Comput. **24**, 386–392 (1986)
112. Winter, K.N., Anderson, D.M., Braun, R.J.: A model for wetting and evaporation of a post-blink precorneal tear film. Math. Med. Biol. **27**, 211–225 (2010)
113. Wong, H., Fatt, I., Radke, C.: Deposition and thinning of the human tear film. J. Colloid Interface Sci. **184**(1), 44–51 (1996)
114. Yanez-Soto, B., Mannis, M.J., Schwab, I., Li, J.Y., Leonard, B.C., Abbott, N.L., Murphy, C.J.: Interfacial phenomena and the ocular surface. Ocul. Surf. **12**, 178–201 (2014)
115. Yokoi, N., Georgiev, G.A.: Tear dynamics and dry eye disease. Ocul. Surf. Dis. London:JP Medical Ltd p. 47 (2013)
116. Yokoi, N., Georgiev, G.A.: Tear-film-oriented diagnosis and therapy for dry eye. Dry eye syndrome: basic and clinical perspectives. London: Future Medicine Ltd pp. 96–108 (2013)
117. Yokoi, N., Georgiev, G.A., Kato, H., Komuro, A., Sonomura, Y., Sotozono, C., Tsubota, K., Kinoshita, S.: Classification of fluorescein breakup patterns: A novel method of differential diagnosis for dry eye. Am. J. Ophthalmol. **180**, 72–85 (2017)
118. Yokoi, N., Takehisa, Y., Kinoshita, S.: Correlation of tear lipid layer interference patterns with the diagnosis and severity of dry eye. Am. J. Ophthalmol. **122**, 818–824 (1996)
119. Zhang, L., Matar, O.K., Craster, R.V.: Analysis of tear film rupture: Effect of non-Newtonian rheology. J. Coll. Interface Sci. **262**, 130–48 (2003)
120. Zhang, L., Matar, O.K., Craster, R.V.: Rupture analysis of the corneal mucus layer of the tear film. Molec. Sim. **30**, 167–72 (2004)
121. Zhong, L., Braun, R.J., Begley, C.G., King-Smith, P.E.: Dynamics of fluorescent imaging for rapid tear thinning. Bull. Math. Biol. **81**, 39–80 (2019)
122. Zhong, L., Braun, R.J., King-Smith, P.E., Begley, C.G.: Mathematical modeling of glob-driven tear film breakup. J. Modeling Ophthalmol. **2**(1), 24–28 (2018)
123. Zhong, L., Ketelaar, C.F., Braun, R.J., Begley, C.G., King-Smith, P.E.: Mathematical modeling of glob-driven tear film breakup. Math. Med. Biol. **36**, 55–91 (2019)
124. Zubkov, V.S., Breward, C.J., Gaffney, E.A.: Coupling fluid and solute dynamics within the ocular surface tear film: a modelling study of black line osmolarity. Bull. Math. Biol. **74**, 2062–2093 (2012)
125. Zubkov, V.S., Breward, C.J.W., Gaffney, E.A.: Meniscal tear film fluid dynamics near marx's line. Bull. Math. Biol. **75**(9), 1524–1543 (2013)

# Part VI
# Cerebrospinal Fluids

# Anatomy and Physiology of the Cerebrospinal Fluid

**David Fleischman and John Berdahl**

**Abstract** The cerebrospinal fluid (CSF) is the primary circulating fluid of the central nervous system. It serves numerous important physiologic and maintenance functions, and its production and movement are highly regulated. Herein, we describe the key anatomic structures of importance in regard to CSF production, circulation, and absorption, followed by the regulatory mechanisms responsible for its proper functioning.

## 1 Introduction

The brain and all components of the central nervous system are bathed in cerebrospinal fluid (CSF). The CSF is a clear, circulating fluid produced by the choroid plexuses. The fluid is responsible for providing mechanical support and homeostatic regulation of the brain's parenchymal interstitial fluid.

The following provides background on the anatomical structures that contain, produce, and drain the CSF. The anatomical description will begin with the site of secretion of the fluid, beginning within the lateral ventricles and concluding its journey to the arachnoid granulations, and the perivascular spaces along the spinal dorsal root ganglia and the cranial nerves (Fig. 1).

D. Fleischman (✉)
Department of Ophthalmology, University of North Carolina at Chapel Hill, Chapel Hill, NC, USA

J. Berdahl
Vance Thompson Vision, Sioux Falls, SD, USA

© Springer Nature Switzerland AG 2019　　　　　　　　　　　　　　　　　　　435
G. Guidoboni et al. (eds.), *Ocular Fluid Dynamics*, Modeling and Simulation in Science, Engineering and Technology, https://doi.org/10.1007/978-3-030-25886-3_18

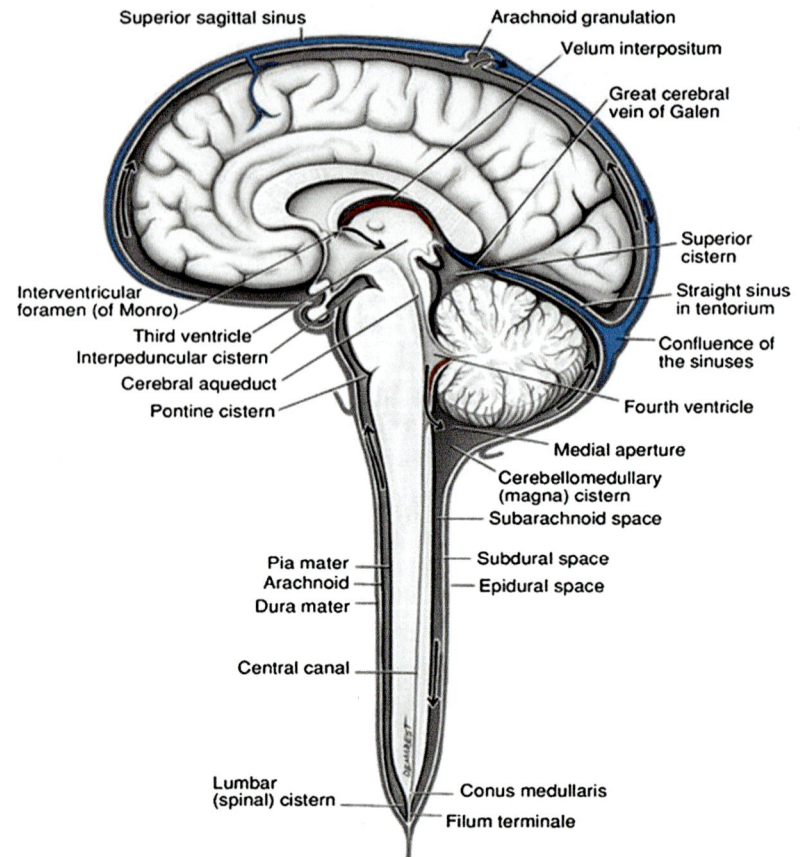

Superior sagittal sinus

Arachnoid granulation

Velum interpositum

Great cerebral vein of Galen

Superior cistern

Straight sinus in tentorium

Confluence of the sinuses

Fourth ventricle

Medial aperture

Cerebellomedullary (magna) cistern

Subarachnoid space

Subdural space

Epidural space

Conus medullaris

Filum terminale

Interventricular foramen (of Monro)

Third ventricle

Interpeduncular cistern

Cerebral aqueduct

Pontine cistern

Pia mater

Arachnoid

Dura mater

Central canal

Lumbar (spinal) cistern

**Fig. 1** CSF-containing spaces within the central nervous system and its circulation. Arrows indicate flow of cerebrospinal fluid from the lateral ventricles entry into the third ventricle via the interventricular foramen of Monro. Arrows indicate the direction of flow to its ultimate drainage into the superior sagittal sinus (figure reprinted from Chap. 5 of Noback's Human Nervous System, Seventh Edition, Structure and Function, Norman L. Strominger, Robert J. Demarest, Lois B. Laemle, Humana Press, Springer Science+Business Media New York 2012)

## 2 Anatomy

### 2.1 Choroid Plexus

The pia lines both the external and internal structures of the brain. The choroid plexus is a highly specialized tissue that is composed of a vascularized modification of the pia and the ependyma. The choroid plexus is a fold of this epithelium with a vascular core tissue. They are located in the roof of the third and fourth ventricles, and the walls of the lateral ventricles. The surfaces of these secretory structures have unique, frond-like, villous processes in the apical membranes. This likely has the function of maximizing surface area for the secretion and post-secretory

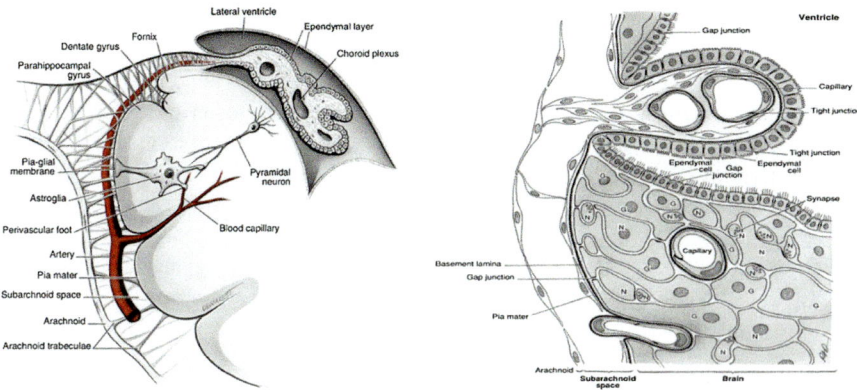

**Fig. 2** Left panel: relations of the pia, arachnoid, subarachnoid, choroid plexus, ventricle, astroglia, and neurons of the CNS. Right panel: ultrastructural features providing the relation of the brain, choroid plexus, leptomeninges, and ventricle. Waste products, ions, and amino acids are transferred from the CSF space within the choroid plexus into the robust capillary system. Ions ($Na^+$, $K^+$, $Cl^-$, $HCO_{3-}$, $Ca^{2+}$, $Mg^{2+}$), vitamins, organic nutrients, and oxygen are released along with CSF from these same capillaries. This system is responsible for its formation and post-secretory regulation (figures reprinted from Chap. 5 of Noback's Human Nervous System, Seventh Edition, Structure and Function, Norman L. Strominger, Robert J. Demarest, Lois B. Laemle, Humana Press, Springer Science+Business Media New York 2012)

modification of the fluid. The vascular supply of the choroid plexus is provided by the anterior and posterior set of choroidal arteries. The plexus is drained by the choroidal vein (Fig. 2).

The material illustrated in the present section is based on references [1–4].

## 2.2 *Ventricles*

The cerebral ventricles are a network of interconnected cavities that contain the fluid-producing tissue of the central nervous system, the choroid plexus. They are lined by ependymal cells. The ventricular system could be thought of as beginning with the lateral ventricles, the largest of the cavities. There is a right and left ventricle, and they are characteristically C-shaped structures. The ventral surface is bordered by the basal ganglia, the dorsal surface by the corpus callosum, and the medial surface by the septum pellucidum. The inferior horn of the lateral ventricle is located in the temporal lobe. The central part, or body, of the lateral ventricle is within the parietal and frontal lobes. The posterior horn extends toward the occipital lobe, while the anterior horn extends into the depths of the frontal lobe. The anterior horn of each lobe of the ventricular system terminates at the interventricular foramina, or the foramen of Monro, which itself empties into the third ventricle.

The third ventricle is a narrow, midline space between the right and left thalamus. The third ventricle then continues caudally with the cerebral aqueduct within the midbrain. The cerebral aqueduct, the aqueduct of Sylvius, is a thin conduit of CSF from the third ventricle to the fourth ventricle.

The aqueduct of Sylvius is clinically an important structure due to its narrow dimensions and tendency to obstruct with compressive mass lesions. In a 3-T MR study, the mean dimensions of the aqueduct of Sylvius were found to be as follows: pars anterior width, 1.1 mm; ampulla width, 1.2 mm; pars posterior width, 1.4 mm, length, 14.1 mm. The narrowest point is 0.9 mm, and the angulation in relation to the third ventricle was 26° and with the fourth ventricle, 18°. Older age tended to be associated with increased width and decreased length of the cerebral aqueduct. These measurements compared favorably to previously published cadaveric studies.

The fourth ventricle is a rhombus-shaped cavity located in the area of the pons, medulla, and cerebellum. CSF flowing through the fourth ventricle will continue either through the foramen of Magendie, the lateral foramina of Luschka, or through the central canal of the spinal cord. The foramina of Magendie establishes a communication with the cerebellomedullary cistern, or the cisterna magna. The foramina of Luschka each open into the prepontine cistern.

The overall volume of the ventricles, as determined by casts in a classic work by Last and Tompsett, was found to be 7.4–56.6 ml (mean 24 ml). Therefore, of the approximately 140 ml of total CSF volume in humans, about 18–20% of the volume is located within the ventricles.

The material illustrated in the present section is based on references [1–3, 5–7].

## 2.3   Dural Sinuses

Originally thought to be the only site of CSF drainage, the sinuses are large openings of the subarachnoid space, which direct into the venous system. We do know now that this is not the only site of CSF drainage, and in fact, in some species the dural sinuses are a minor outflow system [8, 9]. The dural sinuses are lined by endothelium; however, they lack the other typical vascular layers as seen in arteries and veins. The sinuses themselves are embedded within the dura mater. These channels receive blood from the internal and external veins of the brain. CSF drains into the venous sinuses from the subarachnoid space through arachnoid granulations. The dural sinuses ultimately drain into the internal jugular vein.

The material illustrated in the present section is based on [1–3, 8, 9].

## 2.4   Arachnoid Granulations (or Arachnoid Villi)

Evaginations of the arachnoid occur within the dural sinuses. The largest are termed arachnoid granulations. Smaller granulations are termed arachnoid villi. Pacchionian bodies are large arachnoid granulations that are calcified. These are known sites of CSF–blood exchange. While the connection is valvular by function in that CSF only enters into the blood and not vice versa, the actual mechanism likely involves the creation of vacuoles of CSF by the endothelial cells lining the arachnoid granulations or by pinocytosis. This will be discussed in further detail below.

The arachnoid granulations in humans have been identified in the superior sagittal sinus, transverse sinus, cavernous sinus, superior petrosal sinus, and straight sinus (in decreasing frequency). The granulations increase in number with age.

Arachnoid villi are also noted along the spinal dorsal root ganglia. Their structures are similar to the arachnoid villi in the brain. As will be discussed below, drainage of CSF through these granulations is not insignificant. What is unknown is whether the drainage is into the venous system or the lymphatics, although studies suggest likely a predominant lymphatic outflow mechanism.

Many have examined the relationship of the aqueous humor and the CSF. While it is not accepted that intraocular pressure is a surrogate for cerebrospinal fluid pressure (CSFP), there is a common drainage point between the two nervous system fluids. The aqueous humor exits the anterior chamber by bulk flow through the trabecular meshwork into Schlemm's canal (interestingly, in a manner similar to CSF leaving the arachnoid granulations). From here, the fluid enters collector channels, aqueous veins, and episcleral veins. The episcleral veins drain through the superior and inferior ophthalmic veins, into the cavernous sinus. From the cavernous sinus, the blood drains into the superior and inferior petrosal sinuses. There are arachnoid granulations present in the cavernous sinus and the superior petrosal sinus as described above, which would be the first point where the aqueous humor and CSF, now contained with blood, meet.

The material illustrated in the present section is based on references [1–3, 10–13] (Fig. 3).

## 2.5   Barriers

The blood–CSF barrier occurs at the choroid plexus. This is the site of nearest communication of blood and the CSF. The presence of these barriers has been known since the studies of Goldmann, in which he infused trypan blue into the bloodstream to investigate the nature of these separations. The dyes only penetrated the choroid plexuses and the dura, indicating the nature and locales of the barriers between CSF and blood.

The other location of blood–CSF contact is within the arachnoid granulations, where CSF drains into the venous sinus. It is believed that the bulk flow of fluid is through a transcellular basal-to-apical vacuolization, maintaining a one-way valve. Pinocytosis is a predominant mechanism in many non-primate species. It is of particular interest to the ophthalmologist that there is a similar mechanism of fluid transport between the trabecular meshwork and the arachnoid granulations.

The relationship between CSF and brain tissue is far more intimate. Goldmann performed a similar trypan blue infusion into the ventricular CSF, and found that the dye passed readily into the brain tissue. This implies an absence of a strict barrier method between CSF and brain tissue. This is understood, as the CSF and the extracellular space of the brain readily exchange water and ions through gradients.

The material illustrated in the present section is based on references [3, 14, 15].

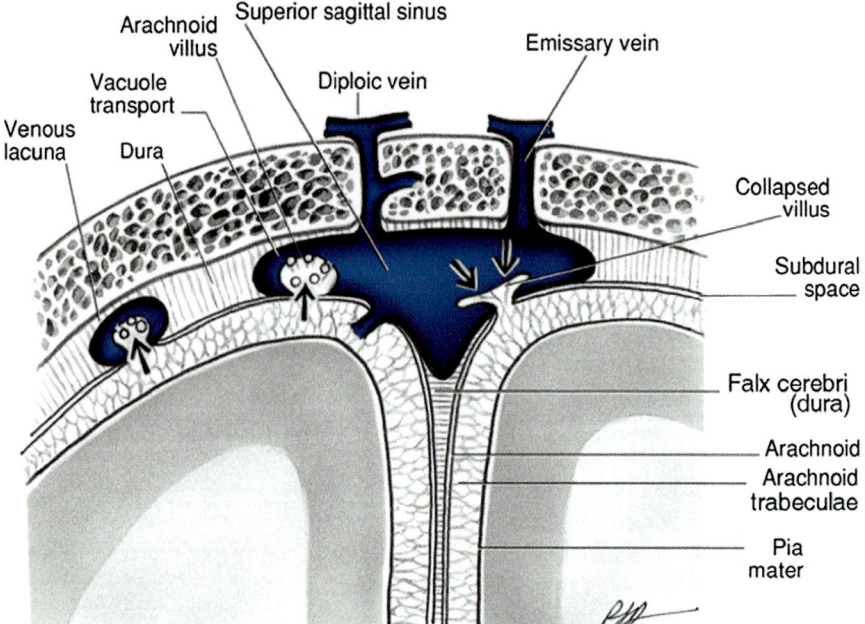

**Fig. 3** Relationship of the arachnoid villi and the dural sinus spaces (figure reprinted from Chap. 5 of Noback's Human Nervous System, Seventh Edition, Structure and Function, Norman L. Strominger, Robert J. Demarest, Lois B. Laemle, Humana Press, Springer Science+Business Media New York 2012)

## 3   Composition and Physical Characteristics of the Cerebrospinal Fluid

The CSF is a clear, watery fluid. The mean density of CSF in adult humans is 1.00059 ($\pm$SD 0.00020), but there are statistically significant differences in the fluid densities among men, premenopausal non-pregnant women, postmenopausal women, and pregnant women, with pregnant women having the least dense (1.00033) and men having the most dense (1.00067). CSF in general is a Newtonian fluid with a viscosity similar to water. Even with increased protein or CSF cellularity, the viscosity of CSF is relatively unaffected, with a range of 0.7–1 mPa.s at 37 °C. The fluid has an osmolarity of 281 mOsm/L at 37 °C. Its specific gravity is 1.006–1.008, and its pH ranges from 7.28 to 7.32.

The CSF is made up predominantly of Na, Cl, and $HCO_3$. $K^+$, $Mg^{++}$, $Ca^{++}$, small RNA, peptides, and micronutrients are found in smaller quantities. It is useful to speak of the composition of the CSF in relation to the plasma. There is nearly 250 times more protein in the plasma compared to the CSF (7 g/100 ml versus

**Table 1** Components of human aqueous humor, vitreous humor, and two intraocular irrigation solutions

| Ingredient | Human aqueous humor | Human vitreous humor | Hartman's lactated ringer's solution | BSS PLUS® intraocular irrigating solution | BSS® intraocular irrigating solution |
|---|---|---|---|---|---|
| Sodium | 162.9 | 144 | 102 | 160.0 | 155.7 |
| Potassium | 2.2–3.9 | 5.5 | 4 | 50 | 10.1 |
| Calcium | 1.8 | 1.6 | 3 | 1.0 | 3.3 |
| Magnesium | 1.1 | 1.3 | – | 1.0 | 1.5 |
| Chloride | 131.6 | 177.0 | – | 130.0 | 128.9 |
| Bicarbonate | 20.15 | 15.0 | – | 25.0 | – |
| Phosphate | 0.62 | 0.4 | – | 3.0 | |
| Lactate | 2.5 | 7.8 | 28 | – | – |
| Glucose | 2.7-3.7 | 3.4 | – | 5.0 | – |
| Ascorbate | 1.06 | 2.0 | – | – | – |
| Glutathione | 0.0019 | – | – | 0.3 | – |
| Citrate | – | – | – | – | 5.8 |
| Acetate | – | – | – | – | 28.6 |
| pH | 7.38 | – | 6.0–7.2 | 7.4 | 7.6 |
| Osmolality (mOsm) | 304 | – | 277 | 305 | 298 |

0.025 g/100 ml, respectively). Most of the protein in the CSF is albumin. Table 1 highlights many of the constituents of the CSF.

The material illustrated in the present section is based on references [14, 16–20].

# 4 Physiology

## 4.1 Secretion of Cerebrospinal Fluid

While there is some controversy about extraventricular production of CSF, it is most accepted that the choroid plexus produces almost all CSF. CSF is a secretion of the choroid plexus, and it is not a simple filtrate or dialysate of the blood. Davson and Segal's monumental work on the CSF and blood–brain barrier systematically review the expected construction of CSF based on a simple membrane with osmotic, ionic, and concentration gradients, versus what is actually appreciated within the CSF. The makeup of CSF is highly regulated, and even with significant changes in plasma concentrations, the CSF makeup tends to be stable. Secretion in adults ranges from 400 to 600 ml per day. These rates are different in a variety of species (Table 1). There is a significant diurnal variation in the production and flow of CSF. There are changes in rate as much as 400%, which one must wonder if a change of this magnitude is an artifact and accentuated from testing methods (Table 2).

**Table 2** Plasma and CSF concentrations and concentration characteristics of common electrolytes and compounds

| Concentration of various substances in human CSF and plasma | | |
|---|---|---|
| Substances | CSF | Plasma |
| • $Na^+$ (meq/kg $H_2O$) | 147.0 | 150.0 |
| • $K^+$ (meq/kg $H_2O$) | 2.9 | 4.6 |
| • $Ca^{2+}$ (meq/kg $H_2O$) | 2.3 | 4.7 |
| • $Cl^-$ (meq/kg $H_2O$) | 113.0 | 99.0 |
| • $HCO_3{-}$ (meq/l) | 25.1 | 24.8 |
| • $PCO_2$ (mm Hg) | 50.2 | 39.5 |
| • pH | 7.33 | 7.40 |
| • Osmolality (mosm/kg $H_2O$) | 289.0 | 289.0 |
| • Protein (mg/dl) | 20.0 | 6000.0 |
| • Glucose (mg/dl) | 64.0 | 100.0 |
| • Cholesterol (mg/dl) | 0.2 | 175.0 |

| Species | $\mu$l/min | Rate, %/min | $\mu$l/min/mg CP |
|---|---|---|---|
| Mouse | 0.325 | 0.89 | – |
| Rat | 2.1 | 0.72 | – |
| | 2.1 | 0.75 | – |
| | 3.0–5.4 | 1.02–1.89 | – |
| | 2.8 | – | – |
| Guinea pig | 3.5 | | 0.875 |
| Rabbit | 10.0 | 0.43 | 0.43 |
| Cat | 20.0 | 0.45 | 0.50 |
| | 22 | 0.50 | 0.55 |
| | 22 | 0.50 | 0.55 |
| Dog | 50.0 | 0.40 | 0.625 |
| | 66 | – | 0.77 |
| | 47 | – | – |
| Goat | 164.0 | 0.65 | 0.36 |
| Sheep | 118.0 | 0.83 | – |
| Monkey | 28.6 | – | – |
| | 41 | – | – |
| Man | 350.0 | 0.38 | – |
| | 370.0 | – | – |
| Chicken | 1.4 | – | – |
| Frog | | | |
| (*R. pipiens*) | 0.3 | – | 0.176 |
| (*R. caiesbiana*) | | 1.7 | – |
| Dogfish | 4.0 | 0.1 | 0.05 |
| Dogfish | 2.0 | 0.15 | 0.025 |
| Nurse shark | 3.0 | 0.15 | – |
| Lemon shark | 4.0 | 0.20 | – |

There is passive filtration of plasma from the choroidal capillaries to the choroid interstitial space in a pressure-dependent mechanism. However, the actual production of CSF is predominantly an active mechanism. The production of CSF has been analogized as the renal tubular system and the production and modification of urine. From the perspective of an ophthalmologist, the production of CSF is not dissimilar to the production of the aqueous humor, with the exception that post-secretion modification does not tend to be as significant as that found in the CSF and urine.

There are a number of transporters that line the choroid plexus that actively shuttle ions, micronutrients, and other molecules against their concentration or electrochemical gradients. Isotope flux studies, electrophysiological, RT-PCR, in situ hybridization, and immunocytochemistry have been used to determine the expression of many of these transporters and channels in the choroid plexus. Many of these transporters have been localized to specific membranes. Choroid plexus carbonic anhydrase produces $H^+$ and $HCO_{3-}$ from water and $CO_2$ in the basolateral membrane. ATP-dependent ion pumps in the apical membrane expel $Na^+$, $Cl^-$, $HCO_{3-}$, and $K^+$ toward the ventricular lumen. Aquaporin I in the apical membrane allows water transport in a passive process reliant of the osmotic gradient generated by active pumps. It is worth noting that *water exchange* actually occurs throughout the entire neuraxis. The pia mater contains gap junctions through which water molecules are continuously displaced throughout the entirety of the CSF-containing spaces. It is this exchange mechanism that has created confusion about the extent and importance of extra-choroidal CSF production.

Vitamins B1, B12, and C, folate, B2-microglobulin, arginine vasopressin, and NO (and countless other molecules, RNA fragments, peptides, and more) are secreted by the plexus into the ventricles through known and as-of-yet uncharacterized transport systems. NaK2Cl cotransporters are located on the apical membrane and transports ions both into the CSF and into the extracellular space of the brain. Bi-directional transporters of this type are important in the secretion and regulation of the CSF composition.

The material illustrated in the present section is based on references [14, 15, 21–24].

## 4.2   Regulation of CSF Secretion

The CSF is remarkably constant in its composition, although it does change as it travels through the neuraxis. For example, micronutrients are found in increased concentration in the ventricles, with concentrations reducing downstream from the site of initial CSF secretion. Therefore, the CSF and extracellular space of the brain are in fluid and constant communication, with the brain extracting these nutrients on a predictable and as-needed basis, considering the relative stability of CSF concentration. We will now look at the factors that affect CSF production and secretion.

An increase in the intraventricular pressure will minimize the hydrostatic imbalance that results in plasma filtration. The choroid plexus is also innervated by cholinergic, adrenergic, serotoninergic, and peptidergic systems which will differentially stimulate production over a spatial cycle. It is believed that Circadian variations of CSF secretion that are present are due to the autonomic nervous system function. An interesting view of this relationship was investigated in male Sprague-Dawley rats by Samuels et al. Chemical stimulation of the dorsomedial and perifornical hypothalamus neurons significantly increased heart rate (69.3 $\pm$ 8.5 beats per min), mean arterial pressure (+22.9 $\pm$1.6 mmHg), intraocular pressure (7.1 mmHg $\pm$1.9 mmHg), and intracranial pressure (3.6 $\pm$0.7 mm Hg) compared to baseline values. However, the change in intraocular pressure was delayed compared to intracranial pressure, resulting in a theorized localized imbalance between intraocular and intracranial pressures across the lamina cribrosa.

Other mechanisms of self-regulation of CSF content are the intrinsic optimal pH range for proper functioning of carbonic anhydrase, aquaporins, and choroid plexus membrane carrier proteins. Dopamine, serotonin, melatonin, atrial natriuretic peptide (ANP), and arginine vasopressin (AVP) receptors are also present in the membrane of the CSF-secreting cells. ANP triggers aquaporin I function, and ANP and AVP decrease the secretion of CSF. AVP concentration within the ventricular fluid increases with older age, and perhaps this is one reason for our findings of reduction of CSF pressure with advancing age.

The material illustrated in the present section is based on references [14, 15, 23–27].

## 4.3   Cerebrospinal Fluid Circulation

CSF is renewed 4–5 times per day in young adults. There is an age-related reduction in CSF production, together with age-related ventricular dilatation, which should result in reduced CSF turnover (May et al. *Neurology* [22]). In female sheep, the turnover rate of CSF changes according to light–dark cycles. It is increased in short day periods, and reduced in long day periods. It is speculated that the difference is responsible for differences in hormonal concentrations in CSF in seasonal species.

The CSF's purpose is that of a nourishing fluid, a system for clearing wastes, and to provide mechanical protection for the brain and central nervous system. As described previously, the fluid is primarily generated by the choroid plexuses located within the ventricles, and therefore CSF migrates unidirectionally in a rostrocaudal direction in the ventricular cavities. The fluid produced in the lateral ventricles traverses through the interventricular foramina of Monro, to the third ventricle. From the third ventricle, the fluid traverses the aqueduct of Sylvius into the fourth ventricle. Finally, the fluid will go through either the lateral foramina of Luschka or the foramen of Magendie into the cisterna magna, which is the principle entrance into the subarachnoid space. Once in the subarachnoid space, however, the fluid flow is far more complex, with characteristic to-and-fro motion as a result of cardiac and respiratory activity. The pulsatility of the CSF is related to the cardiac cycle and respiration.

The subcommisural organ in rats and lower mammals produces a peptide responsible for the formation of Reissner fibers. These fibers serve the purpose of helping the CSF circulate through the cerebral aqueduct. Evolutionarily, this structure was found to be unnecessary, given that the organ disappears early in human development.

An important variable in CSF fluid circulation is a gravitational and positional component. An MRI study by Alperin et al. identified changes in venous outflow, total cerebral blood flow, and intracranial compliance based on changes in posture. In the upright position, venous outflow is less pulsatile and occurs mostly through the vertebral plexus. In the supine position, venous outflow occurs predominantly through the internal jugular veins. In the sitting position, there is lower total cerebral blood flow, much smaller CSF volume oscillating between the cranium and spinal canal, and a significantly greater intracranial compliance. Intracranial pressures in this study were derived by MRI, which is not wholly accepted as an accurate measure of CSFP. Regardless, these changes readily apparent on MRI positional studies underscore the importance of variability in the fluid measures depending on posture.

There is a fluid conductivity between the brain matter and the CSF throughout the neural axis. As such, the CSF also serves as a drainage mechanism for brain interstitial fluid with likely waste and inflammatory byproducts. This is modeled to occur, most actively, within perivascular spaces that permeate sensitive areas of the brain. Perivascular spaces, also known as Virchow-Robin spaces, are the fluid-filled spaces surrounding perforating vessels within the CNS. They are not directly connected to the subarachnoid space, and its fluid content is similar but different than the CSF. They are suspected of playing an immunologic role within the brain. There are three main types of Virchow-Robin spaces, categorized by their typical location:

Type 1: found in the basal ganglia—lenticulostriate arteries
Type 2: found in the cortical gray matter—perforating medullary arteries
Type 3: found in the midbrain

These spaces are theorized to serve as an entry point for CSF to enter into the brain parenchyma. With the help of the pulsatility created by arterial contraction and respiration, the fluid creates a complex extracellular fluid system. This interstitial fluid ultimately reaches perivenuolar spaces through which collected fluid and waste products are excreted, ultimately through the cervical lymphatic system. This process appears to be facilitated by supine and lateral positioning, and accentuated during sleep, where it is described that there is an expansion of the interstitial space during sleep as a clear indication of the process of waste clearance and the rejuvenating properties of sleep. It is to be noted, however, that this mechanism is still at an early stage and not all have accepted this theory to date.

The material illustrated in the present section is based on references [14, 22–24, 28–32].

## 4.4    Cerebrospinal Fluid Absorption

CSF is absorbed into the venous system and the lymphatic system, and the location of this drainage is found throughout the neural axis, from the intracranial compartment to through spinal roots. The ratios of absorption, both via the venous and lymphatic, and the intracranial and extracranial routes, vary dependent on species. Tracer studies have found approximately 30% of CSF produced drains into the deep cervical lymphatics in rabbits, while the value is closer to 10–15% in cats. In humans it is approximately 10%. In Sprague-Dawley rats, lymphatic CSF drainage in the olfactory region is likely a predominant CSF drainage mechanism. This is consistent with the fact that there is a relative paucity of arachnoid granulations in this species. Similarly, in sheep, external ethmoidectomies were performed to block CSF transport through the cribriform plate. This resulted in a nearly twofold increase in resting CSF pressure in the operated animals. In primates, surgical isolation of the olfactory region does not result in hydrocephalus, implying that the lymphatics may have a much lesser role—at least intracranially.

In humans, the arachnoid granulations located in the dura mater are principle sites of CSF absorption into the venous system. The movement of CSF into the dural venous sinuses is a one-way system. Fluid from the subarachnoid space exists at a higher pressure than that in the dural venous sinus, a difference of approximately 3–5 mmHg. From here, the majority of intracranial CSF drains into the internal jugular vein. Another site of CSF drainage exists in the spinal arachnoid villi. The CSF either flows into the epidural venous plexus or into the lymphatic system.

The nerve sheaths of cranial and spinal nerves are also responsible for absorption of CSF. Even in the early twentieth century, it was noted that the dorsal root of the spinal nerves contained clusters of arachnoidal cells. The overall contribution of this mechanism of CSF drainage is unknown. The process of fluid absorption is a dynamic process dependent on pressure. This is important in humans, who are upright, since this means the majority of the day is spent with the CSF fluid column under greatest hydrostatic pressure near the lumbar spine. Therefore, CSF absorption throughout the nerve roots is likely another important mechanism of CSF drainage, with most of this entering the lymphatic system.

As described above, the olfactory nerve is known to serve as a site for CSF drainage into the lymphatics, but other cranial nerves such as the trigeminal, cochlear, and even the optic nerve may have a contributory role for CSF absorption. The cribriform plate of the ethmoid bone, through with the extension of the olfactory nerve traverse, is a site of CSF absorption. As described above, although approximately 10% CSF is found in the cervical lymphatic collection at normal intracranial pressure, when CSF pressure increases, this value can increase up to nearly 80% indicating the dynamic nature of CSF regulation.

The material illustrated in the present section is based on references [8, 14, 15, 23, 24, 33].

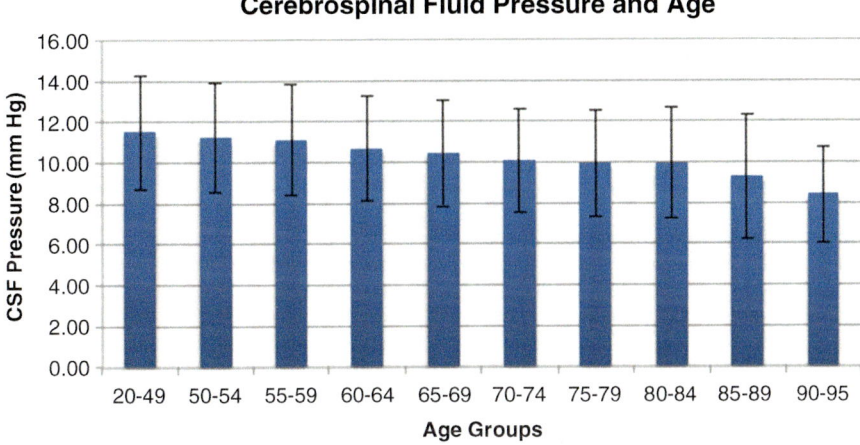

**Fig. 4** Mean CSFP within specific age groups. Bars represent one standard deviation. (Fleischman D et al. PLoS One. 2012)

## 4.5 Cerebrospinal Fluid Pressure

CSF pressure is typically defined by the intracranial pressure in the prone position. The pressure measured in the lateral decubitus position accurately describes the pressure in the lateral ventricles. Overall, pressure is a result of the equilibrium between CSF secretion, absorption, and resistance to flow. Abnormalities in any of these functions will result in changes in the CSF pressure. However, there occur natural changes in the production and reabsorption of CSF that also affect CSF ranges, outside of pathologic instances. The normal range of CSF pressure falls between 8 and 15 mmHg (10.9–20.4 cm $H_2O$). In humans, CSF pressure tends to be stable in the first five decades of life. Thereafter, there is a steady reduction in CSF pressure, so that by the age of 90–95, the pressure, on average, tends to be nearly 33% less compared to the mean at 20–50 years of age (Fig. 4). While there is a known decrease in CSF production with healthy aging, there is a matched increase in CSF outflow through both the arachnoid granulations and the lymphatic outflow channels. Reduction in CSF pressure with aging must be multifactorial and further work should be performed to corroborate and explain these findings.

There is also an independent, linear increase in CSF pressure with increase in body mass index. Whether this is artifact due to compression of the central veins and Valsalva due to positioning in the lateral decubitus position is debated, but this does not explain the consistent and linear relationship at lower BMI (Fig. 5).

Many of these studies discuss CSF pressure measured from the lumbar spine as a surrogate for the pressure within the cranium. As mentioned above, this has been studied and verified by Lenfeldt. However, in fields such as ophthalmology and the increasing interest in CSF pressure regulation from microgravity, the fluid pressure within the perioptic subarachnoid space may not be represented properly

OD                              OS

before surgery

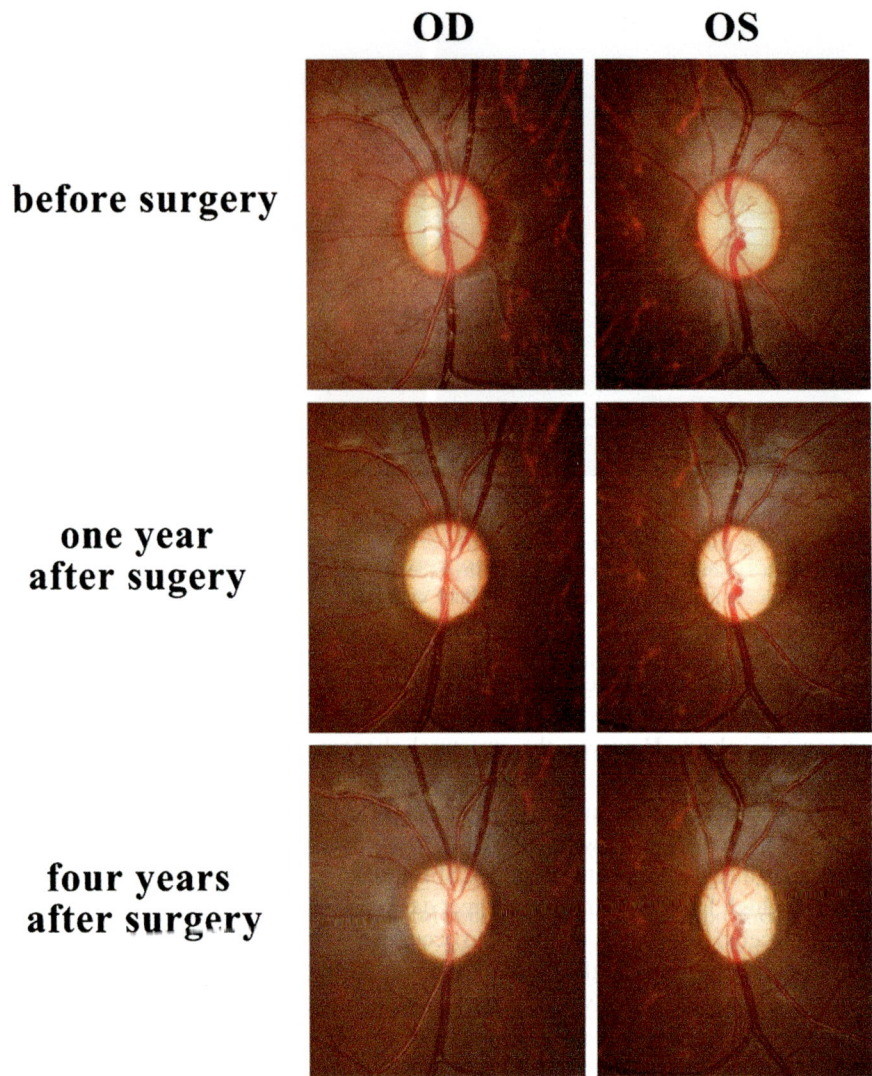

one year
after sugery

four years
after surgery

**Fig. 2** Follow-up fundus photograph of 1# monkey. The bilateral fundus were normal before surgery; RNFLT was decreased in the first year after surgery, but no further change after 1 year of surgery

## 2.2 Retinal Oxygen Saturation

Oxygen is critical to the retina and normal oxygen levels in ocular tissues are vital for healthy eyes [15]. Many ocular diseases, such as age-related macular degeneration (AMD) [16], diabetic retinopathy [17, 18], central retinal vein

## OD                    OS

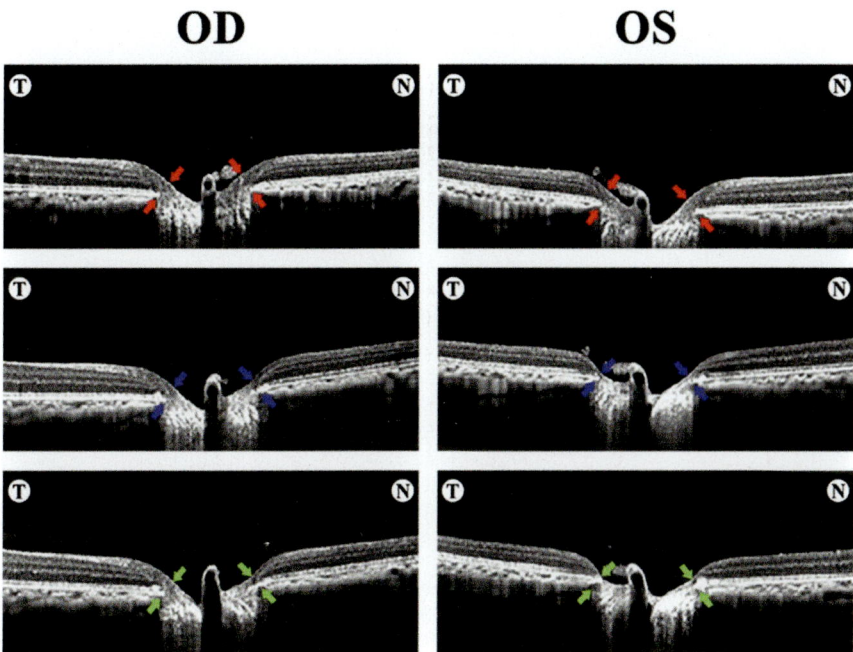

**Fig. 3** OCT images in the both eyes of 1# monkey. 1# monkey presented with a decrease in the retinal nerve fiber layer thickness (RNFLT), rim area (RA), and rim volume (RV) and the enlargement of cup-to-disc area ratio (CDA) in the first year after surgery. RA, RV, and CDA were still in the progression in the following 3 years of follow-up. No further changes were found in RNFLT in the following 3 years. However, no backward displacement of LC was detected in the whole period of follow-up

occlusion (CRVO) [19, 20], branch retinal vein occlusion (BRVO) [21], and glaucoma [22, 23], have shown abnormal oxygen saturation. With the development of modern spectrophometric retinal oximetry techniques, which allow measurement of differences in absorption wavelengths of oxygenated or deoxygenated hemoglobin, oxygen saturation ($SO_2$) in retinal blood vessels can be assessed noninvasively [24]. An instrument that has been developed for this purpose is the Oxymap T1 Retinal Oximeter (Oxymap, Reykjavik, Iceland) [25].

Though oxygen saturation of people (of different ethnicities) with various ocular diseases has been reported many times [16–23, 26–31], the retinal oxygen saturation in low cerebrospinal fluid pressure rhesus monkeys is still lacking. So, we carried out this research to assess the normal values of retinal oxygen saturation and study the retinal oxygen saturation in low cerebrospinal fluid pressure (CSFP) rhesus monkeys. Eighteen normal adult Rhesus macaque monkeys and two low-CSFP monkeys were included in this experimental study. An Oxymap T1 retinal oximeter (Oxymap, Reykjavik, Iceland) was used to perform oximetry on all subjects. Global arterial ($SaO_2$) and venous oxygen saturation ($SvO_2$), and arteriovenous difference

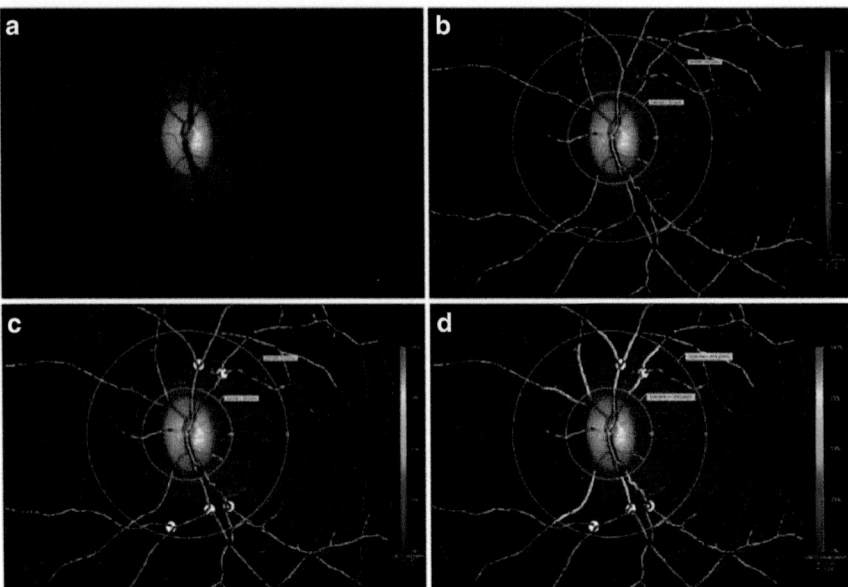

**Fig. 4** Method of analyzing images. (**a**) 570 nm image; (**b**) The vessels were minimum 50 pixels from the optic disc margin and at least 100 pixels from the edge of the image; (**c**) Masking of vessel branchings and A–V crossings along with marking of two concentric circles within which vessel segments would be analyzed; (**d**) Vessel segment markings (thickest arteriole and venule per quadrant).(Reprinted with permission from Li J, Yang Y, Yang D, et al. Normative Values of Retinal Oxygen Saturation in Rhesus Monkeys: The Beijing Intracranial and Intraocular Pressure (iCOP) Study[J]. PLoS One, 2016,11(3):e0150072)

in $SO_2$ were measured (Fig. 4). For both the normal monkeys and the low-CSFP monkeys, each eye was imaged three times, and then mean values were calculated. We found that the average $SaO_2$ and $SvO_2$ were 89.48 ± 2.64% and 54.85 ± 2.18%, respectively. The global A–V difference was 34.63 ± 1.91%. For two low-CSFP monkeys, retinal oxygen saturation measurements were all in the normal range (Fig. 5). Our study is the first to describe retinal $SO_2$ in healthy and low-CSFP Rhesus monkeys. Retinal oxygen saturation was normal in low-CSFP monkeys, the reason for which maybe the autoregulation and homeostasis of retinal vessels. However, there were some limitations in our study. First, anesthesia may cause a decrease in oxygen saturation; in other words, the actual oxygen saturation of Rhesus macaque monkeys may be higher than the data determined in our study. Second, the number of low-CSFP monkeys was only two; more low-CSFP monkeys need to be studied in the future.

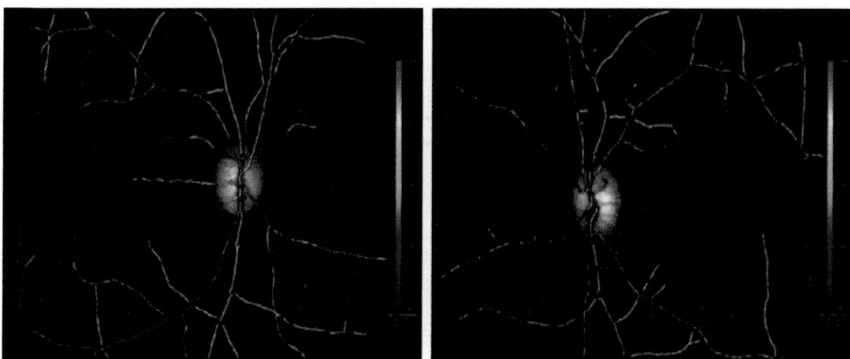

**Fig. 5** Retinal oxygen saturation in a low cerebrospinal fluid pressure rhesus monkey was normal. (Reprinted with permission from Li J, Yang Y, Yang D, Liu X, Sun Y, Wei S, Wang N. Normative Values of Retinal Oxygen Saturation in Rhesus Monkeys: The Beijing Intracranial and Intraocular Pressure (iCOP) Study[J]. PLoS One, 2016, 11(3):e0150072)

## 2.3   Perfusion of Optic Disc

Till now, the mechanism underlying low-CSFP-induced optic neuropathy is not clearly understood. There are a growing number of evidences showing that glaucoma is associated with vascular dysfunction [32–35]. More recently, prospective trials have shown that peripapillary atrophy and optic disc hemorrhage are both related to accelerated glaucoma progression [36, 37]. A method of measuring local circulation has been recently developed by using high-speed OCT to perform quantitative angiography [38]. Using a split-spectrum amplitude decorrelation angiography (SSADA) algorithm, flow in the optic disc and macula can be quantified [39–45].

We therefore conducted this study to investigate the influence of experimentally reduced CSFP on optic disc and macular perfusion compared to elevated IOP. Eighteen healthy monkeys (18 eyes), two low-CSFP monkeys (four eyes), and two sham group monkeys (four eyes) were subjected to optic disc flow index measurements via a high-speed and high-resolution spectral-domain OCT (RTVue XR) with a split-spectrum amplitude decorrelation angiography algorithm. Each eye of all monkeys was imaged three times by a high-speed 840-nm wavelength OCT instrument. The split-spectrum amplitude decorrelation angiography (SSADA) algorithm was used to compute 3-dimensional optic disc angiography. Flow index of both optic disc and macula was computed from four registered scans. We found that the average flow indexes of the four optic disc area layers were $0.171 \pm 0.009$ (optic nerve head, ONH), $0.015 \pm 0.004$ (vitreous), $0.052 \pm 0.009$ (radial peripapillary capillary, RPC), and $0.167 \pm 0.011$ (choroid). In the low-CSFP group as compared to the sham group, flow index of RPC layer was markedly decreased ($p < 0.05$) (Fig. 6 and Table 1). Flow index of RPC layer was significantly correlated with IOP, CSFP, and TLPD (Fig. 7). So, we concluded that experimental models with chronic

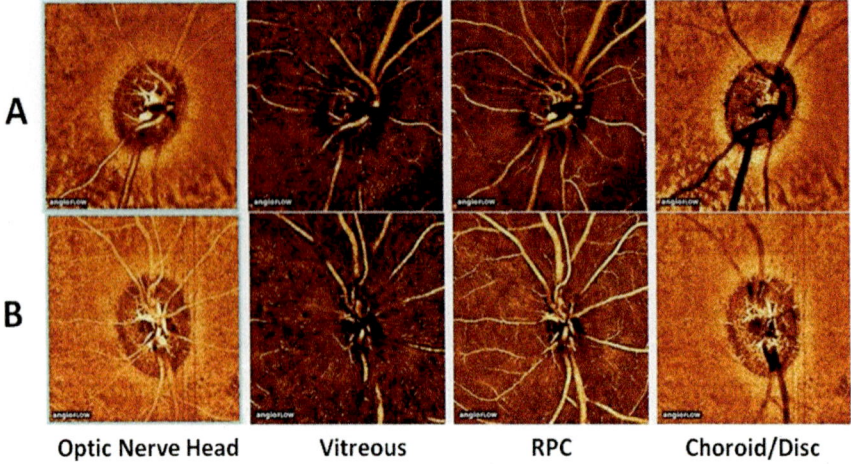

**Optic Nerve Head**        **Vitreous**        **RPC**        **Choroid/Disc**

**Fig. 6** OCT angiograms of optic disc region. (**a**), low-CSFP group; (**b**), sham group. In the low-CSFP group as compared to the sham group, flow index of RPC layer was markedly decreased

**Table 1** CSFP group as compared to the sham group, flow index of RPC layer was markedly decreased ($p < 0.05$)

| Layer | Low-CSFP group ($n = 4$) | Sham group ($n = 4$) |
|---|---|---|
| ONH | $0.16 \pm 0.01$ | $0.17 \pm 0.01$ |
| Vitreous | $0.02 \pm 0.01$ | $0.03 \pm 0.01$ |
| RPC | $0.04 \pm 0.01*$ | $0.07 \pm 0.01$ |
| Choroid | $0.15 \pm 0.01$ | $0.17 \pm 0.01$ |

$*p < 0.05$

**Fig. 7** Plots of mechanical parameters (IOP/CSFP/TLCPD) versus flow index of RPC. Flow index of RPC layer was significantly negatively correlated with IOP and TLPD, and positively correlated with CSFP

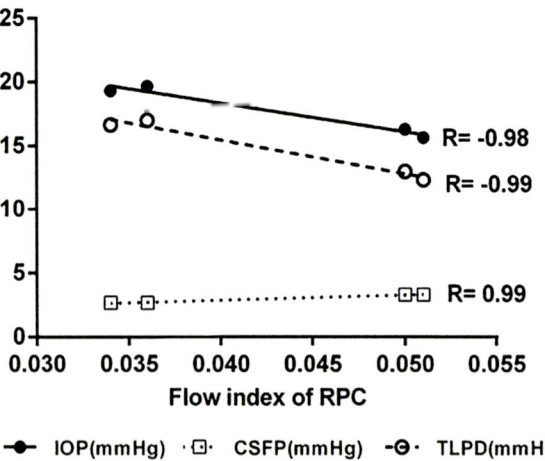

CSFP reduction showed blood flow changes in the optic disc which mainly affect the radial peripapillar capillaries. It suggests that an experimental model with chronic reduction in CSFP may be different in the process of optic nerve damage.

## 2.4   Rat Model

### 2.4.1   Axonal Transport

Previous studies showed that an abnormal TLPD due to an experimentally elevated IOP was associated with an impediment of both, the orthograde and the retrograde, axoplasmic flows in animals [46–52]. By including the CSFP as one of the two determinants of TLPD into the discussion on the pathogenesis of glaucomatous optic neuropathy, Zhang et al. conducted an experimental investigation in rats to further explore whether an abnormally low CSFP is associated with abnormalities of the axoplasmic flow, similar to the situation with an elevated IOP [14]. The orthograde axoplasmic flow was assessed by examining the distribution rhodamine-ß-isothiocyanate (RITC) within optic nerve axons after injection of RITC into the vitreous, and the retrograde axoplasmic flow was assessed by examining the distribution of fluorogold within optic nerve axons after the injection of fluorogold into the superior colliculi. At 24 h after baseline, the intensity of RITC staining was significantly lower in both the high-IOP and low-CSFP groups than in the control group at the retinal pigment epithelium-Bruch's membrane complex line at the optic nerve head and at the measurement points of 50-, 100-, 150-, 200-, 250-, and 450-$\mu$m posterior to Bruch's membrane line (Fig. 8). The density of fluorogold fluorescent retinal ganglion cells somas as measured at 6 h after the fluorogold injection was significantly lower in the high-IOP and low-CSFP groups than in the control group (Fig. 9). The findings of the study provide new information with respect to a disturbance of both the orthograde and retrograde axonal transports in retinal ganglion cell axons with an acute experimental reduction in CSFP. These findings support the hypothesis of an association between abnormally low CSFP and optic nerve damage.

The axonal morphology and axonal motor proteins in retinal ganglion cells as influenced by an experimentally reduced CSFP or elevated IOP were examined in another study [13]. In both groups of rats with high IOP or low CSFP as compared with a control group of animals, the retinal ganglion cell axons became abnormally dilated and accumulated organelles including vesicles, dense bodies, and mitochondria, as well as apparent disorganization of microtubules and neurofilaments (Fig. 10). Both groups as compared to the control group showed an accumulation of dynein intermediate chain (IC) at the optic nerve head and retina and a reduction in kinesin heavy chain (HC) immunoreactivity in the optic nerve fiber axons. As a corollary, Western blot analysis revealed an elevation of dynein IC protein levels in the optic nerve head and retina and a decrease in kinesin HC protein levels in the optic nerve. This supports the notion that experimental models with an acute reduction in CSFP and an acute rise in IOP show similarities in the process of optic nerve damage.

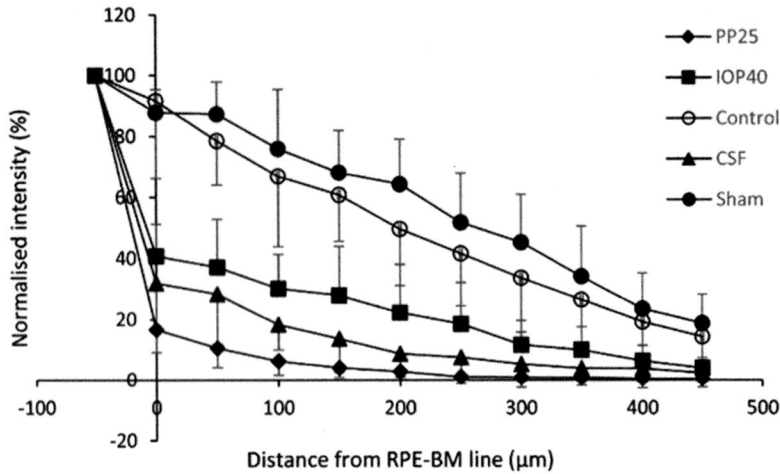

**Fig. 8** Relationship between the mean intensity of rhodamine isothiocyanate (RITC) in the optic nerve and distance from the retinal pigment epithelium-Bruch's membrane complex at the optic nerve head in rats of a control group ("control"), in rats of a sham control group ("sham"), in rats with a short-term (6 h) elevation of intraocular pressure to 40 mmHg ("IOP40"), in rats with a short-term (6 h) reduction of the ocular perfusion pressure to 25 mmHg ("PP25"), and in rats with an experimental short-term (6 h) reduction in cerebrospinal fluid pressure ("CSF"), imaged at 1 day after baseline. The RITC intensity was normalized and expressed as a percentage of the intensity measured close to the inner limiting membrane. The RITC intensity did not differ significantly between the control group and the sham control group nor between the IOP40 study group and the low-CSFP study group nor between the PP25mmHg study group and the low-CSFP study group, at any measurement point along the optic nerve. (Reprinted with permission from Zhang, Z., D. Liu, et al. (2015). "Axonal Transport in the Rat Optic Nerve Following Short-Term Reduction in Cerebrospinal Fluid Pressure or Elevation in Intraocular Pressure." Invest Ophthalmol Vis Sci 56(8): 4257–4266)

### 2.4.2 Early Glial Reactivity in Visual Pathway

Glial reactivity has been reported in a variety of retinal pathological conditions such as ischemia, inflammation, trauma, and so on [53–55]. It has been also demonstrated in diverse higher IOP models, which involved in the whole visual pathway [56–60]. In particular, glial cells reactivation is notable during the initial phases of glaucoma precede optic neuropathy, even as a hallmark of neural injury in central nervous system (CNS) [56, 60, 61]. We propose a hypothesis that early glial reactivity may also participate in the damage of RGCs in the acute cerebrospinal fluid pressure (CSFP) reduction rat model.

We observe glial reactivity in the whole visual pathway, including retina, lateral geniculate nucleus (LGN), and superior colliculus (SC). Both glial fibrillary acidic protein (GFAP) and glutamine synthetase (GS) were seen as indicators for Muller cell activation in retina and CNS, especially GS is specifically expressed in Muller cells [62, 63]. We found that retinal GFAP and GS immunolabeling

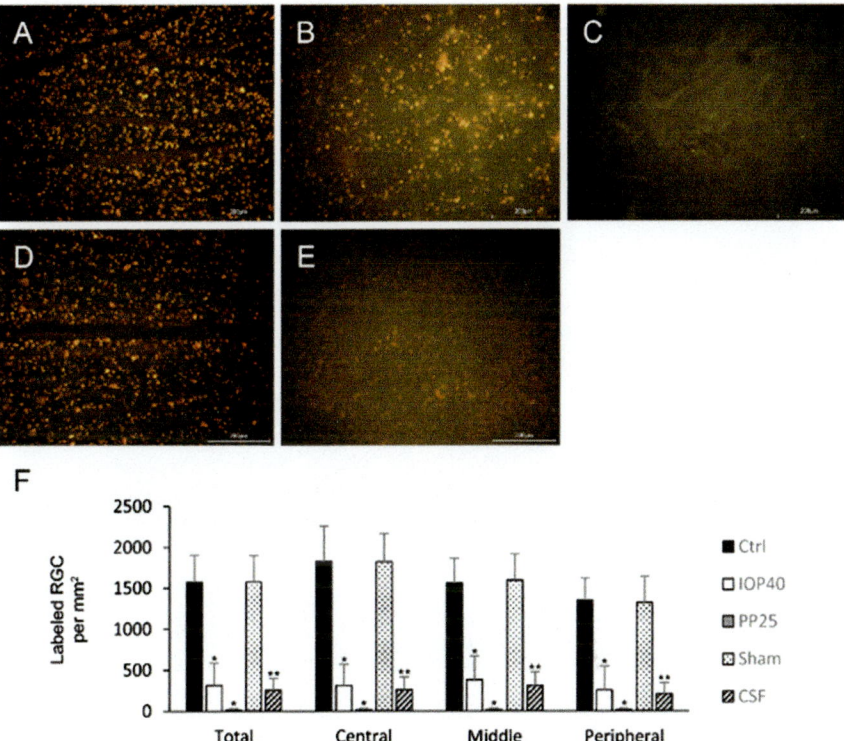

**Fig. 9** Staining of retinal ganglion cells by fluorogold on retinal flat mounts after injection of fluorogold in both superior colliculi in rats of the control group (**a**), in rats with a short-term (6 h) elevation of intraocular pressure to 40 mmHg (**b**), in rats with a short-term (6 h) reduction in the ocular perfusion pressure to 25 mmHg (**c**), in rats of a sham control group (**d**), and in rats with an experimental short-term (6 h) reduction in cerebrospinal fluid pressure (**e**), imaged at 6 h after baseline; (**f**) Quantitative analysis of the retrograde axonal transport assay. *Error bars*: Standard deviation; * Significant at $p < 0.05$ when high-intraocular pressure group or the low-ocular perfusion pressure group were compared with the control group; ** Significant at $p < 0.05$ when low cerebrospinal fluid pressure group was compared with the sham control group; $n = 6$ retinal flat mounts per group. *Scale bars*: (**a–e**) 200 μm. (Reprinted with permission from Zhang, Z., D. Liu, et al. (2015). "Axonal Transport in the Rat Optic Nerve Following Short-Term Reduction in Cerebrospinal Fluid Pressure or Elevation in Intraocular Pressure." Invest Ophthalmol Vis Sci 56(8): 4257–4266)

were consistently more intense obviously over time in lower CSFP group compared with control after 3 days of acute CSFP reduction (Fig. 1). GFAP and GS were $1.78 \pm 0.01$ and $1.71 \pm 0.05$ fold, respectively, with significant changes compared with control group (Figs. 11 and 12).

In LGN and SC, GFAP were $1.84 \pm 0.07$, $2.23 \pm 0.08$ folds at 3 days post-surgery, respectively, compared with control group (Fig. 13), which were determined to be significantly changed. And GS expressions were $0.85 \pm 0.04$,

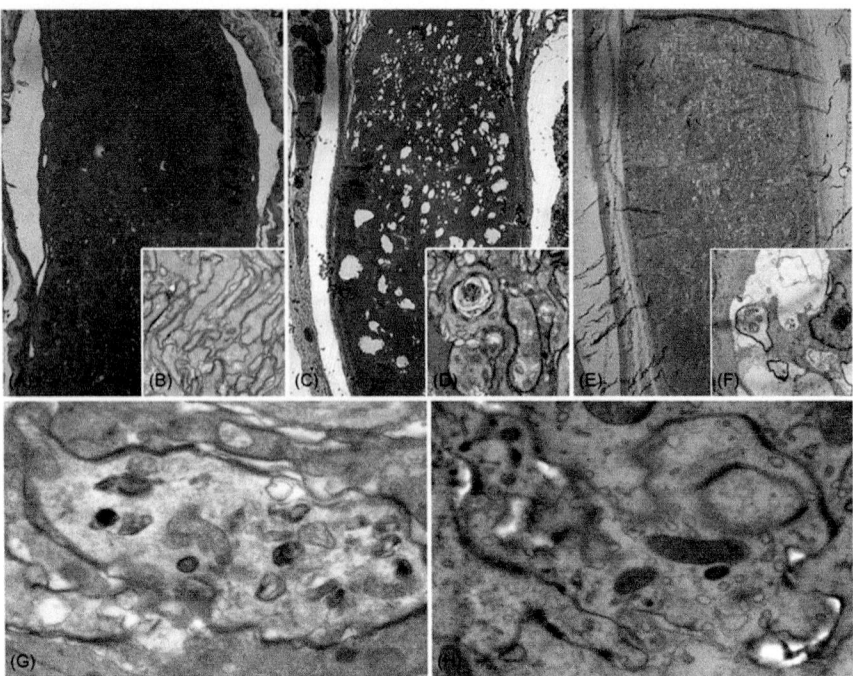

**Fig. 10** Morphology of rat optic nerve at normal, elevated intraocular pressure and reduced cerebrospinal fluid pressure. (**a**) Light micrograph of normal rat optic nerve (azur-methylene blue stain; magnification, ×20). (**b**) Electron micrograph of normal axons from the myelinated optic nerve (magnification, ×20,000). (**c**) Light micrograph of reduced cerebrospinal fluid pressure optic nerve shows clear, swollen axons behind the scleral canal (azur-methylene blue stain; magnification, ×20). (**d**) Clear zones within myelinated axonal bundles when viewed by electron microscopy are dilated nerve fibers (magnification, ×20,000). (**e**) Light micrograph of elevated IOP optic nerve shows slightly dilated RGC axons behind the scleral canal (azur-methylene blue stain; magnification, ×20). (**f**) Clear zones within myelinated axonal bundles when viewed by electron microscopy are dilated nerve fibers (magnification, ×20,000). (**g**) Electron micrograph demonstrating axons of the myelinated optic nerve of rat eyes with reduced CSFP swollen by accumulations of vesicles, dense bodies, and mitochondria. The lamellar structure of optic nerve myelin sheath began to distort (magnification, ×40,000). (**h**) Electron micrograph demonstrating axons of the myelinated optic nerve of rat eyes with elevated IOP. Organelles seen in swollen fiber include mitochondria, smooth vesicles, and dense bodies. The lamellar structure of optic nerve myelin sheath was broken and distorted (magnification, ×40,000). (Reprinted with permission from Zhang, Z., S. Wu, et al. (2016). "Dynein, kinesin and morphological changes in optic nerve axons in a rat model with cerebrospinal fluid pressure reduction: the Beijing Intracranial and Intraocular Pressure (iCOP) study." Acta Ophthalmol 94(3): 266–275)

$1.25 \pm 0.12$ folds at 3 days post-surgery, respectively, without any significant difference (Fig. 14).

To exclude the influences of cerebrospinal fluid dynamics change, we tested glial reactivity in hippocampus which is highly susceptible to stress as astrocytes reactivity [64, 65]. GFAP expressions of western blot analysis revealed $0.97 \pm 0.14$

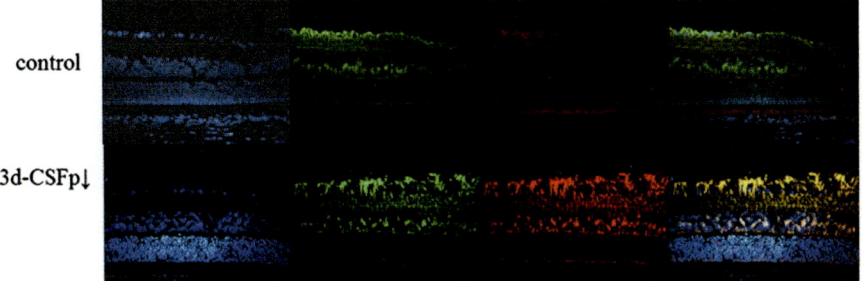

**Fig. 11** Changes in the expression of GFAP (red) and GS (green) in Muller cells of the control, CSFP reduction in rat retinas. A1, B1, immunofluorescence labeling show DAPI in rat retinal vertical slices taken from control, and those obtained at 3D after CSFP reduction. A2, B2, immunofluorescence staining show GS expression in the same slices in A1 and B1, respectively; note that GS changes in Muller cells. A3, B3, immunofluorescence staining show GFAP expression; note that GFAGP changes in Muller cells. D4, B4, merged images of A1–A1, B1–B3. Scale bar, for all images 100 $\mu$m

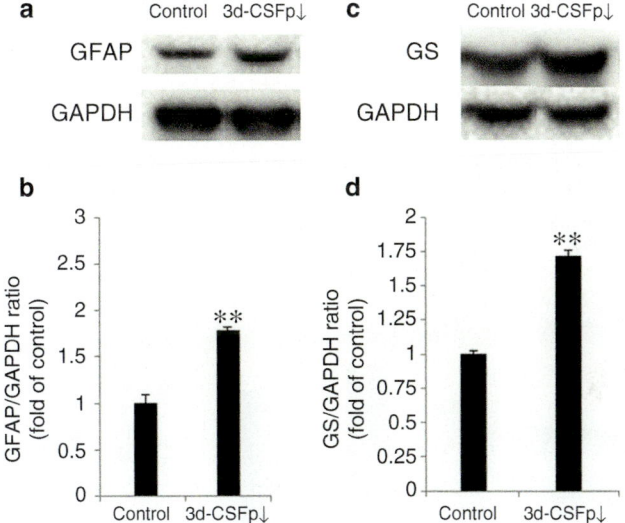

**Fig. 12** Western blot analysis of retinal GFAP (**a, b**) and GS (**c, d**) expressions in CSFP reduction group. (**a–d**) The relative intensity of the chemiluminescence for each protein band was normalized using GAPDH, and the data have been presented as mean ± SD of the fold increase compared with controls. **: vs. control and $p \leq 0.01$

fold at 3 days compared to control group (Fig. 15a, b). And GS expressions were 1.06 ± 0.22 fold at 3 days (Fig. 15a, c), and both changes were not statistically significant.

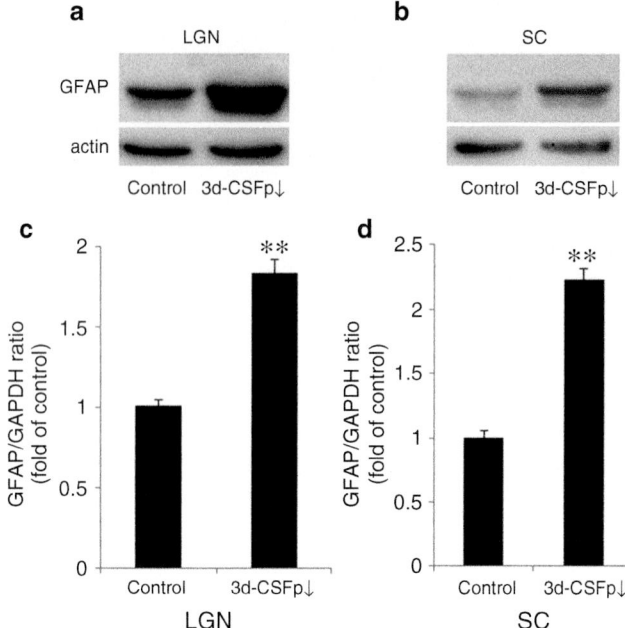

**Fig. 13** Western blot analysis of LGN (**a, c**) and SC (**b, d**) GFAP expressions in CSFP reduction group. The relative intensity of the chemiluminescence for each protein band was normalized using actin, and the data have been presented as mean ± SD of the fold increase compared with controls. **: vs. control and $p \leq 0.01$

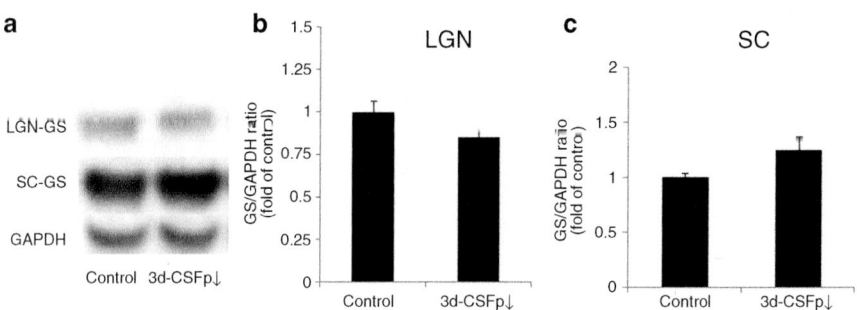

**Fig. 14** Western blot analysis of LGN (**a, b**) and SC (**a, c**) GS expressions in CSFP reduction group. The relative intensity of the chemiluminescence for each protein band was normalized using actin, and the data have been presented as mean ± SD of the fold increase compared with controls

In conclusion, glial reactivity occurred in the whole visual pathway in CSFP reduction group without influence hippocampus. These results may contribute to understanding the underlying mechanism of CSFP or trans-laminar pressure gradient (TLPD) for the physiology of glaucoma or glaucoma-like disease.

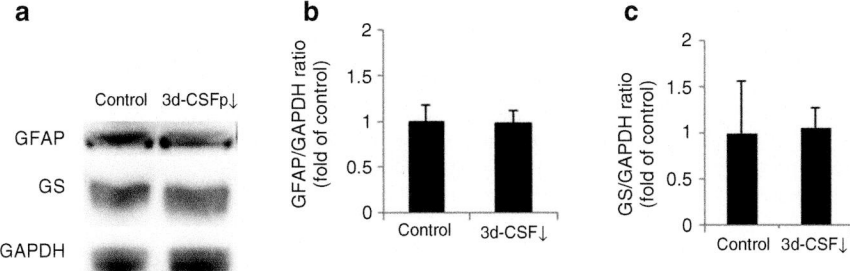

**Fig. 15** Western blot analysis of GFAP (**a, b**) and GS (**a, c**) expressions of hippocampus in CSFP reduction group. The relative intensity of the chemiluminescence for each protein band was normalized using actin, and the data have been presented as mean ± SD of the fold increase compared with controls

# References

1. Morgan WH, Yu DY, Cooper RL, et al. The influence of cerebrospinal fluid pressure on the lamina cribrosa tissue pressure gradient. Invest Ophthalmol Vis Sci. 1995;36:1163–1172.
2. Morgan WH, Chauhan BC, Yu DY, et al. Optic disc movement with variations in intraocular and cerebrospinal fluid pressure. Invest Ophthalmol Vis Sci. 2002;43:3236–3242.
3. Jonas JB, Berenshtein E, Holbach L. Anatomic relationship between lamina cribrosa, intraocular space, and cerebrospinal fluid space. Invest Ophthalmol Vis Sci. 2003;44:5189–5195.
4. Morgan WH, Yu DY, Balaratnasingam C. The role of cerebrospinal fluid pressure in glaucoma pathophysiology: the dark side of the optic disc. J Glaucoma. 2008;17:408–413.
5. Berdahl JP, Allingham RR, Johnson DH. Cerebrospinal fluid pressure is decreased in primary open-angle glaucoma. Ophthalmology. 2008;115:763–768.
6. Ren R, Jonas JB, Tian G, et al. Cerebrospinal fluid pressure in glaucoma: a prospective study. Ophthalmology. 2010;117: 259–266.
7. Hiraoka M, Inoue K, Ninomiya T, Takada M. Ischaemia in the Zinn-Haller circle and glaucomatous optic neuropathy in macaque monkeys. Br J Ophthalmol. 2012;96:597-603.
8. Križaj D, Ryskamp DA, Tian N, et al. From mechanosensitivity to inflammatory responses: new players in the pathology of glaucoma. Curr Eye Res. 2014;39:105-119.
9. Cherecheanu AP, Garhofer G, Schmidl D, Werkmeister R, Schmetterer L. Ocular perfusion pressure and ocular blood flow in glaucoma. Curr Opin Pharmacol. 2013;13:36-42.
10. Leske MC. Ocular perfusion pressure and glaucoma: clinical trial and epidemiologic findings. Curr Opin Ophthalmol. 2009;20:73-78.
11. Chrysostomou V, Rezania F, Trounce IA, Crowston JG. Oxidative stress and mitochondrial dysfunction in glaucoma. Curr Opin Pharmacol. 2013;13:12-15.
12. Zhang Z, Liu D, Jonas JB, et al. Glaucoma and the Role of Cerebrospinal Fluid Dynamics. Invest Ophthalmol Vis Sci. 2015;56:6632.
13. Zhang Z, Wu S, Jonas JB, et al. Dynein, kinesin and morphological changes in optic nerve axons in a rat model with cerebrospinal fluid pressure reduction: the Beijing Intracranial and Intraocular Pressure (iCOP) study. Acta Ophthalmol. 2016;94:266-75.
14. Zhang Z, Liu D, Jonas JB, et al. Axonal Transport in the Rat Optic Nerve Following Short-Term Reduction in Cerebrospinal Fluid Pressure or Elevation in Intraocular Pressure. Invest Ophthalmol Vis Sci. 2015;56:4257-4266.
15. Wangsa-Wirawan ND, Linsenmeier RA. Retinal oxygen: fundamental and clinical aspects. Arch Ophthalmol 2003; 121:547–557.

16. Geirsdottir A, Hardarson SH, Olafsdottir OB, Stefánsson E. Retinal oxygen metabolism in exudative age-related macular degeneration. Acta Ophthalmologica 2014; 92: 27–33.
17. Hardarson SH, Stefansson E. Retinal oxygen saturation is altered in diabetic retinopathy. Br J Ophthalmol 2012; 96:560–563.
18. Hammer M, Heller T, Jentsch S, Dawczynski J, Schweitzer D, Peters S, et al. Retinal Vessel Oxygen Saturation under Flicker Light Stimulation in Patients with Nonproliferative Diabetic Retinopathy. Invest Ophthalmol Vis Sci 2012; 53: 4063–4068.
19. Hardarson SH, Stefansson E. Oxygen saturation in central retinal vein occlusion. Am J Ophthalmol 2010; 150:871–875.
20. Eliasdottir TS, Bragason D, Hardarson SH, Kristjansdottir G, Stefánsson E. Venous oxygen saturation is reduced and variable in central retinal vein occlusion. Graefes Arch Clin Exp Ophthalmol 2015;253:1653–1661.
21. Hardarson SH, Stefánsson E. Oxygen saturation in branch retinal vein occlusion. Acta Ophthalmologica 2011; 90: 466–470.
22. Tobe LA, Harris A, Schroeder A, Gerber A, Holland S, Amireskandari A, et al. Retinal oxygen saturation and metabolism: how does it pertain to glaucoma? An update on the application of retinal oximetry in glaucoma. Eur J Ophthalmol 2013; 23:465–472.
23. Vandewalle E, Abegao Pinto L, Olafsdottir OB, De Clerck E, Stalmans P, Van Calster J, et al. Oximetry in glaucoma: correlation of metabolic change with structural and functional damage. Acta Ophthalmol 2014; 92:105–110.
24. Hardarson SH, Harris A, Karlsson RA, Halldorsson GH, Kagemann L, Rechtman E, et al. Automatic retinal oximetry. Invest Ophthalmol Vis Sci 2006; 47:5011–5016.
25. Hardarson SH. Protocol for Analysis of Oxymap T1 Oximetry Images. Reykjavik, Iceland: Oxymap;2011.
26. Jani PD, Mwanza J-C, Billow KB, Waters AM, Moyer S, Garg S. Normative values and predictors of retinal oxygen saturation. Retina 2014; 34: 394–401.
27. Geirsdottir A, Palsson O, Hardarson SH, Olafsdottir OB, Kristjansdottir JV, Stefánsson E. Retinal Vessel Oxygen Saturation in Healthy Individuals. Invest Ophthalmol Vis Sci 2012; 53: 5433–5442.
28. Yip W, Siantar R, Perera SA, Milastuti N, Ho KK, Tan B, et al. Reliability and determinants of retinal vessel oximetry measurements in healthy eyes. Invest Ophthalmol Vis Sci 2014; 55:7104–7110.
29. Mohan A, Dabir S, Yadav NK, Kummelil M, Kumar RS, Shetty R. Normative Database of Retinal Oximetry in Asian Indian Eyes. PLoS ONE 2015; 10: e0126179.
30. Man REK, Sasongko MB, Kawasaki R, Noonan JE, Lo TC, Luu CD, et al. Associations of retinal oximetry in healthy young adults. Invest Ophthalmol Vis Sci 2014; 55:1763–1769.
31. Türksever C, Orgül S, Todorova MG. Reproducibility of retinal oximetry measurements in healthy and diseased retinas. Acta Ophthalmol 2015; 93: e439–e445.
32. Burgoyne CF, Downs JC. Premise and prediction-how optic nerve head biomechanics underlies the susceptibility and clinical behavior of the aged optic nerve head. J Glaucoma 2008,17(4):318-328.
33. Fechtner RD, Weinreb RN. Mechanisms of optic nerve damage in primary open angle glaucoma. Surv Ophthalmol 1994, 39(1):23-42.
34. Leske MC. Ocular perfusion pressure and glaucoma: clinical trial and epidemiologic findings. Curr Opin Ophthalmol 2009, 20(2):73-78.
35. Schmidl D, Garhofer G, Schmetterer L. The complex interaction between ocular perfusion pressure and ocular blood flow - relevance for glaucoma. Exp Eye Res. 2011;93(2):141-55.
36. Suh MH, Park KH, Kim H, Kim TW, Kim SW, Kim SY, Kim DM. Glaucoma progression after the first-detected optic disc hemorrhage by optical coherence tomography. J Glaucoma. 2012;21(6):358-66.
37. Seidensticker F, Reznicek L, Mann T, Hübert I, Kampik A, Ulbig M, Hirneiss C, Neubauer AS, Kernt M. Assessment of β-zone peripapillary atrophy by optical coherence tomography and scanning laser ophthalmoscopy imaging in glaucoma patients. Clin Ophthalmol. 2014;8:1233-9.

38. Jia Y, Tan O, Tokayer J, Potsaid B, Wang Y, Liu JJ, Kraus MF, Subhash H, Fujimoto JG, Hornegger J, Huang D. Split-spectrum amplitude-decorrelation angiography with optical coherence tomography. Opt Express. 2012;20(4):4710-25.
39. Jia Y, Morrison JC, Tokayer J, Tan O, Lombardi L, Baumann B, Lu CD, Choi W, Fujimoto JG, Huang D. Quantitative OCT angiography of optic nerve head blood flow. Biomed Opt Express. 2012;3(12):3127-37.
40. Wei E, Jia Y, Tan O, Potsaid B, Liu JJ, Choi W, Fujimoto JG, Huang D. Parafoveal retinal vascular response to pattern visual stimulation assessed with OCT angiography. PLoS One. 2013 Dec 2;8(12):e81343.
41. Jia Y, Wei E, Wang X, Zhang X, Morrison JC, Parikh M, Lombardi LH, Gattey DM, Armour RL, Edmunds B, Kraus MF, Fujimoto JG, Huang D. Optical coherence tomography angiography of optic disc perfusion in glaucoma. Ophthalmology. 2014;121(7):1322-32.
42. Jia Y, Bailey ST, Wilson DJ, Tan O, Klein ML, Flaxel CJ, Potsaid B, Liu JJ, Lu CD, Kraus MF, Fujimoto JG, Huang D. Quantitative optical coherence tomography angiography of choroidal neovascularization in age-related macular degeneration. Ophthalmology. 2014;121(7):1435-44.
43. Kuehlewein L, Sadda SR, Sarraf D. OCT angiography and sequential quantitative analysis of type 2 neovascularization after ranibizumab therapy. Eye (Lond). 2015;29(7):932-5.
44. Yu J, Jiang C, Wang X, Zhu L, Gu R, Xu H, Jia Y, Huang D, Sun X. Macular perfusion in healthy Chinese: an optical coherence tomography angiogram study. Invest Ophthalmol Vis Sci. 2015;56(5):3212-7.
45. Pechauer AD, Jia Y, Liu L, Gao SS, Jiang C, Huang D. Optical coherence tomography angiography of peripapillary retinal blood flow response to hyperoxia. Invest Ophthalmol Vis Sci. 2015;56(5):3287-91.
46. Vrabec JP, Levin LA. The neurobiology of cell death in glaucoma. Eye (Lond). 2007;21(suppl 1):S11–S14.
47. Johnson EC, Guo Y, Cepurna WO, Morrison JC. Neurotrophin roles in retinal ganglion cell survival: lessons from rat glaucoma models. Exp Eye Res. 2009;88:808–815.
48. Pease ME, McKinnon SJ, Quigley HA, Kerrigan-Baumrind LA, Zack DJ. Obstructed axonal transport of BDNF and its receptor TrkB in experimental glaucoma. Invest Ophthalmol Vis Sci. 2000;41:764–774.
49. Martin KR, Quigley HA, Valenta D, Kielczewski J, Pease ME. Optic nerve dynein motor protein distribution changes with intraocular pressure elevation in a rat model of glaucoma. Exp Eye Res. 2006;83:255–262.
50. Minckler DS, Bunt AH, Johanson GW. Orthograde and retrograde axoplasmic transport during acute ocular hypertension in the monkey. Invest Ophthalmol Vis Sci. 1977;16: 426–441.
51. Johansson JO. Inhibition of retrograde axoplasmic transport in rat optic nerve by increased IOP in vitro. Invest Ophthalmol Vis Sci. 1983;24:1552–1558.
52. Abbott CJ, Choe TE, Lusardi TA, Burgoyne CF, Wang L, Fortune B. Evaluation of retinal nerve fiber layer thickness and axonal transport 1 and 2 weeks after 8 hours of acute intraocular pressure elevation in rats. Invest Ophthalmol Vis Sci. 2014;55: 674–687.
53. Zhang S, Wang H, Lu Q, et al. Detection of early neuron degeneration and accompanying glial responses in the visual pathway in a rat model of acute intraocular hypertension. Brain research 2009;1303:131-143.
54. Sapienza A, Raveu AL, Reboussin E, et al. Bilateral neuroinflammatory processes in visual pathways induced by unilateral ocular hypertension in the rat. Journal of neuroinflammation 2016;13:44.
55. Mac Nair CE, Schlamp CL, Montgomery AD, Shestopalov VI, Nickells RW. Retinal glial responses to optic nerve crush are attenuated in Bax-deficient mice and modulated by purinergic signaling pathways. Journal of neuroinflammation 2016;13:93.
56. Sun D, Qu J, Jakobs TC. Reversible reactivity by optic nerve astrocytes. Glia 2013;61:1218-1235.
57. Dai Y, Sun X, Yu X, Guo W, Yu D. Astrocytic responses in the lateral geniculate nucleus of monkeys with experimental glaucoma. Veterinary ophthalmology 2012;15:23-30.

58. Tehrani S, Davis L, Cepurna WO, et al. Astrocyte Structural and Molecular Response to Elevated Intraocular Pressure Occurs Rapidly and Precedes Axonal Tubulin Rearrangement within the Optic Nerve Head in a Rat Model. PloS one 2016;11:e0167364.
59. Trost A, Motloch K, Bruckner D, et al. Time-dependent retinal ganglion cell loss, microglial activation and blood-retina-barrier tightness in an acute model of ocular hypertension. Experimental eye research 2015;136:59-71.
60. Ramirez AI, Salazar JJ, de Hoz R, et al. Macro- and microglial responses in the fellow eyes contralateral to glaucomatous eyes. Progress in brain research 2015;220:155-172.
61. Tezel G, Fourth APORICWG. The role of glia, mitochondria, and the immune system in glaucoma. Investigative ophthalmology & visual science 2009;50:1001-1012.
62. Wang X, Su J, Ding J, et al. alpha-Aminoadipic acid protects against retinal disruption through attenuating Muller cell gliosis in a rat model of acute ocular hypertension. Drug design, development and therapy 2016;10:3449-3457.
63. Xue LP, Lu J, Cao Q, Hu S, Ding P, Ling EA. Muller glial cells express nestin coupled with glial fibrillary acidic protein in experimentally induced glaucoma in the rat retina. Neuroscience 2006;139:723-732.
64. Kim EJ, Pellman B, Kim JJ. Stress effects on the hippocampus: a critical review. Learning & memory 2015;22:411-416.
65. Butenko O, Dzamba D, Benesova J, et al. The increased activity of TRPV4 channel in the astrocytes of the adult rat hippocampus after cerebral hypoxia/ischemia. PloS one 2012;7:e39959.

# Instruments to Measure and Visualize Geometrical and Functional Parameters Related to the Fluid Dynamics of Cerebrospinal Fluid in the Eye

**Ingrida Januleviciene and Lina Siaudvytyte**

**Abstract** Intracranial and intraocular pressures are interrelated and relatively independent pressure systems, which keeps themselves in a relatively stable state through aqueous and cerebrospinal fluid circulations. Recently, researchers have focused on intracranial pressure role in eye diseases. This chapter summarizes various instruments to measure and visualize geometrical and functional parameters related to the fluid dynamics of cerebrospinal fluid in the eye.

## 1 Cerebrospinal Fluid Dynamics

Cerebrospinal fluid (CSF) plays an essential role in homeostasis and metabolism of central nervous system (CNS). It regulates balance of the electrolytes, circulation of active molecules, and elimination of catabolites. CSF transports the choroidal plexus secretion products to their sites of action, and modulates the activity of certain regions of the brain by impregnation. Furthermore, CSF returns interstitial fluid and proteins to the circulation, as there are no lymphatic channels in the brain [1].

CSF is a clear, colorless fluid, which consists mostly of water (99%) and only 1% of other substances, such as glucose, proteins, lipids, amino acids, creatinine, hormones, and vitamins. The CSF is separated from blood by the blood–brain barrier. Only lipid soluble substances can easily cross this barrier, and this is important in maintaining the compositional differences [2]. CSF has higher concentrations of sodium, chloride, and magnesium, while potassium, calcium, and glucose concentrations are lower than those of plasma [3]. The protein content is very low (0.2 g/l) resulting in a low buffering capacity. The $pCO_2$ is higher (50 mmHg) resulting in a lower CSF pH (7.33) [4].

I. Januleviciene (✉) · L. Siaudvytyte
Eye Clinic, Lithuanian University of Health Sciences, Kaunas, Lithuania
e-mail: ingrida.januleviciene@kaunoklinikos.lt

© Springer Nature Switzerland AG 2019                                                    469
G. Guidoboni et al. (eds.), *Ocular Fluid Dynamics*, Modeling and Simulation in
Science, Engineering and Technology, https://doi.org/10.1007/978-3-030-25886-3_20

CSF is produced by the choroid plexus in the lateral ventricles and travels through interventricular foramina to the third ventricle, and then the fourth ventricle via the cerebral aqueduct and finally to the subarachnoid spaces (SAS) via the median aperture of the fourth ventricle. In the cranial SAS, CSF circulates rostrally to the villous sites of absorption or caudally to the spinal SAS. CSF is absorbed into the blood by the arachnoid villi (90%) or directly into cerebral venules (10%) [2].

CSF circulation is a dynamic phenomenon. The total volume of CSF ranges from 140 to 270 ml in adults. The volume of the ventricles is about 25 ml. CSF production rate is 0.2–0.7 ml per minute or 600–700 ml per day. CSF turnover is about four to five times per day in young adults and three times per day at the age of 77 years, therefore wastes of brain metabolism, peroxidation products, and glycosylated proteins accumulate with aging [2].

Cerebrospinal fluid pressure (CSFP) is the result of a dynamic equilibrium between CSF secretion, absorption, and resistance to flow. Intracranial pressure (ICP) is the pressure inside the skull and thus in the brain tissue and CSF. It is established that the CSFP measured by lumbar puncture corresponds to ICP in the lateral decubitus position [5]. Therefore, ICP in the prone position is by classic definition the CSFP and these terms are used interchangeably in this text, as it is carried out in clinical practice. Physiological values of ICP varies with age and body posture but is generally considered to be 5–15 mmHg in healthy supine adults, 3–7 mmHg in children, and 1.5–6 mmHg in infants [6–11]. ICP varies with the systolic pulse wave, respiratory cycle, abdominal pressure, jugular venous pressure, state of arousal, physical activity, and posture. Head elevation decreases ICP by displacing CSF into the spinal canal and by improving cerebral venous drainage by opening alternative venous channels in the posterior circulation which remain closed while patients remain recumbent. It remains unclear whether the CSFP, measured by lumbar puncture, corresponds to the CSFP in the orbit around the optic nerve. As part of the CNS, the optic nerve is surrounded by the meninges and CSF. The CSF dynamics of the retrolaminar space probably have unique properties since there are numerous septae present which could limit tree flow of CSF [12]. In addition, unlike in other areas, the dura of optic nerve sheath contains atypical meningeal tissue with lymphoid characteristics [13]. Killer et al. found reduced CSF exchange between the basal cisterns and the SAS surrounding the optic nerve in normal tension glaucoma (NTG) patients, but not in control subjects. Lower ICP in NTG could explain the reduced density of the contrast-loaded CSF in the SAS of the optic nerve [14]. Experimental studies showed that ICP in the sitting position at the level of the occipital prominence, equivalent to eye level, ranges between 0 and $-10$ mmHg [15]. However, the method depends on the optic nerve path at SAS between the orbital and intracranial parts. It is not known what happens when the optic nerve canal is blocked (for example, in cases of suprasellar meningioma, tuberculous meningitis, or intracanalicular ophthalmic artery (OA) aneurysm).

## 2  Cerebrospinal Fluid in the Eye

Intracranial and intraocular pressures are interrelated and relatively independent pressure systems, which keeps themselves in a relatively stable state through aqueous and CSF circulations. These two circulating fluids have many similarities as they both are produced by carbonic anhydrase-catalyzed reactions, generally represent an ultrafiltrate of blood and have nearly identical chemical composition, with more proteins and less ascorbates in CSF. Physiologically ICP and intraocular pressure (IOP) are dynamic parameters, both have circadian variations (24-h) and similar response to changes in posture, intraabdominal or intrathoracic pressures [16]. IOP circadian cycle is quite well known [17]; however, circadian pattern of ICP is less clear, suggesting a nocturnal elevation in ICP [18, 19]. Zhang et al. found that the Valsalva maneuver-associated short-term increase in CSFP was significantly higher than increase in IOP. It led to a Valsalva maneuver-associated decrease or reversal of the translaminar pressure difference (TPD), which was associated with decreased optic cup-related parameters and enlarged neuroretinal rim-related parameters [20]. Other studies suggested that the risk of glaucoma is higher in patients with frequent Valsalva efforts [21, 22]. Wostyn et al. hypothesized that fluctuations in ICP could result in TPD fluctuations and thus fluctuations in the shear stress in the retinal ganglion cell axons, ultimately leading to glaucomatous damage [23].

Discussions about the importance of the CSFP in the pathogenesis of glaucomatous optic neuropathy started in 1908 by Russian ophthalmologist Noishevsky and were confirmed later experimentally with dogs [24, 25]. Hedges et al. presented experimental evidence of direct influence of raised intracranial and systemic arterial pressures on the optic nerve tissue [26]. Volkov pointed out that low CSFP could pathogenetically be associated with glaucomatous optic neuropathy [27]. In 1995, Morgan et al. published in IOVS their ideas about the influence of CSFP on the translaminar pressure gradient (TPG). According to their observations—there is a large number of patients with typical glaucomatous optic neuropathy in whom the IOP measurements have always been in the normal range [26]. Recently, researchers have focused on ICP and TPD role in glaucomatous optic neuropathy [28–34]. The optic nerve is exposed not only to IOP in the eye but also to ICP, as it is surrounded by CSF in the SAS. The lamina cribrosa demarcates these two pressurized zones and the pressure difference between them is called TPD (TPD = IOP–ICP) [35]. Physiologically, the difference between IOP ($14.3 \pm 2.6$ mmHg) and ICP ($12.9 \pm 1.9$ mmHg, in the supine position) is small [34]. A higher TPD may lead to abnormal function and damage of the optic nerve due to changes in axonal transportation, deformation of the lamina cribrosa, altered blood flow, or a combination thereof leading to glaucomatous damage [27, 36, 37, 38]. Furthermore, it is considered that the TPD can be a primary pressure-related factor for glaucoma, as the optic nerve head is located at the junction between the intraocular and retrobulbar spaces [38]. Experimental studies showed that chronic reduction in CSFP was associated with the development of an optic neuropathy in some monkeys

[39]. Growing evidence suggests a possible link between ICP and the pathogenesis of glaucomatous optic neuropathy, a low CSFP in the retrobulbar region of the orbit may theoretically have a similar effect as an increased IOP on TPD [29–34]. Retrospective analysis of patients who had lumbar puncture revealed that ICP was 3–4 mmHg lower in primary open-angle glaucoma (POAG) [40], and its subset NTG, compared with age-matched control subjects and patients with ocular hypertension. Further, they reported that the amount of glaucomatous damage to the optic nerve correlated with the TPD [41]. Recently, prospective studies revealed similar results to those in the retrospective studies, with the control group having the highest ICP and the smallest TPD [29–34].

ICP can affect the optic nerve in diseases in which the ICP becomes higher than the IOP, resulting in a swelling of the optic nerve head (papilledema). Possible conditions causing high ICP and papilledema include intracerebral mass lesions, cerebral hemorrhage, head trauma, meningitis, hydrocephalus, spinal cord lesions, impairment of cerebral sinus drainage, anomalies of the cranium, and idiopathic intracranial hypertension [42]. Similar swelling occurs in ocular hypotony, where the TPD is altered not by elevated ICP but by low IOP. Either way, the optic nerve swells secondary to an alteration of the TPD. The main mechanism of visual loss is likely due to axoplasmic flow stasis. High ICP produces a rise in CSFP surrounding the optic nerves, which disturbs the normal gradient between IOP and retrolaminar pressure, leading to high tissue pressure within the nerves. The increased tissue pressure within the nerves interrupts the metabolic processes that mediate axoplasmic flow [43–45].

Furthermore, researchers in the longitudinal Beijing Eye study have started to analyze ICP role in other ocular diseases, such as central retinal vein occlusion (CRVO), hypertensive retinopathy, and diabetic retinopathy [46–48]. They used estimated ICP formula for measuring CSFP; however, they did not include radiographical parameters that were previously tested and verified by Xie et al. [49]. Data showed that higher estimated ICP was associated with a higher incidence of CRVO ($p = 0.004$) [46], supporting the idea that ICP may have an influence on the retinal vein pressure. Jonas et al. found that larger retinal vein diameters and higher vein-to-artery diameter ratios were significantly associated with higher estimated ICP ($p = 0.001$). In contrast, retinal arterial diameters were not significantly associated with estimated ICP nor were other microvascular abnormalities such as arteriovenous crossing signs [47]. In another report, diabetic retinopathy was significantly associated with higher estimated ICP ($p = 0.04$) after adjusting for a higher hemoglobin A1c value, longer duration of diabetes mellitus, higher serum concentration of creatinine, lower educational level, and shorter axial length [48]. The reason for these associations may be that an elevated retinal venous pressure in patients with higher ICP may be associated with a higher retinal capillary blood pressure, potentially explaining the increased incidence and prevalence of retinal hemorrhages, edema, and lipid exudates as part of diabetic retinopathy.

# 3   Intracranial Pressure Measuring Instruments

## 3.1   General Requirements, International Standards, Pros and Cons

Currently, direct measurement of CSFP via lumbar puncture or via implantation of the pressure sensor into the brains ventricle is considered to be the "gold standard" of ICP measurement [15, 50, 51]. Direct measurement of ICP is not without risk due to its invasiveness and potential risk of intracranial hemorrhage, or even cerebral herniation and hence cannot be widely used as a matter of safety concerns [52]. Overview of advantages and disadvantages of possible invasive/non-invasive methods to evaluate ICP is shown in Table 1.

The optimal ICP monitoring device should be accurate, reliable, cost effective, and cause minimal patient morbidity, as stated by Lundberg [53]. International standard requirements for technologies are accuracy and precision. According to ISO 5725-1, the general term "accuracy" is used to describe the closeness of a measurement to the true value; precision is the closeness of agreement among a set of results: the repeatability or reproducibility of the measurement. Standard deviation may serve as a measure of uncertainty showing how far typical values tend to be from the mean. Furthermore, the use of correlation coefficient to compare devices against the gold standard is inappropriate, and agreement should be evaluated using the method described by Bland and Altman [54].

ICP monitoring device should have the following specifications [55, 56]:

- pressure range from 0 to 100 mmHg;
- accuracy $\pm 2$ mmHg in the 0–20 mmHg range;
- maximum error not exceed $\pm 10\%$ in the 20–100 mmHg range.

**Table 1** Pros and Cons of ICP measurement devices

|  | Pros | Cons |
|---|---|---|
| Invasive ICP devices | 1. Gold standard of accuracy (high accuracy)<br>2. Can be used for drainage of CSF<br>3. Administering of drugs intrathecally<br>4. Allows ICP control | 1. Invasive<br>2. Risk of infection (low or moderate)<br>3. Risk of hemorrhages (low)<br>4. High/relative high cost |
| Non-invasive ICP devices | 1. Non-invasive<br>2. Risk of infection (almost zero)<br>3. Risk of hemorrhages (almost zero)<br>4. Low cost | 1. Low accuracy<br>2. High percentage of unsuccessful measurements<br>3. Clinically not applicable |

According to the Brain Trauma Foundation guidelines for management of severe traumatic brain injury, further improvement in ICP monitoring technology should focus on developing multiparametric ICP devices that can provide simultaneous measurement of ventricular CSF drainage, parenchymal ICP, and other advanced monitoring parameters [57].

Consensus summarizing the results of clinical studies using different sensors and locations showed the following [58]:

- uncertainty (U = ± 2 standard deviations (SD)) of invasive ICP measurement techniques varies from ±4.2 mmHg to ±19.2 mmHg;
- invasive ICP sensors cannot be calibrated in situ after implantation into the brain and systematic error (bias) cannot be eliminated;
- required uncertainty U of non-invasive absolute ICP measurement method must be equal or less than that of most accurate invasive ICP measurement technique (U = 2 SD ≤ 4.2 mmHg).

## 3.2 Invasive Intracranial Pressure Measuring Methods

Direct measurements of CSFP are considered to be the "gold standard" of ICP measurement [15, 50, 51]. Different invasive ICP measuring methods are used in clinical practice, as they have relative accuracy and precision; however, these methods require sophisticated technology and complicated procedures.

Currently, the external ventricular drainage (EVD) is the most accurate, low-cost, and reliable method of monitoring ICP [59]. This method has been proven to be reliable, permits periodic recalibration, and allows therapeutic CSF drainage. Nevertheless, obstruction of the system and the requirement to consistently maintain the external transducer at a fixed reference point can lead to inaccuracy during clinical use [60]. Importantly, this invasiveness includes the potential risk for intracranial hemorrhages, infection, cerebral herniation, or leakage of CSF [53, 60]. It led to the development of alternative intracranial sites and technologies for ICP monitoring. The most common alternative location for ICP monitoring is the cerebral parenchyma. Microtransducers are very accurate at the time of placement; they have been reported to drift over the time [59, 61]. Precision of parenchymal ICP monitors has been assessed by comparing the measurement value at the time of ICP monitor removal with atmospheric pressure [62]. The cost of these microtransducers is higher than EVD. Subdural and epidural monitors and externally placed anterior fontanelle monitors are less accurate than the above-mentioned devices. Moreover, invasive methods lend themselves only to a small portion of pathological disorders in which ICP measurement can be used and consequently many patients who might benefit from ICP measurement, including glaucoma patients, do not do so.

Depending on the technique, invasive ICP measuring methods can be undertaken in different intracranial anatomical locations: intraventricular, intraparenchymal, epidural, subdural, and subarachnoidal (Fig. 1) [63]. Technology evaluation is

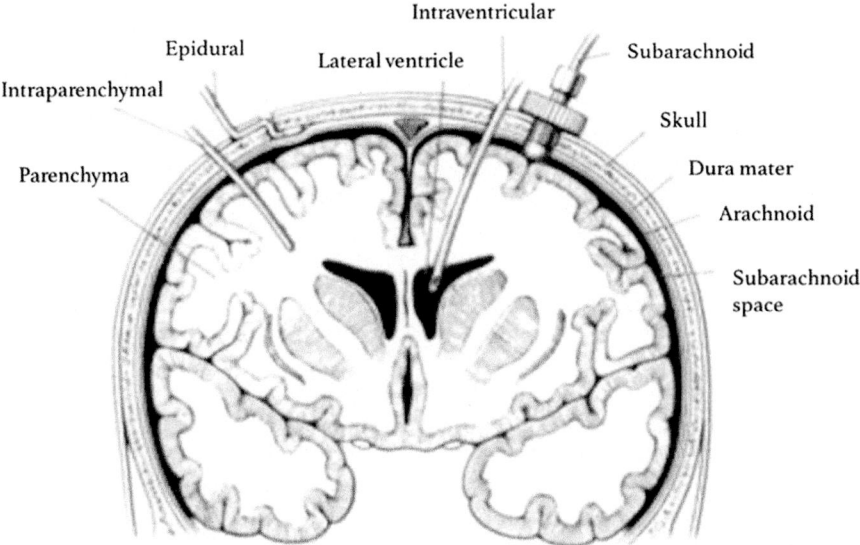

**Fig. 1** Different intracranial pressure monitors. (Reprinted with permission from Lyons et al. [63])

not ideally suited to evidence-based medicine, but the literature suggests that a ventricular catheter connected to an external strain gauge, catheter tip strain gauge devices, or catheter tip fiber optic technology inserted into the ventricles or brain parenchyma is accurate, whereas fluid-coupled or pneumatic devices placed in the subarachnoid, subdural, and epidural compartments are less accurate [56].

### 3.2.1   External Ventricular Drainage or Ventriculostomy

EVD (also called a ventriculostomy) is considered to be the gold standard of ICP measurement in neurosurgical intensive care unit when ICP is within limits of 0– 100 mmHg [5, 64–68]. During this procedure, a ventricular catheter is placed into one of the ventricles by the way of a twist-drill craniostomy. This technique can also be used for drainage of CSF and administering of drugs intrathecally [56, 64]. Drainage of CSF has been used in routine clinical practice for lowering of ICP. However, acute CSF drainage can displace the intracerebral structures, and in severe cases even provoke subincarceration [67]. EVD has the disadvantage that it penetrates the meninges and brain, which gives the risk of bacterial transmission through the fluid coupling. Complications related to ventricular catheters include 0–27% risk of infection and 0.61–41% risk of hemorrhage [69, 70]. There are factors that predispose toward higher infection rates such as frequent CSF sampling, duration of usage, catheter irrigation, and site leaks [71]. The rate of nonfunctioning EVD has been found to be 6.3% [59, 72, 73]. The cost for placing an EVD includes around 200$ in materials [59].

### 3.2.2 Microtransducer Intracranial Pressure Monitoring Devices

The first ICP microtransducers were introduced in the 1980 with the first commercial design of the Honeywell Microtransducer Catheter MTC-P5F [74]. The first widely used microtransducer was the fiberoptic Camino OLMICP Monitor (Camino Laboratories, San Diego, CA, USA).

Microtransducers are mostly designed for intraparenchymal ICP measurement and do not allow drainage of CSF. The accuracy of microtransducers is lower than EVD and remains reliable [63, 75]. Furthermore, microtransducers require calibration and zeroing only once before insertion to the brain. Accuracy of readings is not dependent upon patient positioning in relation to the transducer [57, 64]. However, without the ability to recalibrate the sensor in situ, parenchymal devices are subject to varying degrees of zero drift over the time [57, 76]. The difference between the starting ICP value when the sensor is calibrated (0 mmHg), and the ICP value that is measured when the sensor is removed is termed zero drift. However, microtransducers perform well during in vitro testing, with drift as low as $0.6 \pm 0.9$ mmHg [77]. Microtransducers carry a lower rate of postoperative infections and bleedings (2%), moreover the rate is lower than with EVD [78]. Microtransducers are defective in 3.14–5.0% of cases [61, 79–81]. These devices are more expensive than EVD, costing at least 400–600\$ each [59].

Microtransducers can be divided into three groups:

1. fiber optic devices,
2. strain gauge devices,
3. pneumatic sensors.

Fiber optic devices, such as the Camino ICP Monitor (Integra Neuroscience, Plainsboro, NJ), transmit light toward a miniature displaceable mirror at the catheter tip [82]. Changes in ICP move the mirror, and the change in reflected light intensity is converted to a measured change in ICP. A very important aspect is that a pressure-sensitive element is placed in the head, so the risk of infection is minimized. This gives it an advantage over prolonged ICP monitoring, which uses a fluid-filled catheter and external transducer [83]. They allow for continuous recording and monitoring of ICP in different brain compartments and give accurate pressure readings, making possible the analysis of waveforms in the compartment where the probe is placed. Camino microtransducers are associated with 1.1–2.5% of hemorrhagic and 4.75–8.5% of infectious complications, and 3.14–4.5% of technical errors [79, 81]. The rate of nonfunctioning microtransducers has been found to be 10% [80]. It is characterized as having very low zero drift over long periods of time, very good frequency response, and stable linearity [84]. Crutchfield et al. reported that the device had an accuracy of $\pm 3$ mmHg [range 0–30 mmHg] in vitro. The maximum daily zero drift was $\pm 2.5$ mmHg, with an average daily drift of $\pm 0.6$ mmHg, and the drift rate over a 5-day period was $\pm 2.1$ mmHg. In vivo, the pressures and waveform characteristics were very similar between the fiberoptic device and a strain gauge transducer connected to a ventriculostomy [85].

Strain gauge devices include the Codman microsensor (Johnson and Johnson, Raynham, MA), the Raumedic Neurovent-P ICP sensor (Raumedic AG, Munchberg, Germany), and the Pressio sensor. The strain gauges are connected via tiny wires that extend through the length of the flexible nylon tube to complete a Wheatstone bridge type circuit located in the connector housing.

The Codman microsensor incorporates two semiconductor strain gauges mounted on a thin diaphragm in titanium housing at the catheter tip. The diaphragm distorts in proportion to the applied pressure and a Wheatstone bridge transducer changes in pressure to changes in resistances that are subsequently displayed as ICP [86]. Studies have found the Codman microsensor to be accurate, with an average difference of $-0.5$ to $+2.6$ mmHg between the microsensor reading and an intraventricular fluid-coupled external strain gauge transducer. It is stable, with a daily drift that varies between $-0.13$ and $+0.11$ mmHg per day. It is flexible and can be tunneled beneath the scalp, preventing it from being easily broken. Its small size is an additional advantage, particularly for pediatric patients. The Codman microsensor is non-fluid coupled therefore does not require *irrigation* and has a low risk of infection [87–89].

The Neurovent-P ICP monitor is also based on an electronic chip strain gauge coated by a thin silicon membrane mounted at the distal tip of the catheter. The incorporation of the Wheatstone bridge into the chip enhances the drift characteristics by reducing temperature sensitivity and the effects of non-pressure-related external strains [90]. Neurovent-P catheters incorporating three monitoring variables (ICP, brain tissue oxygen partial pressure, and temperature) are now available, although clinical data with this device are limited [91]. Neurovent-P microsensor is associated with almost no complications (hemorrhages 2.0%, infection 0%) [89–91].

The relatively new Pressio sensor has yet to be tested thoroughly in vivo. Lescot et al., looking at the accuracy of Pressio sensors and Codman MicroSensors in comparison to EVD, found that two types of sensors performed much alike, with a mean difference of $\pm 7$ mmHg compared to ICP values acquired by the EVD. Complications were not recorded [76].

Pneumatic sensors (Spiegelberg GmbH, Hamburg, Germany) use a small air pouch balloon at the end of the catheter to register changes in pressure, and additionally allow quantitative measurement of intracranial compliance [92]. Depending on the technique, monitoring can be done in the intraventricular, intraparenchymal, epidural, subdural, or subarachnoidal compartment. Lang et al. reported no incidences of hemorrhage or infection, and just 3.45% of incorrect measurements [61]. The Spiegelberg parenchymal ICP measurement system is unique—it is the first such system capable of performing regular automatic zeroing in situ and can be recalibrated.

It is worth mentioning that Neurovent-P sensor, Spiegelberg sensor, and Codman microsensor are compatible with magnetic resonance imaging (MRI) without any danger to the patient. The Camino monitor and Pressio sensor contain ferromagnetic components, and therefore patients with these devices cannot undergo MRI [91, 93].

Looking over accuracy and precision on invasive ICP measuring techniques:

- Meta-analysis of microtransducers vs EVD [94]:

    - fiberoptic/microstrain gauge probes vs EVD–mean difference 0.9 mmHg (95% confidence interval (CI) 0.4–1.5 mmHg);
    - fiberoptic vs microstrain gauge probes—mean difference 1.8 mmHg (95% CI 1.5–2.2 mmHg);

- Pressio and Codman microtransducers approximates EVD with an accuracy of ±7 mmHg [76]:

    - Pressio microtransducer vs EVD—mean difference 0.6 mmHg (limits of agreement (1.96 SD) −8.1 to 6.9 mmHg);
    - Codman microtransducer vs EVD—mean difference 0.3 mmHg (limits of agreement (1.96 SD) −6.7 to 7.1 mmHg);

- Spiegelberg microtransducer vs EVD—mean difference $0.1 \pm 4.9$ mmHg [95];
- Strain gauge microtransducer vs EVD—mean difference $0.5 \pm 2.6$ mmHg, ($r = 0.97, p < 0.001$) [96];
- Neurovent-P microsensor vs EVD—mean difference $0.8 \pm 2.2$ mmHg (range −4 to 8 mmHg) [90];
- Codman microtransducer vs EVD—mean difference 0.7 mmHg (ICP measured with Codman $19.0 \pm 0.2$, with EVD was $18.3 \pm 0.3$ mmHg ($r = 0.79$, $p < 0.0001$)) [97].
- Intraparenchymal ICP monitoring vs EVD—mean difference in pulse pressure amplitude −0.13 mmHg (95% CI −0.13 to −0.12 mmHg), single wave pressure −0.71 mmHg (95% CI −0.74 to −0.68 mmHg), latency −0.01 s (95% CI −0.01 to −0.01 s) [98];
- Intraparenchymal ICP monitoring vs EVD, using the Hummingbird® Synergy Ventricular System, which is a novel device allowing multiparametric neurological monitoring, including both ventricular and parenchymal ICP—mean difference −0.95 mmHg (95% CI −0.97 to −0.93 mmHg) [99].

### 3.2.3  Lumbar Puncture

The lumbar puncture technique was introduced in 1891 and resulted in new options to investigate the intrathecal environment [100]. Although not explicitly proven, the technique is believed to accurately estimate ICP when CSF circulates freely [101]. CSFP measured by lumbar puncture varies depending on body position. Most variations are due to coughing or internal compression of jugular veins in the neck. When lying down, the CSF as estimated by lumbar puncture is similar to the ICP. In newborns, CSFP ranges from 8 to 10 cmH$_2$O (4.4–7.3 mmHg or 0.78–0.98 kPa), in adults—ranges from 10 to 18 cmH$_2$O (8–15 mmHg or 1.1–2.0 kPa) with the patient lying on the side and from 20 to 30 cm H$_2$O (16–24 mmHg or 2.1–3.2 kPa) in a sitting position [102]. There are many indications for lumbar puncture, but obtaining CSF may be the only way of confirming or refuting subarachnoid hemorrhage, meningitis, and neuro-inflammatory diseases. The most common complication (32%) is post lumbar puncture headache [103]. The pain is usually diffuse, global, or bitemporal headache, which can be accompanied by nausea, altered hearing, tinnitus, photophobia, or neck stiffness. Low pressure may produce diplopia due to traction on the fourth or sixth cranial nerve [104]. Cortical vein thrombosis [105] and reversible cerebral vasoconstriction syndrome [106] have been reported as very rare complications of low CSFP. Other risks include failure to obtain CSF, localized bruising, bleeding, and local discomfort at the injection site. Iatrogenic meningitis and nerve root injury are exceptionally rare [107].

## 3.3   Non-invasive Intracranial Pressure Measuring Methods

The idea of non-invasive ICP measuring method is captivating, as complications seen in relation to the invasive ICP measuring methods, such as hemorrhage or infection, are avoidable. These considerations have driven investigators to develop non-invasive approaches for ICP estimation [108–112]. All approaches of non-invasive ICP estimation methods are based on mathematical model or correlation of some anatomical or physiological parameters of the human head or brain with ICP:

- Fundoscopy and papilledema (correlation-based approach);
- MRI & computer tomography (mathematical model-based approach);
- Optic nerve sheath diameter (ONSD) (correlation-based approach);
- Transcranial Doppler (TCD) ultrasonography (correlation or mathematical model-based approach);
- Tympanic membrane displacement (TMD) (correlation-based approach);
- Visual-evoked potentials (VEP) (correlation-based approach);
- Brain stem auditory-evoked potentials (BAEP) (correlation-based approach);
- Ophthalmodynamometry (ODM) (correlation-based approach);
- Scalp blood flow (SBF)—laser Doppler flowmetry (correlation-based approach);
- Impedance audiometry (correlation-based approach).

Unfortunately, correlation-based approaches cannot be accurately used for quantitative ICP value measurement (mmHg or mmH$_2$O) because patient-specific calibration is needed for these methods. Thus, there was a need for a non-invasive absolute ICP measurement device with accuracy similar to that of the invasive "gold standard" ICP meters [113]. Seeking to measure absolute ICP value researchers created non-invasive ultrasound-based measurement—a two-depth TCD device which uses an OA as a natural pressure sensor. Accuracy and precision of this device have been previously investigated and shown to be clinically useful [112, 114].

The non-invasive techniques require only the single expense of purchasing the device, after which the devices can be applied multiple times without further costs apart from wages and maintenance. The current non-invasive techniques are simply not accurate enough to replace the traditional invasive techniques.

### 3.3.1 Fundoscopy and Papilledema

Raised ICP typically manifests as papilledema or swelling of the prelaminar portion of the optic nerve head, which can result (10%) in permanent visual impairment [115]. Papilledema can be visualized by fundoscopy and graded by the Frisen Scale into five categories depending on signs of disturbed axoplasmic transport (reproducibility 88–96% and sensitivity between 93 and 100%) [116]. However, even though fundoscopy is often used as a screening method in cases of suspected increase in ICP, the grading scale is not widely applicable or accepted. Furthermore, since the process of optic disc swelling in cases of raised ICP takes time, the technique cannot be applied in emergency situations with sudden increases in ICP, such as trauma [115]. Besides, optic disc swelling may occur in a large variety of conditions (e.g., papillitis and optic neuropathies). Moreover, studies showed no association between severity of papilledema and CSFP, IOP, blood pressure (BP), or any other physiologic parameter [117].

### 3.3.2 Magnetic Resonance Imaging and Computer Tomography

Ideas of using MRI as a non-invasive method of ICP measurement started in 2000 [118]. Researchers considered to small changes in intracranial volume and ICP which occur naturally with each cardiac cycle. The pressure change during the cardiac cycle is derived from the CSFP gradient waveform calculated from the CSF velocities. The intracranial volume change is determined by the instantaneous differences between arterial blood inflow, venous blood outflow, and CSF volumetric flow rates into and out of the cranial vault. An elastance index was derived from the ratio of pressure and volume changes and found to correlate well with invasively measured ICP ($r^2 = 0.96$; $p < 0.005$). However, the technique is very sensitive to differences in heart rate measured in the circulation contra the CSF flow rate as well as CSF measurements, besides it is not easy to choose the representative blood vessels [119]. MRI-based ICP measurements could play a role in diagnosis and evaluation of several chronic disorders potentially associated with increased ICP values, i.e., hydrocephalus, pseudotumor cerebri, intracranial mass lesions, and so forth [120]. Wang et al. found decreased optic nerve SAS width measured by MRI with fat-suppressed fast recovery fast spin echo T2-weighted sequence in NTG patients, compared to high tension glaucoma patients and healthy controls in Chinese population [121]. They measured the orbital SAS width by subtracting the optic nerve diameter from the optic nerve sheath diameter (ONSD). In another study, Xie et al. showed a linear relationship between the orbital SAS width determined by MRI and direct lumbar CSFP measurements [49].

Eide reported no significant correlation between actual size of cerebral ventricles by cranial computer tomography (CT) scans and invasively monitored ICP in consecutive patients [122]. A linear but ultimately non-predictive relationship between baseline ICP and initial head CT scan characteristics was found by Miller and colleagues [123]. Similar results were observed by Hiler et al. who concluded that mean ICP values in the first 24-h cannot be predicted by using the Marshall CT scan classification [124]. Overall, no method of estimating ICP on the basis of cranial CT scans currently exists.

### 3.3.3 Optic Nerve Sheath Diameter

Recently, research has extended into ultrasonography of ONSD (Fig. 2) and its relation with ICP [125]. The optic nerve is a part of the CNS and therefore surrounded by the dural sheath and has 0.1–0.2 mm SAS, which communicates with the SAS surrounding the brain. Changes in ONSD can be visualized using transocular ultrasound. The technique is cheap, efficient, and takes around 5 min per patient [126]. However, it requires training and has intra-observer (± 0.1–0.2 mm) and inter-observer (± 0.2–0.3 mm) variances [127–129]. In 10% of patients investigated using ONSD, no valid measurements could be made [125, 127]. Moreover, several conditions can affect the diameter of the optic nerve (for example, tumors, inflammations, sarcoidosis, Grave's disease, and trauma [127, 129]), which makes ONSD measurements impossible. In patients with optic atrophy and a reduced optic nerve diameter, a normal sheath diameter could be associated with an elevated CSFP.

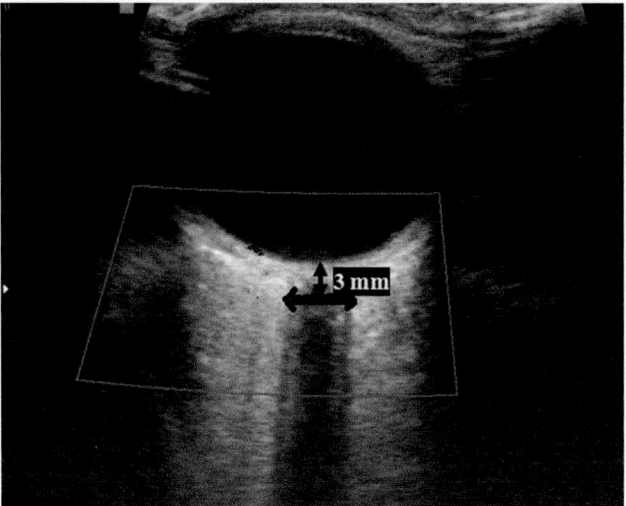

**Fig. 2** Optic nerve sheath diameter measurement 3 mm behind the globe

ONSD measurement method can potentially be used as a screening method for detecting raised ICP—condition when the sheaths expand. Several studies found correlations between invasively measured ICP and ONSD [110, 126, 127, 129–132], some of them even showed 96% sensitivity and 94% specificity for increased ICP (>20 mmHg) for ONSD cutoff value of 4.8 mm [130]. However, other authors found poor reliability of this method [133].

Analyzing glaucoma patients, Jaggi et al. found that NTG patients had increased ONSD, obtained by CT, similar to those of patients with increased ICP [134]. They concluded that increased ONSD in NTG patients might have been due to an optic nerve sheath compartmentation or due to the thinning of the optic nerve. However, Pinto et al. found no statistically significant differences in ONSD, obtained by ultrasonography, between POAG, NTG, and healthy controls [135]. That finding contradicted the hypothesis of an abnormally low ICP in patients with NTG. ONSD is a promising tool for estimation of increased ICP [125, 136, 137], but not for measurement of normal or decreased ICP.

### 3.3.4 Transcranial Doppler Ultrasonography

TCD ultrasonography is a simple non-invasive method used to measure blood flow velocity in the middle cerebral artery (MCA). The use of TCD ultrasonography as a predictor of ICP was first described by Klingerhofer et al. in a pilot study on patients with dissociated brain death. They reported that flow patterns in MCA measured by TCD ultrasonography were related to changes in ICP [138]. Several studies found that MCA pulsatility index (PI), which is calculated as difference between systolic and diastolic flow velocities, divided by the mean flow velocity, correlates with ICP; however, correlation range was from 0.43 to 0.93 [138–140]. In contrast, other researchers failed to find a relationship between MCA PI and ICP [141–144]. Behrens et al. announced that an ICP of 20 mmHg found by using PI had 95% CI of −3.8 to 43.8 mmHg [143]. It must be mentioned that PI depends on several factors such as arterial pressure pulsatility, heart rate, cerebral perfusion pressure, arterial carbon dioxide concentration, cerebral resistance, and compliance of the big vessels. Furthermore, there is intra-observer and inter-observer variations [145] and the technique cannot be used on 10–15% of patients due to the ultrasound not being able to penetrate the skull [146].

### 3.3.5 Tympanic Membrane Displacement

TMD technique requires a patent cochlear aqueduct, normal middle ear pressure, and an intact stapedial reflex. The principle of this technique is through the communication of the CSF and the perilymph via the perilymphatic duct. Stimulation of the stapedial reflex causes a movement of the tympanic membrane, which is shown to correlate with ICP [147, 148]. This membrane is flexible, meaning that the pressure of the fluid in the cochlea affects how the membrane and stapes are positioned and how they move. Shimbles et al. found a correlation between the invasively measured ICP and ICP measured by TMD. However, the method has high inter-subject variability, thus precluding the method for clinical use [148]. In 60% of patients investigated using TMD, no valid measurements could be made [149]. Moreover, perilymphatic duct is less passable with age, especially after the age of 40, thus TMD measurements have a relatively low practicality.

### 3.3.6  Visual-Evoked Potentials

VEP is a method for recording the electrical signals at the scalp over the occipital cortex in response to light stimulus. The mechanical effects of intracranial hypertension on the central visual projection pathways, together with the repercussions on brain perfusion, may slow neuronal conduction, which could be demonstrated through VEP. The technique has advantages of being simple, rapid, and that may be performed in the supine position and does not require strict cooperation and visual fixation, allowing this procedure to be performed even in comatose patients [150, 151]. Studies showed a relationship between a prolongation of the latency of waves from the diffuse light flash VEP test and increased ICP in patients with hydrocephalus, traumatic brain injury, drug-related cerebral edema, and cryptococcal meningitis [152–159]. However, VEP in head trauma is biased by other variables as arterial pO2, sagittal sinus venous pO2, and regional cerebral flow and is not an adequate substitute for direct ICP measures [156–158].

### 3.3.7  Brain Stem Auditory-Evoked Potentials

BAEP is a very important neurophysiological examination in neurotraumatology, giving functional information about the severity and clinical prognosis of brain injury, sensitive to pontomesencephalic integrity, transtentorial brain herniation, and increased ICP [160–163]. Continuous monitoring of BAEP was carried out in 57 comatose patients for periods ranging from 5 h to 13 days. In 53 cases, ICP was also simultaneously monitored. The simultaneous study of BAEP and ICP showed that apparently significant (greater than 40 mmHg) acute rises in ICP were not always followed by BAEP changes. The stability of BAEP despite significant ICP rises was associated with a high probability of survival, while prolongation of central latency of BAEP in response to ICP modifications was almost invariably followed by brain death [164]. Continuous monitoring of brainstem responses provided a useful physiological counterpart to physical parameters such as ICP.

### 3.3.8  Ophthalmodynamometry

ODM is a useful method for measuring the venous outflow pressure of the central retinal vein (CRV) (Fig. 3) [108, 165–168]. Physiologically, the pressure in the CRV is equal to or higher than ICP. CRV passes through the optic nerve head and the

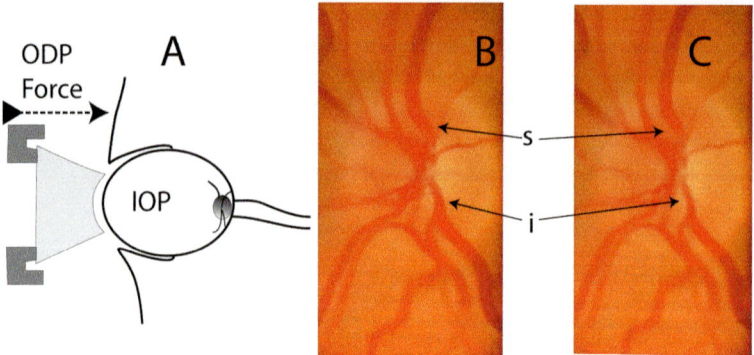

**Fig. 3** Ophthalmodynamometry. (**a**) Schematic diagram of ophthalmodynamometric pressure (ODP) being applied through an ophthalmodynamometer to the eye with baseline, post-dilation intraocular pressure (IOP). (**b**) Superior (s) and inferior (i) hemi-veins are pulsating with showing them at dilation and (**c**) during collapse phase of pulsation. (From Morgan et al. [26]; image made available under CC-BY 4.0)

first 10 mm of the anterior portion of the optic nerve before it pierces through the orbital SAS and the optic nerve meninges. First, the optic nerve head is examined ophthalmoscopically with a non-contact ophthalmoscopic lens. In 80–90% of all healthy subjects, a spontaneous pulsation of the CRV can be observed [166]. If there is no spontaneous vein pulsation, a Goldmann contact lens with a pressure sensor in its holding grip is put onto the cornea and a slight pressure is applied until the vein shows early pulsations. At that point, the pressure applied onto the contact lens plus the IOP at baseline gives the diastolic CRV pressure. The pressure value at the point of collapse is termed the venous outflow and is found to linearly predict ICP [108, 167–169]. Firsching et al. found that an increased pressure of the CRV indicated an elevated ICP, with a probability of 84.2%, whereas a normal pressure of the CRV indicated a normal ICP in 92.8% of patients [8]. Querfurth et al. showed that ODM is able (area under curve 0.89; 95% CI 0.73–1.05) to predict raised ICP [167]. However, ODM cannot be applied in cases of ocular trauma or conditions that selectively affect the optic nerve and gives erroneously high readings in the presence of a papilledema, which may persist long after ICP has returned to normal. ODM is useful for momentary assessment of the ICP, can easily be repeated, and may be used whenever an elevated ICP is suspected. However, ODM cannot detect an abnormally low ICP, since the technique can be applied only if there are no spontaneous pulsations of the CRV.

### 3.3.9 Scalp Blood Flow: Laser Doppler Flowmetry

Intracranial and extracranial circulations are interrelated by emissary veins as well as other collateral veins and arteries [169]. The same direct effect of elevated ICP on brain venous system could be responsible for the changes in SBF. Experimental study with rabbits showed SBF rapid response to ICP manipulations, correlating well up to pressures of about 18 mmHg. A negative correlation between SBF and ICP was found ($r = -0.839$) [170]. The mechanism of changes in SBF due to changes in ICP is unclear at present. One may speculate that blood flow changes are due to an alteration in BP. BP elevation is probably a compensatory mechanism to correct brain blood flow impairments induced by high ICP values.

### 3.3.10  Impedance Audiometry (Otoacoustic Emission)

Otoacoustic emission (OAE) is a sound generated by subtle oscillations of the endolymph and perilymph caused by contractions of the outer hair cells of the inner ear in response to a loud sound. The sound is transmitted to the stapes and further through the ossicles, to the tympanic membrane from which it can be detected with a sensitive microphone inserted into the ear canal. OAE is used in clinical practice to test for hearing deficits in babies and children who are too young to cooperate. The equipment can be made portable, and is relatively easy to use. Researchers proposed formula that relates ICP to the intensity or phase of the measured OAE signal, and described how the other physiological signals or behaviors that are known to affect ICP such as small oscillations of ICP with each heartbeat, respiration, or posture changes can be used to confirm the validity of the obtained measurements (e.g., the absence of modulation of the measured OAE phase with respiration may indicate occlusion of the cochlear aqueduct, in which case OAE cannot provide any information about ICP). There is little data up to date about the clinical utility or accuracy of OAE as a measure of ICP. Frank et al. in a pilot study revealed that increased ICP or conditions known to increase ICP (e.g., posture changes, abdomen compression, and coughing) were associated with notable decreases in intensity of the evoked OAE. All results were however reported only as group averages, and no attempt was made to derive a quantitative one-to-one relation between the OAE intensity and ICP. This method as all other correlation-based approaches cannot be used for absolute ICP value measurement because of impossibility of individual calibration [171].

### 3.3.11  Other Methods

Xie et al. estimated mathematical CSFP formula based on MRI-assisted measurement of the orbital SAS width, mean BP, and body mass index (BMI) [49]. In their study, measured CSFP via lumbar puncture ($13.6 \pm 5.1$ mmHg) did not differ significantly from the calculated MRI-derived CSFP (accordingly, $12.7 \pm 4.2$ mmHg when measuring orbital SAS width 3 mm behind the globe ($p = 0.07$), $13.4 \pm 5.1$ mmHg—9 mm behind the globe ($p = 0.35$), and $14.0 \pm 4.9$ mmHg—15 mm behind the globe ($p = 0.87$)) [49]. Jonas et al. created a surrogate corrective factor to account for population-based tendencies of the orbital SAS width that was not described in any paper, but used in many subsequent studies

Two-depth TCD device had better reliability and relationship with CSFP than ONSD method [178]. To estimate its validity in glaucoma patients, researchers have made a pilot study and found that TPD was higher in glaucoma patients compared to healthy controls ($p < 0.001$). Additionally, reduction of neuroretinal rim area was related to higher TPD in NTG patients [31]. The TCD-based non-invasive absolute ICP value measurement method is suitable for more than 96% individuals with a normal OA anatomy, when OA is a branch of the internal carotid artery, and it has an intracranial segment compressed by ICP. The technique depends on a free communication between the intracranial compartments to the orbital portion of the CSF space. The method may thus not be applicable if the optic nerve canal is blocked.

## 4 Conclusions

Currently, the "gold standard" for ICP measurement remains invasive measurement of the pressure in the CSF via lumbar puncture or via implantation of the pressure sensor into the brains ventricle. However, the direct measurements of ICP are not without risk due to its invasiveness and potential risk of intracranial hemorrhage or infection, and cannot be widely used as a matter of safety concerns. Therefore, development of non-invasive absolute ICP measurement method could be a tool to overcome the high risk to benefit ratio of current invasive ICP measurement methods, leading to possible multidisciplinary application. Further investigations are needed to prove and find the agreement between invasive and non-invasive ICP measurement methods, processing to implementation to everyday clinical practice.

## References

1. Veening JG, Barendregt HP. The regulation of brain states by neuroactive substances distributed via the cerebrospinal fluid: a review. Cerebrospinal Fluid Res 2010;7:1.
2. Sakka L, Coll G, Chazal J. Anatomy and physiology of cerebrospinal fluid. European Annals of Otorhinolaryngology. Head and Neck diseases 2011;128:309–316.
3. Harrington MG, Salomon RM, Pogoda JM et al. Cerebrospinal fluid sodium rhythms. Cerebrospinal Fluid Res 2010;7:3.
4. Siesjo BK. The regulation of cerebrospinal fluid pH. Kidney International 1972;1:360–374.
5. Lenfeldt N, Koskinen L, Bergenhein A, et al. CSF pressure assessed by lumbar puncture agrees with intracranial pressure. Neurology 2007;68(2):155–158.
6. Gjerris F, Brennum J. The cerebrospinal fluid, intracranial pressure and herniation of the brain. Clinical Neurology and Neurosurgery. Paulson OB, Gjerris F, Sørensen PS, Eds., FADL's Forlag Aktieselskab, Copenhagen, Denmark, 2004;179–196.
7. Smith M. Monitoring intracranial pressure in traumatic brain injury. Anesth Analg 2008;106:240–248.
8. Gilland O. Normal cerebrospinal-fluid pressure. N Engl J Med 1969;280:904–905.
9. Gilland O, Tourtellotte WW, O'Tauma L, et al. Normal cerebrospinal fluid pressure. J Neurosurg 1974;40:587–593.

10. Chapman PH, Cosman ER, ArnoldMA. The relationship between ventricular fluid pressure and body position in normal subjects and subjects with shunts: a telemetric study. J Neurosurg 1990:26:181–189.
11. Albeck MJ, Børgesen SE, Gjerris F, et al. Intracranial pressure and cerebrospinal fluid outflow conductance in healthy subjects. J Neurosurg 1991;74:597–600.
12. Jaggi GP, Mironov A, Huber AR, et al. Optic nerve compartment syndrome in a patient with optic nerve sheath meningioma. Eur J Ophthalmol 2007;17(3):454–458.
13. Killer HE, Jaggi G, Miller NR, et al. Does immunohistochemistry allow easy detection of lymphatics in the optic nerve sheath? J Histochem Cytochem 2008;56(12):1087–1092.
14. Killer HE, Miller NR, Flammer J, et al. Cerebrospinal fluid exchange in the optic nerve in normal-tension glaucoma. Br J Ophthalmol 2012;96(4):544–548.
15. Magnaes B. Body position and cerebrospinal fluid pressure. Part 2: Clinical studies on orthostatic pressure and the hydrostatic indifferent point. J Neurosurg 1976;44(6):698–705.
16. Dickerman RD, Smith GH, Langham-Roof L. Intra-ocular pressure changes during maximal isometric contraction: does this reflect intra-cranial pressure or retinal venous pressure? Neurol Res 1999;21(3):243–246.
17. Sit AJ, Liu JHK. Pathophysiology of glaucoma and continuous measurements of intraocular pressure. Mol Cell Biomech 2009;6:57–69.
18. Maurel D, Ixart G, Barbanel G. Effects of acute tilt from orthostatic to head-down antiorthostatic restraint and of sustained restraint on the intra-cerebroventricular pressure in rats. Brain Res 1996;736:165–173.
19. Morrow BA, Starcevic VP, Keil LC. Intracranial hypertension after cerebroventricular infusions in conscious rats. Am J Physiol 1990;258: R1170–1176.
20. Zhang Z, Wang X, Jonas JB, et al. Valsalva manoeuver, intra-ocular pressure, cerebrospianl fluid pressure, optic disc topography: Beijing Intracranial and Intra-ocular Pressure Study. Acta Ophthalmol 2014;92(6):e475–80.
21. Schuman JS, Massicotte EC, Connolly S et al. Increased intraocular pressure and visual field defects in high resistance wind instrument players. Ophthalmology 2000;107:127–133.
22. Krist D, Cursiefen C, Junemann A. Transitory intrathoracic and abdominal pressure elevation in the history of 64 patients with normal pressure glaucoma. Klin Monatsbl Augenheilkd 2001;218 209–213.
23. Wostyn P, De Groot V, Audenaert K, et al. Are intracranial pressure fluctuations important in glaucoma? Med Hypotheses 2011;77 598–600.
24. Рейтузов ВА, Кириллов ЮА. Ноишевский и его вклад в офтальмологию и неврологию. Конференция Глаукома: теория и практика. Горизонты нейропротекции Сборник научных статей: Под редакцией: проф. В.Н. Алексеева, доц. В.И. Садкова – СПб.: Изд-во «Человек и его здоровье» 2014;9:92–96.
25. Noishevsky KI. Stagnant nipple and excavation of the optic nerve by lowering the intracranial pressure. Vestn Ophthalmol 1912;29(2):117–125.
26. Hedges TR, Zaren HA. The Relationship of optic nerve tissue pressure to intracranial and systemic arterial pressure. Am J Ophthalmol. 1973;75:90–98.
27. Volkov VV. Essential element of the glaucomatous process neglected in clinical practice. Oftalmol Zh 1976;31:500–504.
28. Morgan WH, Yu DY, Cooper RL, et al. The influence of cerebrospinal fluid pressure on the lamina cribrosa tissue pressure gradient. Invest Ophthalmol Vis Sci 1995;36(6):1163–1172.
29. Ren R, Wang N, Zhang X, et al. Trans-lamina cribrosa pressure difference correlated with neuroretinal rim area in glaucoma. Graefes Arch Clin Exp Ophthalmol 2011:249(7):1057–1063.
30. Jonas JB. Role of cerebrospinal fluid pressure in the pathogenesis of glaucoma. Acta Ophthalmol 2011:89(6):505–514.
31. Siaudvytyte L, Januleviciene I, Ragauskas A, et al. The difference in translaminar pressure gradient and neuroretinal rim area in glaucoma and healthy subjects. J Ophthalmol 2014;2014:937360.

32. Siaudvytyte L, Januleviciene I, Daveckaite A, et al. Literature review and meta-analysis of translaminar pressure difference in open-angle glaucoma. Eye (Lond). 2015;29(10):1242–1250.
33. Siaudvytyte L, Januleviciene I, Daveckaite A, et al. Neuroretinal rim area and ocular haemodynamic parameters in patients with normal-tension glaucoma with differing intracranial pressures. Br J Ophthalmol 2016;100(8):1134–1138.
34. Jonas JB, Nangia V, Wang N, et al. Trans-lamina cribrosa pressure difference and open-angle glaucoma. The Central India Eye and Medical Study. PLoS One 2013;8(12):e82284.
35. Morgan WH, Yu DY, Alder VA, et al. The correlation between cerebrospinal fluid pressure and retrolaminar tissue pressure. Invest Ophthalmol Vis Sci 1998;39(8):1419–1428.
36. Ren R, Jonas JB, Tian G, et al. Cerebrospinal fluid pressure in glaucoma: a prospective study. Ophthalmology 2010;117(2):259–266.
37. Morgan WH, Chauhan BC, Yu DY, et al. Optic disc movement with variations in intraocular and cerebrospinal fluid pressure. Invest Ophthalmol Vis Sci 2002;43(10):3236–3242.
38. Burgoyne CF, Downs JC, Bellezza AJ, et al. The optic nerve head as a biomechanical structure: a new paradigm for understanding the role of IOP-related stress and strain in the pathophysiology of glaucomatous optic nerve head damage. Prog Retin Eye Res 2005;24(1):39–73.
39. Yang D, Fu J, Hou R et al. Optic neuropathy induced by experimentally reduced cerebrospinal fluid pressure in monkeys. Invest Ophthalmol Vis Sci 2014;55:3067–3073.
40. Berdahl JP, Allingham RR, Johnson DH. Cerebrospinal fluid pressure is decreased in primary open-angle glaucoma. Ophthalmology 2008;115(5):763–768.
41. Berdahl JP, Fautsch MP, Stinnett SS, et al. Intracranial Pressure in Primary Open-Angle Glaucoma, Normal Tension Glaucoma, and Ocular Hypertension: A Case-Control Study. Invest Ophthalmol Vis Sci 2008;49(12):5412–5419.
42. Rigi M, Almarzouqi SJ, Morgan ML, et al. Papilledema: epidemiology, etiology, and clinical management. Eye Brain 2015;7:47–57.
43. Hayreh SS. Pathogenesis of optic disc oedema in raised intracranial pressure. Trans Ophthalmol Soc UK 1976;96:404–407.
44. Hayreh SS. Optic disc edema in raised intracranial pressure. V Pathogenesis. Arch Ophthalmol 1977;95:1553–1565.
45. Hayreh SS, March W, Anderson DR. Pathogenesis of block of rapid orthograde axonal transport by elevated intraocular pressure. Exp Eye Res 1979;28:515–523.
46. Jonas JB, Wang N, Wang YX, et al. Incident retinal vein occlusions and estimated cerebrospinal fluid pressure. The Beijing Eye Study. Acta Ophthalmol. 2015;93:e522–e526
47. Jonas JB, Wang N, Wang S, et al. Retinal vessel diameter and estimated cerebrospinal fluid pressure in arterial hypertension. The Beijing Eye Study. Am J Hypertens 2014;27:1170–1178.
48. Jonas JB, Wang N, Xu J, et al. Diabetic retinopathy and estimated cerebrospinal fluid pressure. The Beijing Eye Study 2011. PLoS One 2014;9(5):e96273.
49. Xie X, Zhang X, Fu J, et al. Non-invasive intracranial pressure estimation by orbital subarachnoid space measurement: the Beijing intracranial and intraocular pressure (iCOP) study. Crit Care 2013;17(4):R162.
50. Andrews P, Citerio G, Longhi L. NICEM consensus on neurological monitoring in acute neurological disease. Intensive Care Med 2008;34:1362–1370.
51. Digre KB, Corbett JJ. Idiopathic intracranial hypertension (pseudotumor cerebri): a reappraisal. Neurologist 2001;7:62–67.
52. Zeng T, Gao L. Management of patients with severe traumatic brain injury guided by intraventricular intracranial pressure monitoring: a report of 136 cases. Chin J Traumatol 2010;13:146–151.
53. Lundberg N. Continuous recording and control of ventricular fluid pressure in neurosurgical practice. Acta Psychiatr Scand 1960;36:1–193.
54. Bland JM, Altman DG. Statistical methods for assessing agreement between two methods of clinical measurement. Lancet 1986;1:307–310.

55. Peter JDA, Citerio G, Longhi L, et al. Neuro–Intensive Care and Emergency Medicine (NICEM) consensus on neurological monitoring in acute neurological disease. Intensive Care Medicine 2008;34:1362–1370.
56. Kawoos U, McCarron RM, Auker CR, et al. Advances in intracranial pressure monitoring and its significance in managing traumatic brain injury. Int J Mol Sci 2015;16(12):28979–28997.
57. Brain Trauma Foundation, American Association of Neurological Surgeons, Congress of Neurological Surgeons. Guidelines for the management of severe traumatic brain injury. J Neurotrauma. 2007;24(Suppl 1):S1–S106.
58. Le Roux P, Menon DK, Citerio G, et al. Consensus summary statement of the International Multidisciplinary Consensus Conference on Multimodality Monitoring in Neurocritical Care: a statement for healthcare professionals from the Neurocritical Care Society and the European Society of Intensive Care Medicine. Intensive Care Med 2014;40(9):1189–209.
59. Brain Trauma Foundation, American Association of Neurological Surgeons, Congress of Neurological Surgeons. Guidelines for the management of severe traumatic brain injury. VII. Intracranial pressure monitoring technology. J Neurotrauma 2007;24(Suppl 1):S45–S54.
60. Citerio G, Andrews PJ. Intracranial pressure. 2. Clinical applications and technology. Intensive Care Med 2004;30:1882–1885.
61. Lang JM, Beck J, Zimmermann M, et al. Clinical evaluation of intraparenchymal Spiegelberg pressure sensor. Neurosurgery 2003;52(6):1455–1459.
62. Koskinen LO, Olivecrona M. Clinical experience with the intraparenchymal intracranial pressure monitoring Codman MicroSensor system. Neurosurgery 2005;56:693–698.
63. Lyons MK, Meyer FB. Cerebrospinal fluid physiology and the management of increased intracranial pressure. Mayo Clin Proc. 1990;65(5):684–707.
64. Zhong J, Dujovny M, Park HK, et al. Advances in ICP monitoring techniques. Neurological Research 2003;25(4):339–350.
65. March K. Intracranial pressure monitoring: why monitor? AACN Clinical Issues 2005;16(4):456–475.
66. Cremer OL. Does ICP monitoring make a difference in neurocritical care? Eur J Anaesthesiol Suppl 2008;42:87–93.
67. Steiner LA, Andrews PJD. Monitoring the injured brain: ICP and CBF. Br J Anaesth 2006;1:26–38.
68. Bhatia A, Gupta AK. Neuromonitoring in the intensive care unit. I. Intracranial pressure and cerebral blood flow monitoring. Intensive Care Med 2007;33(7):1263–1271.
69. Beer R, Lackner P, Pfausler B, et al. Nosocomial ventriculitis and meningitis in neurocritical care patients. J Neurol 2008;255(11):1617–1624.
70. Lozier AP, Sciacca RR, Romagnoli MF, et al. Ventriculostomy-related infections: a critical review of the literature. Neurosurgery 2002;51(1):170–182.
71. Hoefnagel D, Dammers R, Ter Laak-Poort MP, et al. Risk factors for infections related to external ventricular drainage. Acta Neurochir 2008;150:209–214.
72. North B, Reilly P. Comparison among three methods of intracranial pressure recording. Neurosurgery 1986;18(6):730–732.
73. Drake JM. Ventriculostomy for treatment of hydrocephalus. Neurosurg Clin N Am 1993;4(4):657–666.
74. Ostrup RC, Luerssen TG, Marshall LF, et al. Continuous monitoring of intracranial pressure with a miniaturized fiberoptic device. J Neurosurg 1987;67:206–209.
75. Sundbarg G, Nordstrom CH, Messeter K, et al. A comparison of intraparenchymatous and intraventricular pressure recording in clinical practice. J Neurosurg 1987;67:841–845.
76. Lescot T, Reina V, LeManach Y, et al. In vivo accuracy of two intraparenchymal intracranial pressure monitors. Intensive Care Med 2011;37:875–879.
77. Czosnyka M, Czosnyka Z, Pickard JD. Laboratory testing of three intracranial pressure microtransducers: technical report. Neurosurgery 1996;38:219–224.
78. Littlejohns LR, Bader MK (eds). AACN-AACN protocols for practice: monitoring technologies in critically Ill neuroscience patients. Jones and Bartlett, Sudbury, MA. 2008

79. Gelabert-Gonzalez M, Ginesta-Galan V, Sernamito-Garcia R, et al. The Camino intracranial pressure device in clinical practice. Assessment in a 1000 cases. Acta Neurochirurg 2006;148(4):435–441.
80. Piper I, Barnes A, Smith D, et al. The Camino intracranial pressure sensor: is it optimal technology? An internal audit with a review of current intracranial pressure monitoring technologies. Neurosurgery 2001;49(5):1158–1164.
81. Bekar A, Dogan S, Abas F, et al. Risk factors and complications of intracranial pressure monitoring with a fiberoptic device. J Clin Neurosci 2009;16(2):236–240.
82. Munch E, Weigel R, Schmiedek P, et al. The Camino intracranial pressure device in clinical practice: reliability, handling characteristics and complications. Acta Neurochir 1998;140:1113–1119.
83. Speck V, Staykov D, Huttner HB, et al. Lumbar catheter for monitoring of intracranial pressure in patients with post-hemorrhagic communicating hydrocephalus. Neurocritical Care 2011;14(2):208–215.
84. Bray RS, Chodroff NG, Narayan RK, et al. A new fiberoptic monitoring device: Development of the ventricular bolt. In: Hoff JT, Betz AL, eds. Intracranial Pressure VII, New York: Springer-Verlag, 1989:45–47.
85. Crutchfield JS, Narayan RK, Robertson CS, et al. Evaluation of a fiberoptic intracranial pressure monitor. J Neurosurg 1990;72:482–487.
86. Fernandes HM, Bingham K, Chambers IR et al. Clinical evaluation of the Codman microsensor intracranial pressure monitoring system. Acta Neurochir Suppl 1998;71:44–46.
87. Gray WP, Palmer JD, Gill J, et al. A clinical study of parenchymal and subdural minature strain-gauge transducer for monitoring intracranial pressure. Neurosurgery 1996;39:927–931.
88. Hong WC, Tu YK, Chen YS, et al. Subdural intracranial pressure monitoring in severe head injury: clinical experience with the Codman MicroSensor. Surg Neurol 2006;66:S8–S13.
89. Koskinen LOD, Olivecrona M. Clinical experience with the intraparenchymal intracranial pressure monitoring Codman microsensor system. Neurosurgery 2005;56(4):693–698.
90. Citerio G, Piper I, Cormio M et al. Bench test assessment of the new Raumedic Neurovent-P ICP sensor: a technical report by the BrainIT group. Acta Neurochir 2004;146:1221–1226.
91. Stendel R, Heidenreich J, Schilling A, et al. Clinical evaluation of a new intracranial pressure monitoring device. Acta Neurochir 2003;145:185–193.
92. Piper I, Dunn L, Contant C, et al. Multi-centre assessment of the Spiegelberg compliance monitor: preliminary results. Acta Neurochir Suppl 2000;76:491–494.
93. Shellock FG. Biomedical implants and devices: assessment of magnetic field interactions with a 3.0-Tesla MR system. J Magn Reson Imaging 2002;16(6):721–732.
94. Zacchetti L, Magnoni S, Corte FD, et al. Accuracy of intracranial pressure monitoring: systematic review and meta-analysis. Critical Care 2015;19:420.
95. Chambers IR, Siddique MS, Banister K, et al. Clinical comparison of the Spiegelberg parenchymal transducer and ventricular fluid pressure. J Neurol Neurosurg Psychiatry 2001;71:383–385.
96. Gopinath SP, Robertson CS, Contant CF, et al. Clinical evaluation of a miniature strain-gauge transducer for monitoring intracranial pressure. Neurosurgery 1995;36(6):1137–1140.
97. Koskinen LD, Olivecrona M. Clinical Experience with the Intraparenchymal Intracranial Pressure Monitoring Codman MicroSensor System. Neurosurgery 2005;56(4):693–698.
98. Brean A, Eide PK, Stubhaug A. Comparison of intracranial pressure measured simultaneously within the brain parenchyma and cerebral ventricles. J Clin Monit Comput 2006;20:411–414.
99. Vender J, Waller J, Dhandapani K, et al. An evaluation and comparison of intraventricular, intraparenchymal, and fluid-coupled techniques for intracranial pressure monitoring in patients with severe traumatic brain injury. J Clin Monit Comput 2011;25:231–236.
100. Quincke H. Lumbalpunktion des hydrocephalus. Berl Klin Wochenschr 1891;929–933.
101. Langfitt TW, Weinstein JD, Kassell NF, et al. Transmission of increased intracranial pressure. I. Within the craniospinal axis. J Neurosurg 1964;21:989–997.

102. Agamanolis D. Cerebrospinal Fluid: The normal CSF. Chapter 14. Neuropathology. Northeast Ohio Medical University. 2011.
103. Armon C, Evans RW. Addendum to assessment: Prevention of postlumbar puncture headaches: report of the Therapeutics and Technology Assessment Subcommittee of the American Academy of Neurology. Neurology 2005;65(4):510–512.
104. Thomke F, Mika-Gruttner A, Visbeck A, et al. The risk of abducens palsy after diagnostic lumbar puncture. Neurology 2000;54(3):768–769.
105. Laverse E, Cader S, de Silva R, et al. Peripartum isolated cortical vein thrombosis in a mother with postdural puncture headache treated with an epidural blood patch. Case Rep Med 2013;2013:701264.
106. Schievink WI, Maya MM, Chow W, et al. Reversible cerebral vasoconstriction in spontaneous intracranial hypotension. Headache 2007;47(2):284–287.
107. Fishman RA. 2nd ed. Philadelphia: WB. Saunders Co.; Cerebrospinal fluid in diseases of the nervous system. 1992.
108. Firsching R, Schultze R, Motschmann M, et al. Venous ophthalmodynamometry: a noninvasive method for assessment of intracranial pressure. J Neurosurg 2000;93:33–36.
109. Bellner J, Romner B, Reinstrup P, et al. Transcranial Doppler sonography pulsatility index (PI) reflects intracranial pressure (ICP). Surg Neurol 2004;62(1);45–51.
110. Bauerle J, Nedelmann M. Sonographic assessment of the optic nerve sheath in idiopathic hypertension. J Neurol 2011;258:2014–2019.
111. Li Z, Yang Y, Lu Y, et al. Intraocular pressure vs intracranial pressure in disease conditions: A prospective cohort study (Beijing iCOP study). BMC Neurol 2012;12:66.
112. Ragauskas A, Matijosaitis V, Zakelis R, et al. Clinical assessment of noninvasive intracranial pressure absolute value measurement method. Neurology 2012;78:1684–1691.
113. Taylor JR. An Introduction to Error Analysis: the Study of Uncertainties in Physical Measurement, 2nd ed. Herndon, VA: University Science Books 1997;128–129.
114. Ragauskas A, Daubaris G, Dziugys A, et al. Innovative non-invasive method for absolute intracranial pressure measurement without calibration. Acta Neurochir Suppl 2005;95:357–361.
115. Hayreh SS. Pathogenesis of optic disc edema in raised intracranial pressure. Prog Retin Eye Res 2016;50:108–144.
116. Sinclair AJ, Burdon MA, Nightingale PG, et al. Rating papilloedema: an evaluation of the Frisén classification in idiopathic intracranial hypertension. J Neurol 2012;259:1406–1412.
117. Fleischman D, Perry JT, Rand Allingham R, et al. Retrospective analysis of translaminar, demographic, and physiologic parameters in relation to papilledema severity. Can J Ophthalmol 2017;52(1):26–29.
118. Alperin NJ, Lee SH, Loth F, et al. MR-intracranial pressure (ICP): a method to measure intracranial elastance and pressure noninvasively by means of MR imaging: baboon and human study. Radiology 2000;217(3):877–885.
119. Marshall I, MacCormick I, Sellar R, et al. Assessment of factors affecting MRI measurement of intracranial volume changes and elastance index. Br J Neurosurg 2008;22(3):389–397.
120. Raksin PB, Alperin N, Sivaramakrishnan A, et al. Noninvasive intracranial compliance and pressure based on dynamic magnetic resonance imaging of blood flow and cerebrospinal fluid flow: review of principles, implementation, and other noninvasive approaches. Neurosurg Focus 2003;14(4):e4.
121. Wang N, Xie X, Yang D, et al. Orbital cerebrospinal fluid space in glaucoma: The Beijing Intracranial and Intraocular Pressure (iCOP) Study. Ophthalmology 2012;119(10):2065–2073.
122. Eide PK. The relationship between intracranial pressure and size of cerebral ventricles assessed by computed tomography. Acta Neurochirurg 2003;145(3):171–179.
123. Miller MT, Pasquale M, Kurek S, et al. Initial head computed tomographic scan characteristics have a linear relationship with initial intracranial pressure after trauma. J Trauma 2004;56(5):967–973.

124. Hiler M, Czosnyka M, Hutchinson P, et al. Predictive value of initial computerized tomography scan, intracranial pressure, and state of autoregulation in patients with traumatic brain injury. J Neurosurg 2006;104(5):731–737.
125. Geeraerts T, Launey Y, Martin L, et al. Ultrasonography of the optic nerve sheath may be useful for detecting raised intracranial pressure after severe brain injury. Intensive Care Med 2007;33(10):1704–1711.
126. Kimberly HH, Shah S, Marill K, et al. Correlation of optic nerve sheath diameter with direct measurement of intracranial pressure. Acad Emerg Med 2008;15(2):201–204.
127. Soldatos T, Karakitsos D, Chatzimichail K, et al. Optic nerve sonography in the diagnostic evaluation of adult brain injury. Critical Care 2008;12(3):R67.
128. Ballantyne SA, O'Neill G, Hamilton G, et al. Observer variation in the sonographic measurement of optic nerve sheath diameter in normal adults. Eur J Ultrasound 2002;15(3):145–149.
129. Bäule J, Lochner P, Kaps M, et al. Intra- and interobserver reliability of sonographic assessment of the optic nerve sheath diameter in healthy adults. J Neuroimaging 2012;22(1):42–45.
130. Soldatos T, Chatzimichail K, Papathanasiou M, et al. Optic nerve sonography: a new window for the non-invasive evaluation of intracranial pressure in brain injury. Emerg Med J 2009;26(9):630–634.
131. Le A, Hoehn ME, Smith ME, et al. Bedside sonographic measurement of optic nerve sheath diameter as a predictor of increased intracranial pressure in children. Ann Emerg 2009;53:785–791.
132. Rajajee V, Vanaman M, Fletcher JJ, et al. Optic nerve ultrasound for the detection of raised intracranial pressure. Neurocrit Care 2011;15(3):506–515.
133. Strumwasser A, Kwan RO, Yeung L, et al. Sonographic optic nerve sheath diameter as an estimate of intracranial pressure in adult trauma. J Surg Res 2011;170 265–271.
134. Jaggi GP, Miller NR, Flammer J, et al. Optic nerve sheath diameter in normal-tension glaucoma patients. Br J Ophthalmol 2012;96:53–56.
135. Pinto LA, Vanderwalle E, Pronk A, et al. Intraocular pressure correlates with optic nerve sheath diameter in patients with normal tension glaucoma. Graefes Arch Clin Exp Ophthalmol 2012;250:1075–1080.
136. Kimberly HH, Shah S, Marill K, et al. Correlation of optic nerve sheath diameter with direct measurement of intracranial pressure. Acad Emerg Med 2008;15:201–204.
137. Dubourg J, Javouhey E, Geeraerts T, et al. Ultrasonography of optic nerve sheath diameter for detection of raised intracranial pressure: a systematic review and metaanalysis. Invest Care Med 2011;37:1059–1068.
138. Klingerhofer J, Conrad B, Benecke R, et al. Intracranial flow patterns at increasing intracranial pressure. J Mol Med 1997;65(12):542–545.
139. Bellner J, Romner B, Reinstrup P, et al. Transcranial Doppler sonography pulsatility index (PI) reflects intracranial pressure (ICP). Surg Neurol 2004;62(1):45–51.
140. Voulgaris SG, Partheni M, Kaliora H, et al. Early cerebral monitoring using the transcranial Doppler pulsatility index in patients with severe brain trauma. Med Sci Monit 2005;11(2):CR49–52.
141. Moreno JA, Mesalles E, Gener J, et al. Evaluating the outcome of severe head injury with transcranial Doppler ultrasonography. Neurosurg Focus 2000;8(1):e8.
142. Figaji AA, Zwane E, Fieggen AG, et al. Transcranial Doppler pulsatility index is not a reliable indicator of intracranial pressure in children with severe traumatic brain injury. Surg Neurol 2009;72:389–394.
143. Behrens A, Lenfeldt N, Ambarki K, et al. Transcranial Doppler pulsatility index: not an accurate method to assess intracranial pressure. Neurosurgery 2010;66(6):1050–1057.
144. Brandi G, Béchir M, Sailer S, et al. Transcranial color-coded duplex sonography allows to assess cerebral perfusion pressure noninvasively following severe traumatic brain injury. Acta Neurochirurg 2010;152(6):965–972.
145. McMahon CJ, McDermott P, Horsfall D, et al. The reproducibility of transcranial Doppler middle cerebral artery velocity measurements: implications for clinical practice. Br J Neurosurg 2007;21(1):21–27.

146. Tsivgoulis G, Alexandrov AV, Sloan MA. Advances in transcranial Doppler ultrasonography. Curr Neurol Neurosci Rep 2009;9(1):46–54.
147. Lang EW, Paulat K, Witte C, et al. Noninvasive intracranial compliance monitoring: technical note and clinical results. J Neurosurg 2003;98(1):214–218.
148. Reid A, Marchbanks RJ, Burge DM, et al. The relationship between intracranial pressure and tympanic membrane displacement. Br J Audiol 1990;24(2):123–129.
149. Shimbles S, Dodd C, Banister K, et al. Clinical comparison of tympanic membrane displacement with invasive intracranial pressure measurements. Physiol Meas 2005;26(6): 1085–1092.
150. American Clinical Neurophysiology Society. Guideline 9B: guidelines on evoked potentials. J Clin Neurophysiol 2006;23(2):138–156.
151. Odom VJ, Bach M, Brigell M, et al. ISCEV standard for clinical visual evoked potentials (2009 update). Doc Ophthalmol 2010;120(1):111–119.
152. Coupland SG, Cochrane DD. Visual evoked potentials, intracranial pressure and ventricular size in hydrocephalus. Doc Ophthalmol 1987;66(4):321–329.
153. Desch LW. Longitudinal stability of visual evoked potentials in children and adolescents with hydrocephalus. Dev Med Child Neurol 2001;43(2):113–117.
154. Gumerlock MK, York D, Durkis D. Visual evoked responses as a monitor of intracranial pressure during hyperosmolar Blood-Brain Barrier Disruption. Acta Neurochir Suppl 1994;60:132–135.
155. Sjöström A, Uvebrant P, Roos A. The light-flash-evoked response as a possible indicator of increased intracranial pressure in hydrocephalus. Childs Nerv Syst 1991;11(7):381–387.
156. Stone JL, Ghaly RF, Hughes JR. Evoked potentials in head injury and states of increased intracranial pressure. J Clin Neurophysiol 1988;5(2):135–160.
157. York D, Legan M, Benner S, et al. Further studies with a noninvasive method of intracranial pressure estimation. Neurosurgery 1984;14(4):456–461.
158. York DH, Pulliam MW, Rosenfeld JG, et al. Relationship between visual evoked potentials and intracranial pressure. J Neurosurg 1981;55(6):909–916.
159. Vieira MA, Cavalcanti Mdo A, Costa DL, et al. Visual evoked potentials show strong positive association with intracranial pressure in patients with cryptococcal meningitis. Arq Neuropsiquiatr 2015;73(4):309–313.
160. Bricolo A, Turazzi S, Alexandre A, et al. Decerebrate rigidity in acute head injury. J Neurosurg 1977;47:680–698.
161. Cant BR, Hume AL, Judson JA, et al. The assessment of severe head injury by short latency somatosensory and brain stem auditory evoked potentials. Electroencephalogr Clin Neurophysiol 1986;65:188–195.
162. Olbrich HM, Nau HE, Lodemann E, et al. Evoked potential assessment of mental function during recovery from severe head injury. Surg Neurol 1986;26:112–118.
163. Ottaviani F, Almadori G, Calderazzo AB, et al. Auditory brain stem and middle-latency auditory responses in the prognosis of severely head injured patients. Electroencephalogr Clin Neurophysiol 1986;65:196–202.
164. Garcia-Larrea L, Artru F, Bertrand O, et al. The combined monitoring of brain stem auditory evoked potentials and intracranial pressure in coma. A study of 57 patients. J Neurol Neurosurg Psychiatry 1992;55(9):792–798.
165. Morgan WH, House PH, Hazelton ML, et al. Intraocular pressure reduction is associated with reduced venous pulsation pressure. PLoS One 2016;11(1): e0147915.
166. Harder B, Jonas JB. Frequency of spontaneous pulsations of the central retinal vein in normal eyes. Br J Ophthamol 2007;91:401e–402e.
167. Querfurth H, Lieberman P, Arms S, et al. Ophthalmodynamometry for ICP prediction and pilot test on the Mt. Everest. BMC Neurol 2010;10:106.
168. Motschmann M, Muller C, Kuchenbecker J. Ophthalmodynamometry: a reliable method for measuring intracranial pressure. Strabismus 2001;9:13e–16e.
169. Hoffman WY. Scalp injuries and their management. In: Youmans JR. Neurological Surgery, 4th ed. WB Saunders 1966;P:1916.

170. Eskandry H, Reihani H, Najafipour H. Relationship between scalp blood flow and intracranial pressure in rabbits. Arc Iranian Med 1999;2(2):71–6.
171. Meyerson SC, Avan PA, Buki B: US20036589189. 2003.
172. Fleischman D, Bicket AK, Stinnett SS, et al. Analysis of cerebrospinal fluid pressure estimation using formulae derived from clinical data. Invest Ophthalmol Vis Sci 2016;57(13):5625–5630.
173. Berdahl JP, Fleischman D, Zaydlarova J, et al. Body mass index has a linear relationship with cerebrospinal fluid pressure. Investig Ophthalmol Vis Sci 2012;53,:1422e–1427e.
174. Fleischman D, Berdahl JP, Zaydlarova J, et al. Cerebrospinal fluid pressure decreases with older age. PLoS One 2012;7(12):e52664.
175. Kashif FM, Verghese GC, Novak V, et al. Model-based noninvasive estimation of intracranial pressure from cerebral blood flow velocity and arterial pressure. Sci Transl Med 2012;129:129–44.
176. Budohoski PK, Schnidt B, Smielewski P, et al. Non-invasive estimated ICP pulse amplitude strongly correlates with outcome after traumatic Brain Injury. Acta Neurochir Suppl Intracranial Pressure and Brain Monitoring XIV, 2012:121–124.
177. Schmidt B, Klingelhofer J, Schwarze JJ, et al. Noninvasive prediction of intracranial pressure curves using transcranial Doppler ultrasonography and blood pressure curves. Stroke 1997;28:2465–2472.
178. Ragauskas A, Bartusis L, Piper I, et al. Improved diagnostic value of a TCD-based non-invasive ICP measurement method compared with the sonographic ONSD method for detecting elevated intracranial pressure. Neurol Res 2014;36(7):607–614.

# Mathematical Modeling of the Cerebrospinal Fluid Flow and Its Interactions

**Lorenzo Sala, Fabrizia Salerni, and Marcela Szopos**

**Abstract** The present chapter provides an overview of mathematical models that describe the cerebrospinal fluid and its possible interactions with other biofluids or neighboring tissues. The description of the underlying mechanisms stems from the basic principles of fluid and solid dynamics and it is translated into systems of partial or ordinary differential equations. The current review aims at specifying a complementary view with respect to other revisions in the field by illustrating some selected studies that may constitute a first step in the connection between cerebral and ocular fluid dynamics.

First we present lumped-parameter models with a simplified mathematical structure that allow fast computations of the average values of the unknowns and capture the main behavior of the cerebral flow. The discussion then moves onto distributed models, which produce a three-dimensional representation of cerebrospinal fluid flows in complex geometries that may derive from medical imaging. These models allow the investigation of the fluid flow at fine scales, constituting a highly valuable tool in providing a better understanding of pathophysiology.

We acknowledge that significant efforts have already been made in the attempt to unify mathematics and medicine in the context of brain diseases caused by cerebrospinal fluid abnormalities; nevertheless, we report a pressing demand for innovative mathematical models that may achieve the complete description and simulation of the biophysical connections between the eye and the brain.

L. Sala
IRMA, UMR CNRS 7501, Université de Strasbourg, Strasbourg, France
e-mail: sala@math.unistra.fr

F. Salerni
Mathematical, Physical and Computer Sciences, University of Parma, Parma, Italy
e-mail: fabrizia.salerni@studenti.unipr.it

M. Szopos (✉)
MAP5, UMR CNRS 8145, Université Paris Descartes, Paris, France
e-mail: marcela.szopos@parisdescartes.fr

© Springer Nature Switzerland AG 2019
G. Guidoboni et al. (eds.), *Ocular Fluid Dynamics*, Modeling and Simulation in
Science, Engineering and Technology, https://doi.org/10.1007/978-3-030-25886-3_21

# 1  Introduction

The aim of the present chapter is to describe mathematical models based on the fundamental laws of physics, for instance, mass and momentum conservation. These models can be multiscale both in space and in time and multiphysics (e.g., involving a fluid–structure coupling).

We focus on the intracranial dynamics, possibly coupled with craniospinal models, since these are the most meaningful compartments in view of modeling ocular fluid dynamics. In addition, we only provide a brief overview of some "stand-alone" spinal interactions models. Note that in the present contribution, we describe macroscopic mathematical models available to study cerebrospinal fluid flow, therefore chemistry, micro- and mesoscale models are not considered.

Whenever possible, we highlight significant contributions to clinical applications for all the models reviewed, in particular in the field of neurological pathologies (hydrocephalus, Chiari malformation, etc.).

# 2  Models and Methods

Cerebrospinal fluid (CSF) dynamics and its complex coupling with the other cerebral fluids and surrounding tissues was recently reviewed in:

- Kurtcuoglu [72].
- Linninger et al. [83].

Our aim is to specify a complementary view to these works by providing a review of selected works from the abundant literature on CSF modeling based on the possible connections between these brain models with existing models in ocular fluid dynamics.

The two following subsections share a similar structure: we first present in a condensed manner a general view of the models included in the review (under the form of three tables) and, in a second stage, we review in detail three most-representative contributions in each topic. Several acronyms are used throughout the text, all the details are provided in Table 7.

The remaining part of the section is organized as follows:

- Section 2.1 is devoted to the description of the reduced modeling approach and to the review of different contributions in this direction;
- Section 2.2 focuses on distributed models and on reviewing several contributions in the literature following this approach;
- Section 2.3 discusses very recent contributions towards modeling the highly complex coupled dynamics of biofluids in the eye-cerebral network.

## 2.1   Reduced Models for CSF Dynamics and Its Interactions

This section reviews several *reduced* models for CSF flow, possibly coupled with cerebral hemodynamics and brain tissue dynamics. We also present two very recent contributions studying the complex interplay between cerebral and ocular biofluids.

Reduced models have the advantage of being a useful tool to provide insights on the interaction between different constituents and/or compartments of the system under investigation. In addition, they are able to capture the main physical phenomena governing the system at a lower computational cost than the full solution of a three-dimensional problem describing the CSF flow by means of a partial differential equation (PDE) model.

Various modeling reduction techniques have been proposed in the literature, see, for instance, [27, Chap.10] and the references cited therein. *One-dimensional models* (1D) take advantage of the basically cylindrical morphology of some biological conduits. The resulting mathematical description is based on 1D hyperbolic PDEs, which can capture wave-propagation phenomena in the biofluids, with a reasonable computational cost. This is not the approach we will follow in the sequel, since we rather focused on an even more simplified description of the flow.

Reduced *zero-dimensional models* (0D), also called *lumped-parameter models* (see chapter "Mathematical and Physical Modeling Principles of Complex Biological Systems"), provide a circuit-based representation of the fluid dynamics in each compartment, based on the analogy between electric and hydraulic networks. By writing *Kirchhoff's laws* for the nodes (conservation of current/flow rate) and for closed loops (conservation of the voltage/pressure), the resulting mathematical model is a system of differential algebraic equations (DAE), potentially nonlinear. These models are used in different ways depending on the specific modeling needs.

In a first approach, many studies implemented them as boundary conditions for three-dimensional simulations in regions of particular interest. Alternatively, stand-alone 0D reduced models have been used to understand the main dynamics of the system, thanks to their simplified mathematical structure, and to compute average values of the unknowns. They are also particularly well-suited for sensitivity analysis studies with respect to changes in the input data that can be used to discriminate the contribution of different factors on flow quantities. In addition, these robust mathematical models have the advantage that their complexity can be dynamically increased and implemented as the understanding of the underlying phenomena evolves.

The main modeling ingredients involved in the mathematical description of 0D reduced models for fluid flows are:

- *Hagen–Poiseuille's law*, according to which the *pressure-driven flux* $Q_{ij}$ between the generic compartments $i$ and $j$ is governed by the hydraulic analogue of Ohm's law

$$Q_{ij} = \frac{P_i - P_j}{R_{ij}} \qquad (1)$$

where $R_{ij}$ denotes the hydraulic resistance between compartment $i$ and compartment $j$;
- *Starling–Landis's law*, describing *filtration-driven flows* in general terms as follows: the flux $Q_{ij}$ due to filtration from compartment $i$ to compartment $j$ is modeled by

$$Q_{ij} = K_{ij}[(P_i - P_j) - \sigma_{ij}(\pi_i - \pi_j)], \tag{2}$$

where $\pi_i$ is the osmotic pressure in $i$, $P_i$ is the mean hydraulic pressure in the compartment $i$, $K_{ij}$ is the filtration coefficient from $i$ to $j$, and $\sigma_{ij}$ is the corresponding reflection coefficient.
- *Darcy's law*, characterizing *diffusive flows* in porous media according to the following expression:

$$Q_{ij} = \frac{kA}{\mu L}(P_i - P_j), \tag{3}$$

where $k$ is the permeability of the medium, $A$ is the cross-sectional area to flow, $\mu$ is the viscosity of the fluid, and $L$ the characteristic length over which the pressure drop $P_i - P_j$ takes place.

For a more in-depth discussion on the above equations, their interpretation and the parameters involved in their description, see [27, Chap.10] and chapter "Mathematical and Physical Modeling Principles of Complex Biological Systems".

The mathematical translation of the *clinical concepts* of CSF production, circulation, absorption, and storage started in the second part of the twentieth century, and early works such as [18, 38] and especially [89, 139] are still used in contemporary clinical neuroscience. These contributions put the basis of the theoretical study of three *basic clinical maneuvers*—bolus CSF withdrawal, addition, and constant rate infusion—and allowed to characterize CSF circulation disorders using parameters from the models, such as resistance to CSF outflow, elasticity, and pressure–volume index (PVI). For example, the servo-controlled constant pressure infusion test [21] based on [89] is used for the assessment of hydrocephalus and progressive dementia; PVI is a parameter utilized to describe CSF compensation in hydrocephalic subjects [66] or traumatic brain injury [120]. Subsequently, the mathematical models describing CSF were enriched to incorporate autoregulation, interactions with blood flow, tissues, or other biofluids and many of these ideas are still actively investigated in current research. We first propose a general overview of these contributions and their clinical relevance in Table 1, then detail the parameters and methods from the mathematical and numerical viewpoints in Table 2, and finally discuss input and output data as well as the validation approach in Table 3. In the remainder of the section we propose three detailed reviews of selected works highlighted in the previous tables.

**Table 1** Reduced models: *general* overview

| Reference | Purpose | Clinical relevance |
|---|---|---|
| *Cerebral CSF hydrodynamics* | | |
| Marmarou [90] | ICP dynamics in terms of nonlinear intracranial compliance [89], dural sinus pressure, CSF formation, and resistance to CSF absorption. | Assessment of the factors leading to elevation of ICP and identification of clinical descriptors useful as therapeutic and prognostic guides. |
| *Cerebral hemo-hydrodynamics* | | |
| **Ursino** [139] | ICP dynamics interacting with cerebral blood volume changes and autoregulation. | Study of ICP in pathological conditions (brain injury, subarachnoid hemorrhage, hydrocephalus, brain tumor) on the basis of data from routine clinical measurements, e.g., PVI tests. |
| *Cerebrospinal hemo-hydrodynamics* | | |
| Ambarki [2] | CSF and blood flows interactions with volume changes during the cardiac cycle. | Study of CSF and blood flow oscillation during cardiac cycle at different sites and their kinetic energy impact at their interface boundary. |
| Gehlen [31] | CSF volume and compliance for different hydrostatic pressure gradients between the CSF and venous system. | Prediction of craniospinal compliance distribution shift and CSF pulsation changes related to postural changes. |
| *Cerebral hemo-hydrodynamics and brain dynamics* | | |
| **Lakin** [76] | ICP, volume, and flow dynamics in a whole body physiology (blood, CSF, interstitial fluid, lymph, cerebrovascular autoregulation and regulation by the sympathetic nervous system). | Study of the effects of CSF infusion, change in body position and pathological conditions (cardiac arrest, hemorrhagic shock). |
| Linninger [84] | ICP dynamics between cerebral vasculature, biphasic brain parenchyma, ventricular system (extended from [81]), and compliant spine. | Prediction of intracranial pressure gradients, blood and CSF flow, and displacements in normal and pathological conditions (hydrocephalus). |
| Buishas [12] | CSF dynamics as a function of both hydrostatic and osmotic pressure gradients between cerebral vasculature, ECS, perivascular space, and CSF. Both the classical and the microvessel hypothesis of CSF production and reabsorption are included. | Assessment of the effects of the osmolarity of ECS, blood, and CSF elucidating the mechanism of water exchange occurring in the brain (osmotic imbalances, hydrocephalus, edema). |
| *Cerebral and eye hemo-hydrodynamics* | | |
| Nelson [96] | Blood and aqueous humor dynamics, lamina cribrosa biomechanics, and IOP-dependent ocular compliance affecting volume/pressure in the eye. | Prediction of the impact on IOP due to increased ocular blood pressures and ICP, associated with short-term effects induced by gravitational variations. |
| **Salerni** [117] | Hemodynamics in the eye (retina, choroid, ciliary body, lamina cribrosa) and brain interacting with the dynamics of aqueous humor and CSF; lamina cribrosa biomechanics influenced by tissue pressure in the optic nerve head due to CSF within the SAS. | Assessment of the role of various factors and the mechanism of their interactions implicated in the loss of visual function in various condition, e.g., ocular pathologies, long head-down tilt experiment, microgravity environment. |

The references in bold are reviewed in detail in the text

**Table 2** Reduced models: *mathematical* overview

| Reference | Parameters | Models and methods |
|---|---|---|
| *Cerebral CSF hydrodynamics* | | |
| Marmarou [90] | Estimated from experimental pressure–volume curves. | General nonlinear equation associated to the equivalent electrical circuit configuration of the CSF system (formation, storage, and absorption) solved through the integrating factor technique. |
| *Cerebral hemo-hydrodynamics* | | |
| **Ursino** [139] | Literature [5, 134, 136, 138] and experimental results. | ICP as solution of a DAEs system obtained by imposing mass conservation principles between compartments. |
| *Cerebrospinal hemo-hydrodynamics* | | |
| Ambarki [2] | Qualitatively estimated from literature [58, 136] and empirically fitted on MRI curves. | DAEs imposing the Kirchhoff's laws to the electric analogue circuit. |
| Gehlen [31] | Estimated from published values from patients with NPH [104–106]. | DAEs system based on volume balance between CSF and cardiovascular system flows. Craniospinal compliance is computed from the slope of CSF and blood pressure–volume curves; jugular vein collapse is included reducing the hydrostatic pressure difference between cranial and spinal veins. |
| *Cerebral hemo-hydrodynamics and brain dynamics* | | |
| **Lakin** [76] | Calibrated on physical data, and other relationships. | DAEs system obtained imposing mass conservation principles at each compartment, and Eqs. (1) and (2) for fluid flows. Deformation of the brain membrane for changes in pressure and compliance related to distensibility and compartmental volume. |
| Linninger [84] | Literature [26, 61, 76, 123, 147] and MRI measurements from normal subjects. | DAEs obtained by discretizing spatial dimensions and imposing continuity, momentum, and distensibility balances in each compartment, and Eqs. (1) and (3) for fluid flows. A linear distensibility model for different cerebral vasculature beds and CSF spaces is incorporated to account for compartment expansion or compression. Implicit Euler scheme for time integration and convergent step-size controlled Newton–Raphson method are used. |
| Buishas [12] | Literature and fitted on experimental results by an optimization routine. | DAEs obtained from mass and molar conservation balances at each node, and constitutive flux equations based on Eqs. (1)–(3) for fluid flows between the compartments. |

(continued)

**Table 2** (continued)

| Reference | Parameters | Models and methods |
|---|---|---|
| *Cerebral and eye hemo-hydrodynamics* | | |
| Nelson [96] | Deduced from published studies [43, 64, 69, 99, 121, 122]. | DAEs system obtained imposing unsteady mass conservation equations or volume changes in terms of rates of pressure changes for each compartment. |
| **Salerni** [117] | Calibrated on published data [11, 35, 65, 75, 87, 131]. | DAEs system obtained imposing balance and constitutive relations for each component (Eqs. (1) and (2) for flow rates). The analogue electric model is implemented in a graphic-oriented design. IOP results from the balance between aqueous humor production and drainage, veins are modeled as Starling resistors, and a compressive stress is exerted by the lamina cribrosa from the combined action between IOP, CSF pressure, and scleral tension. |

The references in bold are reviewed in detail in the text

**Table 3** Reduced models: *clinical* overview

| Reference | Input | Output | Validation |
|---|---|---|---|
| *Cerebral CSF hydrodynamics* | | | |
| Marmarou [90] | Changes of CSF volume. | Static and dynamic response of CSF pressure useful in characterizing the physiological mechanism responsible for elevated ICP. | Comparison with experimental pressure response measured in the cisterna magna of 6 adult cats by changing the animal's CSF volume with an infusion of saline into the cisterna magna or ventricles. |
| *Cerebral hemo-hydrodynamics* | | | |
| **Ursino** [139] | Arterial pressure and its time derivative, venous sinus pressure, and an amount of mock CSF injected into or subtracted from the cranial cavity. | ICP reflecting the volume stored in the craniospinal compliance, and the arterial–arteriolar compliance influenced by the action of cerebrovascular control mechanisms. | Simulation results are compared with clinical literature [70, 109]. |
| *Cerebrospinal hemo-hydrodynamics* | | | |
| Ambarki [2] | Cardiac pulsation as a sinusoidal arterial pressure reproducing physiological conditions. | Arterial, venous, ventricular, and spinal CSF flow responses. | Reproduction of arterial, venous, and CSF flows distribution (amplitude and phase shift) observed in 7 healthy volunteers by PC-MRI velocity measurements. |

(continued)

504

L. Sala et al.

**Table 3** (continued)

| Reference | Input | Output | Validation |
|---|---|---|---|
| Gehlen [31] | Cranial arterial volume pulsation based on arterial blood flow rates recorded by averaged PC-MRI of 16 NPH patients, and an imposed constant rate of CSF formation. | CSF volume, cranial CSF pressure, and blood and CSF craniospinal flow rates during the cardiac cycle. | Comparison of craniospinal flow rates with clinical MRI measurements recorded in supine and sitting posture. |

*Cerebral hemo-hydrodynamics and brain dynamics*

| | | | |
|---|---|---|---|
| **Lakin** [76] | Time variation of cardiac uptake and output, blood flow into the cerebral capillaries and in the choroid plexus autoregulated by a pressure dependent fluidity. | Pressure, flow, and volume dynamics of blood and CSF. | Comparison of simulations assuming normal physiology with clinically measured pressure pulsations. |
| Linninger [84] | Carotid pulsatile pressure based on MRI measurements, a zero amplitude venous pressure. Constant CSF production through the choroid plexus, capillary diffusion through the brain parenchyma and absent flow at the end of the spinal canal are imposed. | Temporal changes in ICP and volumes of brain components. | Comparison of computational prediction of brain area change and CSF flow velocity with MRI measurements in 11 subjects (6 NPH and 5 with hydrocephalus). |
| Buishas [12] | Changes in hydrostatic and osmotic pressures. | Water flux related to compartments volume accumulation. | Comparison with clinical data. |

*Cerebral and eye hemo-hydrodynamics*

| | | | |
|---|---|---|---|
| Nelson [96] | Change of gravitational field and assuming a constant aqueous humor production. | Changes of IOP and blood volume in eye, biomechanical strains in the tissues of the posterior eye. | Validation against 4 existing datasets on parabolic flight, body inversion, and head-down tilt. |
| **Salerni** [117] | Changes in hydrostatic and osmotic pressures, alterations of the blood–brain barrier. Constant active secretion of aqueous humor and CSF is imposed. | Determination of IOP and changes of flow and pressures distribution within the eye and the brain. | Qualitatively and quantitatively reproduction of the relationship among intraocular, CSF, and blood pressures reported in major clinical and population-based studies [9, 20, 94, 107, 146]. |

The references in bold are reviewed in detail in the text

**Detailed Reviews**

### 2.1.1   A Simple Mathematical Model of the Interaction Between Intracranial Pressure and Cerebral Hemodynamics (Ursino et al. [139])

Ursino et al. [139] proposed a simple mathematical model to simulate the interactions between CSF and cerebral blood flow, which accounts for cerebral venous pressure instability, cerebral blood volume, and autoregulation, with the *aim* of improving the comprehension of physiological ICP time patterns. As acknowledged by the authors, the simplicity of the model is a consequence of the fact that it arises from the reduction of complexity of previously proposed models by the first author and collaborators in [136–138].

Intracranial dynamics is inferred from the connection between brain fluids circulation and craniospinal storage capacity. The electrical analogue of the *model* is shown in Fig. 1. Arterial–arteriolar vascular bed from large intracranial arteries down to cerebral capillaries is represented by a Windkessel-type model [27, Sec. 10.2.3] consisting of a nonlinear compliance $C_a$ and a nonlinear hydraulic resistance $R_a$ controlled by the cerebral autoregulation mechanism. Any decrease in the arterial blood volume $V_a$ caused by passive drop or active vasoconstriction is reflected in resistance increase, and vice versa. Veins from cerebral capillaries down to the lateral lacunae and bridging veins are represented by means of a proximal

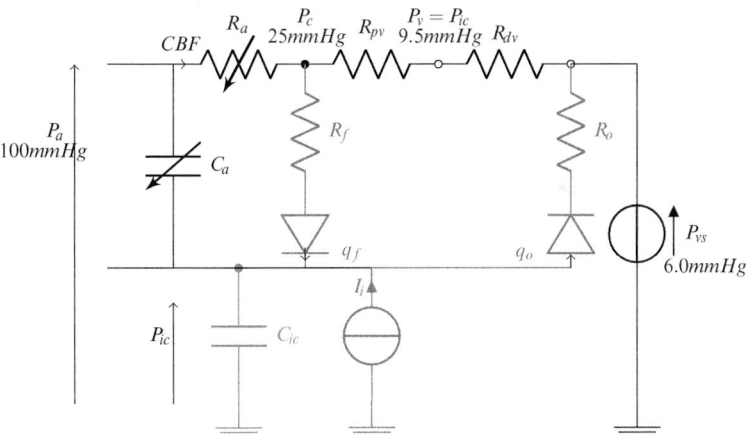

**Fig. 1** Network model for the intracranial vasculature (black portion), CSF production and drainage (gray portion), and craniospinal storage capacity (tan portion). The vasculature comprises CBF in the arterial–arteriolar ($a$), capillaries ($c$), and venous cerebrovascular bed (vs), respectively. CSF fluid is passively produced at capillary level ($q_f$) and drained at venous sinus level ($q_o$). Intracranial pressure (tan filled node) results from the balance between CSF inflow and outflow, blood volume changes in arterial capacity, and mock CSF injection rate ($I_i$). Figure adapted from Ursino et al. [139]

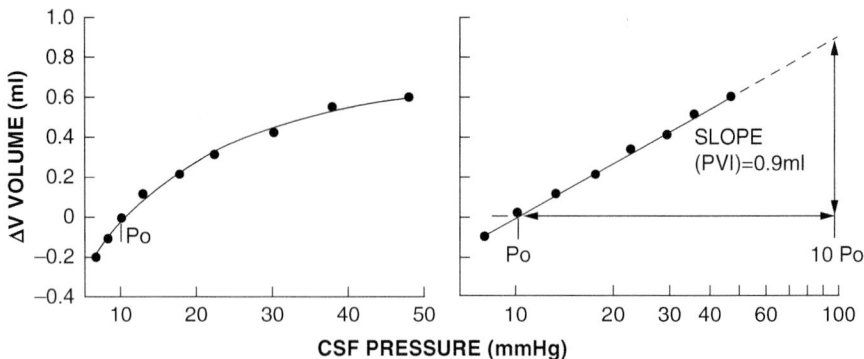

**Fig. 2** The CSF volume–pressure curve plotted on a linear axis (left) and on semilogarithmic axis (right). The compliance given by the slope $\Delta V/\Delta P$ decreases as pressure increases, as a result of the hydrodynamics of the system. Figure taken from Marmarou [89]

hydraulic resistance $R_{pv}$ and a downstream Starling resistor $R_{dv}$ that is assumed to collapse at its entrance into the dural sinuses. Therefore, cerebral venous pressure $P_v$ is always approximately equal to intracranial pressure $P_{ic}$. $R_f$ and $R_o$ represent CSF formation and outflow resistances, respectively, $C_{ic}$ is the nonlinear intracranial compliance which stores the amount of blood and CSF volume and follows a mono-exponential pressure–volume relationship depicted in Fig. 2, as in Marmarou [89]:

$$C_{ic} = \frac{1}{K_E \cdot P_{ic}} \tag{4}$$

where $K_E$ is the elastance coefficient of the craniospinal system, inversely proportional to PVI (Fig. 2). This index is the slope of the volume-log pressure curve, corresponding to the amount of volume necessary to raise opening pressure by a factor of 10. PVI is calculated by injecting into the CSF space, raising the pressure from an initial level of $P_0$ to a peak pressure of $P_p = 10P_0$: $PVI = \Delta V / \log \frac{P_p}{P_0}$. The total PVI of a system of several compartments is equal to the algebraic sum of the PVI of each compartment.

Systemic arterial pressure and its time derivative, dural sinus pressure $P_{vs}$ and the amount of mock CSF possibly injected into or subtracted from the cranial cavity $I_i$ are the *input* quantities of the model.

Interestingly, the model accounts for the coupling between blood and CSF dynamics in the following way: the Monro–Kellie doctrine [62, 95] states that changes in arterial–arteriolar blood volume

$$V_a = C_a \cdot (P_a - P_{ic}) \tag{5}$$

or in CSF volume cause a compression of the remaining intracranial volumes which means an increase of ICP. However, since the described framework aspired to be used directly in a clinical setting, some inherent simplifications were needed (for instance, neglecting the role of venous blood volume change). Therefore, the model cannot be accurately used to study more complex phenomena, such as the effect of ICP pulsating waves synchronous with the cardiac beat or with respiration and perturbation arising in the extracranial venous drainage pathways.

By imposing the flow rate conservation law at the filled nodes in Fig. 1 and by recalling the constitutive equation for the nonlinear compliance $C_a$, the main equations to solve are

$$C_{ic} \cdot \frac{dP_{ic}}{dt} = \frac{dV_a}{dt} + \frac{P_c - P_{ic}}{R_f} - \frac{P_{ic} - P_{vs}}{R_0} + I_i \tag{6a}$$

$$\frac{dC_a}{dt} = \frac{1}{\tau} \left(-C_a + \sigma(G \cdot x)\right) \tag{6b}$$

resulting in a DAE system with first order ODEs with two state variables:

- $P_{ic}$ denoting the ICP;
- $C_a$, which is influenced by the action of cerebrovascular control mechanisms.

In particular $\tau$ represents the time constant of the autoregulation, $x = \frac{q - q_n}{q_n}$ where $q$ is the CBF, $\sigma$ is a sigmoidal static function with lower and upper saturation, and $G$ is the maximum autoregulation gain. The full formulation with all the parameters involved can be found in the Appendix A of the paper [139].

In normal steady-state conditions the model is *validated* by computing the pattern of CBF and arteriolar volume as a function of systemic arterial pressure and for different values of autoregulation gain: according to experimental data on cats in [70], the maximum cerebral blood volume increase occurs close to autoregulation lower limit.

Model dynamics is represented by bifurcation diagrams in the phase-space plane, a mathematical description of the mutual dependence of the two state variables, converging towards a stable equilibrium point. Instability is characterized by the loss of steady-state equilibrium and by the occurrence of self-sustained waves representing the periodic oscillations of ICP correlated with periodic oscillations in arterial–arteriolar blood volume.

In *conclusion*, simulations of ICP response to changes in main model parameters during acute arterial pressure changes and PVI tests in patients with preserved autoregulation pointed out the importance of studying the effects of small alterations close to the stability boundary. In *clinical practice*, PVI tests are performed to estimate intracranial parameters monitoring ICP during mock CSF injection; moreover, sensitivity model simulations of PVI, reported in Fig. 3, showed that in a subject with efficient autoregulation, ICP increases after the injection phase and returns monotonically towards baseline. The model predicts that the *main* alterations, which determine intracranial instability, are due to a decrease in intracranial compliance and an inverse proportional increase in the CSF outflow resistance. In addition, the

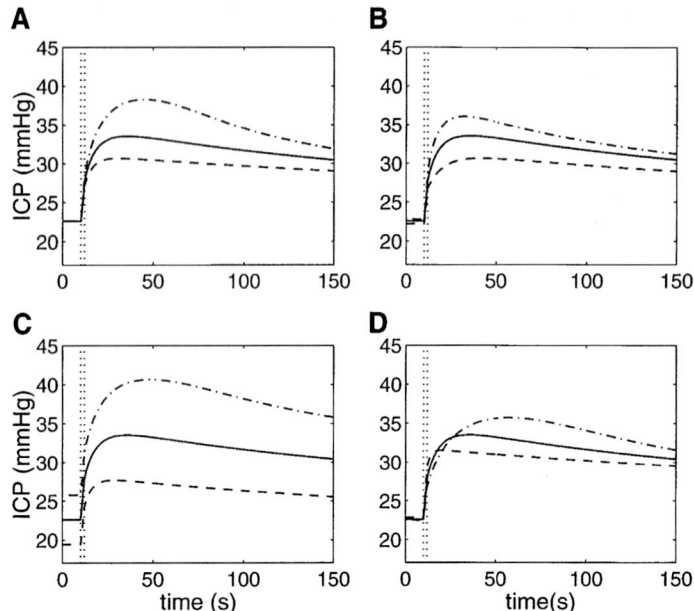

**Fig. 3** ICP response to PVI tests in a subject with preserved autoregulation (solid curves) by injecting 2 ml of mock CSF between 10 and 12 s (vertical dotted lines). Four different sensitivity analyses have been performed for decreased (dashed curves) and increased (dot-dashed curves) values of: (**a**) $k_E$, (**b**) the central autoregulation gain, (**c**) $R_0$, and (**d**) the arteriolar–arterial compliance, respectively. Figure taken from Ursino [139]

presented model constitutes a valuable support for more rigorous management of arterial pressure in head-injured patients, as described in the companion paper [50].

### 2.1.2  A Whole Mathematical Model for Intracranial Pressure Dynamics (Lakin et al. [76])

A complex model of the intracranial system (tissues, vasculature, CSF) embedded in an extensive whole-body physiology was developed by Lakin et al. [76] for the *study* of intracranial pressures, volumes, and flow dynamics.

The *lumped-parameter model* accounts for cerebrovascular autoregulation, regulation of systemic vascular pressures by the sympathetic nervous system, regulation of CSF production in the choroid plexus, an extracranial lymphatic system, the colloid osmotic pressure effects, and a realistic description of cardiac output.

Physical constituents are subdivided into 16 distinct spatially limited, linked, interacting compartments, each one containing a single physical constituent (blood, CSF, or tissue and the associated interstitial fluid). Intracranial and extracranial systemic vasculature is subdivided into compartments representing fluid in the arteries, capillaries, choroid plexus, veins, and venous sinus. Ventricular CSF is contained in an intracranial compartment connected to a bridging compartment; the latter contains extra-ventricular CSF partially outside of the cranial vault which represents the subarachnoid and spinal portions of CSF space. The deformation of the membrane between adjacent compartments is a function of the change in pressure difference between these compartments

$$\frac{dV}{dt} = C\,\frac{dP}{dt}.\tag{7}$$

Intracranial autoregulation mechanisms provide constant blood flow to the cerebrovascular capillaries and the choroid plexus that is placed in a separate compartment from the rest of the intracranial capillary bed providing a constant CSF production.

Biofluid dynamics in each compartment is described by lumped, time-dependent functions of pressures ($P$). Pressure differences are related to changes in flows ($Q$) and compartmental volumes ($V$) by assuming resistance ($R$) and compliance ($C$) between adjacent compartments. Direct flows between adjacent compartments are driven by compartmental pressure differences, Eq. (1); on the contrary, fluid transfer between capillaries and tissue by filtration is governed by Eq. (2). Interaction with the external environment is allowed through flows representing inflow and outflow via the central body.

Cardiac uptake and output are the *forcing forces* of the system, their realistic representation being based on a mean cardiac output of 6900 ml and a heart rate of 76 beats per minute [127]. The model assumes that inflow to the intracranial arteries equals outflow from the jugular bulb producing a system that conserves the total intracranial volume (Kellie–Monro doctrine [95]). In each compartment the fluids flow conservation law is imposed and the final model system of differential equations is solved numerically using the symbolic mathematical software package Mathematica employing maximum accuracy settings (NDSolve). The model was *validated* by comparing simulation results with clinical data. As an example, Fig. 4 describes the model output for a CSF infusion test into the lower lumber space.

Figure 4 (top) depicts the pressure of the extracranial CSF space associated with the total CSF volume change. The slope of the curve describes the elastance—inverse of the compliance—of the intra and extracranial CSF space. The curve has a lower pressure plateau—large compliance—near the pressure where CSF production is balanced by CSF absorption: venous system compression accommodates increases in the volume of the CSF space at these relatively low pressures. Increasing pressures by adding CSF volume decreases the ability of the system to adjust, yielding to a reduction in the compliance of the system and a compression of the veins. The slope of the pressure–volume relationship continues to increase for larger infusion volumes until the resulting CSF pressures are high enough that the intracranial arteries can begin to be compressed. There is an upper pressure plateau at the diastolic pressure of the intracranial arteries indicating additional

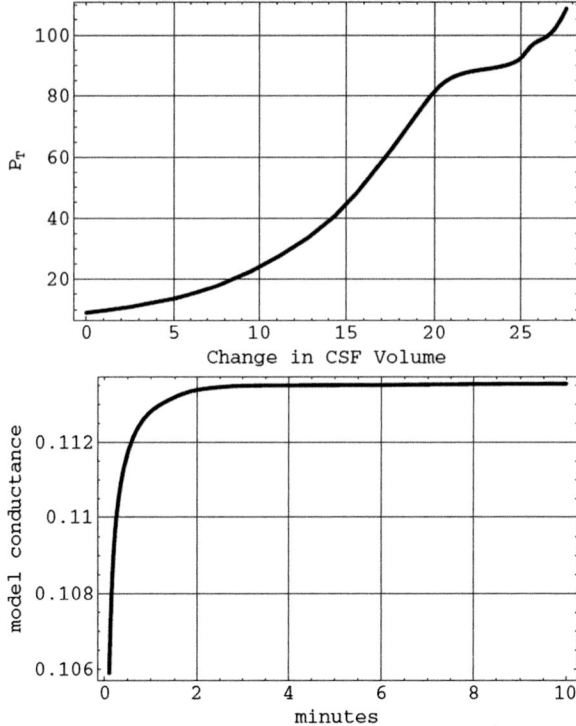

**Fig. 4** Top: CSF pressure–volume curve. The slope describes the elastance (inverse of the compliance) of the entire CSF space, including extracranial portions. Bottom: model conductance to CSF output. Figure taken from Lakin [76]

compliances to buffer additional volume increases, but an increase of CSF pressures to the systolic pressure leads the CSF compliance to fall to zero.

Figure 4 (bottom) reports the values of the simulated conductance of CSF outflow $C_{out}$, corresponding to the slope of the relationship between CSF pressure differences and CSF absorption. Experimentally, in Albeck [1] a mean value for $C_{out}$ of 0.11 (ml/min)/mmHg was measured for a sample of eight healthy volunteers, and in Sullivan [129] a value greater than 0.10 is reported as being in physiological ranges. In Fig. 4 (bottom) the values change with time, but stay within 0.004 units of the mean value of 0.11 (ml/min)/mmHg as in [1]. They are also greater than 0.10 as suggested by [129].

In *conclusion*, the present closed-system model incorporating the influence of important extracranial factors provides a valuable complex systemic view of the interaction between CSF and other biofluids. Simulated CSF infusion tests allow to recover experimentally derived intracranial pressure–volume relationships. The model can also be utilized to examine the effect on cerebral blood flow of a change in body position, and for simulating cardiac arrest and hemorrhagic shock, thus demonstrating its predictive capabilities in pathological conditions.

### 2.1.3 A Mathematical Model of Ocular and Cerebral Hemo-Fluid Dynamics: Application to VIIP (Salerni et al. [117])

Salerni et al. [117] developed a mathematical model describing the *coupled dynamics* of fluid flows in the eye (blood and aqueous humor) and brain (blood, cerebrospinal, and interstitial fluid) embedded in a simplified description of the whole-body circulation. The *aim* of the model is to theoretically investigate the complex interplay between the numerous factors that influence the biofluid dynamics in the eye-cerebral system, since their individual contribution is difficult or even impossible to measure or isolate in a clinical setting. These factors include arterial blood pressure (BP), IOP, ICP, and CSF pressure. The model also targets prediction of fluid redistribution in the upper body vasculature by inducing changes in the plasma colloid osmotic pressure and hydrostatic pressure, and its implication on IOP and ICP.

The electric analogue representation of the model and the included compartments are described in Fig. 5. In summary, the *brain model* consists of three fluid networks: the blood vasculature, the cerebrospinal fluid network, and the interstitial fluid network. An extra-ventricular compartment filled with CSF bridges the intracranial and extracranial regions and includes the SAS in the optic nerve, posterior to the lamina cribrosa. In each of the *two eyes*, the interaction between blood and aqueous humor (AH) is incorporated and the lamina cribrosa is represented, loaded by IOP on the one side and by the pressure within the optic nerve tissue on the other side, due to the CSF pressure in the SAS. The model accounts for cerebral hemodynamics, in the intracranial arteries, capillaries, and intracranial venous sinuses and for blood flow within three major ocular vascular beds, namely retina, choroid, and ciliary body. The fluid dynamics brain and eye models are *coupled* and *linked* to the rest of the body through use of a highly simplified model consisting of two compartments: central arteries and central veins. The flow is *driven* by the following mechanisms: (i) the pressure drop between the central arteries and the central veins; (ii) the active secretion of AH in the eyes, and (iii) the production of CSF in the brain.

The lumped-parameter circuit for the brain is adapted from the work described and validated in Lakin and Stevens [75]; the eye model originates from the approach proposed in Guidoboni et al. [35] for the study of retinal circulation and has been extended to account for the three ocular vascular beds (retina, choroid, and ciliary body) on the basis of Kiel et al. [63]; finally, ocular hemodynamics is coupled with the AH dynamics as in Szopos et al. [131].

Two main types of flow are included in the model. The flux due to *filtration* is modeled by Eq. (2), the electric analogue of which is an element with a resistor and a current generator arranged in parallel, as depicted in Fig. 5; this constitutive law is used for filtration of interstitial fluid from the capillaries to the interstitial space (tan portion), and filtration of aqueous humor from the ciliary body capillaries into the posterior/anterior chambers (cyan portion). All the other flows in the model are simply driven by the *hydraulic pressure difference* between compartments, governed by the hydraulic analogue of Ohm's law Eq. (1).

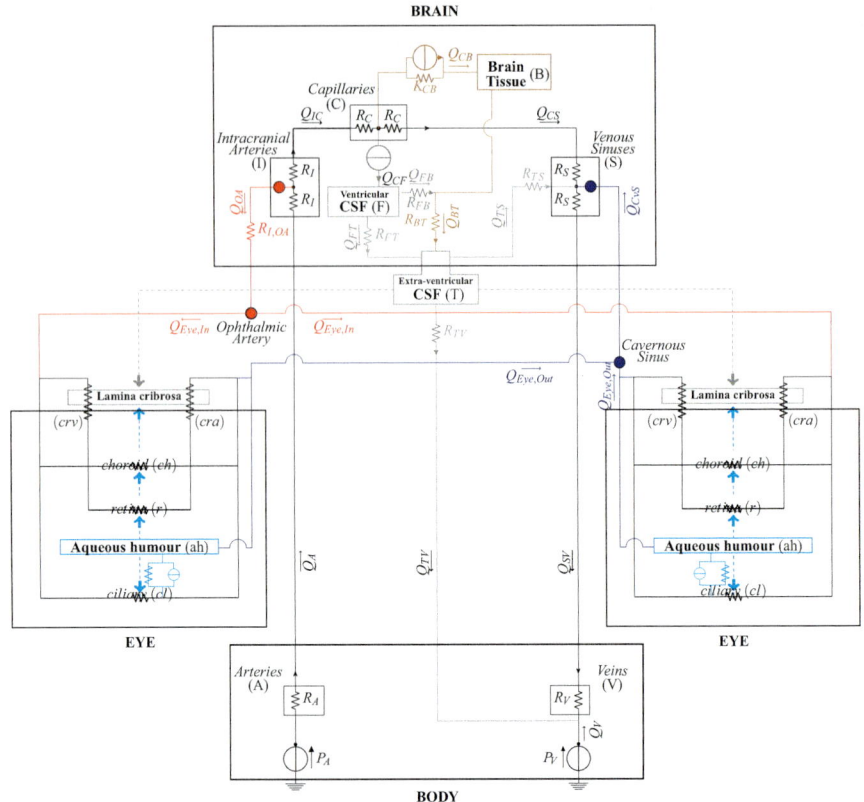

**Fig. 5** Network model of Body-Brain-Eyes system: blood vasculature (black portion), cerebrospinal fluid network (grey portion), interstitial fluid network (tan portion), and aqueous humor network (cyan portion). The nodes correspond to the connection between the brain and eye models. The red connection *Intracranial Arteries-Ophthalmic Artery* represents arterial supply; the blue connection *Venous Sinuses-Cavernous Sinus* represents the venous drainage; the gray and cyan arrows represent the pressures acting on both sides of the lamina cribrosa. Lower case letters denote compartments in the eyes and upper case letters in the brain and upper part of the body. Figure taken from Salerni [117]; for a detailed representation of the eye see [116]

By formulating Eqs. (1) and (2) for all compartments and by writing the Kirchhoff's law of currents at all circuit nodes, a set of nonlinear algebraic equations is obtained (Tables 1 and 2 in [116]). The nonlinearity is a consequence of the fact that, in some compartments, resistances are assumed to depend on the pressures, therefore deformable tubes are modeled as Starling resistors, reflecting the physiological high collapsibility of these vessels when the transmural pressure becomes negative. Time variations occurring on the time scale of heartbeat are neglected, as well as autoregulation mechanisms of small vessels, since the aim was to keep the model relatively simple in order to understand its basic behavior.

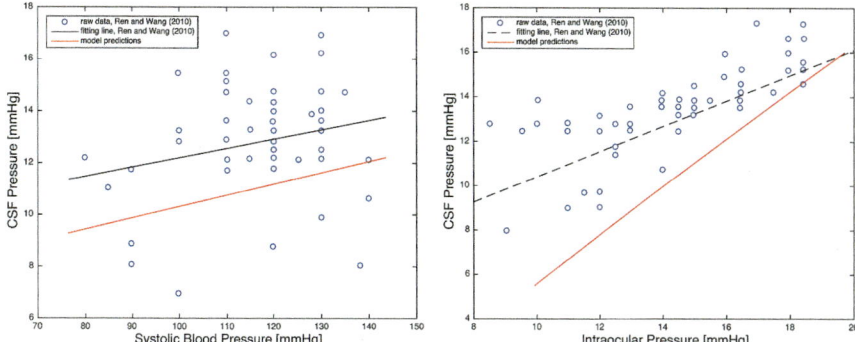

**Fig. 6** Comparison between regression lines obtained by CSF pressure measured via lumber puncture in [107] (black lines) and computed by the model in extra-ventricular CSF compartment (red lines). CSF pressure was correlated with systolic blood pressure (left) and IOP (right). Figure taken from [37]

The free, open-source modeling and simulation environment OpenModelica [29] is used to obtain an accurate numerical solution of the resulting nonlinear algebraic system.

In order to assess the *validity* of the assumptions related to the coupling between the various elements, the whole model is compared qualitatively and quantitatively with several clinical findings. The model shows satisfactory agreement with experimental data regarding IOP variations due to changes in blood pressure [20, 94, 146], and choroidal venous pressure equals to IOP [9]. Moreover, the model is able to capture the main trends of CSF pressure variations due to change in blood pressure and IOP. As an illustration, Fig. 6 depicts the regression lines (black lines) alongside the data scatter plots of a clinical study by Ren et al. [107] about CSF pressure via lumber puncture related with blood pressure and IOP. The pressure computed by the model in the extra-ventricular CSF compartment, namely the spinal region (red lines in Fig. 6) is within the ranges reported in clinical studies.

After this thorough validation process, the model was employed to investigate changes in flow and pressure distributions associated with long-term exposure to microgravity conditions. Simulation results point at a purely mechanical, i.e., purely passive, feedback mechanism due to the collapsibility of the venous segments in controlling pressures and fluxes in the ocular circulation; in particular, retinal venules exposed to IOP help maintaining a relatively constant level of blood flow through the retina despite microgravity induced changes in intracranial and blood colloid osmotic pressures.

In *conclusion*, the present mathematical model describes the complex relationship between different biofluids in the brain and in the eye at a level of detail that was not previously available in the literature to the best of our knowledge. The advantage of such a network-based model is that it provides a systemic view, able to capture the overall behavior of the interwoven physiology of blood, CSF, ocular humors and interstitial fluids in the eye and in the brain, while maintaining a relatively accessible mathematical complexity and low computational costs. In addition to the reported findings, the proposed approach could serve as a sound framework to theoretically investigate other pathological states, most importantly glaucoma, in which the interaction between ocular and cerebral hemo-fluid dynamics is thought to play a crucial role.

## 2.2 Distributed Models for CSF Dynamics and Its Interactions

The goal of this part is to discuss the main challenges in the mathematical and computational modeling of CSF flow in realistic geometries and to review several contributions within this framework. We therefore focus in this section on the so-called *distributed* (or macro-scale) *models* that are suitable when aiming for a three-dimensional representation of the flow (see chapter "Mathematical and Physical Modeling Principles of Complex Biological Systems"). The description of the underlying mechanisms stems from the basic principles of fluid dynamics and it is translated into systems of PDEs, supplied with suitable initial and boundary conditions of particular importance. Specifically, CSF is often considered as an incompressible Newtonian fluid, therefore the *Navier–Stokes equations* are employed (see, for instance, [27, Chap.2, Chap.3]). In addition, the fluid dynamics description can be enriched to take into account the combined effects of flow and different structures from a *multiphysics* perspective including *poroelasticity* [27, Chap.7] and *fluid–structure interaction* [27, Chap.3, Chap.9]. For a detailed presentation of the mathematical models and different approaches in their numerical solution, see chapter "Mathematical and Physical Modeling Principles of Complex Biological Systems".

The issue of boundary conditions when modeling biological fluids is of major importance and matter of intense research, from the mathematical and computational viewpoint. When performing large-scale three-dimensional simulations, the domain has to be reduced to a region of interest because of the prohibitive computational costs. Therefore, its boundary is composed of two parts: a *physical* boundary—corresponding to the geometrical description of the compartments—and an *artificial* boundary—at the level where the domain is truncated. On the one hand, two possibilities can be considered for the physical boundaries: either assume them rigid, in which case the no-slip Dirichlet condition is considered, or incorporate their ability to deform, and thus include a much more complex fluid–structure behavior in the model. On the other hand, at the artificial boundaries, the formulation should be able to take into account the rest of the closed circuit representing the CSF circulation, which is still a very challenging issue, as discussed, for instance, in the recent review [103]. Moreover, in order to achieve physiological simulations, data used to impose the boundary conditions should be taken from clinical measurements.

In this perspective, the geometrical representation of the CSF space is very complex and CSF flow quantification requires special care. Recent progress in segmentation of vascular volumes from MRI data, design of a vascular models and construction of computational meshes of the vascular structures, as well as PC-MRI techniques for blood flow measurements are translated into use for CSF flows. However, the CSF compartments have their own features that are contributing to the difficulty of deriving accurate information, among which: (i) complex geometric shapes, in contrast to the (locally) near-cylindrical form of blood vessels; (ii) smaller magnitudes in the velocity field compared to blood flow; (iii) difficulties in finding planes perpendicular to the flow direction; (iv) presence of reversal flow, etc.

The understanding of CSF behavior at *fine scales* is of major importance, since abnormal flow dynamics or pressures imbalances in different compartments may lead to several pathologies. A recent compilation of studies on abnormal CSF dynamics in central nervous system (CNS) diseases can be found in [83], with a focus on some of the most common ones: hydrocephalus, Chiari malformation type 1, syringomyelia, pseudotumor cerebri, idiopathic intracranial hypertension, and benign intracranial hypertension. The pathophysiology and the etiology of these syndromes are still poorly understood and trigger a lot of open research questions. In addition, the clinical manifestations, diagnosis, and management strategies are currently a matter of active discussion. In perspective, a better understanding of both normal and pathological CSF flow thanks to mathematical and computational models encompassing the whole 3D dynamics is very appealing, since it may be able to provide not only a qualitative assessment of the dynamics, but also a more quantitative knowledge about CSF velocities, flow rates, and pressures.

The long-term goal is to devise a complete in silico model of CNS dynamics, accounting for the full coupling between CSF, blood and brain tissues and incorporating various physical properties and settings (e.g., variations of body position, patient-specific parameters and geometries, etc.). However, to the best of our knowledge, a full 3D mathematical and computational modeling of the complex connections between biofluids in the coupled eye–brain system is still not available and only a few recent reduced models tackled this challenging issue (see Sects. 2.1 and 2.3). Such a very complex formulation would require large-scale availability of data, sophisticated mathematical and numerical methods, and high computational costs. Several recent contributions that we will review in the sequel made significant steps in this direction and provided new insights to the understanding and better management of CSF flow conditions: Chiari malformation (Bunck et al. [13], Støverud et al. [128], Jain et al. [59]), hydrocephalus and cerebral edema (Vardakis et al. [141]), drug delivery in leptomeningeal dissemination, epilepsy, Parkinson (Howden et al. [53]) to give just a few examples.

A more detailed description of several contributions is proposed in the sequel: first, we give in Table 4 a general overview of the models; next, we focus in Table 5 on the geometrical, mathematical, and numerical ingredients; finally, we review clinically meaningful information about input/output data and the validation process in Table 6. The last part of the section is dedicated to detailed reviews of three selected works from the literature.

**Table 4** Spatially distributed models: *general* overview

| Reference | Purpose | Clinical relevance |
|---|---|---|
| Jacobson [56] | Computation of CSF flow in idealized cylindrical domains. | Estimation of the pressure drop required to achieve normal physiological CSF flow rate within the aqueduct of Sylvius. |
| Kurtcuoglu [73] | Investigation of subject-specific CSF flow characteristics by combining MRI scans and CFD simulations. | Quantitative prediction of the subject-specific flow field in the third ventricle and the aqueduct of Sylvius. |
| Gupta [40] | Prediction of subject-specific CSF flow in the cranial SAS using in vivo data. | Quantitative and qualitative description of cranial CSF transient flow, potentially useful to compare healthy with pathological CSF dynamics. |
| **Howden** [53] | Prediction of the CSF flow in the CNS, the SAS, and in the ventricular system, with highlights in the aqueduct of Sylvius, the exit of Magendie, and entrance to the Spinal SAS. | Investigation of the role of the cardiac cycle, of the pulsatility, and of the porous arachnoid trabeculae within the SAS on the fluid motion of the CSF by a global 3D model of the CNS. |
| Sweetman [130] | Prediction of 3D CSF velocity and pressure fields in the CNS, comparison with in vivo measurements from a healthy human subject. | Quantification of normal intracranial dynamics, potentially significant to analyze diseased intracranial dynamics. |
| Hadzri [42] | Investigation of 3D CSF flow dynamics in a stenosed aqueduct. | Inclusion of a stenosis inducing significantly increased pressure drop; possible clinical application to monitor hydrocephalus. |
| Bunck [13] | Statistical analysis on 3D data for a population base study: 20 Chiari malformation patients and 10 healthy volunteers. | Comparison between healthy subjects and Chiari I patients, showing qualitative and quantitative alterations of CSF flow, in particular for patients with associated syringomyelia. |
| Hsu [54] | Construction of a drug delivery model through the SAS and spinal cord tissue. | Identification of patient-specific key variables (frequency and magnitude of CSF pulsations) on drug distribution at the spine level. |
| Rutkowska [110] | Visualization of CSF flow simulation in patient-specific models of the SAS. | Comparison of CSF dynamics between healthy subjects and Chiari I patients both with CFD simulations and PC-MRI results. |
| **Tangen** [133] | Quantification of the contribution of micro-anatomical aspects on CSF flow patterns and flow resistance within the entire CNS. | Investigation of complex CSF dynamics in a subject-specific model of the entire CNS, including micro-anatomical aspects and their impact on drug distribution. |
| Vardakis [141] | Development of a novel spatio-temporal model of fluid regulation and cerebral tissue displacement in the framework of Multiple-Network Poroelastic Theory. | Investigation of interstitial edema formation and its alleviation, for a better understanding of diseases like hydrocephalus or cerebral edema. |

(continued)

**Table 4** (continued)

| Reference | Purpose | Clinical relevance |
|---|---|---|
| Støverud [128] | Construction of a subject-specific model to compare CSF flow variations in the cervical SAS and in the SAS of the posterior cranial fossa. | Comparison of CSF flow differences between Chiari I patients and healthy individuals. |
| **Jain** [59] | Quantification of CSF hydrodynamics down to very fine scales to investigate the possible occurrence of transitional and turbulent CSF flows. | Comparison between CSF flow in healthy and Chiari malformation I patients, with a special focus on the craniovertebral junction. |
| Garnotel [30] | Construction of a simplified model to understand the dynamics of the CSF, its impact on ICP and CBF. | Description of CSF distribution in cerebral ventricles, cerebral SAS and spinal SAS, accounting for the confined environment of the skull and the synchronization with the cardiac cycle. |

The references in bold are reviewed in detail in the text

**Table 5** Spatially distributed models: *mathematical* overview

| Reference | Geometry | Models and methods |
|---|---|---|
| Jacobson [56] | Three different geometries, from an ideal structure to a realistic representation, following the description in [77, 93]. | 3D NS equations for an incompressible Newtonian fluid solved with FVM for steady and unsteady conditions using CFDS-FLOW3D [119]. |
| Kurtcuoglu [73] | MRI of the 3rd ventricle and aqueduct of a 27-year-old healthy male converted to NURBS surfaces for high-quality mesh. | 3D NS equations for an incompressible Newtonian fluid with imposed domain wall motion solved by the FEM with second-order accuracy in space and first-order in time, using PISO [55] pressure correction method within FLUENT [3]. |
| Gupta [40] | PC-MRI of the SAS of a 25-year-old healthy male manipulated via AMIRA [126]. | 3D NS/Brinkman equations [10] solved with FVM, using FLUENT [3]. |
| **Howden** [53] | MRI of the entire CNS, manipulated via MIMICS [97] and GAMBIT [98]. | 3D NS equations for an incompressible Newtonian fluid solved with FVM and laminar solver using FLUENT [3]. |
| Sweetman [130] | PC-MRI of the entire ventricular system and SAS. | 3D FSI system solved with FEM using ADINA-FSI 8.6 [8]. |
| Hadzri [42] | MRI of 3rd and 4th ventricles and aqueduct of a healthy female volunteer, manipulated via AMIRA [126]. | 3D NS equations for incompressible Newtonian fluids solved with FEM using EFD (Engineering Fluid Dynamics). |

(continued)

**Table 6** (continued)

| Reference | Input parameters | Output | Validation |
|---|---|---|---|
| Hsu [54] | CSF pulsatile flow velocities via CINE-MRI and other parameters from literature. | CSF flow and pressure distribution in the CNS, drug distribution within the CSF. | Results showing good match between the simulation and experiments with a radiotracer [50], CSF dynamics is in agreement with clinical measurements [49] and drug infusion [118]. |
| Rutkowska [110] | BCs data from PC-MRI and cardiac cycle information from the electrocardiogram. | CSF flow and pressure distribution in the SAS, in particular systolic and diastolic snapshots at each vertebra in the foramen magnum. | Simulation results showing increased CSF velocities, pressure distribution and synchronous bidirectional flow between Chiari I patients and controls, in agreement with PC-MRI data. |
| **Tangen** [133] | Velocity data at vertebra C4, T6 and L4 from MRI. | CSF flow and pressure in the geometry, vorticity adjacent nerve roots in the spinal SAS, local macroscopic Reynolds number, CSF phase lag at C4, flow resistance attributable to micro-anatomy. | Simulated CSF phase lag comparable to literature data. Simulated CSF flows in similar ranges as MRI data measured by the authors. |
| Vardakis [141] | Model parameters from literature [79, 123, 135, 142, 143, 145]. | Volume dilation and CSF/ISF fluid content variation within the parenchymal tissue, radial and tangential components of transparenchymal effective stress, ventricular displacement, capillary filtration velocity, CSF/ISF pressure, arterial, venous and capillary pressure. | Volume dilation due to FVOO and CSF/ISF increment: *qualitative* comparison with [124, 143] and [78], respectively. Displacement fields: *qualitative* comparison with [67]. Similar findings of capillary network resistance to the flow as in [135, 141, 143]. |
| Støverud [128] | BCs: velocity flow data at caudal end and at the aqueduct of Sylvius from PC-MRI. | Velocity flow virtual image and pressure distribution in the geometry (streamlines and values), in particular the pressure drop between cervical SAS and pontine cistern, and between aqueduct and pontine cistern. | Comparable values with previous literature studies and with experiments. |
| **Jain** [59] | BCs for cranial SAS and aqueduct from PC-MRI. | Velocity and pressure distribution in the geometry, turbulent intensities, Kolmogorov micro-scales. | Main results comparable with other studies in literature. |

(continued)

**Table 6** (continued)

| Reference | Input parameters | Output | Validation |
|---|---|---|---|
| Garnotel [30] | Wall BCs and model parameters estimated via PC-MRI. | CSF flow distribution and pressure within cerebral SAS and ventricles. | Comparison with assessed mathematical models and with physiological results both in pathological [17] and healthy case [1, 105], and with experimental results obtained via MRI. |

The references in bold are reviewed in detail in the text

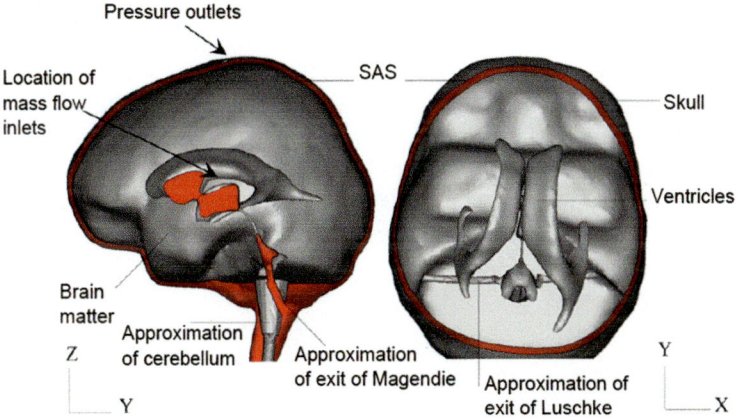

**Fig. 7** The left image highlights the exit of Magendie in a side view of a section of the CNS geometry. On the right, the exits of Luschke are displayed, thanks to an aerial view of a section via the CNS geometry. Image taken from [53]

## Detailed Reviews

### 2.2.1 Three-Dimensional Cerebrospinal Fluid Flow Within the Human Central Nervous System (Howden et al. [53])

The aim of the work of Howden et al. [53] in 2011 is to develop a virtual laboratory to study the CSF flow motion in the entire CNS. The *major interest* is to use numerical simulations to address the limitations of MRI and try to solve the ambiguities about CSF motion.

The *geometrical domain* (Fig. 7) used for their simulation is retrieved by MRI slices, manipulated via the softwares MIMICS [97] in order to produce a three-dimensional surface model of the entire volume of the CNS, the SAS, and ventricular system. Moreover, the same software is employed to generate the flow exit apertures, the inlets and to connect the ventricles to the SAS. Highly detailed MRI data are used to highlight the importance of the aqueduct of Sylvius, the exit of Magendie and the entrance to the spinal SAS. The final volumetric geometry is obtained by conversion adopting the CFD pre-processing software GAMBIT [98].

The *mathematical model* computes the CSF flow dynamics via incompressible Navier–Stokes (NS) equations for a laminar viscous fluid, numerically solved by the software FLUENT with the finite volume method (FVM). The model follows the hypothesis of Bering, which means that the pulsatile motion of the CSF is a result of the expansion and contraction of the choroid plexus, due to the pulsing of arterial blood around the site during the cardiac cycle. Velocity inlet BCs are applied to the choroid plexus, which is treated as the source of pulsatility of the CSF [81], whereas the SAS is modeled with rigid walls, associated with no-slip velocity BCs. The moderate value of the Reynolds number within the region of highest velocity magnitude allows the numerical simulation to be run with the laminar solver over 5 cardiac cycles.

The model provides as *output* the CSF velocity and pressure within the aqueduct of Sylvius, the ventricles, at the exit of Magendie and entrance to the SAS (Fig. 8); moreover it is also able to produce directional vectors of velocity field and average velocity magnitude throughout the spinal column and laterally through the CNS (Fig. 9). To *validate* the accuracy of the model, the results have been compared to medical literature values [6, 23, 33, 100, 125] and other computational outcomes [15, 56, 81, 82]. For instance, the maximum pressure within the lateral ventricles is

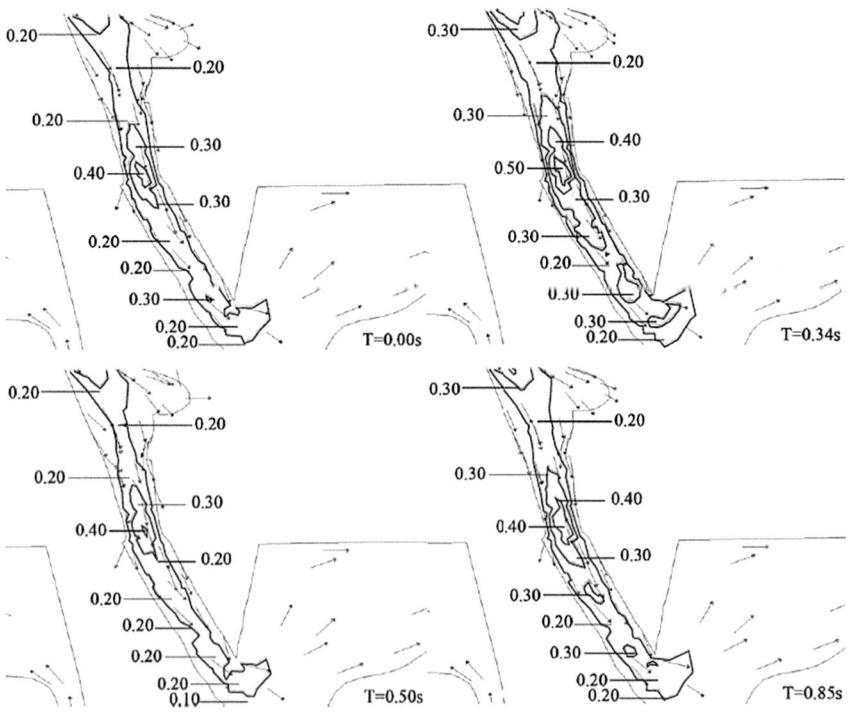

**Fig. 8** Simulated velocity magnitudes in mm/s on a plane through the fourth ventricle at 0.0, 0.34, 0.50, and 0.85 s during one cycle. Image taken from [53]

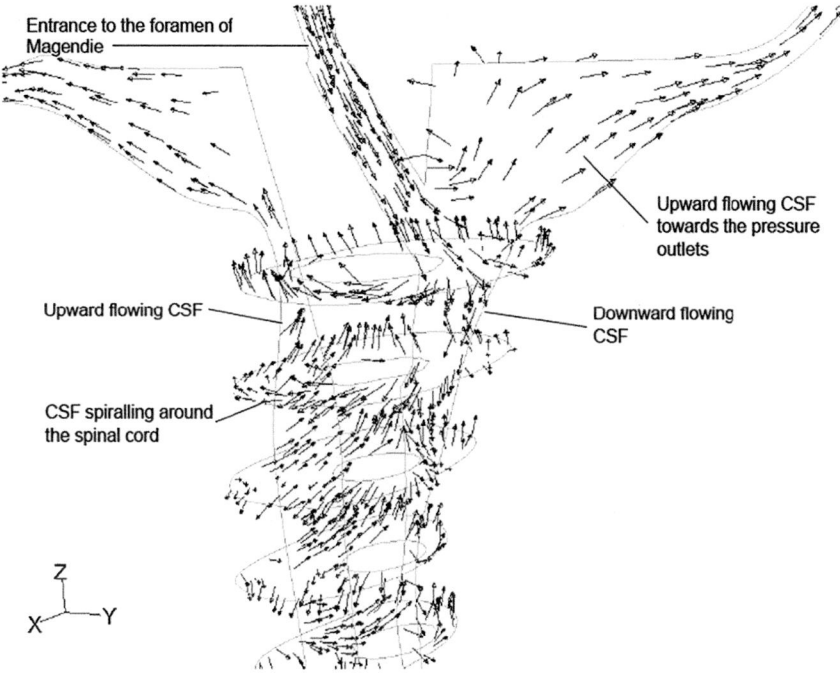

Entrance to the foramen of Magendie

Upward flowing CSF towards the pressure outlets

Upward flowing CSF

Downward flowing CSF

CSF spiralling around the spinal cord

**Fig. 9** Direction of the CSF flow illustrated by these vectors as it enters the spinal SAS and moves around the spinal cord. Image taken from [53]

close to the value by Cheng et al. [15] and the upward trend of the pulsatile CSF motion within the spinal SAS has the same behavior as in the findings by Greitz et al. [33].

In *conclusion* this work provides a systemic view of the CNS, including ventricular system and SAS by presenting a simplified MRI-based 3D CFD numerical simulation. The model is able to predict the flow field within the ventricular system and cranial SAS, as well as partially describe the flow field within the spinal column. A future development of the mathematical framework presented by the authors is to study the effect of respiratory cycle as a possible driving force on the spinal flow motion. The authors also envision the possibility to employ the model to study drug delivery within the CNS.

### 2.2.2 CNS Wide Simulation of Flow Resistance and Drug Transport Due to Spinal Microanatomy (Tangen et al. [133])

Using a computational fluid mechanics approach, the work of Tangen et al. [133] presents a model able to quantify the contribution of micro-anatomical aspects on CSF flow patterns and flow resistance within the entire CNS. The *major interest* is to investigate the effect of micro-anatomical aspects on flow, due to the fact that their length scale is below medical image resolution. Besides, this is a first attempt to perform CFD simulations on the entire CNS including micro-anatomical trabeculae in the spinal SAS and towards a complete in silico model of CNS dynamics with full coupling between CSF, blood and brain tissue. The adoption of a systemic view eliminates the burden of introducing uncertain assumptions about artificial BCs.

The simulation domain consists in the entire CNS including nerve roots that induce CSF mixing patterns. The critical point of the work is the need of computing the solution within the entire CNS including the cranial and spinal SAS with their microscopic aspects. The *geometry* of the spinal canal of a 29-year-old subject has been reconstructed from MRI and manipulated using the software MIMICS [97]. The data are collected on a 3T Signa ® HDx MR Scanner (GE Healthcare, Waukesha, WI, US) equipped with a high-density CTL spine coil. Velocity data at three planes, the 4th cervical vertebra(C4), 6th thoracic vertebra (T6), and 4th lumbar vertebra (L4), are acquired through 16 cardiac cycles and linearly interpolated to reconstruct time averaged velocity vectors over one cardiac cycle.

The *mathematical model* employs the NS equations for an incompressible Newtonian fluid, imposing subject-specific anatomical BCs. Thus, the cranial pial surface and the epithelial surfaces of the choroid plexi in the lateral ventricles are periodically displaced to account for pulsate expansion and contraction of the cerebral vasculature. To assess the impact of nerve roots, intrathecal drug infusion in the lumbar region is simulated in multiple scenarios, and modeled by an advection–diffusion system of bupivacaine, solved for a period of 12 h post injection. In particular, the species transport is described by the following equation:

$$\frac{\partial C}{\partial t} + \nabla \cdot \left[ \underline{u}(\underline{x}, t) \, C \right] = \nabla \cdot (D \, \nabla C)$$

where $\underline{u}(\underline{x}, t)$ is the fluid velocity, $C(\underline{x}, t)$ is the drug concentration, and $D = 2.1 \cdot 10^{-10} \text{ m}^2 \text{ s}^{-1}$ is the diffusivity coefficient, whereas drug metabolism and uptake are considered negligible. The fluid motion is solved with ANSYS 14.0 FLUENT on Blacklight at Pittsburgh Supercomputing Center. It exploits the fact that the geometry is reasonably symmetrical along the mid-sagittal plane: symmetric BCs are therefore adopted to reduce size of the computational domain and, consequently, computational time.

The *outputs* of the model are the CSF flow and pressure in the simulation domain described above. In addition, all the computational results about pressure and volumetric flow are condensed into an easy-to-use Fourier-approximation formula (Fig. 10). Other significant outcomes are the vorticity adjacent to nerve roots in the spinal SAS, local macroscopic Reynolds number, and drug distribution contours in the full spinal SAS (Fig. 11). All these results have been validated via comparison with literature data and MRI measurements obtained by the authors.

In *conclusion*, this study tests the hypothesis that micro-anatomical structures generate significant vortex phenomena in spinal CSF flow which increase flow resistance, raise pressure drop, and induce complex mixing patterns responsible for

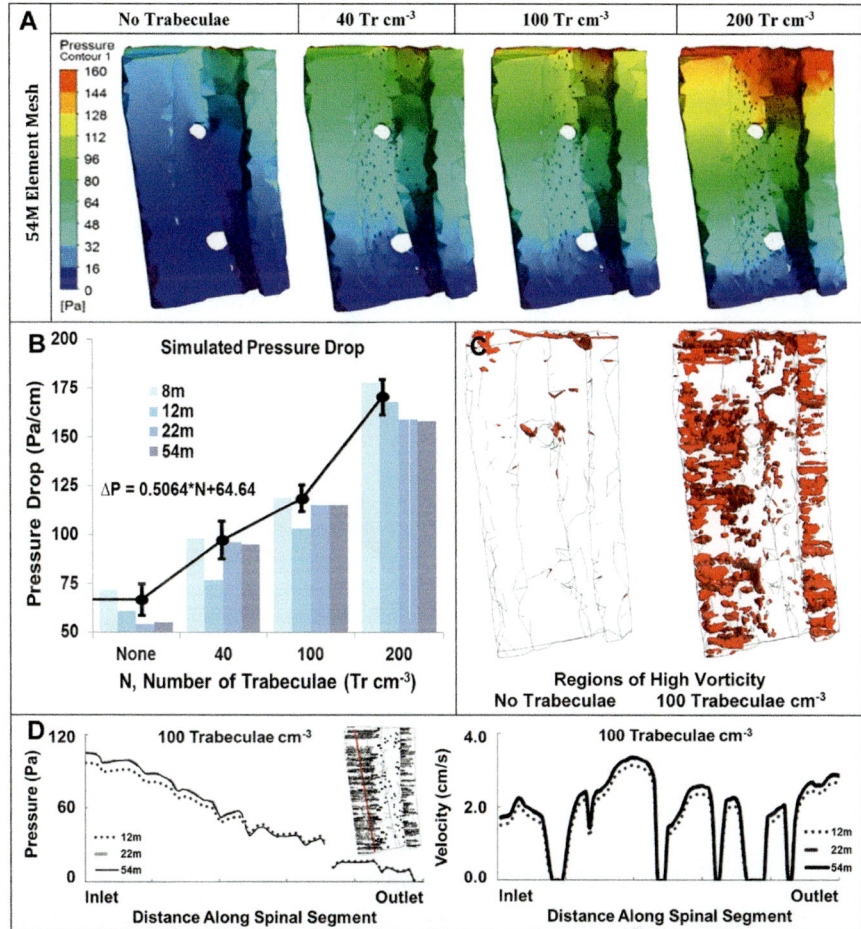

**Fig. 10** (**a**) Pressure contours for spinal SAS segment models with 40–200 Trabeculae cm$^{-3}$. (**b**) Simulated pressure drop, proportional to the number of trabeculae, independently from the chosen mesh. (**c**) High vorticity regions that are around trabeculae, almost absent in meshes with no trabeculae (vorticity threshold = 6 s$^{-1}$). (**d**) Mesh independency study, along a cut for different mesh densities. Image taken from [133]

the rapid biodispersion of moieties administered to the spinal SAS. The simulations confirm the prior prediction implicating CSF pulse and frequency as the main driving forces of solute dispersion in slow continuous CSF drug administration. In particular, the model proves that drug biodistribution depends strongly on CSF pulsatility, but only weakly on drug diffusivity. Simulated drug rise towards the cranial compartment is considerably slower in idealized annular models without micro-anatomy. Through their work, the authors provided a deeper understanding of the mechanisms connecting vascular pulsations to CSF dynamics. However,

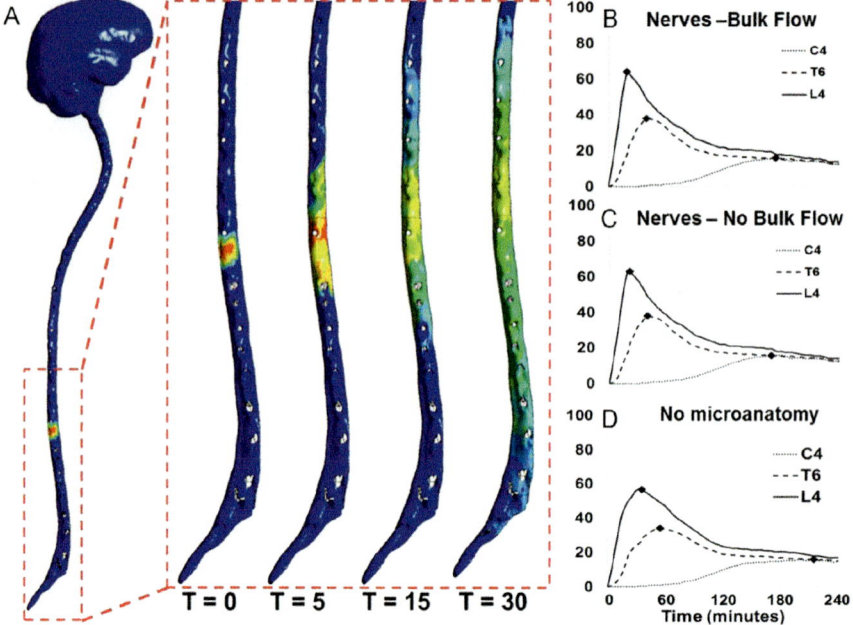

**Fig. 11** (**a**) Time evolution of the caudocranial rise of the injected drug: CSF bulk flow, due to arachnoid absorption, shown not to alter the speed of rostral drug motion in a significant way. (**b**), (**c**), and (**d**). One minute bolus injection in a model with nerves, without bulk flow and in an idealized annular model, respectively. Image taken from [133]

despite these recent progresses, the exact source location and type of force coupling between the expanding and contracting vasculature, brain movement, and induced CSF dynamics are still uncertain.

### 2.2.3 Direct Numerical Simulation of Transitional Hydrodynamics of the Cerebrospinal Fluid in Chiari I Malformation: The Role of Cranio-Vertebral Junction (Jain et al. [59])

The work presented by Jain et al. [59] investigates in detail the possibility of transitional CSF hydrodynamics, by performing fully resolved direct numerical simulations (DNS) based on the NS model on three subjects: one control and two Chiari malformation I patients. The *major interest* of the study is to observe and quantify CSF hydrodynamics at the finest scales possible, in particular in the craniovertebral junction.

To fetch the patient-specific feature, the *geometry* adopted in the model is retrieved from MR images obtained at Oslo University Hospital on a 3T Siemens scanner (Skyra) for the healthy individual and on a 3T Philips scanner (Achieva 2.5.3) for the two patients. 32 images were obtained per cardiac cycle using retrospective cardiac gating. For the segmentation of the SAS and ventricles, the authors used T2-heavily fluid weighted 3D steady-state echo images with a spacing of 0.5 mm × 0.5 mm or 1 mm × 1 mm in the sagittal plane and slice thickness of 1 mm. The anatomical reconstruction of the cervical SAS, pontine cistern, and the 4th ventricle including the aqueduct was performed using the VMTK via level set methods. VMTK was also used to generate the computational meshes. Besides the geometry, the volume flux at the cranial SAS and the aqueduct are retrieved by the PC-MRI and employed as BCs data. Other meaningful input in the model is the kinematic viscosity of the fluid defined by

$$\nu = \frac{\partial x^2}{\partial t} \, \frac{(1/\Omega) - 0.5}{3}$$

where $\Omega$ is the relaxation parameter in the Lattice Boltzmann method (LBM) algorithm in order to solve the overall system.

The *mathematical model* aims at assessing the impact of the assumption of laminar flow on the CSF dynamics by quantifying the Kolmogorov energy cascade. To achieve this goal the authors employ the LBM to numerically solve the NS system, due to its suitability in DNS of moderate Reynolds flows in complex anatomical geometries and its appreciable computational efficiency on modern supercomputers. Particularly the model utilizes the mesh generator Seeder with 700 to 800 millions cells for each model and the solver Musubi [45, 46, 68, 102]. It requires around 20–25 min on two supercomputers based in Munich, Germany (Leipzig Supercomputer Center) and Stuttgart, Germany (High Performance Computing Center), respectively, using more than 12,000 cores.

The *output* of this model are (i) velocity traces, (ii) pressure distribution, in particular the drop between pontine cistern and cervical spine and between aqueduct and pontine cistern, (iii) turbulent intensities, and (iv) Kolmogorov micro-scales (Figs. 12, 13, and 14). In particular the images in Fig. 13 show a remarkable flow impingement in Patient1 and Patient2, and flow fluctuations in Patient2. Main results are comparable with other studies in literature; in addition, the contribution confirms a previous prediction of possible transitional CSF flow already described in [48].

In *conclusion*, this work intends to quantify the sophisticated CSF hydrodynamics down to the Kolmogorov scales. The outcomes have prospects for future flow simulations and several implications from fluid mechanical, modeling, and, potentially significant clinical impact. For the future, the authors envision a further analysis on a larger cohort of healthy subjects and Chiari patients in order to have a better understanding of the presence of transitional/turbulent CSF flow.

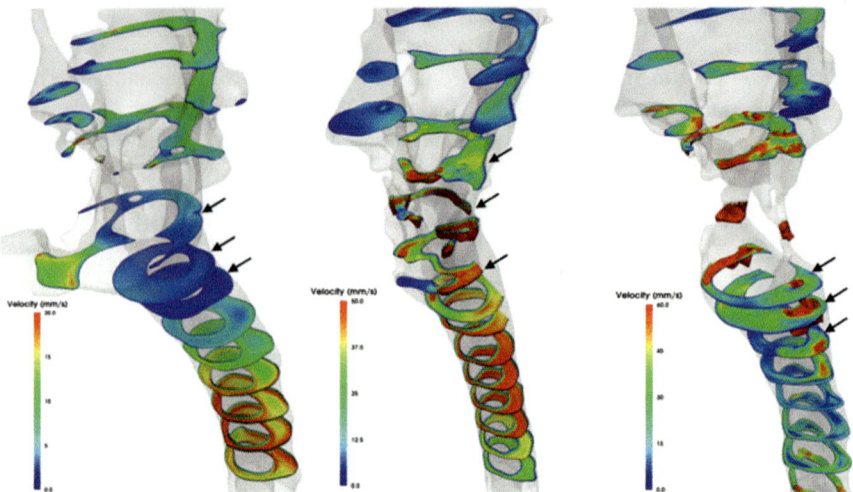

**Fig. 12** Peak systolic velocity field across axial planes in the SAS, Control, Patient1, and Patient2, respectively. Black arrows indicate the planes on which velocity field in Fig. 13 is analyzed. Image taken from [59]. The color scale corresponds to the velocity magnitude in mm/s. The lowest (blue) value is 0.0 mm/s for the three panels. The highest value (red) is 30, 50, and 60 mm/s for left, central, and right panel, respectively

## 2.3 Current Developments in Modeling the Coupled Cerebral-Ocular Flows Dynamics

The review of the mathematical and computation models that we have presented in Sects. 2.1 and 2.2 shows a substantial lack of works that investigate the link between the cerebral and ocular fluids, especially in the case of spatially distributed models. This deficiency can be traced back to the complexity of the problem from a physiological viewpoint and the fact that the mechanisms responsible for the interaction between the two fluids are still a matter of debate in medicine (see chapter "Anatomy and Physiology of the Cerebrospinal Fluid"). The multiplicity of different aspects in the physiology of the eye and brain calls for advanced mathematical and computational tools that can increase our understanding of these complex phenomena, in synergy with other approaches. Up to here, we focused on the works related to the CSF dynamics, we list in the sequel some interesting recent contributions that are approaching the cranial fluid dynamics starting from ocular modeling.

One of the main points of interest in the cerebral and ocular connection is the balance between the intracranial and intraocular pressure. Inner and outer factors may influence this delicate equilibrium causing different diseases, for example, glaucoma or spaceflight associated neuro-ocular syndrome (SANS).

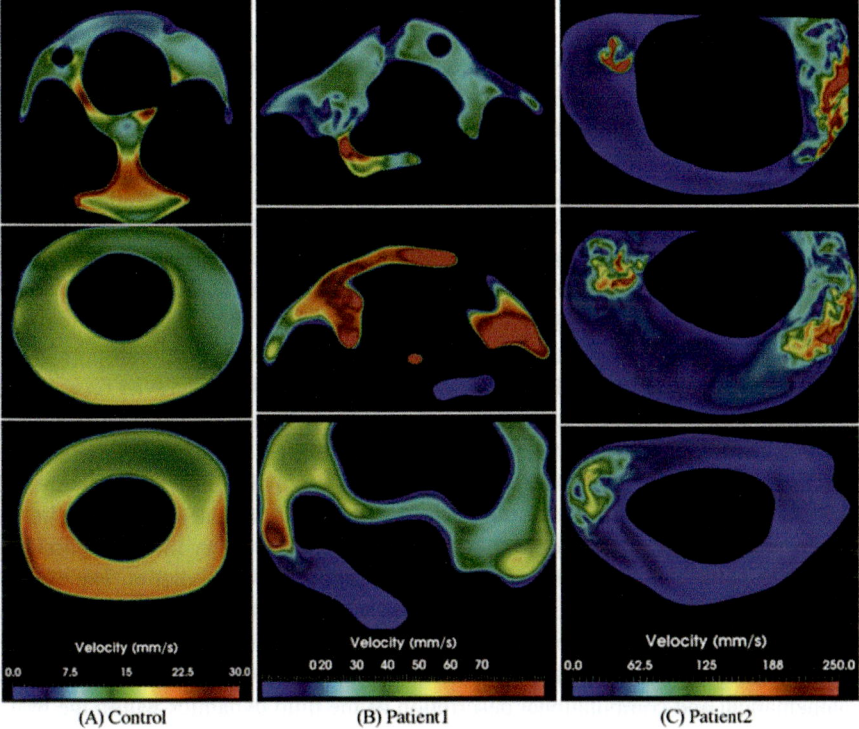

**Fig. 13** Velocity magnitude during peak diastole of the 8th cardiac cycle for Control (left), 50th cardiac cycle of Patient1 (middle) and Patient2 (right). Image taken from [59]

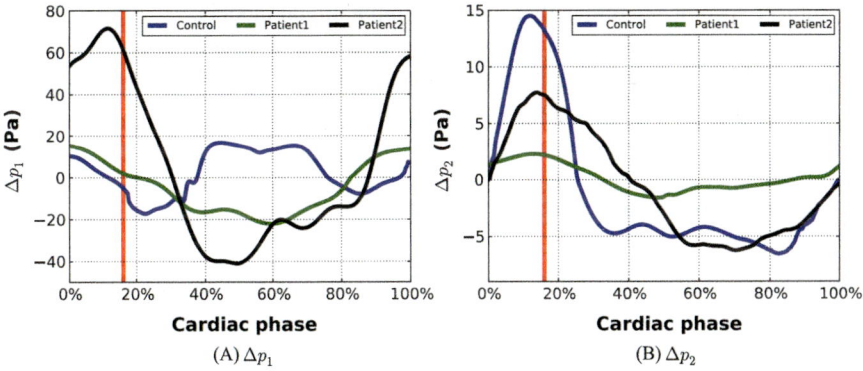

**Fig. 14** Pressure drop $\Delta p_1$ between pontine cistern and cervical SAS (left); $\Delta p_2$ between aqueduct and pontine cistern (right). The $\Delta p$ is ensemble averaged for the last $n = 15$ cardiac cycles in Patient1 and Patient2, while it is shown for the last (eighth) cycle of Control. The vertical red line indicates the systolic peak. Image taken from [59]

18. H Davson, G Hollingsworth, and MB Segal. The mechanism of drainage of the cerebrospinal fluid. *Brain*, 93(4):665–678, 1970.

19. Jurre den Haan, Aleid van de Kreeke, Johannes de Boer, Femke Bouwman, and Frank D Verbraak. Optical coherence tomography angiography in amyloid proven alzheimers disease; a non-invasive biomarker? *Investigative Ophthalmology & Visual Science*, 59(9):724–724, 2018.

20. Ida Dielemans, Johannes R Vingerling, Douwe Algra, Albert Hofman, Diederick E Grobbee, and Paulus TVM de Jong. Primary open-angle glaucoma, intraocular pressure, and systemic blood pressure in the general elderly population: the rotterdam study. *Ophthalmology*, 102(1):54–60, 1995.

21. Anders Eklund, Bo Lundkvist, L-OD Koskinen, and J Malm. Infusion technique can be used to distinguish between dysfunction of a hydrocephalus shunt system and a progressive dementia. *Medical and Biological Engineering and Computing*, 42(5):644–649, 2004.

22. DR Enzmann and NfsJ Pelc. Normal flow patterns of intracranial and spinal cerebrospinal fluid defined with phase-contrast cine mr imaging. *Radiology*, 178(2):467–474, 1991.

23. D.R. Enzmann and N.J. Pelc. Normal flow patterns of intracranial and spinal cerebrospinal fluid defined with phase-contrast cine mr imaging. *Radiology*, 178(2):467–474, 1991.

24. Andrew J Feola, Jerry G Myers, Julia Raykin, Lealem Mulugeta, Emily S Nelson, Brian C Samuels, and C Ross Ethier. Finite element modeling of factors influencing optic nerve head deformation due to intracranial pressure. *Investigative ophthalmology & visual science*, 57(4):1901–1911, 2016.

25. Stefan E Fischer, GC McKinnon, SE Maier, and P Boesiger. Improved myocardial tagging contrast. *Magnetic resonance in medicine*, 30(2):191–200, 1993.

26. Robert A Fishman et al. *Cerebrospinal fluid in diseases of the nervous system*. Saunders Philadelphia, 1992.

27. L. Formaggia, A. Quarteroni, and A. Veneziani. Cardiovascular mathematics, volume 1 of ms&a. modeling, simulation and applications, 2009.

28. Sigrid Friese, Uwe Klose, and Karsten Voigt. Zur pulsation des liquor cerebrospinalis. *Klinische Neuroradiologie*, 12(2):67–75, 2002.

29. Peter Fritzson, Peter Aronsson, Håkan Lundvall, Kaj Nyström, Adrian Pop, Levon Saldamli, and David Broman. The OpenModelica modeling, simulation, and development environment. In *46th Conference on Simulation and Modelling of the Scandinavian Simulation Society (SIMS2005), Trondheim, Norway, October 13–14, 2005*, 2005.

30. Simon Garnotel, Stéphanie Salmon, and Olivier Balédent. Numerical modeling of the intracranial pressure using windkessel models. *MathS In Action*, 8:9–25, 01 2017.

31. Manuel Gehlen, Vartan Kurtcuoglu, and Marianne Schmid Daners. Is posture-related craniospinal compliance shift caused by jugular vein collapse? a theoretical analysis. *Fluids and Barriers of the CNS*, 14(1):5, 2017.

32. Katuhiko Goda. A multistep technique with implicit difference schemes for calculating two-or three-dimensional cavity flows. *Journal of Computational Physics*, 30(1):76–95, 1979.

33. D. Greitz, A. Franck, and B. Nordell. On the pulsatile nature of intracranial and spinal csf-circulation demonstrated by mr imaging. *Acta radiologica*, 34(4):321–328, 1993.

34. Giovanna Guidoboni, Alon Harris, Lucia Carichino, Yoel Arieli, and Brent A Siesky. Effect of intraocular pressure on the hemodynamics of the central retinal artery: a mathematical model. *Mathematical Biosciences & Engineering*, 11(3):523–546, 2014.

35. Giovanna Guidoboni, Alon Harris, Simone Cassani, Julia Arciero, Brent Siesky, Annahita Amireskandari, Leslie Tobe, Patrick Egan, Ingrida Januleviciene, and Joshua Park. Intraocular pressure, blood pressure, and retinal blood flow autoregulation: A mathematical model to clarify their relationship and clinical relevance. *Investigative Ophthalmology and Visual Science*, 55(7):4105–4118, 2014.

36. Giovanna Guidoboni, Fabrizia Salerni, Alon Harris, Christophe Prud'homme, Marcela Szopos, Peter M Pinsky, and Rodolfo Repetto. Ocular and cerebral hemo-fluid dynamics in microgravity: a mathematical model. *Investigative Ophthalmology & Visual Science*, 58(8):3036–3036, 2017.

37. Giovanna Guidoboni, Fabrizia Salerni, Rodolfo Repetto, Marcela Szopos, and Alon Harris. Relationship between intraocular pressure, blood pressure and cerebrospinal fluid pressure: a theoretical approach. *Investigative Ophthalmology & Visual Science*, 59(9):1665–1665, 2018.
38. JE Guinane. An equivalent circuit analysis of cerebrospinal fluid hydrodynamics. *American Journal of Physiology-Legacy Content*, 223(2):425–430, 1972.
39. Sumeet Gupta, Michaela Soellinger, Peter Boesiger, Dimos Poulikakos, and Vartan Kurtcuoglu. Three-dimensional computational modeling of subject-specific cerebrospinal fluid flow in the subarachnoid space. *Journal of biomechanical engineering*, 131(2):021010, 2009.
40. Sumeet Gupta, Michaela Soellinger, Deborah M Grzybowski, Peter Boesiger, John Biddiscombe, Dimos Poulikakos, and Vartan Kurtcuoglu. Cerebrospinal fluid dynamics in the human cranial subarachnoid space: an overlooked mediator of cerebral disease. i. computational model. *Journal of the Royal Society Interface*, 7(49):1195–1204, 2010.
41. GyroTools. Gtflow: software solution that delivers all the necessary functionality for visualization, assessment and interpretation of multidimensional mri phase-contrast flow datasets. http://www.gyrotools.com/products/gt-flow.html, 2012.
42. Edi Azali Hadzri, Kahar Osman, Mohamed Rafiq Abdul Kadir, and Azian Abdul Aziz. Computational investigation on csf flow analysis in the third ventricle and aqueduct of sylvius. *IIUM Engineering Journal*, 12(3), 2011.
43. John E Hall. *Guyton and Hall textbook of medical physiology e-Book*. Elsevier Health Sciences, 2015.
44. A. Harris, J. Gross, D. Prada, B. Siesky, A. C. Verticchio Vercellin, L. Saint, and G. Guidoboni. The relationship between cerebrospinal fluid pressure and blood flow in the retina and optic nerve. In *Intraocular and Intracranial Pressure Gradient in Glaucoma (N. Wang)*. Springer, Singapore, 2019.
45. Manuel Hasert, Kannan Masilamani, Simon Zimny, Harald Klimach, Jiaxing Qi, Jörg Bernsdorf, and Sabine Roller. Complex fluid simulations with the parallel tree-based lattice boltzmann solver musubi. *Journal of Computational Science*, 5(5):784–794, 2014.
46. Manuel Hasert and Sabine P Roller. *Multi-scale lattice Boltzmann simulations on distributed octrees*. Number RWTH-CONV-144063. German Research School for Simulation Sciences GmbH, 2014.
47. Frédéric Hecht. New development in freefem++. *Journal of numerical mathematics*, 20(3-4):251–266, 2012.
48. Anders Helgeland, Kent-Andre Mardal, Victor Haughton, and Bjørn Anders Pettersson Reif. Numerical simulations of the pulsating flow of cerebrospinal fluid flow in the cervical spinal canal of a chiari patient. *Journal of biomechanics*, 47(5):1082–1090, 2014.
49. MC Henry-Feugeas, I Idy-Peretti, B Blanchet, D Hassine, G Zannoli, and E Schouman-Claeys. Temporal and spatial assessment of normal cerebrospinal fluid dynamics with mr imaging. *Magnetic resonance imaging*, 11(8):1107–1118, 1993.
50. HDM Hettiarachchi, Ying Hsu, Timothy J Harris, and Andreas A Linninger. The effect of pulsatile flow on intrathecal drug delivery in the spinal canal. *Annals of biomedical engineering*, 39(10):2592, 2011.
51. David W Holman, Vartan Kurtcuoglu, and Deborah M Grzybowski. Cerebrospinal fluid dynamics in the human cranial subarachnoid space: an overlooked mediator of cerebral disease. ii. in vitro arachnoid outflow model. *Journal of The Royal Society Interface*, page rsif20100032, 2010.
52. RM Hoogeveen. Quantitative flow measurement: accuracy and precision, spatial and temporal resolution. In *Proceedings of the International Workshop on Flow and Motion*. Zurich, 2004.
53. L. Howden, D. Giddings, and H. Power. Three-dimensional cerebrospinal fluid flow within the human central nervous system. *Discrete and continuous dynamical systems*, 2011.
54. Ying Hsu, HD Madhawa Hettiarachchi, David C Zhu, and Andreas A Linninger. The frequency and magnitude of cerebrospinal fluid pulsations influence intrathecal drug distribution: key factors for interpatient variability. *Anesthesia & Analgesia*, 115(2):386–394, 2012.

95. Alexander Monro. Observations on the structure and functions of the nervous system. 1783.
96. Emily S Nelson, Lealem Mulugeta, Andrew Feola, Julia Raykin, Jerry G Myers, Brian C Samuels, and C Ross Ethier. The impact of ocular hemodynamics and intracranial pressure on intraocular pressure during acute gravitational changes. *Journal of Applied Physiology*, 123(2):352–363, 2017.
97. Materialise NV. Mimics: advanced segmentation toolbox for patient specific-device design or medical image-based research and development. http://www.materialise.com/en/medical/software/mimics, 2010.
98. University of North Carolina at Chapel Hill. Gambit: end-to-end application allowing group-wise automatic mesh-based analysis of cortical thickness as well as other surface area measurements. https://www.nitrc.org/projects/gambit/, 2010.
99. Hae-Young Lopilly Park, Hye-Young Shin, Kyoung In Jung, and Chan Kee Park. Changes in the lamina and prelamina after intraocular pressure reduction in patients with primary open-angle glaucoma and acute primary angle-closure. *Investigative ophthalmology & visual science*, 55(1):233–239, 2014.
100. R.K. Parkkola, M.E. Komu, T.M. Aarimaa, M.S. Alanen, and C. Thomsen. Cerebrospinal fluid flow in children with normal and dilated ventricles studied by mr imaging. *Acta radiologica*, 42(1):33–38, 2001.
101. R Price, S Gady, K Heinemann, ES Nelson, L Mulugeta, CR Ethier, BC Samuels, A Feola, J Vera, and JG Myers. An integrated model of the cardiovascular and central nervous systems for analysis of microgravity induced fluid redistribution. 2015.
102. Jiaxing Qi, Kartik Jain, Harald Klimach, and Sabine Roller. Performance evaluation of the lbm solver musubi on various hpc architectures. In *PARCO*, pages 807–816, 2015.
103. A. Quarteroni, A. Veneziani, and C. Vergara. Geometric multiscale modeling of the cardio-vascular system, between theory and practice. *Computer Methods in Applied Mechanics and Engineering*, 302:193 – 252, 2016.
104. Sara Qvarlander, Khalid Ambarki, Anders Wåhlin, Johan Jacobsson, Richard Birgander, Jan Malm, and Anders Eklund. Cerebrospinal fluid and blood flow patterns in idiopathic normal pressure hydrocephalus. *Acta neurologica Scandinavica*, 135(5):576–584, 2017.
105. Sara Qvarlander, Bo Lundkvist, Lars-Owe D Koskinen, Jan Malm, and Anders Eklund. Pulsatility in csf dynamics: pathophysiology of idiopathic normal pressure hydrocephalus. *J Neurol Neurosurg Psychiatry*, 84(7):735–741, 2013.
106. Sara Qvarlander, Nina Sundström, Jan Malm, and Anders Eklund. Postural effects on intracranial pressure: modeling and clinical evaluation. *Journal of Applied Physiology*, 115(10):1474–1480, 2013.
107. Ruojin Ren, Jost B Jonas, Guohong Tian, Yi Zhen, Ke Ma, Shuning Li, Hongtao Wang, Bin Li, Xiaojun Zhang, and Ningli Wang. Cerebrospinal fluid pressure in glaucoma: a prospective study. *Ophthalmology*, 117(2):259–266, 2010.
108. Catherine Bowes Rickman. Brain and retinal degenerative diseases: Is there a common thread? Paper presentation at the Annual Meeting of ARVO, 2017.
109. Jarl Risberg, Nils Lundberg, and David H Ingvar. Regional cerebral blood volume during acute transient rises of the intracranial pressure (plateau waves). *Journal of neurosurgery*, 31(3):303–310, 1969.
110. G Rutkowska, V Haughton, S Linge, and K-A Mardal. Patient-specific 3d simulation of cyclic csf flow at the craniocervical region. *American Journal of Neuroradiology*, 33(9):1756–1762, 2012.
111. Gabriela Marta Rutkowska. Computational fluid dynamics in patient-specific models of normal and chiari i geometries. Master's thesis, 2011.
112. Riccardo Sacco, Aurelio G Mauri, Alessandra Cardani, Brent A Siesky, Giovanna Guidoboni, and Alon Harris. Multiscale modeling and simulation of neurovascular coupling in the retina. *Journal for Modeling in Ophthalmology*, 2(2):30–35, 2018.

113. Lorenzo Sala, Christophe Prud'homme, Giovanna Guidoboni, Marcela Szopos, Brent A Siesky, and Alon Harris. Analysis of iop and csf alterations on ocular biomechanics and lamina cribrosa hemodynamics. *Investigative Ophthalmology & Visual Science*, 59(9):4475–4475, 2018.
114. Lorenzo Sala, Christophe Prud'Homme, Marcela Szopos, and Giovanna Guidoboni. Towards a full model for ocular biomechanics, fluid dynamics, and hemodynamics. *Journal for Modeling in Ophthalmology*, 2(2):7–13, 2018.
115. Lorenzo Sala, Riccardo Sacco, and Giovanna Guidoboni. Multiscale nature of ocular physiology. *Journal for Modeling in Ophthalmology*, 2(1):12–18, 2018.
116. F. Salerni, R. Repetto, A. Harris, P. Pinsky, C. Prud'homme, M. Szopos, and G. Guidoboni. Biofluid modeling of the coupled eye-brain system and insights into simulated microgravity conditions, *PLoS One*, 14(8):e0216012, 2019.
117. Fabrizia Salerni, Rodolfo Repetto, Alon Harris, Peter Pinsky, Christophe Prudhomme, Marcela Szopos, and Giovanna Guidoboni. Mathematical modeling of ocular and cerebral hemo-fluid dynamics: application to viip. *Journal for Modeling in Ophthalmology*, 2(2):64–68, 2018.
118. B Sallerin-Caute, Y Lazorthes, B Monsarrat, J Cros, and R Bastide. Csf baclofen levels after intrathecal administration in severe spasticity. *European journal of clinical pharmacology*, 40(4):363–365, 1991.
119. Flow Science. Flow3d: complete and versatile cfd simulation platform for engineers investigating the dynamic behavior of liquids and gas in a wide range of industrial applications and physical processes. https://www.flow3d.com/, 1980.
120. Kenneth Shapiro and Anthony Marmarou. Clinical applications of the pressure-volume index in treatment of pediatric head injuries. *Journal of neurosurgery*, 56(6):819–825, 1982.
121. David M Silver and Orna Geyer. Pressure-volume relation for the living human eye. *Current eye research*, 20(2):115–120, 2000.
122. Arthur J Sit, Cherie B Nau, Jay W McLaren, Douglas H Johnson, and David Hodge. Circadian variation of aqueous dynamics in young healthy adults. *Investigative ophthalmology & visual science*, 49(4):1473–1479, 2008.
123. Alan Smillie, Ian Sobey, and Zoltan Molnar. A hydroelastic model of hydrocephalus. *Journal of Fluid Mechanics*, 539:417–443, 2005.
124. Ian Sobey and Benedikt Wirth. Effect of non-linear permeability in a spherically symmetric model of hydrocephalus. *Mathematical Medicine and Biology*, 23(4):339–361, 2006.
125. F. Staglberg, J. Mogelvang, C. Thomsen, B. Nordell, M. Stubgaard, and A. Ericsson. A method for mr quantification of flow velocities in blood and csf using interleaved gradient-echo pulse sequences. *Magnetic resonance imaging*, 7(6):655–667, 1989.
126. Detlev Stalling, Malte Westerhoff, and Hans-Christian Hege. Amira: a highly interactive system for visual data analysis, 2005.
127. Scott A Stevens, William D Lakin, and Wolfgang Goetz. A differentiable, periodic function for pulsatile cardiac output based on heart rate and stroke volume. *Mathematical Biosciences*, 182(2):201–211, 2003.
128. Karen-Helene Støverud, Hans Petter Langtangen, Geir Andre Ringstad, Per Kristian Eide, and Kent-Andre Mardal. Computational investigation of cerebrospinal fluid dynamics in the posterior cranial fossa and cervical subarachnoid space in patients with chiari i malformation. *PloS one*, 11(10):e0162938, 2016.
129. HG Sullivan and JD Allison. Physiology of cerebrospinal fluid. *Neurosurgery*, 3:2125–2135, 1985.
130. B. Sweetman and A.A. Linniger. Cerebrospinal fluid flow dynamics in the central nervous system. *Annals of biomedical engineering*, 39(1):484–496, 2011.
131. M. Szopos, S. Cassani, G. Guidoboni, C. Prud'homme, R. Sacco, B. Siesky, and A. Harris. Mathematical modeling of aqueous humor flow and intraocular pressure under uncertainty: towards individualized glaucoma management. *Journal for Modeling in Ophthalmology*, 1(2):29–39, 2016.

132. Yukio Tada and Tatsuya Nagashima. Modeling and simulation of brain lesions by the finite-element method. *IEEE Engineering in Medicine and Biology Magazine*, 13(4):497–503, 1994.
133. Kevin M Tangen, Ying Hsu, David C Zhu, and Andreas A Linninger. Cns wide simulation of flow resistance and drug transport due to spinal microanatomy. *Journal of biomechanics*, 48(10):2144–2154, 2015.
134. Minoru Tomita, H Naritomi, and WD Heiss. *Cerebral Hyperemia and Ischemia, from the Standpoint of Cerebral Blood Volume: Proceedings of the Statellite Symposium, Brain Section, Fourth World Congress for Microcirculation, Osaka, Japan, 1-2 August 1987*. Elsevier, 1988.
135. B. Tully and Y. Ventikos. Cerebral water transport using multiple-network poroelastic theory: application to normal pressure hydrocephalus. *Journal of Fluid Mechanics*, 2011.
136. Mauro Ursino. A mathematical study of human intracranial hydrodynamics part 1 – the cerebrospinal fluid pulse pressure. *Annals of biomedical engineering*, 16(4):379–401, 1988.
137. Mauro Ursino. A mathematical study of human intracranial hydrodynamics part 2 – simulation of clinical tests. *Annals of biomedical engineering*, 16(4):403–416, 1988.
138. Mauro Ursino and Patrizia Di Giammarco. A mathematical model of the relationship between cerebral blood volume and intracranial pressure changes: The generation of plateau waves. *Annals of Biomedical Engineering*, 19(1):15–42, Jan 1991.
139. Mauro Ursino and Carlo Alberto Lodi. A simple mathematical model of the interaction between intracranial pressure and cerebral hemodynamics. *Journal of Applied Physiology*, 82(4):1256–1269, 1997.
140. Kristian Valen-Sendstad, A Logg, Kent-A Mardal, H Narayanan, and M Mortensen. A comparison of some common finite element schemes for the incompressible navier-stokes equations. *Automated Scientific Computing (Logg, Mardal, Wells et al), preliminary draft*, 2010.
141. J.C. Vardakis, D. Chou, B.J. Tully, C.C. Hung, T.H. Lee, P.H. Tsui, and Y. Ventikos. Investigating cerebral oedema using poroelasticity. *Medical engineering and physics*, 2016.
142. J.C. Vardakis, B.J. Tully, and Y. Ventikos. *Computer Models in biomechanics*, chapter Multicompartmental poroelasticity as a platform for the integrative modeling of water transport in the brain. Netherlands, Springer, 2013.
143. J.C. Vardakis, B.J. Tully, and Y. Ventikos. Exploring the efficacy of endoscopic ventriculostomy for hydrocephalus treatment via a multicompartmental poroelastic model of csf transport: a computational perspective. *PLoS One*, 2013.
144. Frank M White. Fluid mechanics, 3rd. *NewYork, McGraw H çœ*, 1994.
145. Benedikt Wirth and Ian Sobey. An axisymmetric and fully 3d poroelastic model for the evolution of hydrocephalus. *Mathematical Medicine and Biology*, 23(4):363–388, 2006.
146. Liang Xu, Han Wang, Yaxing Wang, and Jost B Jonas. Intraocular pressure correlated with arterial blood pressure: the beijing eye study. *American journal of ophthalmology*, 144(3):461–462, 2007.
147. Mokhtar Zagzoule and Jean-Pierre Marc-Vergnes. A global mathematical model of the cerebral circulation in man. *Journal of biomechanics*, 19(12):1015–1022, 1986.
148. David C Zhu, Michalis Xenos, Andreas A Linninger, and Richard D Penn. Dynamics of lateral ventricle and cerebrospinal fluid in normal and hydrocephalic brains. *Journal of Magnetic Resonance Imaging*, 24(4):756–770, 2006.

# Part VII
# Perspectives

# Image Analysis for Ophthalmology: Segmentation and Quantification of Retinal Vascular Systems

Kannappan Palaniappan, Filiz Bunyak, and Shyam S. Chaurasia

**Abstract** The retina is directly connected to the central nervous system and the vascular circulation, which uniquely enables three-dimensional retinal tissue structures and blood flow dynamics to be imaged and visualized from the exterior using non-invasive imaging modalities. Rapid advances in the types of diagnostic imaging modalities, combined with image processing, computer vision, artificial intelligence, and machine learning algorithms for quantitative image analytics are opening up a host of new possibilities for early diagnosis and treatment of a broad range of eye and systemic diseases with clinical impact. Incorporating patient-specific imaging to estimate geometric structures of vessel morphology and boundary conditions as input to the mathematical and computational fluid-dynamics modeling frameworks described in earlier chapters will enable new ways to predict treatment outcomes and model physiological effects at the systemic level. This chapter describes a set of widely used retinal imaging modalities, including fundoscopy, fluorescein angiography (FA), and optical coherence tomography (OCT), along with emerging modalities to measure retinal blood flow dynamics like optical coherence tomography angiography (OCTA) and laser speckle flowgraphy (LSFG). We use vessel segmentation and quantification as a prototypical ophthalmology image analysis pipeline that can be applied across imaging modalities, to describe processing techniques for measuring geometrical vascular structures. Current challenges and future opportunities especially in using artificial intelligence and deep learning architectures for patient optimized precision medicine and clinical efficacy are highlighted.

K. Palaniappan (✉) · F. Bunyak · S. S. Chaurasia
Department of Electrical Engineering and Computer Science, University of Missouri, Columbia, MO, USA
e-mail: palaniappank@missouri.edu; bunyak@missouri.edu; chaurasias@missouri.edu

© Springer Nature Switzerland AG 2019     543
G. Guidoboni et al. (eds.), *Ocular Fluid Dynamics*, Modeling and Simulation in Science, Engineering and Technology, https://doi.org/10.1007/978-3-030-25886-3_22

# 1 Introduction

Imaging modalities are an essential component of clinical practice and research for non-invasive assessment of all the different organ systems of the body, including the eye. The neurovascular physiology of the eye offers a clinical window to assess the overall health of the patient and offers a rich range of opportunities for diagnosis and monitoring of not only eye diseases but also systemic diseases, including cardiovascular, neurological, and autoimmune pathologies. Common ophthalmological imaging techniques described in this section include color fundoscopy, fluorescein angiography, optical coherence tomography, and optical coherence tomography angiography. These modalities image different structures of the retina and are widely used in the diagnosis and follow-up of patients experiencing vision degradation or loss due to chronic health conditions, congenital diseases, age related degeneration, or as a result of environmental exposure and accidents.

Over the past half century, there has been great progress in quantitative characterization of biomedical and clinical images, such as facilitating computer-aided diagnosis, identifying disease state and severity using automated grading of diagnostic imagery, recommending treatment type and therapeutic medication dosage or titration. Recently, the first artificial intelligence-based software for retinal disease diagnosis has received US FDA approval.

However, there is a wide gap between disease specific image acquisition and image analysis, and physiologically driven numerical models incorporating anatomical structure and function of tissues, organs, and whole body systems across scale, from the molecular to the macroscopic level. Combining quantitative clinical image analysis of the ocular vascular structure and dynamics, with physics- and omics-based models that capture the interaction of the local physiology with the systemic state of the patient, will offer new ways to model patient health and predict treatment outcomes within a mathematical modeling framework that holds great promise for therapeutics.

Vascular geometric structure and functional flow analysis is common across multiple retinal imaging modalities and serves as a case study to illustrate the computational frameworks used to transform images from very high-dimensional pixel space to low-dimensional interpretable patient-centric measurements for quantitative clinical decision support. Quantitative image analysis, of the vessel and capillary architecture, can then provide a framework for initializing patient-specific ocular physiology. This would provide accurate boundary conditions for simulating ocular biomechanics, mesoscale fluid flow, and nutrient and gas flux [1] to predict disease progression under alternative treatment options including watchful waiting. The area of combining image analysis for patient-specific vessel geometry with mathematical modeling of the microcirculation [2–4] is ripe with opportunity as 3-D imaging modalities continue to improve to measure geometric structures and flow at finer scales. Additional enabling technologies include new image analytics and machine learning algorithms for characterizing anatomical structures, dynamics, and disease conditions, to reach expert levels of performance, while at the same time computational resources for micron resolution volumetric numerical models of ocular tissues and blood flow become more affordable. There are a number of challenges to enable clinical application of quantitative image analysis for dynamical flow modeling which are outlined in this chapter.

We describe a set of quantitative algorithms for extracting the vessel structures from retinal imagery that can be used to characterize the physiological state of the tissue. Artificial intelligence and deep learning methods applied to imagery are emerging for ophthalmological applications including diagnosing and treating diseases of the retina and systemic diseases manifested in the eye such as age-related macular degeneration (AMD), diabetic retinopathy (DR), glaucoma, maculopathies, choroidal neovascularization, hypertensive retinopathy, and arteriosclerosis of the retinal vessels [5–13]. Challenges in processing multimodal retinal imagery for clinical and research applications include: (1) variations in anatomy, color, size, shape, curvature and tortuosity of retinal vessels, vessel brightness profile, and weak boundaries; (2) complex patient-specific three-dimensional anatomical structures and topology such as branching points, vessel crossings, centerline reflex, optic disc, and macula; (3) variability in blood flow regulation and transport with age, gender, variable physiological conditions and medications; (4) presence of signs of pathological conditions such as microaneurysms, exudates, leakage, hemorrhages, and sign of infective processes; and (5) imaging instrument limitations including resolution, limited field of view, illumination variation, limited quantitative flow measurement, patient tolerance, and artifacts due to eye motion. The application of image analysis with deep learning methods to quantitatively identify and characterize disease progression and longitudinal patient response to treatment is in its infancy. There are a wide range of opportunities for incorporating modern deep learning methods to the clinically beneficial image analytics pipelines. We describe image segmentation and quantification of retinal vascular systems as an example pipeline with a wide range of applications across image modalities and pathological conditions.

Characterizing the vascular structure and microcirculation in the retina is essential for a range of eye diseases including AMD, DR, and glaucoma, which are the leading causes of vision loss associated with aging. AMD is the most common cause of visual loss in the USA affecting over half of all legally blind Americans [14]. AMD damages the macula causing loss of central vision with the dry AMD form progressing slowly and accounting for 80% of patients. Wet AMD, also called choroidal neovascularization (CNV), involves formation of abnormal blood vessels in the retina (ingrowth into the macula) which can leak blood and cause inflammation, is faster progressing, and without early intervention can lead to vision loss. Treatments for wet AMD include intravitreal injections of agents such as anti-vascular endothelial growth factor (anti-VEGF) that stops the growth and formation of abnormal blood vessels. Diabetic retinopathy due to alterations in the retinal microvasculature from diabetes mellitus is a leading cause of blindness affecting over 100 million people globally [15]. Hyperglycemia (excess blood glucose) damages the retinal vessel walls leading to ischemia and growth of new blood vessels (proliferative DR) or breakdown of the blood-retinal barrier leading to damage of photoreceptors from fluid leakage into the retina and diabetic macular edema [16]. Clinical manifestation of DR in fundus images includes microaneurysms, hard exudates (proteins leaking through retinal vessels), soft exudates or cotton wool spots (cellular damage from occlusion of arterioles), hemorrhages from weakened capillaries, neovascularization with leaky vessels that bleed into the vitreous, and macular edema (swelling of the retina due to permeability changes in retinal capillaries) [17]. Glaucoma is the third most common cause of blindness in the USA and is primarily an optic neuropathy, caused by damage to the retinal nerve fibers. The major known risk factor for glaucoma disease onset and progression is due to an increased intraocular pressure arising from an impairement in the acqueous humor fluid dynamics [18]. The vertical cup to disc ratio of the optic nerve head in fundus images using quantitative image analysis is used as a structural indicator for detecting and monitoring progression of glaucoma [19].

Novel treatment methods for previously intractable ophthalmological diseases are emerging for a variety of retinal degenerative diseases including genetic disorders such as retinitis pigmentosa and choroideremia [20]. Experimental treatment approaches include retinal prosthetics, biocompatible synthetic replacements, transplants, restoring photoreceptors using optogenetics, stem cell therapy, and gene therapy among others. New treatment methods to prolong or restore vision will spur the development of new imaging modalities with higher cellular level resolution and deeper tissue penetration. This in turn will require better artificial intelligence-based image analysis techniques combined with a more complete mathematical model of ocular geometric structures and fluid dynamics to characterize multi-scale physiological responses and predict outcomes. The process of coupling high resolution quantitative image analysis with numerical models of the ocular system is currently in its infancy. It will usher in a more rigorous approach for improved health care and personalized medicine by matching existing and emerging treatments with patients, monitoring outcomes, and quantitatively measuring physiological and omic changes in tissue structures due to medication, surgical intervention, or introduction of biological or synthetic materials within the eye and retina that are all patient specific processes.

## 2  Diagnostic Retinal Imaging Modalities

### 2.1  Fundoscopy for Ophthalmic Examination

**Background** The retina has a unique position in the central nervous system, as it is the only neural tissue that can be imaged and visualized from the exterior using non-invasive devices. Also, it is the only neuronal tissue where the microvasculature can be imaged externally [21, 22]. A fundus (or retinal) camera optical system uses a low power microscope with a high intensity (xenon) flash and specialized optics to capture the reflected light that is electronically recorded using a high resolution

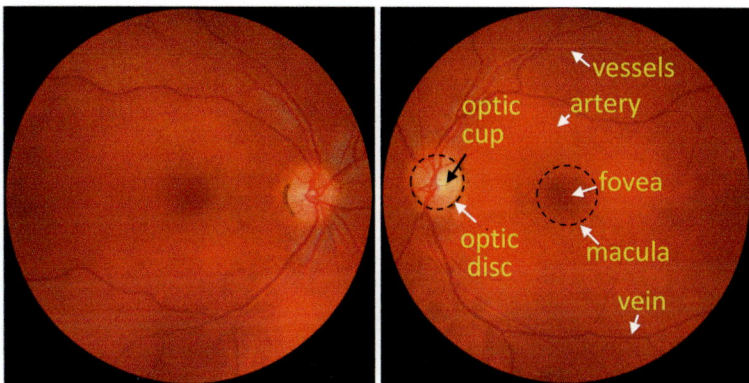

**Fig. 1** Fundus photographs using light illumination to image the retina without the use of dyes, of a normal right (OD) and left eye (OS) (left and right images respectively) with the macula near the center of each image and the optic disc located towards the nose

digital color camera, to examine and photograph the interior of the eye including the retina, retinal vasculature, optic nerve head, and posterior pole [14] (see Fig. 1). In the normal fundus, arteries and veins emerge from the optic nerve head and the veins appear darker than the arteries. The vascular walls are not visible, but the column of blood within the walls is visible [23, 24]. Although fundoscopy has been in use since the early part of the twentieth century, the first automatic and quantitative retinal image analysis methods were initiated in the 1970s [14], and are only recently reaching performance levels with high enough sensitivity and specificity suitable to consider computer-aided diagnosis for selective disease categories like DR and severity, using deep learning and a very large corpus of labeled fundus imagery to train the learning algorithms [25]. But many challenges remain, including screening for multiple disease conditions, validating accuracy on larger patient cohorts and communities with greater heterogeneity in disease manifestation, best use of artificial intelligence (AI) software in clinical workflows such as initial assessment or second opinion, and explainable AI to build clinician patient trust in automated diagnostic systems [15].

**Clinical Applications** The routine use of fundoscopy has led to the early recognition of several retinal diseases and to the prevention of ophthalmic complications and vision loss including DR, glaucoma, and AMD [26, 27] (see Fig. 2a–d). Damage to the eye including macular edema and retinal detachment can be observed in fundus images. The fundoscopic examination can provide diagnostic information related to hypertension, diabetes, and infections. In pathological conditions such as AMD and DR, fundus examinations can be used for detection of hard exudates comprised of fat-laden macrophages and serum lipids [26]. Cotton wool spots, sometimes also called soft exudates, are localized in the retinal nerve fiber layer recognized by white lesions with fuzzy edges. Roth spots can be visualized as retinal

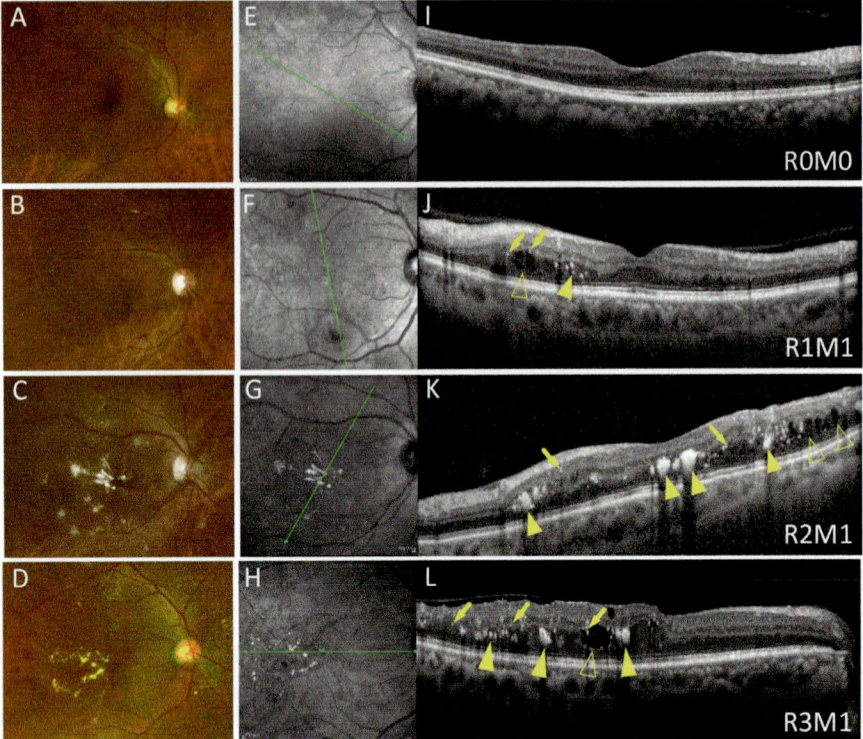

**Fig. 2** Multimodal clinical retinal imaging of diabetic retinopathy patients. Retinal fundus color fundoscopy (Col 1, **a–d**), fluorescein angiography (Col 2, **e–h**), and optical coherence tomography (Col 3, **i–l**) imaging of the same eye in four type 2 diabetic patients. Diabetic retinopathy was graded according to the Gloucestershire classification. The degree of retinopathy (shown in Col 3) is represented as R0 = no retinopathy, R1 = mild NPDR (non-proliferative diabetic retinopathy), R2 = moderate to severe NPDR, R3 = PDR (proliferative diabetic retinopathy). Maculopathy is denoted as M0: absent, M1: present. *Filled triangle*, hard exudate; *Up arrow*, microaneurysm; *Open triangle*, intra-retinal edema. Figure adapted from [27]

hemorrhages with a white or yellow center. Cholesterol emboli can be found in the arteriolar lumen and can cause vessel occlusion and infarct of the tissue. Fundoscopy can also be used to detect optic disc edema depicted by elevated optic disc covered by cotton wool spots and hemorrhages [28–31].

## 2.2 Fluorescein Angiography (FA)

**Background** Fluorescein angiography (FA) is a diagnostic technique which allows the study of the retinal and choroidal circulation of blood vessels in physiologic and pathologic conditions. It is an invasive technique where patients are intravenously injected with an orange-red dye called sodium fluorescein with a molecular weight of 376 daltons that diffuses through most of the body fluids. Sodium fluorescein disperses into the retinal and choroidal vasculature which can be visualized using FA microscopy. The dye is relatively inert and safe as it does not cross the blood-retinal barriers and eliminated within 24–36 h via the urine [32–35].

**Clinical Applications** FA is an essential test used to classify several retinal diseases and to follow-up on patients with diseases at different stages of progression [36–39]. FA in patients can show blockage or leakage in the retinal vessels, characterized in terms of decreased (hypo-) and increased fluorescence (hyperfluorescence). Hypofluorescence or vascular filling defect on FA characterizes various stages of the retinal diseases clinically represented by intraretinal hemorrhage, subretinal exudate, pigment proliferation/clumping, sub-RPE (retinal pigment epithelium) hemorrhage, vascular atrophy, and retinal capillary occlusion (Fig. 2e–h). On the contrary, hyperfluorescence can be detected with FA in patients with proliferative DR and wet AMD indicated by the presence of RPE rip or detachment, soft drusen, cystoid macular edema, choroidal neovascularization, and retinal angiogenesis [27].

## 2.3   Optical Coherence Tomography (OCT)

**Background** Optical coherence tomography (OCT) is becoming a standard of care imaging modality that uses light waves to visualize the 3-D structure of different layers in the retina at high resolution without harming the tissue similar to ultrasound imaging but using light [40]. OCT is a non-invasive (label-free) technology which uses optical back-scattering or back-reflection to measure in real-time and in situ cross-sectional depth slices of the patient's retina at micron scale resolution. Unlike traditional histopathology where tissue (biopsy) specimens are surgically excised, stained, and processed for microscopy-based pathology examination, OCT, that was introduced in 1991 and now widely adopted clinically, offers a non-invasive optical approach. OCT enables a visualization of the tissue comparable to a biopsy, allowing for monitoring of disease progression, including pre- and post-treatment evaluation. In conventional OCT imaging, a short pulse or continuous wave laser (with 840 nm wavelength, for example) is used as the light source because of its extremely short coherence lengths and high output power enabling high-resolution and high-speed imaging by measuring the backscattering of light as it passes through the tissue [41]. Figure 3 shows the process of acquiring a 3-D OCT volume of the retinal tissue by rapidly generating a patterned sequence axial laser scans covering the region of clinical interest. The figure illustrates the A- and B-scan patterns used to image the tissue volume and several ways in which the retinal data cube can be visualized using different available software. There are several advantages of OCT technology for biomedical imaging of the retina and many other tissues. OCT image slices are characterized by high axial resolutions

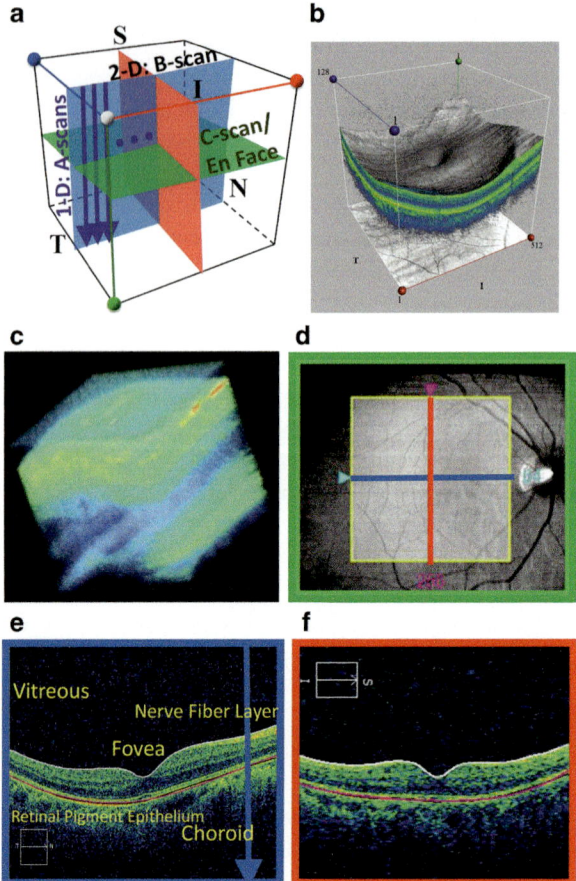

**Fig. 3** Volumetric 3-D OCT transverse scan of the retina using raster scanning modes (**a**) OCT laser backscatter scanning in depth through the tissue at a single location along optical axis is called an A-scan or axial scan ($z$-axis); several A-scans along a line form a B-scan or 2-D optical cross-sectional slice ($x-z$ plane) or transverse scan. A set of adjacent B-scans form a 3-D data cube ($x-y-z$ volume). Extracting a (virtual) slice from the volume at a single depth is known as a C-scan ($x-y$ plane); multiple C-scans can be combined into a depth slice known as an *en face* image. (**b**) Volume rendering showing the top surface of a normal human retina (512 × 128 × 1024 voxels). (**c**) Volume rendering of a full 3-D OCT cube (rat retina) showing multiple layers rendered transparently using computer graphics techniques. (**d**) Sample en face image with the macula region outlined by the yellow box (human retina OD, right eye). (**e**) Sample B-scan slice (TN blue-plane) with several retinal layers labeled. (**f**) Sample SI red-plane ($y-z$ plane) is a virtual slice created from the 3-D volume scan

of 1–15 μm and real-time imaging. OCT supports in situ imaging on excised specimens, is compact and portable, and is compatible with a wide range of surgical instruments. The major drawbacks to the OCT imaging modality are its limitation in imaging tissues deeper than 2–3 mm and its limited ability to penetrate through opaque tissues including dense cataracts and excessive bleeding in the vitreous.

**Clinical Applications** OCT is a powerful imaging platform that provides a real-time, non-contact, and non-invasive imaging of the anterior and posterior chambers of the eye for diagnosis and monitoring of disease progression and clinical response to treatment. OCT imaging has made a substantial clinical impact in the field of ophthalmology [42–46]. In the last two decades, imaging algorithms have been developed to investigate the use of OCT in diagnosis and monitoring of ocular diseases including, AMD, DR, glaucoma, diabetic macular edema, macular hole, macular pucker, central serous chorioretinopathy, epiretinal membranes, and vitreous traction [47–50]. We characterized the diabetic retinopathy patients using OCT imaging technique in our recent study (Fig. 2i–l; [27]). Figures 4 and 5 show how clinical visualization software present OCT volumetric data to the ophthalmologist for diagnosis and treatment in terms of 2-D slices, 2.5-D surfaces, and linked data visualization sub-windows. The linked window paradigm that is now in pervasive use in different fields was pioneered by the NASA Interactive Image SpreadSheet (IISS) visualization system for examining large volumes of time-varying 2-D imagery and 3-D model data [51, 52]. In recent years, several developments have been made in OCT technology, which might lead to the development of OCT interfaced catheter and endoscopes imaging gastrointestinal, pulmonary, urinary tracts, and arterial imaging. In the future, OCT technology can be utilized for cellular level imaging as well as a tool for guiding surgical intervention making it a potential tool in the diagnosis and clinical management of many diseases [53–58].

## 2.4   Optical Coherence Tomography Angiography (OCT-A)

**Background** OCT angiography is an emerging imaging modality that extends OCT to non-invasively measure the vessel blood flow dynamics within the layers of the human retina and choroid at unprecedented resolution without dyes or contrast agents. It was recently approved by the FDA in 2016 [59]. OCT-A uses the intensity and phase of backscattered light off the surface of moving red blood cells to measure flow or motion-contrast in the microcirculation within the retinal and choroid tissues [60, 61]. OCT-A captures dynamics by extending OCT to repeatedly image the same tissue area (i.e., transverse slice), registering slices to compensate for eye-movement and using image background modeling and differencing of multiple B-scans to characterize regions of high and low blood flow rates. OCT-A is becoming a more popular alternative to visualize vascular flow changes since it is fast, avoids intravenous dye injection which takes more time, and can have adverse clinical effects in some patients. The dye-based angiographic techniques in clinical practice include fluorescein angiography (FA) for imaging retinal blood flow and indocyanine green angiography (ICGA) to visualize blood flow in the choroidal vasculature. Both FA and ICGA provide primarily 2-D angiogram imagery of dye staining and leakage. OCT-A, on the other hand, provides vessel flow dynamics and

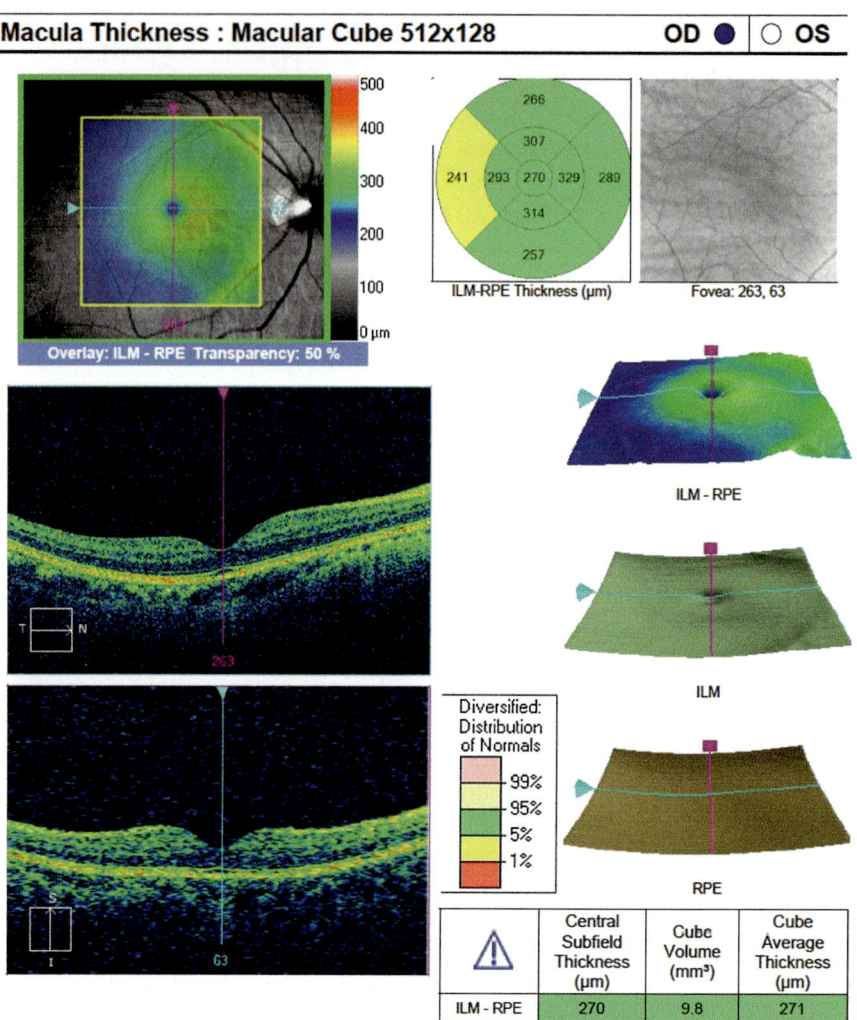

**Fig. 4** Macula cube from a normal patient showing clinical data visualization of the OCT volume as slices and surfaces that are linked together so that user manipulation of one visual object updates the other sub-windows. OCT visualization using the Zeiss Cirrus HD-OCT v9.5.1 software

motion-contrast structures across both the retinal and choroidal tissues as well as visualization of 3-D microvasculature flow relationships as seen in Fig. 6.

**Clinical Applications** OCT-A aids the clinician in visualizing the anatomical structures with vessel blood flow and also provides imaging of small blood vessel tissues across retinal layers that can be missed by OCT. OCT-A facilitates identification of vascular changes such as capillary dropout or abnormal new vessel growth in AMD or diabetic retinopathy that can lead to vision loss. In diseased

**Fig. 5** Macula cube from a normal patient showing slices 59, 61, 63, and 65 that is visualized as linked side-by-side windows, which is a data visualization technique that is well suited to look at three-dimensional data, anatomical structures and relationships, from different viewing perspectives. Visualization using the Zeiss Cirrus HD-OCT v9.5.1 software

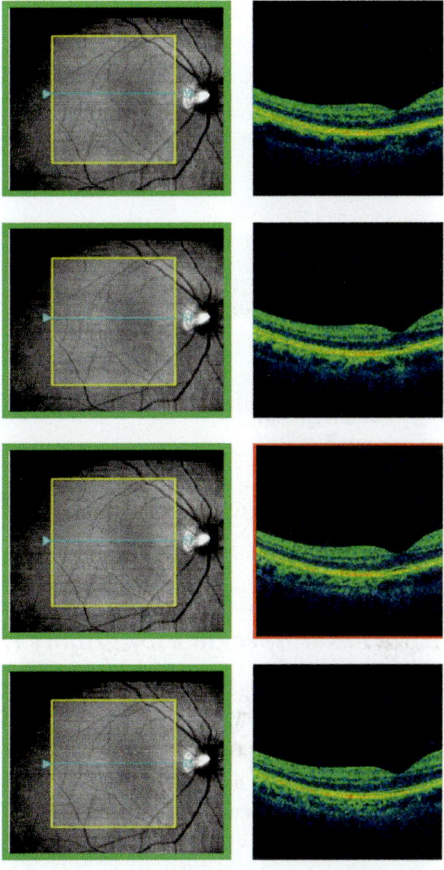

eyes, pathological signs such as drusen, intraretinal cysts, edema, or subretinal fluid can be visualized by OCTA. This novel technique also allows an improvement in the quality of the automated segmentation. Quantitative analysis of blood flow is done using image analysis of en face images and measuring skeleton density, vessel density, fractal dimension, vessel diameter index, flow index, and neovascularization area. OCT-A is well suited for characterizing diseases with abnormal presence of flow (neovascularization), anomalous vessel geometry (dilated vessels, aneurysms), macular telangiectasia, microaneurysms, capillary remodeling, or the absence of flow (nonperfusion/capillary dropout) which are characteristic pathologic hallmarks of many retinal and choroidal vascular diseases [59]. The limitations of OCT-A include limited clinical availability compared to FA, limited field of view, need to compensate for eye-motion and for noise reduction in order to detect small vessels, and difficulty in assessing changes in a vessel's permeability and leakage [60, 61].

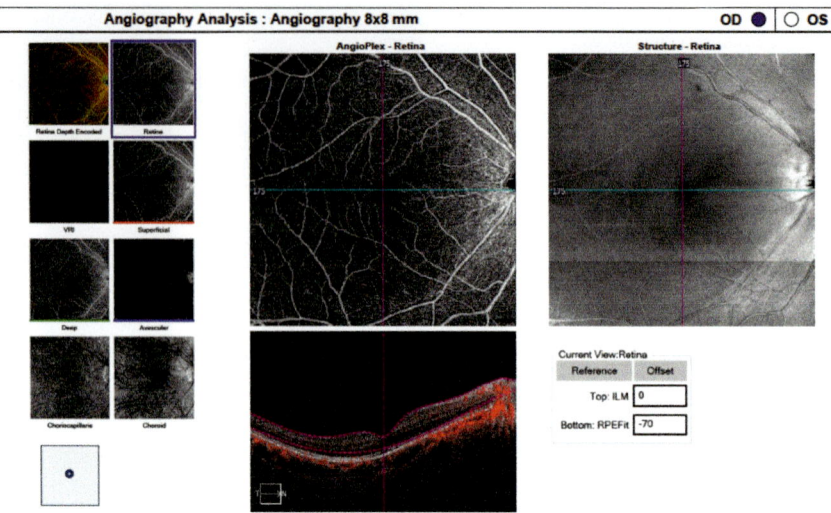

**Fig. 6** OCT angiography showing flow in a normal patient across different retinal layers

## 2.5 Laser Speckle Flowgraphy (LSFG)

**Background** Laser speckle flowgraphy (LSFG) is an emerging non-invasive imaging technique to quantitatively estimate the real-time relative blood flow volume for the assessment of hemodynamics and ocular perfusion in retinal vessels, choroid, and optic nerve head in living eyes. LSFG exploits a laser speckle phenomenon to measure in real-time 2-D fundus blood flow changes to quantify the movement of red blood cells (RBCs) without requiring pupillary dilation or contrast media. LSFG gathers interference obtained through ray scattering through the application of laser speckle to the blood flow creating a random speckled pattern [62–64]. The rate of ocular blood flow is directly proportional to the changes in the speckle phenomenon, which can be expressed numerically and the flow rate can be assessed as mean blue rate (MBR). Waveform analysis of the range of MBR values allows for the calculation of the blowout score (BOS). The retinal vasculature is known to self-regulate blood flow dynamics using ocular perfusion pressure and vascular resistance and maintain blood homeostasis under pathological conditions.

**Clinical Applications** LSFG is a non-invasive, precise, and fast technique to assess patients with ocular conditions associated with vascular dysregulation such as glaucoma, age-related macular degeneration, or diabetic retinopathy. Previous studies have suggested a potential correlation between the basal blood flow and neuronal activity-dependent alterations in ocular diseases. In fact, the retinal glial cells are known to regulate the activity-dependent blood flow in the pathogenesis of many ocular diseases, including proliferative diabetic retinopathy (PDR). In our recent report [65], we measured the relative retina flow volume (RFV) derived by filtering the background choroidal flow from the overall blood flow value in a region of interest centered on a retinal vessel, measuring vessel diameter and retinal flow velocity (Fig. 7i–l) and compared with fundus (Fig. 7a–d) and fluorescein angiography (Fig. 2e–h). A double transgenic (Akimba) mouse model representing proliferative diabetic retinopathy (PDR) showed a significant decrease in the RFV compared to their parental mouse strains in the study (Akita, Kimba or wild-type retina). A clinical study by Shiba et al. [62] showed a positive correlation between the incidence of metabolic syndrome (MeTS) and ocular defects in the microcirculation of the optic nerve head and the choroid, detected using LSFG in sleep apnea patients in a hospital setting [62]. A recent study indicated that

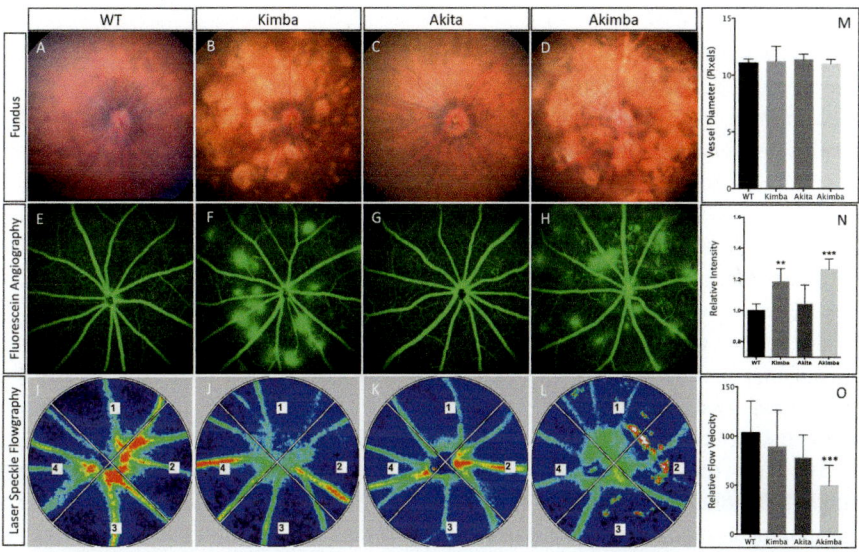

**Fig. 7** Colored fundus photographs and fundus fluorescein angiograms (FFA) of wild type (WT) mice and mice models of type 1 diabetes (Kimba, Akita, and Akimba). WT mouse (**a, e**) displayed normal arterial and venular branching and caliber in the retina. Kimba transgenic mice overexpressing VEGF in the photoreceptors showed multiple spots of hyperfluorescein throughout the retina (**b, f**). Akita hyperglycemic mice exhibited an increase in tortuosity without leaky vessels (**c, g**). Double transgenic Akimba mice showed severe vasculature abnormalities with leakages and significant tortuosity (**d, h**). Next, we used laser speckle flowgraphy (LSFG) to measure relative flow volume between WT and various animal models. All the images were acquired continuously for 4 s to produce a composite map of ocular blood flow (**i–l**). Relative retinal flow volume was calculated by first subtracting the choroidal background flow from the overall blood flow value centered on a retinal vessel. The double transgenic Akimba mice representing proliferative diabetic retinopathy depicted significant attenuation of blood flow volume (**l, o**). No difference was seen between the Kimba (**j**) and Akita (**k**) when compared to WT retina (**i**). Measurement of vessel diameter showed no difference in caliber among all four groups (**m**). Next, we used laser speckle flowgraphy (LSFG) to measure the speckle pattern produced by random laser interference revealing the decrease in relative flow volume compared to WT mice. ***$p < 0.001$ compared to WT. Figure adapted from [65].

pulse waveform analysis of choroidal microcirculation measured by LSFG has a significant correlation with left ventricle systolic and left ventricle diastolic dysfunction [66]. Also, LSFG has been used to evaluate the foot blood flow in patients with peripheral arterial disease before and after surgical intervention [64]. Thus, LSFG could become a tool in clinical practice, to diagnose and monitor the progression of ocular flow-related diseases and their relation to systemic diseases, as well as to evaluate the postoperative visual outcomes in patients undergoing surgical or pharmacological interventions.

## 2.6   Retinal Fundoscopy Benchmark Datasets

A non-exhaustive list of benchmark (fundoscopy) datasets used for developing, testing, and validating retinal image analysis pipelines is given in Table 1. The most popular datasets in the literature used to test image analysis algorithms for fundus imagery include DRIVE, STARE, and CHASE. A large retinal fundus image EyePACS-1 dataset for automated detection of diabetic retinopathy and diabetic macular edema with 128,175 images for training and 9963 images for testing has been used to train a deep learning network [25]. Other datasets, particularly for different imaging modalities such as fluorescein angiography [67], or retinal optical coherence tomography [68, 69], are also available.

There are a number of good review papers focusing on different aspects of retinal image analysis or vessel image analysis more broadly: in [88], Patton et al. discuss techniques used to automatically detect landmark features of the fundus, such as the optic disc, fovea and blood vessels, use of image analysis in diagnosis of pathology (particularly retinopathy), and quantitative measurements of vascular topography; in [17], Mookiah et al. review retinal feature extraction methods with a focus on developing computer-aided diagnosis methods for diabetic retinopathy; in [89], Fraz et al. examine techniques and algorithms for blood vessel segmentation in two-dimensional fundus retinal images.

## 3   Image Analysis Pipelines for Retinal Vascular Analysis

Image processing pipelines for vascular morphological and fluid dynamics flow analysis involve a sequence of algorithm modules or workflow tasks as illustrated in Fig. 8. The upper pipeline is based on traditional computer vision approaches wherein each module is often independently optimized using expert knowledge and extensive manual testing and tuning. The pipelines, shown in the lower part of the figure, are characteristic of modern supervised deep learning methods that perform end-to-end joint optimization across multiple tasks using a vast amount of labeled training data. The final result of quantitative image analysis is to provide patient specific structural and functional physiological and pharmacokinetics parameters

**Table 1** Some benchmark fundoscopy datasets with ground-truth annotations for developing and testing retinal image analysis pipelines

| Name | Source | Content |
| --- | --- | --- |
| DRIVE (digital retinal images for vessel extraction) [70] | A diabetic retinopathy screening program in Netherlands, https://www.isi.uu.nl/Research/Databases/DRIVE/download.php | 40 diabetic subjects between 25 and 90 years of age. **40 fundus images**, 33 with no sign of diabetic retinopathy, 7 with signs of mild early diabetic retinopathy |
| STARE (structured analysis of the retina) [71, 72] | University of California, San Diego, and Veterans Administration Medical Center in San Diego, USA, http://cecas.clemson.edu/~ahoover/stare/ | **20 fundus images**, 10 with some pathologies. Manually segmented by two observers |
| CHASE DB1 retinal image database | Kingston University London, https://blogs.kingston.ac.uk/retinal/chasedb1/ | **28 images** with a resolution of $1280 \times 960$ pixels. Subset of retinal images of multiethnic children from the Child Heart and Health Study in England |
| ARIA (automated retinal image analysis) [73–75] | St. Paul's Eye Unit, Royal Liverpool University Hospital Trust and Department of Ophthalmology, Clinical Sciences, University of Liverpool, Liverpool, UK, https://eyecharity.weebly.com/aria_online.html | **92 images** with age-related macular degeneration, **59 images** with diabetes and a control group consists of **61 images** |
| ImageRet DIARETDB0 (standard diabetic retinopathy database calibration level 0) [76, 77] | Lappeenranta University of Technology, Finland, http://www.it.lut.fi/project/imageret/diaretdb1/ | **130 color fundus images**. 20 are normal and 110 contain signs of the diabetic retinopathy (hard exudates, soft exudates, microaneuyrysms, hemorrhages, and neovascularization) |
| ImageRet DIARETDB1 (standard diabetic retinopathy database calibration level 1) [76] | Lappeenranta University of Technology, Finland, http://www.it.lut.fi/project/imageret/diaretdb1/ | **89 images**, 5 images with healthy retina, 84 with some signs of mild proliferative diabetic retinopathy. Independent markings from 4 medical experts |
| Messidor-2 (methods to evaluate segmentation and indexing techniques in the field of retinal ophthalmology within the scope of diabetic retinopathy) [78, 79] | Messidor program partners and LaTIM laboratory Brest University Hospital, France, http://www.adcis.net/en/third-party/messidor/, http://latim.univ-brest.fr/ | **1748 images**, 874 patients. Messidor-Original: 529 examinations (1058 images). Messidor-Extension: 345 examinations (690 images), diabetic patients from Ophthalmology Department of Brest University Hospital (France) |
| REVIEW (retinal vessel image set for estimation of widths) [80] | Department of Computing and Informatics at the University of Lincoln, Lincoln, UK, http://www.aldiri.info/REVIEWDB/REVIEWDB.aspx | **16 mydriatic images** (8 high resolution, 4 vascular disease, 2 central light reflex, 2 kickpoint) manually marked by three experts |

(continued)

**Table 1** (continued)

| Name | Source | Content |
|------|--------|---------|
| VICAVR [81, 82] | Varpa Research Group, University of Coruña, Spain, www.varpa.es/research/ophtalmology.html#databases | **58 images**. The database includes the caliber of the vessels measured at different radii from the optic disc as well as the vessel type (artery/vein) labeled by three experts |
| HRF (high-resolution fundus) [83, 84] | Pattern Recognition Lab, Department of Ophthalmology, Friedrich-Alexander University Erlangen-Nuremberg, Germany, and Brno University of Technology, Department of Biomedical Engineering, Brno, Czech Republic, https://www5.cs.fau.de/research/data/fundus-images/ | **45 images** (15 healthy, 15 diabetic retinopathy, 15 glaucomatous patients). Binary gold standard vessel segmentation images are available for each image |
| DR HAGIS (diabetic retinopathy, hypertension, age-related macular degeneration and glacuoma images) [85, 86] | UK diabetic retinopathy screening programme. Copyright Faculty of Biology, Medicine and Health, University of Manchester, UK, https://personalpages.manchester.ac.uk/staff/niall.p.mcloughlin/ | **39 fundus images** from diabetic patients. Four co-morbidity subgroups: glaucoma, hypertension, diabetic retinopathy, age-related macular degeneration |
| Kaggle diabetic retinopathy detection challenge dataset | Images captured in multiple primary care sites throughout California and elsewhere and uploaded to EyePACS, a free platform for diabetic retinopathy screening. https://www.kaggle.com/c/diabetic-retinopathy-detection | **88,702 color fundus photographs** from 44,351 patients: one photograph per eye. Image sizes ranging from $433 \times 289$ pixels to $5184 \times 3456$ pixels [87] |

that can be integrated into mathematical models including initializing boundary conditions and state estimation for computational fluid-dynamics.

This section explores image analysis methods for quantification of retinal vessel morphology and function.

## 3.1 Image Enhancement

Preprocessing of retinal images involves processes such as color conversion, noise removal, and background normalization. The aim of this step is to increase the separability of vessel and background classes for the following steps. Some processing pipelines such as those involving intensity thresholding may be more sensitive to preprocessing than others such as those involving shape-based processing.

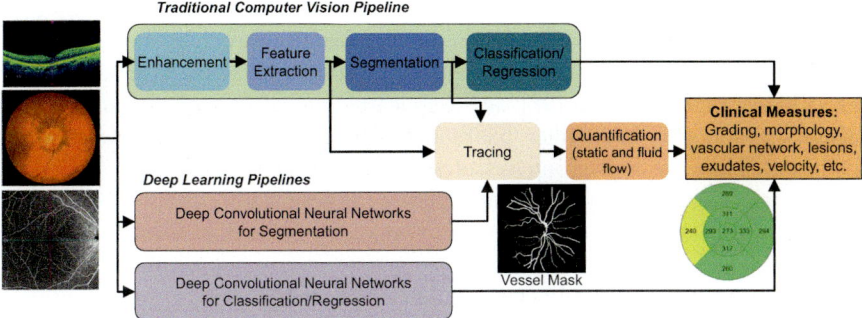

**Fig. 8** Image processing pipelines for vascular morphological and fluid dynamics flow analysis based on traditional computer vision modules are typically optimized individually one at a time (top), while more recent deep learning architectures (bottom) are usually jointly optimized using back propagation type algorithms from end-to-end

Generic and custom-designed or hand-crafted image processing methods have been developed to enhance retinal images and improve their visual quality for human viewing and automated image analysis. In [90], Zhou et al. overview retinal image enhancement methods and propose a luminosity and contrast adjustment approach designed for color retinal images. Hessian matrix-based filters capturing second order local structure of an image [91], directional filters [92], morphological filters such as top-hat filter [93, 94] and various anisotropic diffusion approaches [95] are among the mostly used vessel enhancement approaches. Prasath et al. [96] evaluated anisotropic diffusion approaches for denoising vascular images and developed a multi-scale tensor with anisotropic diffusion model that progressively and adaptively updates the amount of smoothing while preserving vessel boundaries. The method was evaluated using epifluorescence images of microvasculature, but can also be applied to retinal images.

## 3.2   Vessel Segmentation

Deformable curvilinear thin structures occur frequently in anatomical and biomedical imagery. Yet their automatic quantitative characterization remains a challenge from retinal vessel characterization for diabetes and glaucoma to neuronal and dendritic structures in computational neuroscience.

Reliable vessel segmentation is crucial for automatic quantitative image analysis of microvasculature dynamics. Retinal vessel extraction approaches can be categorized as segmentation-based where binary masks are generated using pixel classification approaches or tracing/tracking-based where vectorized representation of vessels are obtained by using local information to extend seed points. This section will review segmentation-based approaches. Vessel tracing methods will be explored in Sect. 3.3.1.

Our group developed automated vascular system segmentation pipelines consisting of expert hand-crafted salient image features associated with vascular structures [93, 97]; machine learning methods including traditional methods such as SVM and random forests [98]; and deep learning methods such as convolutional neural networks to learn filter banks optimized for thin curvilinear structures [99, 100]. Here we will group vessel segmentation approaches as unsupervised versus supervised and explore recent papers in the field.

### 3.2.1 Unsupervised Vessel Segmentation Approaches

Various unsupervised segmentation methods have been proposed to segment retinal vessels from background. These methods rely on inherent intensity, color, shape, or texture patterns of vascular structures and do not require labeled training data. Unsupervised methods used in retinal image segmentation range from simple intensity thresholding to sophisticated methods involving clustering, mathematical morphology, active contours, etc. on more complex image features. Some of these vessel segmentation approaches are compared in [101].

Vessel segmentation pipelines often involve filtering or processing to extract features and to enhance curvilinear structures. These filters exploit shape-based properties that are robust to image contrast and intensity variations. Since vessels produce ridges/creases in the intensity map, ridge detection methods are widely used in vessel segmentation. Various definitions and associated detection methods for ridges/creases can be found in [102]. These methods can be roughly classified as: curvature-based [102, 103], directional derivative-based [102–105], and height definitions or Hessian-based [70, 91, 106–109]. Principal curvatures and directions of a surface $L$ correspond to the eigenvalues $\kappa_1 \geq \ldots \geq \kappa_{n-1}$ and corresponding eigenvectors $\xi_1, \ldots, \xi_{n-1}$ of the shape operator matrix on the tangent space $W$ defined in Eq. (1),

$$W = \mathbf{I}^{-1}\mathbf{II} = \begin{bmatrix} L_x \cdot L_x & L_x \cdot L_y \\ L_x \cdot L_y & L_y \cdot L_y \end{bmatrix}^{-1} \begin{bmatrix} L_{xx} \cdot c & L_{xy} \cdot c \\ L_{xy} \cdot c & L_{yy} \cdot c \end{bmatrix} \tag{1}$$

where $\mathbf{I}$ and $\mathbf{II}$ are the first and second fundamental forms, $L_x$ and $L_{xy}$ are the first and second partial derivatives of the surface, respectively, and $c = \frac{L_x \times L_y}{|L_x \times L_y|}$ [102]. Ridges can be defined as local extrema of principal curvatures, where the differentiation is taken along the principal directions [102]. Ridges should be invariant to translations, rotations and uniform scaling in the spatial variables, and monotonic transformations of the intensity function to ensure that the ridge operator commutes is unit consistent and insensitive to image intensity quantization range [102]. Since computation of eigenvalues, the individual principal curvatures are expensive, mean curvature $H$ (Eq. (2)) is often used to classify a surface patch as a ridge, planar (flat), or valley [102, 103].

$$H = \frac{(\kappa_1 + \kappa_2)}{2} = \frac{trace(W)}{2} = \frac{L_y^2 L_{xx} + L_x^2 L_{yy} - 2L_x L_y L_{xy}}{2(L_x^2 + L_y^2)^{\frac{3}{2}}} \quad (2)$$

The *height* definition for creases (ridges and valleys) is a generalization of local extrema for real-valued functions of a vector variable [102]. The height definition is based on computing local maxima for the intensity or height function in special directions, which is a local process with the desired geometric invariances except for monotonic transformations of the height function. A point $x_0$ is classified as maximum if $\nabla L(x_0) = 0$ (critical point) and the Hessian of $L$, $\mathcal{H}(L(x_0))$, is negative definite (all eigenvalues $\lambda_i$ are negative). The Hessian matrix, $\mathcal{H}$, defined in Eq. (3), captures the second order structure of local intensity variations around each point of the image $L(x, y)$ modeled as a manifold surface,

$$\mathcal{H} = \begin{bmatrix} L_{xx} & L_{xy} \\ L_{xy} & L_{yy} \end{bmatrix}. \quad (3)$$

Table 2 shows possible orientation patterns based on the value of the eigenvalues $\lambda_{1,2}$ (Eq. (4)) of the Hessian matrix $\mathcal{H}$,

$$\lambda_{1,2} = \frac{1}{2}(L_{xx} + L_{yy} \pm \sqrt{(L_{xx} - L_{yy})^2 + (2L_{xy})^2}). \quad (4)$$

When the height or surface condition holds (critical points, $\nabla L = 0$), then using a Taylor series expansion and curvature definitions, eigenvalues $\lambda_i$ and eigenvectors $\mathbf{v_i}$ of the Hessian matrix correspond to principal curvatures $\kappa_i$ and principal directions $\xi_i$, respectively [110].

Eigenvalues and eigenvectors of the Hessian matrix have been used in many medical image processing applications as a ridgeness measure for linear or tubular structure enhancement and detection [70, 91, 93, 106–109, 111, 112]. Other filters relying on shape-based information include variants of matched filters [72, 113–115], other second order derivatives (i.e., amplitude modified second-order Gaussian filter [116]), 2-D Gabor wavelets [117], directional mathematical morphology filters [118], etc. A matched filter describes the expected profile of a signal. Detection is performed by comparative matching [72] of the defined filter. Chaudhuri et al. [113] note that the gray-level profiles of the cross-sections of retinal vessels have an intensity profile which can be approximated by a Gaussian. First, a 2-D

| | $\lambda_1$ | $\lambda_2$ | Orientation pattern |
|---|---|---|---|
| **Table 2** Possible orientation patterns based on the eigenvalues $\lambda_1$, $\lambda_2$ of the Hessian matrix, $\mathcal{H}$, with H = high, L = low magnitude, and $|\lambda_1| \geq |\lambda_2|$) [91] | L | L | Flat or noise no preferred direction |
| | H− | L | Bright tubular structure |
| | H+ | L | Dark tubular structure |
| | H− | H− | Bright blob-like structure |
| | H+ | H+ | Dark blob-like structure |

matched filter kernel matching this profile is proposed, then vessel segments at various orientations are detected by convolving the image with rotated versions of the matched filter kernel and retaining the maximum response.

Mathematical morphology is a framework developed for the analysis of the geometrical structures and relationships in an image. Mathematical morphology provides powerful methods for extracting information from images and segmenting binary and grayscale images based on set theory, lattice theory, topology, and random functions [119]. In [120], Zana and Klein present an algorithm that combines morphological filters and cross-curvature evaluation to segment vessel-like patterns. In [121], Mendonca and Campilho use a set of morphological operators with increasing structuring element size to generate several enhanced representations of the vascular network. Morphological reconstruction is then used to segment vessels from these enhanced representations.

Active contours evolve/deform a curve $\mathscr{C}$ subject to constraints from a given image [122]. Active contours are classified as parametric [123, 124] or geometric [125, 126] according to their representation. Parametric active contours (i.e., classical snakes) are represented explicitly as parametrized curves, geometric active contours are implicitly represented as level sets [127]. In [128], Al-Diri et al. use oriented morphological filters similar to top-hat filters for initial detection followed by four linked active contours for vessel segmentation. In [129], Zhao et al. propose an active contour model with hybrid region information and a novel regularization term that allows for better detection of small branching structures. In [115], Oliveira et al. first enhance the retinal images with combined matched filter, Frangi's filter, and Gabor wavelet filter. Then image is segmented using deformable models by region-scalable fitting (RSF) energy approach which is a variation of Chan and Vese active contours [126].

### 3.2.2 Supervised Vessel Segmentation Approaches

Supervised classification or regression methods such as random forests are also popular for retinal vessel segmentation. These methods produce segmentation masks by classifying each image pixel as vessel or nonvessel, based on corresponding feature vectors and based on manually segmented ground-truth images that are used for training. Our group developed a supervised random forest (RF) classifier for segmenting thin vessel structures using multi-scale features based on Hessian, oriented second derivatives, Laplacian of Gaussian and line features [98]. This approach was used for segmentation of epifluorescence images of dura mater. Random forests belong to a class of techniques that use an ensemble learning method for classification and regression. For classification they operate by constructing a collection of decision trees at training time and pixels are labeled using the mode of the votes from the set of decision trees. In [130], Wang et al. use convolutional neural network for hierarchical feature extraction and combine it with ensemble of random forests for a trainable classifier. In [131], Fraz et al. use an ensemble system of bagged and boosted decision trees along with a feature vector based on the orientation analysis of gradient vector field, morphological transformation, line strength measures, and Gabor filter responses. Other supervised classification methods are also popular for retinal vessel segmentation. In [132], Ricci and Perfetti combine a line detector with support vector machines (SVM). In [117], Soares et al. combine Gabor wavelet transform based features with a Bayesian classifier with class-conditional probability density functions described as Gaussian mixtures. While there are various classical approaches for supervised processing of retinal vessels, in this review we want to concentrate on more recent deep learning based approaches.

### 3.2.3 Deep Learning-Based Vessel Segmentation

Image analytics pipelines traditionally involve a series of independent steps including enhancement (calibration, registration, transformation, filtering), feature extraction, segmentation, classification, tracking (when spatio-temporal data is involved), and measurement. These steps typically involve carefully tuned features. The performance of these distinct components is optimized separately, even though the stages of the pipeline are highly dependent on each other, and require a great deal of engineering expertise. Overall accuracy depends on each preceding step. Such a loosely connected pipeline is difficult to optimize jointly since modules are nonlinear operators. In recent years, deep artificial neural network architectures that tightly connect together multiple stages of visual image processing have shown outstanding performance in computer vision and pattern recognition tasks and benchmark challenges, by jointly optimizing all adjustable parameters across all stages of the hierarchical deep network through a series of back propagation steps (i.e., chain rule for efficiently computing partial derivatives through the network function graph) by minimization of the misclassification error or other metric loss function over the training set [133, 134]. Beside classical vision problems, these deep learning methods have also shown great performance in biomedical applications [135].

Deep learning provides a systematic approach for automated bottom-up learning of feature sets adapted to particular domains and specialized tasks without requiring manual design and/or feature selection, which is often a critical performance limiting step of traditional image processing pipelines. Convolutional neural networks (CNNs) are one of the most popular and widely used types of deep learning models used in biomedical image analysis and computer vision. CNNs have been used in various image or image block/window classification tasks and recently have been adapted to semantic image segmentation tasks using block-based or superpixel-based architectures [136, 137].

A convolutional neural network is a function $y = g(x)$ mapping input tensors $x$ (i.e., a multispectral image or volume) to an output vector $y$. The function $g(\cdot)$ is the composition of a nested sequence of simpler functions $\{f_l\}$, which are called computational blocks or layers, $g = \{f_L \circ \ldots \circ f_l \circ \ldots \circ f_1\}$. Assuming the network input is $x_0 = x$, and each of internal network output layers are labeled, $x_1, x_2, \ldots, x_L$. Then each internal output layer,

$$x_l = f_l(x_{l-1}; w_l) \qquad (5)$$

is computed from the previous output $x_{l-1}$ by applying the function $f_l$ with weight parameters $w_l$. The network is called a convolutional neural network (CNN), because the functions $f_l$ act as a local translation invariant operator. Backpropagation methods for gradient estimation using optimization techniques like stochastic gradient descent (SGD) and adaptive momentum (Adam) from first and second moments of the gradients are used to learn from scratch or fine-tune the deep neural network architecture weight parameters. Segmentation and classification CNNs typically output a vector of class probabilities $y = g(x)$ for all learned image classes or labels based on the supervised labeled training data [138].

Our group explored deep convolutional neural network architectures for segmenting blood vessel microvasculature [99, 100] and demonstrated that this same deep learning pipeline, shown in Fig. 9, exhibits remarkable resilience for segmenting retinal fundoscopy imagery (brightfield imaging of the retinal blood vessel tree) and had the best performance compared to 12 other published methods [99].

In [139], Liskowski and Krawiec propose several convolutional neural networks to classify small retinal image matches (i.e., size $27 \times 27$). The paper also describes a number of image preprocessing and data augmentation steps. In [140], Fu et al. formulate the retinal vessel segmentation problem as a boundary detection task and solve it using a novel deep learning architecture. The proposed method applies a multi-scale and multi-level convolutional neural network (CNN) with a side-output layer to learn a rich hierarchical representation, and utilizes a conditional random field (CRF) to model the long-range interactions between pixels. In [141], Li et al. propose a wide and deep neural network to map a retinal image patch to the corresponding vessel map. The network uses a specifically designed two-stage training scheme consisting of pre-training of the first layer and overall training. In [142], Maninis et al. propose a unified framework called deep retinal image understanding

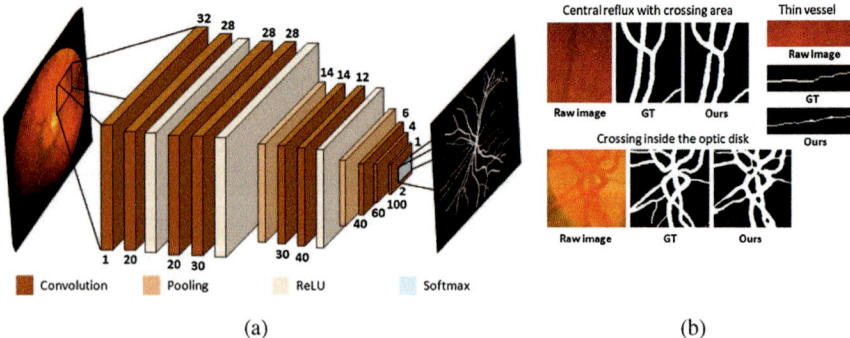

(a)                                                                                            (b)

**Fig. 9** Convolutional neural network (CNN) used for segmentation of retinal vessel structures in fundus imagery: (**a**) CNN computational architecture, (**b**) sample results showing image patches of challenging cases with corresponding ground-truth (GT). Figure adapted from [99]

(DRIU) for retinal vessel and optic disc segmentation. The proposed system uses a base shared CNN network and per-task specialized layers to perform both vessel and optic disc segmentation within the same framework (Fig. 10a). In [143], Lin et al. propose a deep learning based vessel segmentation method, named as deeply supervised and smoothly regularized network (DSSRN). DSSRN combines the

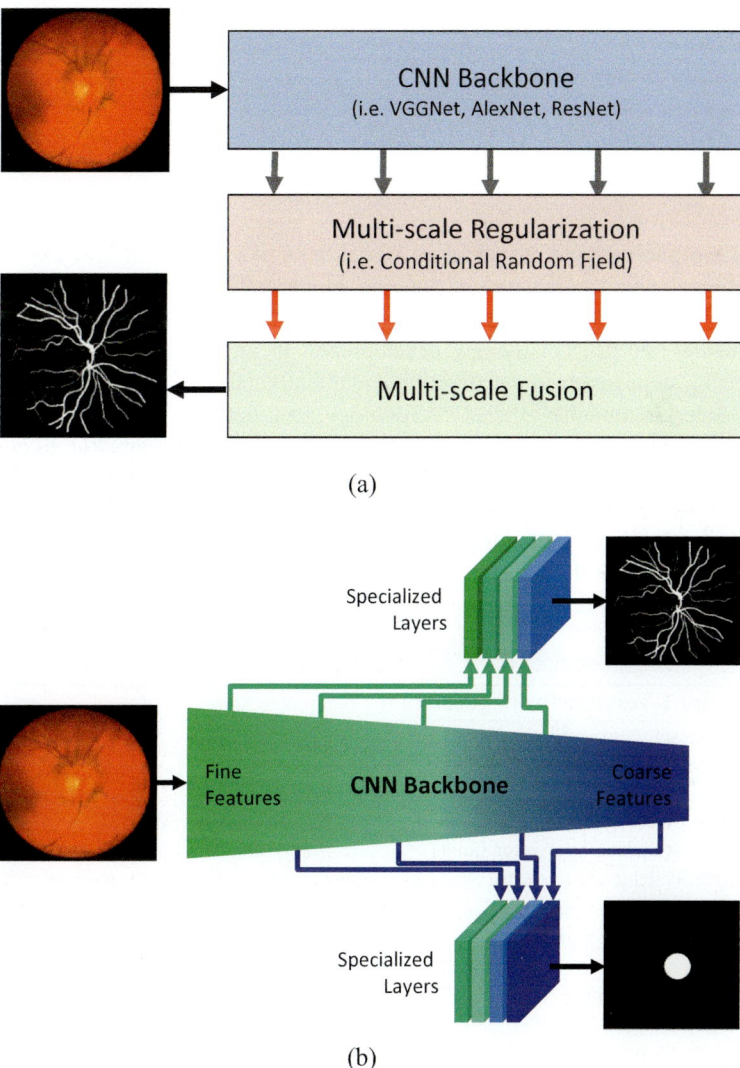

(a)

(b)

**Fig. 10** Example of deep learning network architecture models that combine one or more CNN backbones with additional image analysis steps have been developed for vessel structure segmentation in retinal fundus images. See text for more details. (**a**) Deep learning architecture with multi-scale analysis. (**b**) Deep learning architecture with regularization

strengths of fully convolutional network (FCN) and conditional random field (CRF) that introduces global smoothness regularization into the network (Fig. 10b). In [144], Yan et al. propose a novel CNN network with joint segment-level and pixel-wise losses that aims more effective feature learning without increasing model complexity. Beside vessel segmentation, deep learning methods are also used in other image segmentation tasks involved in ophthalmology. In [145], multiple U-Net structures are combined with a conditional generative adversarial network for detection of exudates in fundus photographs.

## 3.3 Quantification of Vessel Morphology and Network Structure

Vessel morphology and properties of vessel networks provide insight for scientific and clinical characterization of function in normal and diseased states. Image analysis tools quantifying these properties are valuable for disease diagnosis, monitoring, and drug or therapy development. In this section, we explore image processing steps involved in extraction and quantification of these properties.

In order to quantify vessel morphology and network structure, first vessel network graphs need to be constructed. A sample processing pipeline to construct the vascular network graph is described in Algorithm 1. Given vessel segmentation masks, some of the major steps involved in the quantitative parameterization are skeletonization, tracing, landmark detection (i.e., bifurcation points), and network graph construction by labeling and linking bifurcation points with the vessel segments joining them.

---

**Algorithm 1** Vessel network generation

**Input** : Vessel segmentation mask $M_{vessel}$
**Output** : Vessel network graph $G = (\mathscr{S}, \mathscr{B}, E)$

1: **Step 1:** Skeleton $\mathscr{S}_{vessel}$ is extracted from vessel mask $M_{vessel}$.
2: **Step 2:** Branching/bifurcation points $\mathscr{B}_{vessel}$ in $\mathscr{S}_{vessel}$ are identified.
3: **Step 3:** Skeleton $\mathscr{S}_{vessel}$ is disconnected at branching points. Connected component labeling is applied on branch points $\mathscr{B}_{vessel}$ and disconnected cell skeleton $\mathscr{S}_{vessel}$ resulting in uniquely labeled branching points $\mathscr{B} = \{b_1, b_2, \ldots, b_m\}$ and skeleton segments $\mathscr{S} = \{s_1, s_2, \ldots s_n\}$.
4: **Step 4:** A vessel network graph $G = (\mathscr{S}, \mathscr{B}, E)$ is constructed with branching points $\mathscr{B}$ and skeleton segments $\mathscr{S}$ as vertices. Each touching branching point $b_i$ and segment $s_j$ are connected with an edge $E_{ij}$.
5: **Step 5:** Short skeleton segments corresponding to spurs are removed and corresponding branching points are updated.
6: **Step 6:** For each branch point $b_i$, set of skeleton segments incident to it $\mathscr{S}_i = \{s_{i,1}, s_{i,2}, ..\}$ are identified.
7: **Step 7:** Labeled skeleton segments $\mathscr{S}_i = \{s_{i,1}, s_{i,2}, ..\}$ are traced and corresponding parametric curves are obtained.

---

### 3.3.1   Vessel Tracing

Vessel tracing or tracking is the critical step in vessel network extraction and graph construction. Tracing should not merge/connect disjoint vessel structures and should not disconnect continuous vessel segments. The process is challenging due to inhomogeneous intensity variations within and around vessels and close proximity of vessel structures, crossing vessels or other complexities of organization, and adverse imaging conditions. There are two main groups of approaches to vessel tracing: (1) segmentation-based, (2) tracking-based. Segmentation-based approaches first segment the vessels from background (as described in the previous section), then form a vascular network graph by analyzing the skeleton of the segmentation mask. Tracking-based approaches start from a set of seed points and trace the curvilinear structures in general, vessels in particular by linking pixels to the traced object (vessel, filament, neuron, etc.) using only local intensity patterns. Some retinal image analysis papers describe vessel tracing methods. In [146], Tolias and Panas propose a fuzzy vessel tracking algorithm based on fuzzy vessel/non-vessel membership functions and special handling of junctions and forks. In [147], Gaussian and Kalman filters are employed to trace retinal vessel. In [148], Can et al. propose a retinal vessel tracing scheme based on recursive processing and directional correlation kernels that act as low-pass differentiators. In [149], Vlachos and Dermatas use a multi-scale line tracking procedure to extract vascular network. In [150], Zhang et al. combine multi-scale line detection with Bayesian theory for improved tracing.

Beside these retinal vessel tracking approaches, other curvilinear structure trackers can be applied to retinal images with some modifications. In [151], Benmansour and Cohen describe an interactive vessel tracking method based on minimal path solved using fast marching algorithm with an anisotropic metric oriented along the direction of the vessel. In [152], Obara et al. propose a contrast-independent approach to identify curvilinear structures based on oriented phase congruency, i.e., the phase congruency tensor (PCT). Recently there has been great interest in tracing neuronal structures. Some automated or semi-automated (i.e., Matlab plugins NeuronJ, Neurite Tracer, etc.) tools designed for tracing neurons can also be used for retinal vessels. A review of neuron tracing tools can be found in [153, 154].

One of the main challenges in the construction of vessel network graphs is resolving whether a junction point corresponds to branching of a single vessel or a crossover of two or more vessels. Methods are described classifying bifurcation versus crossover classification approaches [155, 156]. In [157] a two step graph-theoretical approach is described to resolve crossover issue using local and global contextual information of the vessel network. After segmentation and skeletonization, the paper first identifies the root nodes (as segments touching the optic disc area) and formulates the vessel tracing problem as label propagation using the matrix-forest theorem. Decision to connect vessel segments is done using an energy function based on the angle between two vessel segments. The intuition is that vessel segments belonging to the same vessel should not bend too much.

### 3.3.2 Quantitative Measures

Once a vascular network graph is constructed, various computational measures characterizing vessel segments such as curvature, tortuosity, ratio of input to output diameters, permeability, number of bifurcation points, bifurcation angles, vessel diameter variation, and network graph properties are computed. Patton et al. [88] and Lau et al. [158] present comprehensive surveys of retinal vessel quantitative measures. Among the most frequently used measures are

(a) *Vessel diameter and related measures*: Length to diameter ratio reflecting retinal arteriolar attenuation [88], calculated using the length from the midpoint of a vascular bifurcation to the midpoint of the preceding bifurcation. Arteriovenous ratio (AVR), ratio between the average diameters of the arterioles with respect to the venules.

(b) *Curvature related measures*: Vessel curvature ($\kappa$) which can be computed locally along a vessel segment or averaged over the full segment, for the medial axis of the vessel or the boundary of the vessel walls, is defined below in both continuous and discrete approximation forms,

$$\kappa = \frac{\dot{x}\ddot{y} - \dot{y}\ddot{x}}{(\dot{x}^2 + \dot{y}^2)^{3/2}} \tag{6}$$
$$= \frac{(x_p - x_{p-1})(y_{p+1} - 2y_p + y_{p-1})}{(\sqrt{(x_p - x_{p-1})^2 + (y_p - y_{p-1})^2})^3}$$
$$- \frac{(y_p - y_{p-1})(x_{p+1} - 2x_p + x_{p-1})}{(\sqrt{(x_p - x_{p-1})^2 + (y_p - y_{p-1})^2})^3}$$

where the dot notation is used to represent spatial derivatives with respect to the length parameter $s$ along the contour, $x(s)$ and $y(s)$ are points on the contour, $p$ is the index of the point. These quantitative network measures are computed to characterize network structures.

(c) *Tortuosity related measures*: Vascular tortuosity ($\tau$) is computed as the ratio of the arc length of a vessel connecting the flow between two points $s_A$ and $s_B$ of the vascular network, and the shortest Euclidean distance between (bifurcation) points $s_A$ and $s_B$ estimated by a straight line [88],

$$\tau = \frac{\text{Arclength}}{\text{Chord}} = \frac{\int_{s_A}^{s_B} \sqrt{\left(\frac{dx}{ds}\right)^2 + \left(\frac{dy}{ds}\right)^2}\, ds}{\sqrt{(x_0 - x_l)^2 + (y_0 - y_l)^2}} \tag{7}$$

where *arc length* is the arc length computed from parametric branch curve $\mathscr{S}(s)$ and *chord* is calculated using the Euclidean distance between the end points of $\mathscr{S}$. The normal vessels are generally smooth and stay straight with low tortuosity values, whereas increased tortuosity represents hypertension of the vessels.

Tortuosity metric defined in Eq. (7) provides a global measure of deviation from a straight line. This metric ignores local properties of vessels that are more in line with an ophthalmologist's perception of tortuosity. A number of extended quantitative vessel curvature and tortuosity related measures have been proposed in the literature. Smedby et al. [159] and Bullitt et al. [160] incorporate local properties to Eq. (7) by multiplying it with the number of inflection points of the curve. In [161], Trucco et al. extend curvature and tortuosity measures that are based solely on vessel skeleton properties with vessel thickness. An evaluation of various vessel curvature and tortuosity measures in terms of scale invariance and positive monotonic response with respect to the amplitude and frequency can be found in [162].

(d) *Branching related measures* such as branching angle computed between two daughter vessels using tangent lines at branching points. Branching angle is thought to be related to blood flow efficiency. Low angles are associated with hypertension, whereas increased angles have been related to decreased blood flow. Branching coefficient is given by the area ratio [158],

$$bc(v) = \frac{w(s_1)^2 + w(s_2)^2}{w(s)^2} \tag{8}$$

where $s$ is the root segment of vessel $v$, $s_1$ and $s_2$ are the two daughter segments of $s$ after the bifurcation point, and $w(s)$ denotes mean width or vessel diameter of segment $s$. The symmetry of the two daughter vessel sizes (diameters) can be measured using the ratio [158],

$$ar(v) = \left( \frac{min\{w(s_1)), w(s_2)\}}{max\{w(s_1)), w(s_2)\}} \right)^2 \tag{9}$$

where $ar(v)$ ranges between zero and one with small values close to zero indicating highly asymmetric branching vessels and high values close to one when the two vessels are nearly equal-sized that is of similar diameters.

While there are a number of papers proposing quantitative measures of vessel properties, relevant measures for clinical use remain an active area of research and motivate the continuing development of objective, quantitative, and high-throughput image analysis and learning methods for ophthalmology.

# 4   Conclusions and Future Work

Non-invasive and label-free imaging of the retina using multiple modalities offers a rich source of information for diagnosing, managing, and treating pathologies of the eye and systemic diseases that manifest in the retina. Quantitative image

14. MD Abramoff, MK Garvin, and M Sonka. Retinal imaging and image analysis. *IEEE Reviews in Biomedical Engineering*, 3:169–208, 2010.

15. TY Wong and NM Bressler. Artificial intelligence with deep learning technology looks into diabetic retinopathy screening. *Journal of the American Medical Association*, 316(22):2366–2367, 2016.

16. H Mir, H Al-Nashash, and UR Acharya. Quantification of diabetic retinopathy using digital fundus images. In EYK Ng, UR Acharya, JS Suri, and A Campilho, editors, *Image Analysis and Modeling in Ophthalmology*, pages 161–169. CRC Press, 2014.

17. MRK Mookiah, UR Acharya, CK Chua, CM Lim, EYK Ng, and A Laude. Computer-aided diagnosis of diabetic retinopathy: A review. *Computers in Biology and Medicine*, 43(12):2136–2155, 2013.

18. MRK Mookiah, UR Acharya, C Chakraborty, LC Min, EYK Ng, and JS Suri. Automated glaucoma identification using retinal fundus images: A hybrid texture feature extraction paradigm. In EYK Ng, UR Acharya, JS Suri, and A Campilho, editors, *Image Analysis and Modeling in Ophthalmology*, pages 9–22. CRC Press, 2014.

19. N Thakur and M Juneja. Survey on segmentation and classification approaches of optic cup and optic disc for diagnosis of glaucoma. *Biomedical Signal Processing and Control*, 42:162–189, 2018.

20. JC Tovey and DP Hainsworth. Artificial vision: Hope for the new millennium. *Missouri Medicine*, 112(1):76–84, 2015.

21. H Schneiderman. The fundoscopic examination. In *Clinical Methods: The History, Physical, and Laboratory Examinations*, chapter 117. Boston: Butterworths, 3rd edition, 1990. Available from: https://www.ncbi.nlm.nih.gov/books/NBK221/.

22. H Kolb. Simple anatomy of the retina. In *Webvision: The Organization of the Retina and Visual System [Internet]*. Salt Lake City (UT): University of Utah Health Sciences Center, 2005. Available from http://www.ncbi.nlm.nih.gov/books/NBK11533/ PubMed PMID: 21413391.

23. TJ MacGillivray, E Trucco, JR Cameron, B Dhillon, JG Houston, and EJ van Beek. Retinal imaging as a source of biomarkers for diagnosis, characterization and prognosis of chronic illness or long-term conditions. *The British Journal of Radiology*, 87(1040), 2014.

24. A Newman, N Andrew, and R Casson. Review of the association between retinal microvascular characteristics and eye disease. *Clinical & Experimental Ophthalmology*, 46(5):531–552, 2018.

25. V Gulshan, L Peng, M Coram, MC Stumpe, Derek Wu, A Narayanaswamy, S Venugopalan, K Widner, T Madams, J Cuadros, R Kim, R Raman, PC Nelson, JL Mega, and DR Webster. Development and validation of a deep learning algorithm for detection of diabetic retinopathy in retinal fundus photographs. *Journal of the American Medical Association*, 316(22):2402–2410, 2016.

26. HK Banda, GK Shah, and KJ Blinder. Applications of fundus autofluorescence and widefield angiography in clinical practice. *Canadian Journal of Ophthalmology*, 2018.

27. RR Lim, T Vaidya, SG Gadde, NK Yadav, S Sethu, DP Hainsworth, RR Mohan, A Ghosh, and SS Chaurasia. Correlation between systemic s100a8 and s100a9 levels and severity of diabetic retinopathy in patients with type 2 diabetes mellitus. *Diabetes & Metabolic Syndrome: Clinical Research & Reviews*, 13:1581–1589, 2019.

28. S Bearelly and SW Cousins. Fundus autofluorescence imaging in age-related macular degeneration and geographic atrophy. In *Retinal Degenerative Diseases*, pages 395–402. Springer, 2010.

29. R Bernardes, P Serranho, and C Lobo. Digital ocular fundus imaging: a review. *Ophthalmologica*, 226(4):161–181, 2011.

30. S Guigui, T Lifshitz, and J Levy. Screening for diabetic retinopathy: Review of current methods. *Hospital Practice*, 40(2):64–72, 2012.
31. N Panwar, P Huang, J Lee, PA Keane, TS Chuan, A Richhariya, S Teoh, TH Lim, and R Agrawal. Fundus photography in the 21st century - a review of recent technological advances and their implications for worldwide healthcare. *Telemedicine and e-Health*, 22(3), 2016.
32. GP Leese, DM Broadbent, SP Harding, and JP Vora. Detection of sight-threatening diabetic eye disease. *Diabetic Medicine*, 13(10), 1996.
33. PFJ Hoyng, AH Rulo, EL Greve, M Astin, and M Gjötterberg. Fluorescein angiographic evaluation of the effect of latanoprost treatment on blood-retinal barrier integrity: A review of studies conducted on pseudophakic glaucoma patients and on phakic and aphakic monkeys. *Survey of Ophthalmology*, 41:S83–S88, 1997.
34. R Brancato and G Trabucchi. Fluorescein and indocyanine green angiography in vascular chorioretinal diseases. In *Seminars in Ophthalmology*, volume 13, pages 189–198, 1998.
35. A Ly, L Nivison-Smith, N Assaad, and M Kalloniatis. Fundus autofluorescence in age-related macular degeneration. *Optometry and Vision Science*, 94(2), 2017.
36. A Wessing. Diabetic retinopathy: update on diagnosis and treatment. *Nephrology, Dialysis, Transplantation: European Dialysis and Transplant Association-European Renal Association*, 12(9), 1997.
37. MA Hochman, CM Seery, and MA Zarbin. Pathophysiology and management of subretinal hemorrhage. *Survey of Ophthalmology*, 42(3):195–213, 1997.
38. KA Klima. Focus on fluorescein angiography. *Insight*, 31(2), 2006.
39. D Schmidt. The mystery of cotton-wool spots - a review of recent and historical descriptions. *European Journal of Medical Research*, 13(6), 2008.
40. JG Fujimoto, C Pitris, SA Boppart, and ME Brezinski. Optical coherence tomography: An emerging technology for biomedical imaging and optical biopsy. *Neoplasia*, 2(1–2):9–25, 2000.
41. ML Gabriele, G Wollstein, H Ishikawa, L Kagemann, J Xu, LS Folio, and JS Schuman. Optical coherence tomography: History, current status, and laboratory work. *Investigative Ophthalmology & Visual Science*, 52(5):2425–2436, 2011.
42. ML Gabriele, G Wollstein, H Ishikawa, J Xu, J Kim, L Kagemann, LS Folio, and JS Schuman. Three dimensional optical coherence tomography imaging: Advantages and advances. *Progress in Retinal and Eye Research*, 29(6):556–579, 2010.
43. T Oshitari and Y Mitamura. Optical coherence tomography for complete management of patients with diabetic retinopathy. *Current Diabetes Reviews*, 6(4):207–214, 2010.
44. W Geitzenauer, CK Hitzenberger, and UM Schmidt-Erfurth. Retinal optical coherence tomography: Past, present and future perspectives. *British Journal of Ophthalmology*, 95(2):171–177, 2011.
45. SY Cohen, A Miere, S Nghiem-Buffet, F Fajnkuchen, EH Souied, and S Mrejen. Clinical applications of optical coherence tomography angiography: What we have learnt in the first 3 years. *European Journal of Ophthalmology*, 28(5):491–502, 2018.
46. H Jiao, LJ Hill, LE Downie, and HR Chinnery. Anterior segment optical coherence tomography: Its application in clinical practice and experimental models of disease. *Clinical and Experimental Optometry*, 2018.
47. DE Baskin. Optical coherence tomography in diabetic macular edema. *Current Opinion in Ophthalmology*, 21(3):172–177, 2010.

48. CE Mendoza-Santiesteban, A Gonzalez-Garcia, TR Hedges, Y Hernandez-Silva, Y Columbie-Garbey, L Fernández-Cherkasova, R Santiesteban-Freixas, and SV Casali. Optical coherence tomography for neuro-ophthalmologic diagnoses. *Seminars in Ophthalmology*, 25(4), 2010.

49. L Van Melkebeke, J Barbosa-Breda, M Huygens, and I Stalmans. Optical coherence tomography angiography in glaucoma: a review. *Ophthalmic Research*, 60(3):139–151, 2018.

50. KD Bojikian, PP Chen, and JC Wen. Optical coherence tomography angiography in glaucoma. *Current Opinion in Ophthalmology*, 30(2):110–116, 2019.

51. A. F. Hasler, K. Palaniappan, M. Manyin, and J. Dodge. A high performance interactive image spreadsheet (iiss). *Computers in Physics*, 8(4):325–342, 1994.

52. K. Palaniappan, A. Hasler, J. Fraser, and M. Manyin. Network-based visualization using the distributed image spreadsheet (diss). In *17th Int. AMS Conf. on Interactive Information and Processing Systems (IIPS) for Meteorology, Oceanography and Hydrology*, pages 399–403, 2001.

53. C. Lamirel, N. Newman, and V. Biousse. The use of optical coherence tomography in neurology. *Rev Neurol Dis*, 6(4):E105–E120, 2009.

54. AP Dhawan, B D'Alessandro, and X Fu. Optical imaging modalities for biomedical applications. *IEEE Reviews in Biomedical Engineering*, 3:69–92, 2010.

55. R Singh, SLCY Mei, W Tam, D Raju, and A Ruszkiewicz. Real-time histology with the endocytoscope. *World Journal of Gastroenterology: WJG*, 16(40), 2010.

56. J Doustar, T Torbati, KL Black, Y Koronyo, and M Koronyo-Hamaoui. Optical coherence tomography in alzheimer's disease and other neurodegenerative diseases. *Frontiers in Neurology*, 8:701, 2017.

57. CJ Chen, JS Kumar, SH Chen, D Ding, TJ Buell, S Sur, N Ironside, E Luther, M Ragosta, MS Park, MY Kalani, KC Liu, and RM Starke. Optical coherence tomography: Future applications in cerebrovascular imaging. *Stroke*, 49(4):1044–1050, 2018.

58. L Wang, O Murphy, NG Caldito, PA Calabresi, and S Saidha. Emerging applications of optical coherence tomography angiography (OCTA) in neurological research. *Eye and Vision*, 2018.

59. AH Kashani, CL Chen, JK Gahm, F Zheng, GM Richter, PJ Rosenfeld, Y Shi, and RK Wang. Optical coherence tomography angiography: A comprehensive review of current methods and clinical applications. *Progress in Retinal and Eye Research*, 60:66–100, 2017.

60. SS Gao, Y Jia, et al. Optical coherence tomography angiography. *Investigative Ophthalmology and Visual Science*, 57:OCT27–OCT36, 2016.

61. RF Spaide, JG Fujimoto, NK Waheed, SR Sadda, and G Staurenghi. Optical coherence tomography angiography. *Progress in Retinal and Eye Research*, 64:1–55, 2018.

62. T Shiba, M Takahashi, T Matsumoto, and Y Hori. Relationship between metabolic syndrome and ocular microcirculation shown by laser speckle flowgraphy in a hospital setting devoted to sleep apnea syndrome diagnostics. *Journal of Diabetes Research*, 2017.

63. X Wei, PK Balne, KE Meissner, VA Barathi, L Schmetterer, and R Agrawal. Assessment of flow dynamics in retinal and choroidal microcirculation. *Survey of Ophthalmology*, 63(5):646–664, 2018.

64. S Kikuchi, K Miyake, Y Tada, D Uchida, A Koya, Y Saito, T Ohura, and N Azuma. Laser speckle flowgraphy can also be used to show dynamic changes in the blood flow of the skin of the foot after surgical revascularization. *Vascular*, 2018.

65. SS Chaurasia, RR Lim, BH Parikh, YS Wey, BB Tun, TY Wong, CD Luu, R Agrawal, A Ghosh, A Mortellaro, E Rackoczy, RR Mohan, and VA Barathi. The NLRP3 inflammasome may contribute to pathologic neovascularization in the advanced stages of diabetic retinopathy. *Nature Scientific Reports*, 8(1), 2018.

66. T Shiba, M Takahashi, T Matsumoto, and Y Hori. Pulse waveform analysis in ocular microcirculation by laser speckle flowgraphy in patients with left ventricular systolic and diastolic dysfunction. *Journal of Vascular Research*, 55(6):329–337, 2018.

67. H Rabbani, MJ Allingham, PS Mettu, SW Cousins, and S Farsiu. Fully automatic segmentation of fluorescein leakage in subjects with diabetic macular edema. *Investigative Ophthalmology & Visual Science*, 56(3):1482–1492, 2015.

68. S Farsiu, SJ Chiu, RV O'Connell, FA Folgar, E Yuan, JA Izatt, and CA Toth. Quantitative classification of eyes with and without intermediate age-related macular degeneration using optical coherence tomography. *Ophthalmology*, 121(1):162–172, 2014.

69. DS Kermany, M Goldbaum, W Cai, CC Valentim, H Liang, SL Baxter, A McKeown, G Yang, X Wu, F Yan, and J Dong. Identifying medical diagnoses and treatable diseases by image-based deep learning. *Cell*, 172(5):1122–1131, 2018.

70. J Staal, MD Abràmoff, M Niemeijer, MA Viergever, and B van Ginneken. Ridge-based vessel segmentation in color images of the retina. *IEEE Transactions on Medical Imaging*, 23(4):501–509, 2004.

71. STARE (Structured Analysis of the Retina). http://cecas.clemson.edu/~ahoover/stare/. Accessed: 2018-07-01.

72. AD Hoover, V Kouznetsova, and M Goldbaum. Locating blood vessels in retinal images by piecewise threshold probing of a matched filter response. *IEEE Transactions on Medical Imaging*, 19(3):203–210, 2000.

73. ARIA (Automated Retinal Image Analysis). https://eyecharity.weebly.com/aria_online.html. Accessed: 2018-07-01.

74. Y Zheng, MHA Hijazi, and F Coenen. Automated "disease/no disease" grading of age-related macular degeneration by an image mining approach. *Investigative Ophthalmology & Visual Science*, 53(13):8310–8318, 2012.

75. DJJ Farnell, FN Hatfield, P Knox, M Reakes, S Spencer, D Parry, and SP Harding. Enhancement of blood vessels in digital fundus photographs via the application of multiscale line operators. *Journal of the Franklin Institute*, 345(7):748–765, 2008.

76. IMAGERET. https://www.it.lut.fi/project/imageret/. Accessed: 2018-07-01.

77. T Kauppi, V Kalesnykiene, JK Kamarainen, L Lensu, I Sorri, H Uusitalo, H Kälviäinen, and J Pietilä. DIARETDB0: Evaluation database and methodology for diabetic retinopathy algorithms. *Machine Vision and Pattern Recognition Research Group, Lappeenranta University of Technology, Finland*, 73, 2006.

78. G Quellec, M Lamard, PM Josselin, G Cazuguel, B Cochener, and C Roux. Optimal wavelet transform for the detection of microaneurysms in retina photographs. *IEEE Transactions on Medical Imaging*, 27(9):1230–41, 2008.

79. E Decencière, X Zhang, G Cazuguel, B Lay, B Cochener, C Trone, P Gain, R Ordonez, P Massin, A Erginay, et al. Feedback on a publicly distributed image database: The Messidor database. *Image Analysis & Stereology*, 33(3):231–234, 2014.

80. B Al-Diri, A Hunter, D Steel, M Habib, T Hudaib, and S Berry. A reference data set for retinal vessel profiles. In *International Conference of the IEEE Engineering in Medicine and Biology Society (EMBC)*, pages 2262–2265, 2008.

81. VICAVR. http://www.varpa.es/research/ophtalmology.html#databases. Accessed: 2018-10-12.

82. M Ortega, N Barreira, J Novo, MG Penedo, A Pose-Reino, and F Gómez-Ulla. Sirius: a web-based system for retinal image analysis. *International Journal of Medical Informatics*, 79(10):722–732, 2010.

83. High-resolution fundus image database (hrf). https://www5.cs.fau.de/research/data/fundus-images/. Accessed: 2018-10-12.

84. A Budai, R Bock, A Maier, J Hornegger, and G Michelson. Robust vessel segmentation in fundus images. *International Journal of Biomedical Imaging*, 2013.
85. DR HAGIS: Diabetic Retinopathy, Hypertension, Age-related macular degeneration and Glacuoma ImageS. https://personalpages.manchester.ac.uk/staff/niall.p.mcloughlin/. Accessed: 2018-10-12.
86. S Holm, G Russell, V Nourrit, and N McLoughlin. DR HAGIS – a fundus image database for the automatic extraction of retinal surface vessels from diabetic patients. *Journal of Medical Imaging*, 4(1):014503, 2017.
87. G Quellec, K Charrière, Y Boudi, B Cochener, and M Lamard. Deep image mining for diabetic retinopathy screening. *Medical Image Analysis*, 39:178–193, 2017.
88. N Patton, TM Aslam, T MacGillivray, IJ Deary, B Dhillon, RH Eikelboom, K Yogesan, and IJ Constable. Retinal image analysis: Concepts, applications and potential. *Progress in Retinal and Eye Research*, 25(1):99–127, 2006.
89. MM Fraz, P Remagnino, A Hoppe, B Uyyanonvara, AR Rudnicka, CG Owen, and SA Barman. Blood vessel segmentation methodologies in retinal images–A survey. *Computer Methods and Programs in Biomedicine*, 108(1):407–433, 2012.
90. M Zhou, K Jin, S Wang, J Ye, and D Qian. Color retinal image enhancement based on luminosity and contrast adjustment. *IEEE Transactions on Biomedical Engineering*, 65(3):521–527, 2018.
91. AF Frangi, WJ Niessen, KL Vincken, and MA Viergever. Multiscale vessel enhancement filtering. *International Conference on Medical Image Computing and Computer-Assisted Intervention (MICCAI)*, 1496:130–137, 1998.
92. PTH Truc, MAU Khan, YK Lee, S Lee, and TS Kim. Vessel enhancement filter using directional filter bank. *Computer Vision and Image Understanding*, 113(1):101–112, 2009.
93. F Bunyak, K Palaniappan, O Glinskii, V Glinskii, V Glinsky, and V Huxley. Epifluorescence-based quantitative microvasculature remodeling using geodesic level-sets and shape-based evolution. In *International Conference of the IEEE Engineering in Medicine and Biology Society (EMBC)*, pages 3134–3137, 2008.
94. M Liao, YQ Zhao, XH Wang, and PS Dai. Retinal vessel enhancement based on multi-scale top-hat transformation and histogram fitting stretching. *Optics & Laser Technology*, 58:56–62, 2014.
95. C Cancio and P Radeva. Vesselness enhancement diffusion. *Pattern Recognition Letters*, 24(16):3141–3151, 2003.
96. VBS Prasath, R Pelapur, OV Glinskii, VV Glinsky, VH Huxley, and K Palaniappan. Multi-scale tensor anisotropic filtering of fluorescence microscopy for denoising microvasculature. In *IEEE International Symposium on Biomedical Imaging (ISBI)*, pages 540–543, 2015.
97. R Pelapur, VBS Prasath, F Bunyak, OV Glinskii, VV Glinsky, VH Huxley, and K Palaniappan. Multi-focus image fusion using epifluorescence microscopy for robust vascular segmentation. In *International Conference of the IEEE Engineering in Medicine and Biology Society (EMBC)*, pages 4735–4738, 2014.
98. YM Kassim, VBS Prasath, R Pelapur, OV Glinskii, RJ Maude, VV Glinsky, VH Huxley, and K Palaniappan. Random forests for dura mater microvasculature segmentation using epifluorescence images. In *International Conference of the IEEE Engineering in Medicine and Biology Society (EMBC)*, pages 2901–2904, 2016.
99. YM Kassim and K Palaniappan. Extracting retinal vascular networks using deep learning architecture. In *IEEE International Conference on Bioinformatics and Biomedicine (BIBM)*, pages 1170–1174, 2017.

100. YM Kassim, VBS Prasath, OV Glinskii, VV Glinsky, VH Huxley, and K Palaniappan. Microvasculature segmentation of arterioles using deep CNN. In *IEEE International Conference on Image Processing (ICIP)*, pages 580–584, 2017.

101. M Niemeijer, J Staal, B van Ginneken, M Loog, and MD Abramoff. Comparative study of retinal vessel segmentation methods on a new publicly available database. In *Medical Imaging: Image Processing*, volume 5370, pages 648–657. International Society for Optics and Photonics, 2004.

102. D Eberly, R Gardner, B Morse, S Pizer, and C Scharlach. Ridges for image analysis. *Journal of Mathematical Imaging and Vision*, 4(4):353–373, 1994.

103. AM Lopez, F Lumbreras, J Serrat, and JJ Villanueva. Evaluation of methods for ridge and valley detection. *IEEE Transactions on Pattern Analysis and Machine Intelligence*, 21(4):327–335, 1999.

104. JBA Maintz, PA van den Elsen, and MA Viergever. Evaluation of ridge seeking operators for multimodality medical image matching. *IEEE Transactions on Pattern Analysis and Machine Intelligence*, 18(4):353–365, 1996.

105. T Lindeberg. Feature detection with automatic scale selection. *International Journal of Computer Vision*, 30(2):77–116, 1998.

106. Y Sato, S Nakajima, H Atsumi, T Koller, G Gerig, S Yoshida, and R Kikinis. 3D multi-scale line filter for segmentation and visualization of curvilinear structures in medical images. *Medical Image Analysis*, 2(2):143–168, 1998.

107. C Lorenz, IC Carlsen, TM Buzug, C Fassnacht, and J Weese. Multi-scale line segmentation with automatic estimation of width, contrast and tangential direction in 2D and 3D medical images. In *Proceedings of CVRMed-MRCAS*, pages 233–242, 1997.

108. K Krissian, G Malandain, N Ayache, R Vaillant, and Y Trousset. Model-based detection of tubular structures in 3D images. *Computer Vision Image Understanding*, 80(2):130–171, 2000.

109. J Zhou, S Chang, D Metaxas, and L Axel. Vessel boundary extraction using ridge scan-conversion deformable model. In *IEEE International Symposium on Biomedical Imaging (ISBI)*, pages 189– 192, Apr. 2006.

110. I. N. Bronshtein and K. A. Semendyayev. *Handbook of mathematics (3rd ed.)*, chapter Chapter 4.3. Springer-Verlag, London, UK, 1997.

111. R Annunziata, A Garzelli, L Ballerini, A Mecocci, and E Trucco. Leveraging multiscale hessian-based enhancement with a novel exudate inpainting technique for retinal vessel segmentation. *IEEE Journal of Biomedical and Health Informatics*, 20(4):1129–1138, 2016.

112. K BahadarKhan, AA Khaliq, and M Shahid. A morphological hessian based approach for retinal blood vessels segmentation and denoising using region based otsu thresholding. *PloS One*, 11(7):e0158996, 2016.

113. S Chaudhuri, S Chatterjee, N Katz, M Nelson, and M Goldbaum. Detection of blood vessels in retinal images using two-dimensional matched filters. *IEEE Transactions on Medical Imaging*, 8(3):263–269, 1989.

114. M Sofka and CV Stewart. Retinal vessel centerline extraction using multiscale matched filters, confidence and edge measures. *IEEE Transactions on Medical Imaging*, 25(12):1531–1546, 2006.

115. WS Oliveira, JV Teixeira, TI Ren, GDC Cavalcanti, and J Sijbers. Unsupervised retinal vessel segmentation using combined filters. *PloS One*, 11(2):e0149943, 2016.

116. L Gang, O Chutatape, and SM Krishnan. Detection and measurement of retinal vessels in fundus images using amplitude modified second-order Gaussian filter. *IEEE Transactions on Biomedical Engineering*, 49(2):168–172, 2002.

117. JVB Soares, JJG Leandro, RM Cesar, HF Jelinek, and MJ Cree. Retinal vessel segmentation using the 2-d gabor wavelet and supervised classification. *IEEE Transactions on Medical Imaging*, 25(9):1214–1222, 2006.

118. EM Sigurdhsson, S Valero, JA Benediktsson, J Chanussot, H Talbot, and E Stefansson. Automatic retinal vessel extraction based on directional mathematical morphology and fuzzy classification. *Pattern Recognition Letters*, 47:164–171, 2014.

119. L Najman and H Talbot. *Mathematical Morphology: From Theory to Applications*. John Wiley & Sons, 2013.

120. F Zana and JC Klein. Segmentation of vessel-like patterns using mathematical morphology and curvature evaluation. *IEEE Transactions on Image Processing*, 10(7):1010–1019, 2001.

121. AM Mendonca and A Campilho. Segmentation of retinal blood vessels by combining the detection of centerlines and morphological reconstruction. *IEEE Transactions on Medical Imaging*, 25(9):1200–1213, 2006.

122. K Palaniappan, F Bunyak, S Nath, and J Goffeney. *Parallel processing strategies for cell motility and shape analysis*. 2009.

123. M Kass, A Witkin, and D Terzopoulos. Snakes: Active contour models. *International Journal of Computer Vision*, 1(4):321–331, 1988.

124. LD Cohen. On active contour models and balloons. *CVGIP: Image Understanding*, 53(2):211–218, 1991.

125. V Caselles, R Kimmel, and G Sapiro. Geodesic active contours. *International Journal of Computer Vision*, 22(1):61–79, 1997.

126. TF Chan and LA Vese. Active contours without edges. *IEEE Transactions on Image Processing*, 10(2):266–277, 2001.

127. JA Sethian. *Level Set Methods and Fast Marching Methods: Evolving Interfaces in Computational Geometry, Fluid Mechanics, Computer Vision, and Materials Science*. Cambridge University Press, 1999.

128. B Al-Diri, A Hunter, and D Steel. An active contour model for segmenting and measuring retinal vessels. *IEEE Transactions on Medical Imaging*, 28(9):1488–1497, 2009.

129. Y Zhao, L Rada, K Chen, SP Harding, and Y Zheng. Automated vessel segmentation using infinite perimeter active contour model with hybrid region information with application to retinal images. *IEEE Transactions on Medical Imaging*, 34(9):1797–1807, 2015.

130. S Wang, Y Yin, G Cao, B Wei, Y Zheng, and G Yang. Hierarchical retinal blood vessel segmentation based on feature and ensemble learning. *Neurocomputing*, 149:708–717, 2015.

131. MM Fraz, P Remagnino, A Hoppe, B Uyyanonvara, AR Rudnicka, CG Owen, and SA Barman. An ensemble classification-based approach applied to retinal blood vessel segmentation. *IEEE Transactions on Biomedical Engineering*, 59(9):2538–2548, 2012.

132. E Ricci and R Perfetti. Retinal blood vessel segmentation using line operators and support vector classification. *IEEE Transactions on Medical Imaging*, 26(10):1357–1365, 2007.

133. Y LeCun, Y Bengio, and G Hinton. Deep learning. *Nature*, 521(7553):436, 2015.

134. J Schmidhuber. Deep learning in neural networks: An overview. *Neural Networks*, 61:85–117, 2015.

135. G Litjens, T Kooi, BE Bejnordi, AAA Setio, F Ciompi, M Ghafoorian, JAWM van der Laak, B van Ginneken, and CI Sánchez. A survey on deep learning in medical image analysis. *Medical Image Analysis*, 42:60–88, 2017.

136. J Long, E Shelhamer, and T Darrell. Fully convolutional networks for semantic segmentation. In *IEEE Conference on Computer Vision and Pattern Recognition (CVPR)*, pages 3431–3440, 2015.

137. Z Al-Milaji, I Ersoy, A Hafiane, K Palaniappan, and F Bunyak. Integrating segmentation with deep learning for enhanced classification of epithelial and stromal tissues in H&E images. *Pattern Recognition Letters*, 119:214–221, 2017.

138. A Vedaldi and K Lenc. MatConvNet: Convolutional neural networks for Matlab. In *ACM International Conference on Multimedia*, pages 689–692, 2015.

139. P Liskowski and K Krawiec. Segmenting retinal blood vessels with deep neural networks. *IEEE Transactions on Medical Imaging*, 35(11):2369–2380, 2016.

140. H Fu, Y Xu, S Lin, DWK Wong, and J Liu. DeepVessel: Retinal vessel segmentation via deep learning and conditional random field. In *International Conference on Medical Image Computing and Computer-Assisted Intervention (MICCAI)*, pages 132–139. Springer, 2016.

141. Q Li, B Feng, L Xie, P Liang, H Zhang, and T Wang. A cross-modality learning approach for vessel segmentation in retinal images. *IEEE Transactions on Medical Imaging*, 35(1):109–118, 2016.

142. KK Maninis, J Pont-Tuset, P Arbeláez, and L Van Gool. Deep retinal image understanding. In *International Conference on Medical Image Computing and Computer-Assisted Intervention (MICCAI)*, pages 140–148. Springer, 2016.

143. Y Lin, H Zhang, and G Hu. Automatic retinal vessel segmentation via deeply supervised and smoothly regularized network. *IEEE Access,* 7:57717–57724, 2018.

144. Z Yan, X Yang, and KTT Cheng. Joint segment-level and pixel-wise losses for deep learning based retinal vessel segmentation. *IEEE Transactions on Biomedical Engineering*, 65(9):1912–1923, 2018.

145. R Zheng, L Liu, S Zhang, C Zheng, F Bunyak, R Xu, B Li, and M Sun. Detection of exudates in fundus photographs with imbalanced learning using conditional generative adversarial network. *Biomedical Optics Express*, 9(10):4863–4878, 2018.

146. YA Tolias and SM Panas. A fuzzy vessel tracking algorithm for retinal images based on fuzzy clustering. *IEEE Transactions on Medical Imaging*, 17(2):263–273, 1998.

147. O Chutatape, L Zheng, and SM Krishnan. Retinal blood vessel detection and tracking by matched Gaussian and Kalman filters. In *International Conference of the IEEE Engineering in Medicine and Biology Society (EMBC)*, volume 6, pages 3144–3149, 1998.

148. A Can, H Shen, JN Turner, HL Tanenbaum, and B Roysam. Rapid automated tracing and feature extraction from retinal fundus images using direct exploratory algorithms. *IEEE Transactions on Information Technology in Biomedicine*, 3(2):125–138, 1999.

149. M Vlachos and E Dermatas. Multi-scale retinal vessel segmentation using line tracking. *Computerized Medical Imaging and Graphics*, 34(3):213–227, 2010.

150. J Zhang, H Li, Q Nie, and L Cheng. A retinal vessel boundary tracking method based on bayesian theory and multi-scale line detection. *Computerized Medical Imaging and Graphics*, 38(6):517–525, 2014.

151. F Benmansour and LD Cohen. Tubular structure segmentation based on minimal path method and anisotropic enhancement. *International Journal of Computer Vision*, 92(2):192–210, 2011.

152. B Obara, M Fricker, D Gavaghan, and V Grau. Contrast-independent curvilinear structure detection in biomedical images. *IEEE Transactions on Image Processing*, 21(5):2572–2581, 2012.

153. E Meijering. Neuron tracing in perspective. *Cytometry Part A*, 77(7):693–704, 2010.

154. L Acciai, P Soda, and G Iannello. Automated neuron tracing methods: An updated account. *Neuroinformatics*, 14(4):353–367, 2016.

155. A Bhuiyan, B Nath, J Chua, and K Ramamohanarao. Automatic detection of vascular bifurcations and cross-overs from color retinal fundus images. In *IEEE International Conference on Signal-Image Technologies and Internet-Based Systems*, pages 711–718, 2007.

156. D Calvo, M Ortega, MG Penedo, and J Rouco. Automatic detection and characterisation of retinal vessel tree bifurcations and cross-overs in eye fundus images. *Computer Methods and Programs in Biomedicine*, 103(1):28–38, 2011.
157. J De, L Cheng, X Zhang, F Lin, H Li, KH Ong, W Yu, Y Yu, and S Ahmed. A graph-theoretical approach for tracing filamentary structures in neuronal and retinal images. *IEEE Transactions on Medical Imaging*, 35(1):257–272, 2016.
158. QP Lau, ML Lee, W Hsu, TY Wong, EYK Ng, UR Acharya, A Campilo, and JS Suri. The Singapore eye vessel assessment system. *Image Analysis and Modeling in Ophthalmology*, pages 143–160, 2014.
159. O Smedby, N Högman, S Nilsson, U Erikson, AG Olsson, and G Walldius. Two-dimensional tortuosity of the superficial femoral artery in early atherosclerosis. *Journal of Vascular Research*, 30(4):181–191, 1993.
160. E Bullitt, G Gerig, SM Pizer, W Lin, and SR Aylward. Measuring tortuosity of the intracerebral vasculature. *IEEE Transactions on Medical Imaging*, 22(9):1163–1171, 2003.
161. Emanuele Trucco, Hind Azegrouz, and Baljean Dhillon. Modeling the tortuosity of retinal vessels: Does caliber play a role? *IEEE Transactions on Biomedical Engineering*, 57(9):2239–2247, 2010.
162. S Lorthois, F Lauwers, and F Cassot. Tortuosity and other vessel attributes for arterioles and venules of the human cerebral cortex. *Microvascular Research*, 91:99–109, 2014.

# The Next Frontier of Imaging in Ophthalmology: Machine Learning and Tissue Biomechanics

Jenna Tauber and Larry Kagemann

**Abstract** Medical imaging has revolutionized the diagnosis and management of disease in healthcare. Early integration of computers into medical imaging brought control of devices and data acquisition to new heights of precision. As computing power and sophistication of software evolve, we have reached an era of computer-based image interpretation. Machine learning approaches have been developed to automate certain quantitative measures derived from these modalities. Prospects for machine learning in ophthalmology address its potential role in approaching some of the most common causes of blindness worldwide: diabetic retinopathy, glaucoma, and age-related macular degeneration. As these analysis techniques evolve, concurrent advancements are seen in ophthalmology imaging technologies themselves, and today, the aqueous outflow tract and the optic nerve head can be visualized in more detail than ever before. The optimization and assimilation of these tools may hold clinically significant answers to screening, diagnosing, and managing disease.

## 1 Introduction: On the Nature of Medical Imaging

Medical images are comprised of visual displays of information. Information is sampled from the body and presented to the physician, technician, or other user. Consider a conventional X-ray. In reality, an X-ray is a two-dimensional (2D) map of the body's ability to block or scatter X-ray energy. Once obtained, something truly mystical occurs: a radiologist is able to examine this 2-D map of radiopacity

*Disclaimer*: All views and opinions express herein are those of the authors, and do not express federal policy or the views of the US Food and Drug Administration.

J. Tauber · L. Kagemann (✉)
Department of Ophthalmology, NYU Langone Medical Center, NYU School of Medicine, New York, NY, USA
e-mail: Lawrence.Kagemann@nyumc.org

© Springer Nature Switzerland AG 2019
G. Guidoboni et al. (eds.), *Ocular Fluid Dynamics*, Modeling and Simulation in Science, Engineering and Technology, https://doi.org/10.1007/978-3-030-25886-3_23

and detect lesions, disease, and malformations—a plethora of information about the health of a patient.[1] Note that three individual steps occur in the process of using medical imaging in healthcare:

1. *Acquisition*: some physical characteristic is sampled and quantified from within a living person.
2. *Presentation*: a pictorial representation of that single physical characteristic is created.
3. *Interpretation*: a trained professional interprets the image.

Sampling of a physical characteristic is accomplished by presenting the body with some manner of energy and detecting the result. For example, ultrasound presents the body with sound waves and detects and maps the strength of echoes and the location of their source. Computed tomography (CT) presents the body with X-rays to produce a 3D map of radiopacity. Magnetic resonance imaging (MRI) presents the body with an electromagnetic pulse (EMP) and creates a 3-D map of the EMP echo.[2] Optical coherence tomography (OCT) presents the body with near-infrared broad-spectrum light and maps the intensity of reflectance.

The physical bases of image acquisition (step 1 in the imaging process) are well known and established, but, by finessing the way in which samples are obtained, a wealth of information can be generated. For example, in MRI, an entire science has arisen based on the creative manipulation of the EMP trains presented to the body. EMP train manipulation has enabled the visualization of brain function and structure by detecting changes in blood flow, blood volume, and blood oxygenation level before and during an activity (functional MRI). Additionally, neural connections may be mapped (e.g., diffusion tensor imaging, Fig. 1) using the EMPs to quiet all reflections within a region, detecting only those hydrogen nuclei that meander into the interrogation field. Similarly, innovative analysis of a sequence of OCT scans obtained in rapid succession has been used to create virtual vascular casting, a technique known as OCT angiography (OCTA). Later in this chapter, advances in the creative processing of OCT signals will be considered, and the resulting measurement abilities will be discussed.

In the area of presentation, the advent of 3D datasets has produced the opportunity for development of efficient ways to convey 3D data (step 2 in the imaging process). There is debate on the best way to achieve this on a screen, and work in this area will continue to advance as computer power increases.

It is step 3, however, the interpretation of imaging data, which promises an explosion of innovation in the near future. As noted in the X-ray example, an almost mystical process occurs by which a trained human observer is able to perceive

---

[1] In discussing this process with radiology residents, I've been told that at some point in the second year of training, something clicks; a veil is lifted from their eyes, and they become aware of the vast hidden features contained in an X-ray.

[2] In the case of MRI, some cleaver finagling is performed with magnetic fields prior to the presentation of the EMP, which is beyond the scope of this discussion.

Blood-oxygenation-level-dependent Activity Map  Color-encoded Fractional Anisotropy Map  Fractional Anisotropy Value Map

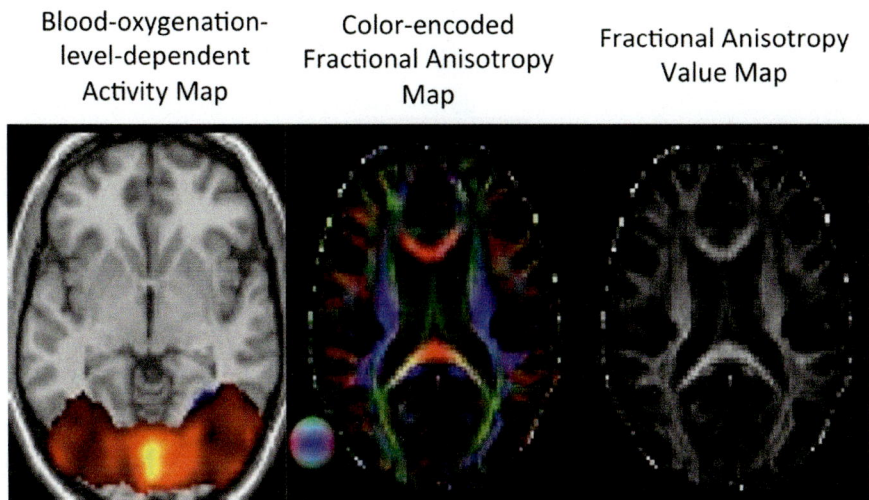

**Fig. 1** Functional magnetic resonance imaging (MRI) (left), and diffusion tensor imaging (middle and right) of a human brain in the axial view using a 3-Tesla MRI scanner. (Left) Blood-oxygenation-level-dependent activity map overlaid on an anatomical T1-weighted image. Hot color represents brain activity during visual presentation with an 8-Hz checkboard pattern; (Middle) Color-encoded fractional anisotropy directionality map. Color representations for the principal diffusion directions: blue, dorsal–ventral; red, left–right; green, caudal–rostral. (Right) Fractional anisotropy value map. Image Courtesy of Dr. Kevin C. Chan, Departments of Ophthalmology and Radiology, NYU School of Medicine, NYU Langone Health, New York University

subtleties within an image, from which he or she can discern signs of health and disease. This connection between sight and perception has been referred to as the "Blink" by Gladwell [1] and has proven to be an unbreachable gap in the automation of the medical imaging process. The growth of artificial intelligence and machine learning technologies may be the key to connecting image content and image interpretation for the non-specialist. This chapter will primarily discuss advances in the application of machine learning techniques to the automated interpretation of ophthalmic images.

## 2 Digital Medical Images and Machine Learning

We will use OCT imaging to serve as a platform for discussion, but the concepts herein generally apply to all imaging modalities.

As stated above, medical imaging modalities sample and quantify some physical characteristic from a portion of tissue. In the case of OCT, the relative reflectance of infrared light is measured. We say "relative" because a number of extraneous factors may influence the signal level, such as lens opacities, floaters, or corneal

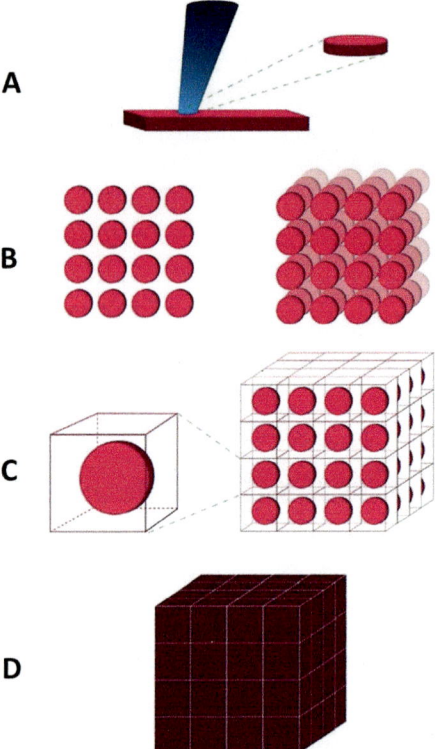

**Fig. 2** The focal point of the OCT beam occurs at the beam-waist (**a**) sampling reflectance in a disk-shaped volume of tissue approximately $20 \times 5$ μm. Sampling occurs in a raster pattern (**b**, left). This raster is repeated at a series of depths to sample a large volume of tissue (**b**, right). The reflectance value measured in the small sample area is inferred across the entire volume of tissue represented by the volume element (voxel, **c**, left). This inference occurs in each of the samples within a scanned tissue volume; the space surrounding each sample attributed the reflectance measured from the volume within (**c**, right). The resulting 3-D image dataset is therefore comprised of a reflectance value obtained from a small sample volume within (**d**). Image courtesy of Laura Cox, graphic artist

aberrations, all of which may reduce the perceived reflectance of an otherwise highly reflective tissue.

At its point of focus, the OCT beam is at its narrowest (beam waist, Fig. 2). When used properly, this will be the location of reflectance sampling. In the case of OCT, the length of the reference arm of the scanner determines the location of sampling along the axial axis, and the coherence length of the light source determines the axial resolution. Physics aside, the result is that reflectance is measured from within a small disk of tissue (Fig. 2a, b). As the beam scans the surface of the tissue, samples are obtained in a grid pattern, and numerous tissue planes are sampled (Fig. 2b). These reflectance samples are given a value. Most medical images present samples on a scale from 0 to 255, with a value of 0 representing black (no signal) and a value

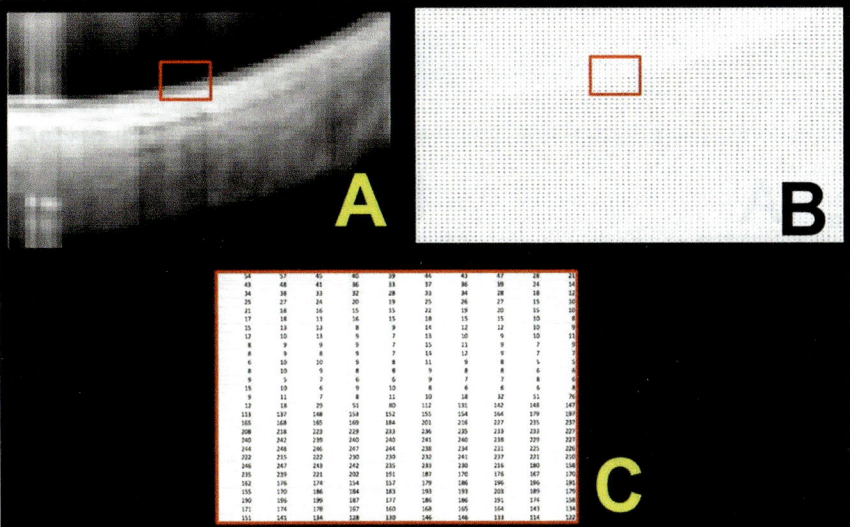

**Fig. 3** A low resolution OCT image of the retina (**a**) is comprised of values representing shades of gray (**b**). The section within the red box in (**a**, **b**) is shown in magnified view (**c**) to display the actual pixel values

of 255 representing white (saturation). These values are displayed in the digital image as continuous blocks of gray, which are usually square in shape (pixels, or picture elements). In the case of 3D data, these measurements are projected in depth so that adjacent image planes touch; i.e., there is no space between images in the virtual stack (Fig. 2b). The spacing of samples will determine whether significant portions of tissue are not represented in the image. Note in Fig. 2d that each block is assigned a color representing the reflectivity of the small sample area (disk) within.

Therefore, the image that we accept to represent tissue (Fig. 3a) consists of an array of numbers (Fig. 3b, c). The challenge is to program a computer to look at that number array and generate "perceptions" matching those of a human expert. Attempts to compose a set of strict rules have failed to come close to producing the most rudimentary machine perception . . . until the advent of machine learning.

## 3   Machine Learning: A Fly-Over View

From self-driving cars to precision medicine, machine learning is an exploding field transforming how we use our technologies. In the late 2000s, traditional central processing units (CPUs) were displaced by the substantially more efficient graphics processing units (GPUs), allowing for tremendous increases in computational power. This critical discovery catalyzed the advances we see today in image recognition with neural networks. An in-depth description of the neural networks that comprise machine learning systems is beyond the scope of this chapter. A

automated algorithms by considering additional factors, such as axial length, macular thickness, visual function, sex, and fasting glucose levels [8–10].

In a unique approach to diagnosis via superficial examination, Khansari et al. [11] used an automatic method to discriminate between stages of diabetic retinopathy on images of conjunctival microvasculature (obtained by digital camera coupled to slit lamp biomicroscope). This method accurately distinguished diseased from healthy eyes at rates higher than experienced human observers.

## 4.2 Glaucoma

Machine learning techniques have been applied to visual function testing and optical imaging and have attained the ability to accurately diagnose and detect glaucoma progression [12, 13]. Asaoka et al. [14] used a deep feed-forward neural network to effectively distinguish glaucomatous from healthy visual fields measured with standard automated perimetry testing. Chen et al. [15] developed a deep learning method using fundus photo inputs with a contextualized training strategy to learn deep glaucoma features and successfully classify disease.

Using standard automated perimetry and OCT, a variety of machine learning classifiers have been explored, including Gaussian [16], support vector machine [17], and random forest [18]. Kim et al. [19] examined C5.0, random forest, support vector machine, and k-nearest neighbor algorithms and concluded that classification accuracy could be improved with ensemble learning. Zilly et al. [20] used ensemble learning convolutional neural network architectures to develop accurate networks to segment the optic cup and optic disc from retinal fundus photos. Importantly, this technique could be applied successfully in glaucoma diagnosis with limited sample data needed for training.

An important goal in glaucoma predictive models has been to improve automatic segmentation of retinal layers; most approaches have been limited in that they are only capable of two-dimensional segmentation. In 2017, Miri et al. [21] developed a machine learning graph-based approach that enabled true three-dimensional segmentation for improved measurement of Bruch's membrane opening-minimum rim width on OCT, an important parameter for open-angle glaucoma diagnosis [22]. Precise identification of parameters such as these implies substantial potential for improvement in recognizing an irreversible disease with devastating consequences.

## 4.3 Age-Related Macular Degeneration

Machine learning has also been applied to the early detection and treatment of age-related macular degeneration (AMD). Mookiah et al. [23] used a supervised machine learning algorithm with features extracted from fundus photos as inputs and applied a variety of classifiers. This approach yielded 91.11% sensitivity and 96.30% specificity in diagnosing dry AMD.

Fraccaro et al. [24] have further demonstrated the clinical utility of machine learning models in AMD diagnosis. Using clinical data such as patient demographics and a number of macular features as identified by ophthalmologists on spectral domain optical coherence tomography (SD-OCT), they applied a supervised algorithm and four different classifiers. These authors illustrated the feasibility of incorporating decision support algorithms into electronic medical records in order to improve diagnostic accuracy and thus enhance patient care.

In a different approach, Bogunovic et al. [25] used a predictive model to identify risk of intermediate AMD progression. In this approach, segmentation data was obtained from longitudinal SD-OCT images at 3-month follow-up visits. A machine learning method was applied, which proved capable of designating risk scores and predicting drusen regression in this vulnerable population.

The same group considered the possibility of predicting anti-VEGF injection requirements in neovascular AMD patient treatment regimens [26]. In a pilot study, automated segmentation of consecutive monthly SD-OCT images starting from initiation of therapy was utilized along with demographics and information about visual acuity. Machine learning methods were applied to predict whether patients would require low, medium, or high injection requirements over an almost two-year maintenance period. The success of this approach illustrated the realistic potential utility of machine learning in managing neovascular AMD.

## 4.4  Ongoing Challenges

Despite the progress that has been made in applying concepts of machine and deep learning in ophthalmology, there are still several limitations.

When working with a paired organ system, within-patient differences must be considered independently from between-patient distinctions. This concept should be addressed when training samples are selected for machine learning models that address ocular conditions. Most of the current predictive approaches have included one eye from each patient, which does not account for distinguishing eye-specific characteristics within each patient [27]. One technique that can be used to address this issue is to use clustering methods, such as the Wilcoxon rank sum procedure, which allows paired eyes to be analyzed as a part of a unit rather than entirely independently [28].

An additional notion will need to be addressed prior to application of most predictive models to the true population; this is the concept that one exclusive diagnosis does not always befit a human eye [27]. For example, if an algorithm has been trained and taught to predict the presence or absence AMD, it may fail to diagnose an eye as having diabetic retinopathy. This is a matter that will need to be addressed in training machine learning models before they can be put to use as reliable clinical tools.

Noise and other acquisition artifacts can also affect accurate prediction of ocular pathology. Approaches to address these barriers are focused on improving data pre-processing and on streamlining input data, which may be obtained from different testing devices.

## 5 Advances in OCT Imaging

### 5.1 Aqueous Outflow Tract

In primary open-angle glaucoma, resistance to aqueous humor movement in the conventional outflow pathway leads to increased intraocular pressure [29]. To investigate this causal pathway, molecular level studies have addressed the variations between the trabecular meshwork, juxtacanalicular tissue, and inner wall endothelium of Schlemm's canal in glaucomatous and control eyes. Contractile cells in Schlemm's canal can stiffen in glaucoma, which may interfere with the sensory input needed to generate an adequate response with pore formation and aqueous humor outflow [30]. Several groups have suggested the role of the extracellular membrane within the juxtacanalicular space in IOP regulation, and imaging tools have been crucial in advancing this study [31]. In order to further improve the understanding of mechanical properties within this tissue, researchers have continued to seek better methods to observe these structural details in vivo (Fig. 5) [31].

Zeng et al. [32] estimated Young's modulus of elasticity in human Schlemm's canal endothelial cells by using magnetic pulling cytometry and finite element

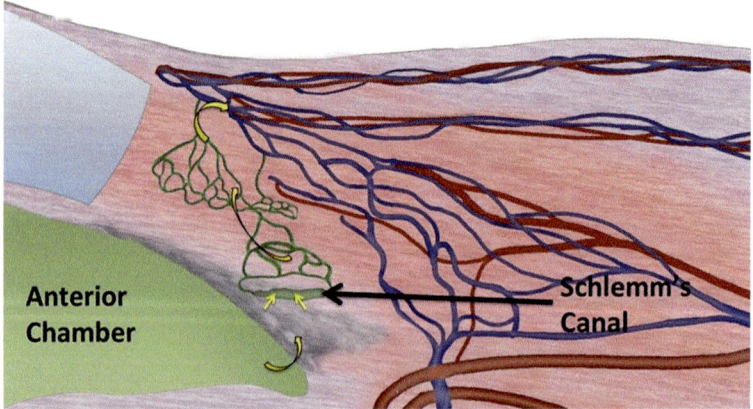

**Fig. 5** In the primary outflow pathway, aqueous humor exits the eye through the trabecular meshwork (gray) into Schlemm's canal, then out via a network of aqueous vessels and drains into scleral veins. Image courtesy of Laura Cox, graphic artist

**Fig. 6** The primary aqueous humor outflow pathway may be visualized, without contrast agent, by mapping the dark scleral voids filled with clear aqueous humor. Above is a composite image comprised of 12 OCT scans of a perfused human cadaver eye

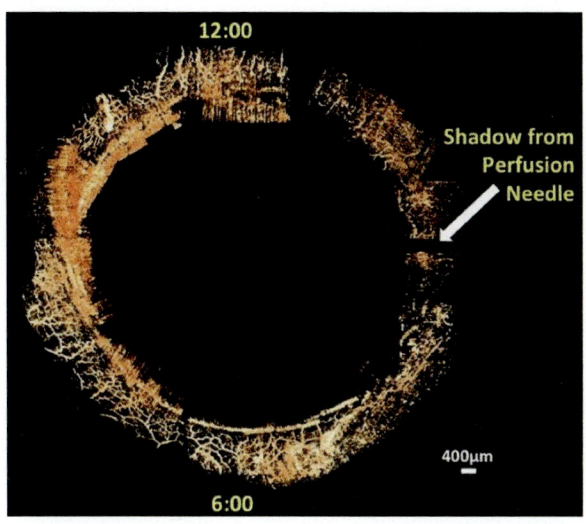

modeling. These authors compared these values at low and high intraocular pressures, noting increasing modulus values with pressure increases. This information allows for prediction of Schlemm's canal cell deformation and thus the amount of resistance that may be generated to flow.

Because of OCT's high-speed, high-resolution three-dimensional imaging capabilities, it has been explored as a tool to acquire this data. In 2010, Kagemann et al. [33] showed Schlemm's canal and the aqueous outflow system could be visualized and measured using OCT (Fig. 6).

Further studies have demonstrated the ability to measure with OCT decreases in Schlemm's canal area with IOP elevation acutely [34] and chronically in glaucoma subjects [35]. The ability to visualize the compression of a tissue resulting from a known perturbation in IOP allows the use of modeling to deduce the Young's modulus of internal tissues non-invasively. Pant et al. [36] demonstrated that a finite element analysis of the visualization of the elastic response of the trabecular meshwork (TM) to pressure elevation calculated a stiffness of 5.75 kPa in the living human eye (Fig. 7).

SD-OCT can be limited by inevitable features such as patient motion, scattering from scleral tissue, nearby vessel shadowing, and poor signal to noise ratio [37]. Hariri et al. [38] designed a setup that could overcome these factors while studying ex vivo primate eyes. Using a cannula to establish steady-state pressures, these authors could identify and measure trabecular meshwork tissue as well as Schlemm's canal lumen, collector channel entrances, and the surrounding structures (Fig. 7).

In 2012, Francis et al. [39] used SD-OCT to image the tract during active aqueous humor outflow in both cadaver and living eyes. After adjusting image contrast to visualize microvasculature, the images were assessed using a 3D viewer. With this method, successful virtual casting of this outflow microvasculature was possible,

**Fig. 7** Simulated results of
the deformation of the
trabecular meshwork of a
living human eye at baseline
(**a**, the marked circle denotes
the penetration of the TM into
the scleral) and during acute
IOP elevation (**b**) achieved
with the application of 30 g of
force applied to the temporal
sclera

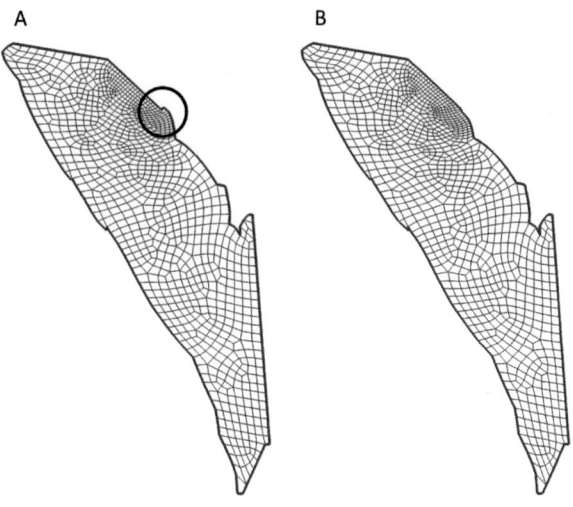

including the intrascleral venous plexus, connecting aqueous veins and Schlemm's canal components.

Phase-sensitive optical coherence tomography (PhS-OCT) can be used to capture tissue motion by comparing adjacent B-scans and can detect movements at a fraction of a nanometer [37]. This technology cannot be used for in vivo imaging alone, but its data can be overlaid with that from SD-OCT to produce an image that reflects structure and function combined. Li et al. [40] successfully implemented this technique by imaging human eyes in vivo and developing a phase compensation algorithm that could remove bulk motion confounders. Sun et al. [41] were able to compare an occluded and a normal outflow system within the same patient due to the presence of an iris cyst in one eye. These authors were able to use PhS-OCT to distinguish a difference in trabecular meshwork motion between these eyes.

Ultimately, clinically useful tools may be developed that will allow for direct observation of the aqueous outflow tract to diagnose disease, even in those without elevated IOP when randomly measured. These tools may aid physicians in making treatment decisions and may even improve predictions about long-term prognosis and recovery. With this new information, therapies can be developed to target biomechanical tissue-level features within this structural pathway.

## 5.2 Optic Nerve Head

Similar goals exist when it comes to biomechanics and imaging in the back of the eye. The optic nerve head is the only discontinuous portion of the globe, making it the weakest part; the connective tissue in this region bears the brunt of mechanical loads imposed upon the eyeball [42]. The lamina cribrosa—the region through which retinal ganglion cell axons pass within the optic nerve head—is considered by many to be a critical pathophysiologic location in the development of the optic neuropathy that characterizes glaucoma. Biomechanical models have also reflected a critical role of the peripapillary sclera in this complex system, in which it informs how sensitive the lamina cribrosa may be to intraocular pressure [43]. Efforts have been directed toward understanding the changes that occur within the laminar and surrounding tissues as they respond to stress acutely with stretch and compression as well as chronically as tissue responds with remodeling. Direct visualization of this tissue in vivo is a critical step in advancing this field of study.

The lamina cribrosa is a structure characterized by its sponge-like arrangement, with pores formed by interlacing beams. Commercial SD-OCT can visualize these features, yet image quality is inadequate. Only a portion of laminar tissue can be seen, as light signals are attenuated and blood vessel shadows further obstruct the viewable field.

Developed by Spaide et al. [44] to improve upon these limitations, enhanced depth imaging (EDI) is an approach in which images are obtained by decreasing the eye's distance from the SD-OCT device in order to obtain inverted fundus photos. These authors centered a $5 \times 15°$ rectangle on the fovea and scanned with seven sections comprised of 100 averaged frames each, which allowed for full laminar and choroidal thickness measurements. Lee et al. [45] demonstrated that EDI, compared directly to SD-OCT, is capable of producing an image with increased signal depth and improved image contrast. In 139 glaucomatous human eyes, Park et al. [46] were able to identify the anterior laminar surface beneath the neuroretinal and scleral rims as well as vascular structures in 65% of eyes, laminar pores in 76% of eyes, features of the central retinal artery and vein in 100% of eyes, and short posterior ciliary arteries in 86% of eyes. Unfortunately, not all brands of commercially available scanners currently offer high-density isotropic laminar sampling.

Swept source OCT (SS-OCT), commercially available in several countries outside of the USA, can more clearly produce images of 3D optical disc structure. This device does not require averaging of multiple B-scans for deep viewing as it uses a high-penetration, long-wavelength laser, allowing for greater tissue density sampling with raster scanning [47]. A raster scanning protocol obtains of consecutive OCT images at equally spaced intervals resulting in increased axial scan number and density, reducing sampling errors, and allowing for 3D data reconstruction. Omodaka et al. [48] developed software that enables the user to manually determine laminar thickness in humans in vivo based upon a three-dimensional evaluation of the tissue. These authors used SS-OCT scans with $3 \times 3$ mm cube areas centered over the optic nerve head and with this tool, reconstructed 12 radial B-scan images from a set of 256 horizontal 3D B-scan images to create a laminar model. This group was able to compare and measure significant differences in parameters such as average lamina cribrosa thickness between groups of normal, preperimetric, and normal tension glaucomatous eyes. Wang et al. [49] used SS-OCT to identify changes in laminar microarchitecture in glaucoma subjects, including reduced pore size and increased pore variability.

**Fig. 8** OCT images of the optic nerve head enhanced using image processing techniques. Visualization of the lamina cribrosa insertion sites (arrows) is improved, and blood vessel shadowing (asterisks) is corrected. Image courtesy of [54]

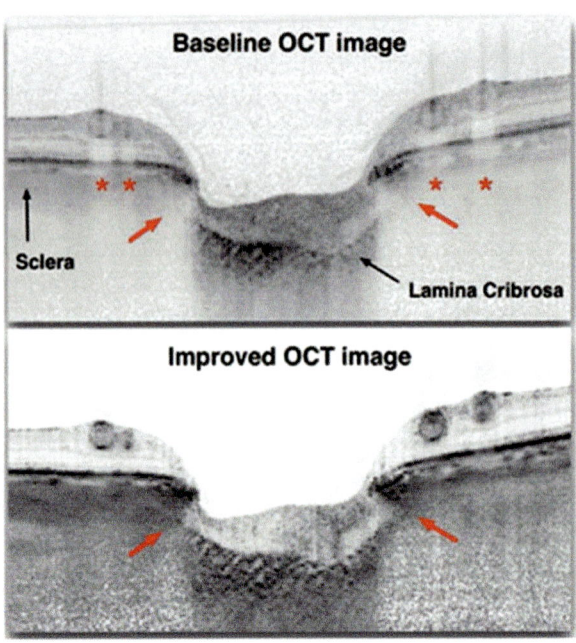

Despite these advances, OCT image contrast continues to be inferior to histology. While these tools can be used to study the lamina in vivo, they cannot be applied in clinical practice at this point, and while OCT may be used to sample the lamina, it cannot image the entire structure consistently [50].

Several additional approaches have been used to further improve upon existing technology. Adaptive optics corrects for aberrations and allows for improved transverse resolution from 20 to 5 μm [51], which also allows for observation of the posterior lamina and trabecular structure. Several groups developed post-image processing algorithms, techniques adapted from ultrasound to improve optic nerve head images by correcting for physical properties of light attenuation. With this approach, blood vessel and peripapillary structural shadowing can be reduced as well as enhanced laminar visualization [52–54] (Fig. 8).

Experimental stress tests of the lamina cribrosa have employed various imaging technologies. Previously, ex vivo studies showed the laminar field to be heterogeneous to strain with localized regions of in-plane strains [55]. Girard et al. [56] were able to study this concept in vivo using SD-OCT and digital volume correction to measure the deformation secondary to intraocular pressure. Similarly, Midgett et al. [57] developed an ex vivo inflation test that used a laser scanning microscope to assess the backscattered second harmonic generation signal of the lamina under pressure in combination with digital volume correction. This allowed the authors to calculate full-field displacement in three dimensions.

The search for enhanced biomechanical data collection continues in the hopes of eventually developing new therapeutics. Collagen cross-linking techniques have already been used to stiffen the sclera in animals in vivo and ex vivo, yet these techniques have not been deemed safe in humans, nor is there evidence to substantiate the claim that they will benefit glaucoma patients [58]. An improved understanding of optic nerve head biomechanics may provide the information needed to facilitate the advancement of this valuable field.

# References

1. Gladwell, M., *Blink: The power of thinking without thinking*. 2005, New York: Little, Brown and Co.
2. Kononenko, I., *Machine learning for medical diagnosis: history, state of the art and perspective*. Artif Intell Med, 2001. **23**(1): p. 89-109.
3. Sajda, P., *Machine learning for detection and diagnosis of disease*. Annu Rev Biomed Eng, 2006. **8**: p. 537-65.
4. Torok, Z., et al., *Combined Methods for Diabetic Retinopathy Screening, Using Retina Photographs and Tear Fluid Proteomics Biomarkers*. J Diabetes Res, 2015. **2015**: p. 623619.
5. Gargeya, R. and T. Leng, *Automated Identification of Diabetic Retinopathy Using Deep Learning*. Ophthalmology, 2017. **124**(7): p. 962-969.
6. Balaratnasingam, C., et al., *Visual Acuity Is Correlated with the Area of the Foveal Avascular Zone in Diabetic Retinopathy and Retinal Vein Occlusion*. Ophthalmology, 2016. **123**(11): p. 2352-2367.
7. Hwang, T.S., et al., *Automated Quantification of Capillary Nonperfusion Using Optical Coherence Tomography Angiography in Diabetic Retinopathy*. JAMA Ophthalmol, 2016. **134**(4): p. 367-73.
8. Linderman, R., et al., *Assessing the Accuracy of Foveal Avascular Zone Measurements Using Optical Coherence Tomography Angiography: Segmentation and Scaling*. Transl Vis Sci Technol, 2017. **6**(3): p. 16.
9. Tan, C.S., et al., *Optical Coherence Tomography Angiography Evaluation of the Parafoveal Vasculature and Its Relationship With Ocular Factors*. Invest Ophthalmol Vis Sci, 2016. **57**(9): p. Oct224-34.
10. Tang, F.Y., et al., *Determinants of Quantitative Optical Coherence Tomography Angiography Metrics in Patients with Diabetes*. Sci Rep, 2017. **7**(1): p. 2575.
11. Khansari, M.M., et al., *Automated fine structure image analysis method for discrimination of diabetic retinopathy stage using conjunctival microvasculature images*. Biomed Opt Express, 2016. **7**(7): p. 2597-606.
12. Bowd, C. and M.H. Goldbaum, *Machine learning classifiers in glaucoma*. Optom Vis Sci, 2008. **85**(6): p. 396-405.
13. Bowd, C., et al., *Bayesian machine learning classifiers for combining structural and functional measurements to classify healthy and glaucomatous eyes*. Invest Ophthalmol Vis Sci, 2008. **49**(3): p. 945-53.
14. Asaoka, R., et al., *Detecting Preperimetric Glaucoma with Standard Automated Perimetry Using a Deep Learning Classifier*. Ophthalmology, 2016. **123**(9): p. 1974-80.
15. Chen, X., Yanwu, X., Yan, S., Wong, D.W.K., Wong, T.Y., Liu, J., *Automatic feature learning for glaucoma detection based on deep learning*, in *International Conference on Medical Image Computing and Computer Assisted Intervention — MICCAI 2015*, N. Navab, Hornegger, J., Wells, W. M., and A.F. Frangi, Editors. 2015, Springer International Publishing: Munich, Germany. p. 669–677.
16. Goldbaum, M.H., et al., *Comparing machine learning classifiers for diagnosing glaucoma from standard automated perimetry*. Invest Ophthalmol Vis Sci, 2002. **43**(1): p. 162-9.

17. Bizios, D., et al., *Machine learning classifiers for glaucoma diagnosis based on classification of retinal nerve fibre layer thickness parameters measured by Stratus OCT.* Acta Ophthalmol, 2010. **88**(1): p. 44-52.

18. Barella, K.A., et al., *Glaucoma Diagnostic Accuracy of Machine Learning Classifiers Using Retinal Nerve Fiber Layer and Optic Nerve Data from SD-OCT.* J Ophthalmol, 2013. **2013**: p. 789129.

19. Kim, S.J., K.J. Cho, and S. Oh, *Development of machine learning models for diagnosis of glaucoma.* PLoS One, 2017. **12**(5): p. e0177726.

20. Zilly, J., J.M. Buhmann, and D. Mahapatra, *Glaucoma detection using entropy sampling and ensemble learning for automatic optic cup and disc segmentation.* Comput Med Imaging Graph, 2017. **55**: p. 28-41.

21. Miri, M.S., et al., *A machine-learning graph-based approach for 3D segmentation of Bruch's membrane opening from glaucomatous SD-OCT volumes.* Med Image Anal, 2017. **39**: p. 206-217.

22. Chauhan, B.C., et al., *Enhanced detection of open-angle glaucoma with an anatomically accurate optical coherence tomography-derived neuroretinal rim parameter.* Ophthalmology, 2013. **120**(3): p. 535-43.

23. Mookiah, M.R., et al., *Decision support system for age-related macular degeneration using discrete wavelet transform.* Med Biol Eng Comput, 2014. **52**(9): p. 781-96.

24. Fraccaro, P., et al., *Combining macula clinical signs and patient characteristics for age--related macular degeneration diagnosis: a machine learning approach.* BMC Ophthalmol, 2015. **15**: p. 10.

25. Bogunovic, H., et al., *Machine Learning of the Progression of Intermediate Age-Related Macular Degeneration Based on OCT Imaging.* Invest Ophthalmol Vis Sci, 2017. **58**(6): p. Bio141-bio150.

26. Bogunovic, H., et al., *Prediction of Anti-VEGF Treatment Requirements in Neovascular AMD Using a Machine Learning Approach.* Invest Ophthalmol Vis Sci, 2017. **58**(7): p. 3240-3248.

27. Caixinha, M. and S. Nunes, *Machine Learning Techniques in Clinical Vision Sciences.* Curr Eye Res, 2017. **42**(1): p. 1-15.

28. Rosner, B., R.J. Glynn, and M.L. Lee, *A nonparametric test for observational non-normally distributed ophthalmic data with eye-specific exposures and outcomes.* Ophthalmic Epidemiol, 2007. **14**(4): p. 243-50.

29. Grant, W.M., *Clinical measurements of aqueous outflow.* AMA Arch Ophthalmol, 1951. **46**(2): p. 113-31.

30. Stamer, W.D., et al., *Biomechanics of Schlemm's canal endothelium and intraocular pressure reduction.* Prog Retin Eye Res, 2015. **44**: p. 86-98.

31. Carreon, T., et al., *Aqueous outflow — A continuum from trabecular meshwork to episcleral veins.* Prog Retin Eye Res, 2017. **57**: p. 108-133.

32. Zeng, D., et al., *Young's modulus of elasticity of Schlemm's canal endothelial cells.* Biomech Model Mechanobiol, 2010. **9**(1): p. 19-33.

33. Kagemann, L., et al., *Identification and assessment of Schlemm's canal by spectral-domain optical coherence tomography.* Invest Ophthalmol Vis Sci, 2010. **51**(8): p. 4054-9.

34. Kagemann, L., et al., *IOP elevation reduces Schlemm's canal cross-sectional area.* Invest Ophthalmol Vis Sci, 2014. **55**(3): p. 1805-9.

35. Wang, F., et al., *Comparison of Schlemm's canal's biological parameters in primary open-angle glaucoma and normal human eyes with swept source optical.* J Biomed Opt, 2012. **17**(11): p. 116008.

36. Pant, A.D., Kagemann, L., Schuman, J.S., Sigal, I.A., Amini, R., *An imaged-based inverse finite element method to determine in-vivo mechanical properties of the human trabecular meshwork.* Journal for Modeling in Ophthalmology, 2017. **3**: p. 100-111.

37. Xin, C., et al., *Aqueous outflow regulation: Optical coherence tomography implicates pressure-dependent tissue motion.* Exp Eye Res, 2017. **158**: p. 171-186.

38. Hariri, S., et al., *Platform to investigate aqueous outflow system structure and pressure-dependent motion using high-resolution spectral domain optical coherence tomography.* J Biomed Opt, 2014. **19**(10): p. 106013.

39. Francis, A.W., et al., *Morphometric analysis of aqueous humor outflow structures with spectral-domain optical coherence tomography.* Invest Ophthalmol Vis Sci, 2012. **53**(9): p. 5198-207.

40. Li, P., et al., *Pulsatile motion of the trabecular meshwork in healthy human subjects quantified by phase-sensitive optical coherence tomography.* Biomed Opt Express, 2013. **4**(10): p. 2051-65.

41. Sun, Y.C., et al., *Pulsatile motion of trabecular meshwork in a patient with iris cyst by phase-sensitive optical coherence tomography: a case report.* Quant Imaging Med Surg, 2015. **5**(1): p. 171-3.

42. Burgoyne, C.F., et al., *The optic nerve head as a biomechanical structure: a new paradigm for understanding the role of IOP-related stress and strain in the pathophysiology of glaucomatous optic nerve head damage.* Prog Retin Eye Res, 2005. **24**(1): p. 39-73.

43. Girard, M.J., et al., *Translating ocular biomechanics into clinical practice: current state and future prospects.* Curr Eye Res, 2015. **40**(1): p. 1-18.

44. Spaide, R.F., H. Koizumi, and M.C. Pozzoni, *Enhanced depth imaging spectral-domain optical coherence tomography.* Am J Ophthalmol, 2008. **146**(4): p. 496-500.

45. Lee, E.J., et al., *Visualization of the lamina cribrosa using enhanced depth imaging spectral-domain optical coherence tomography.* Am J Ophthalmol, 2011. **152**(1): p. 87-95.e1.

46. Park, S.C., et al., *Enhanced depth imaging optical coherence tomography of deep optic nerve complex structures in glaucoma.* Ophthalmology, 2012. **119**(1): p. 3-9.

47. Nadler, Z., et al., *Automated lamina cribrosa microstructural segmentation in optical coherence tomography scans of healthy and glaucomatous eyes.* Biomed Opt Express, 2013. **4**(11): p. 2596-608.

48. Omodaka, K., et al., *3D evaluation of the lamina cribrosa with swept-source optical coherence tomography in normal tension glaucoma.* PLoS One, 2015. **10**(4): p. e0122347.

49. Wang, B., et al., *In vivo lamina cribrosa micro-architecture in healthy and glaucomatous eyes as assessed by optical coherence tomography.* Invest Ophthalmol Vis Sci, 2013. **54**(13): p. 8270-4.

50. Sigal, I.A., et al., *Recent advances in OCT imaging of the lamina cribrosa.* Br J Ophthalmol, 2014. **98 Suppl 2**: p. ii34-9.

51. Hermann, B., et al., *Adaptive-optics ultrahigh-resolution optical coherence tomography.* Opt Lett, 2004. **29**(18): p. 2142-4.

52. Girard, M.J., et al., *Shadow removal and contrast enhancement in optical coherence tomography images of the human optic nerve head.* Invest Ophthalmol Vis Sci, 2011. **52**(10): p. 7738-48.

53. Mari, J.M., et al., *Enhancement of lamina cribrosa visibility in optical coherence tomography images using adaptive compensation.* Invest Ophthalmol Vis Sci, 2013. **54**(3): p. 2238-47.

54. Kim, T.W., et al., *Imaging of the lamina cribrosa in glaucoma: perspectives of pathogenesis and clinical applications.* Curr Eye Res, 2013. **38**(9): p. 903-9.

55. Sigal, I.A., et al., *Eye-specific IOP-induced displacements and deformations of human lamina cribrosa.* Invest Ophthalmol Vis Sci, 2014. **55**(1): p. 1-15.

56. Girard, M.J., et al., *In vivo optic nerve head biomechanics: performance testing of a three-dimensional tracking algorithm.* J R Soc Interface, 2013. **10**(87): p. 20130459.

57. Midgett, D.E., et al., *The pressure-induced deformation response of the human lamina cribrosa: Analysis of regional variations.* Acta Biomater, 2017. **53**: p. 123-139.

58. Strouthidis, N.G. and M.J. Girard, *Altering the way the optic nerve head responds to intraocular pressure-a potential approach to glaucoma therapy.* Curr Opin Pharmacol, 2013. **13**(1): p. 83-9.

# Statistical Methods in Medicine: Application to the Study of Glaucoma Progression

**Alessandra Guglielmi, Giovanna Guidoboni, Alon Harris, Ilaria Sartori, and Luca Torriani**

**Abstract** Statistical models provide a variety of powerful methods for data analysis in medicine. In this chapter, we aim at illustrating the insights that statistical models can provide regarding the study of disease progression. In particular, we analyze a unique dataset on glaucoma progression by means of mixed-effects statistical models, where the form of the probability distribution for the multiple measurements is assumed to be the same for each individual in the study, but the parameters of that distribution can vary over individuals. Two illustrative case studies are presented in the context of structural and functional progression in glaucoma.

## 1  Introduction

Glaucoma is an optic neuropathy characterized by progressive retinal ganglion cell death, structural changes to the retina and optic nerve head, and irreversible visual field loss. Currently, elevated intraocular pressure (IOP) is the only treatable risk factor for glaucoma, despite overwhelming evidence that many additional factors are involved in the disease process, including alterations in blood pressure, cerebrospinal fluid pressure, intracranial pressure, and vascular regulation [10]. Currently, the only therapeutic strategies available to treat glaucoma are directed at lowering IOP, even though optimal target IOP levels for individual patients

A. Guglielmi (✉) · I. Sartori · L. Torriani
Dipartimento di Matematica, Politecnico di Milano, Milano, Italy
e-mail: alessandra.guglielmi@polimi.it

G. Guidoboni
Department of Electrical Engineering and Computer Science, Department of Mathematics, University of Missouri, Columbia, MO, USA
e-mail: guidobonig@missouri.edu

A. Harris
Icahn School of Medicine at Mount Sinai, New York, NY, USA
e-mail: alharris@indiana.edu

© Springer Nature Switzerland AG 2019
G. Guidoboni et al. (eds.), *Ocular Fluid Dynamics*, Modeling and Simulation in Science, Engineering and Technology, https://doi.org/10.1007/978-3-030-25886-3_24

Other measurements included in the dataset and utilized for our analysis are:

1. *age* (years) of the patient at the first visit (baseline) (sample average and sd are 64.94 and 10.98, respectively);
2. *gender*: indicator of the gender of the patient, set to be equal to 1 if the patient is a male (sample proportion is 38.9%) and equal to 0 if the patient is female;
3. *cardio*: indicator of cardiovascular diseases, set to be equal to 1 if the patient is affected by cardiovascular diseases (sample proportion =17%) and equal to 0 otherwise;
4. *diab*: indicator of diabetes, set to be equal to 1 if the patient is affected by diabetes (sample proportion is 19.5%) and equal to 0 otherwise;
5. *fh*: indicator of the patient family history of glaucoma, set to be equal to 1 if the patient relatives suffered from glaucoma (sample proportion=40.7%) and equal to 0 otherwise;
6. *hypt*: indicator of hypertension, set to be equal to 1 if the patient is hypertensive (sample proportion=19.5%) and equal to 0 otherwise;
7. *race*: categorical variable of the ethnic group of the patient; this variable assumes values *White*, *Black*, *Asian*, or *Hispanic*, with sample proportion equal to 71%, 26%, 2%, and 1%, respectively;
8. *YoG*: years from diagnosis, up to the time of the visit; the sample average and sd of years from diagnosis up to the baseline visit are 11.12 and 8.56, respectively;
9. *IOP*: intraocular pressure (measured in mmHg); elevated IOP is a recognized risk factor for glaucoma; the overall sample mean of this variable is 15.08 and the corresponding sd is 4.13;
10. *MAP*: mean arterial pressure (measured in mmHg); it is a convex linear combination of systolic blood pressure, denoted by *SYS*, and diastolic blood pressure, denoted by *DIA*, so that $MAP = (1/3) SYS + (2/3) DIA$; overall sample average and sd are 98.69 and 12.17, respectively;
11. *OPP*: ocular perfusion pressure; it is a synthetic index, defined as $OPP = (2/3) MAP - IOP$, which combines the effects of elevated IOP and low blood pressure; overall sample average and sd are 15.08 and 4.12, respectively.

The dataset contains *fixed-time measurements*, which remain constant at each visit, and *longitudinal measurements*, which may vary at each visit. In the list above, items 1, 2, and 7 are fixed-time measurements and items 8–11 are longitudinal measurements. We notice that the variable *YoG* is longitudinal since it increases linearly with time. In principle, items 3–6 should be considered as longitudinal measurements because they could vary at each visit. For example, a patient who does not suffer from cardiovascular problems when first joining the study may develop them as years go by. However, we have verified that items 3–6 remain constant for every patient in the particular dataset under consideration and so we treat them as fixed-time measurements.

*Remark 1* The IGPS dataset includes additional measurements that were not included in the present analysis, such as *vertical integrated rim area (volume) (VIRA)*, assessed via OCT, measuring the total volume of the retinal nerve fiber layer in the neuroretinal border, and *RNFL*, also assessed via OCT, measuring the average of the retinal nerve fiber layer thickness, see [19, 24]. The variables included in the analysis were selected on the basis of preliminary statistical results obtained on a similar dataset from the same clinical study [19], and from the literature [12, 24]. In addition, the selected variables, namely HIRWA and MD, present the minimum number of missing measurements, which is 13% for HIRWA and 7% for MD in this dataset.

*Remark 2* Clinical datasets are often incomplete, since patients might skip some visits or drop the study due to personal or medical reasons. How to deal with data incompleteness constitutes one of the biggest challenges in mathematical statistics. Several strategies have been proposed in the theoretical literature to deal with this issue, including simple random imputation, regression prediction for deterministic imputation (see [6, Chapter 25]), and case deletion. In this chapter, we have imputed all missing values using deterministic imputation via simple regression.

## 3 Methods: Longitudinal Linear Mixed-Effect Models for Glaucoma Progression

In this section, we describe the longitudinal linear mixed-effect models utilized in this chapter to study glaucoma structural and functional progression.

Linear mixed models for longitudinal data are statistical models that provide a description of the variability between different subjects and within the single subject of the dataset for repeated measurements over time. In this case, data for each subject is denoted by $y_i = (y_{i1}, \ldots, y_{i n_i})$, where $y_{ij}$ is the measurement of interest in the patient $i$ at the visit $j$ taken at time $t_{ij}$ (starting from the time when the patient joined the study), with $j = 1, \ldots, n_i$. In the study of glaucoma, $y_{ij}$ is the value of one of the variables that are hypothesized to drive disease progression, such as HIRWA or MD. The data vector $y_i$ is a particular realization of a vector $Y_i$ of random variables, which, following the notation used in [9] and [14], is assumed to satisfy the following statistical model:

$$Y_i = X_i \boldsymbol{\beta} + Z_i b_i + \epsilon_i, \quad i = 1, \ldots, M \tag{1}$$

$$\epsilon_i \stackrel{\text{ind}}{\sim} \mathcal{N}_{n_i}(\mathbf{0}, \sigma^2 I) . \tag{2}$$

Here, $\stackrel{\text{ind}}{\sim}$ means that we assume the error vectors $\epsilon_1, \ldots, \epsilon_M$ to be (stochastically) independent, as customary in statistical inference; moreover, the vector $Y_i$ is also referred to as the $n_i$-dimensional response vector for the subject $i$, where $n_i$ is the total number of visits available for the patient $i$ and $M$ is the number of subjects. Equation (1) assumes that $Y_i$ is determined by three main contributions:

- a contribution $X_i\boldsymbol{\beta}$, which characterizes the average effect of the covariates on the response. Assuming that $p$ covariates contribute to this average effect, $X_i$ is a $(n_i \times p)$ matrix containing the subject-specific values of the $p$ covariates and $\boldsymbol{\beta}$ is a $p$-dimensional column vector of coefficients. Notice that $\boldsymbol{\beta}$ does not have the subscript $i$, thereby indicating that this column vector is the same for each patient $i$, with $i = 1, \ldots, M$;
- a contribution $Z_i\boldsymbol{b}_i$, which characterizes the subject-specific effect of the covariates on the response, to be added to the average effect. Assuming that $q$ covariates contribute to this subject-specific effect, $Z_i$ is a $(n_i \times q)$ matrix containing the subject-specific values of the $q$ covariates and $\boldsymbol{b}_i$, with $i = 1, \ldots, M$, is a $q$-dimensional column vector of coefficients that may vary from subject to subject;
- a contribution $\boldsymbol{\epsilon}_i$, which characterizes the random error. Equation (2) assumes that $\boldsymbol{\epsilon}_i$ is an $n_i$-dimensional vector whose random components are described by a Gaussian distribution with mean equal to 0 and variance equal to $\sigma^2$.

The statistical model (1)–(2) is also known as *likelihood function*, that is the probability of observing the data given the specified value of the parameters; it quantifies information contained in the data. Since the parameters $\boldsymbol{\beta}, \boldsymbol{b}_1, \ldots, \boldsymbol{b}_M$, and $\sigma^2$ are unknown, they need to be estimated from the data $\{y_i, X_i, Z_i, i = 1, \ldots, M\}$. Under the frequentist framework, usual estimates are those obtained via the maximum likelihood estimation (MLE) method or its generalization [14]. In the Bayesian approach, the components of $\boldsymbol{\beta}$ and $\boldsymbol{b}_1, \ldots, \boldsymbol{b}_M$ are assumed to be random parameters described by certain probability distributions. These parameter distributions could be of various forms, such as uniform or Gaussian; in the Bayesian approach, such forms are *assumed* based on *a priori* information, coming from other studies or evidences available in the literature. This is the reason why the parameter distribution is called *prior distribution*. Upon evaluation of the likelihood of the dataset under investigation, the parameter distributions can be updated via a formula known as Bayes theorem (for instance, see [5, Chapter 1]), resulting in the *posterior distribution* of the parameters.

The inference analysis is based on summary statistics of this posterior distribution utilizing, for instance, credible intervals of each unidimensional parameter. In general, the posterior distribution is not available in closed analytic form; typical approximations are based on a class of simulation methods known as Markov chain Monte Carlo (MCMC), which can be easily built using specific software such as BUGS or JAGS, see [17]. Typical choices for prior distributions in this context are

$$\boldsymbol{b}_i \overset{iid}{\sim} \mathcal{N}_q(\boldsymbol{0}, \boldsymbol{\Psi}), \quad i = 1, \ldots, M \tag{3}$$

$$\boldsymbol{\Psi} \sim \pi_\Psi(\boldsymbol{\Psi}) \tag{4}$$

$$\sigma^2 \sim inv - gamma(a, b), \quad \boldsymbol{\beta} \sim \mathcal{N}_p(\boldsymbol{0}, \boldsymbol{\Sigma}) \tag{5}$$

where $\mathcal{N}_q(\boldsymbol{0}, \boldsymbol{\Psi})$ denotes the $q$-dimensional Gaussian distribution with mean given by the vector of all 0's and covariance matrix $\boldsymbol{\Psi}$, and $\boldsymbol{\Psi}$ is random with prior distribution $\pi_\Psi(\boldsymbol{\Psi})$. Parameter $\sigma^2$ is given the typical *inverse-gamma* (see [5,

Table A.2]) prior; hyperparameters $a$ and $b$ specify the distribution, i.e., the expectation of this distribution is $b/(a-1)$. The acronym $iid$ in (3) means that the parameter vectors $b_1, \ldots, b_M$ are assumed *independent and identically distributed*, according to the distribution specified after the symbol $\overset{\text{iid}}{\sim}$. Even though there has been little application of Bayesian methodology to integrate data in different domains in glaucoma research, some interesting examples can be found in [18, 23] and [12].

Note that model (1)–(2) allows us to take into account an unequal number of measurements for each subject, as well as the correlation between repeated measurements within subjects. Since random effects represent a natural heterogeneity between subjects, this assumption is justified for data where the between-subject variability is large in comparison to the within-subject variability; see [14].

For instance, the simplest form of model (1) is

$$Y_{ij} = \beta_0 + \beta_1 t_{ij} + b_{i1} + b_{i2}t_{ij} + \epsilon_{ij}, \ j = 1, \ldots, n_i \qquad (6)$$

where $Y_{ij}$ is the measurement of HIRWA in patient $i$ at time $t_{ij}$. Notice that (6) is an example of Eq. (1), where $X_i$ is an $n_i \times 2$ matrix, with the first column containing all 1's, and the second column containing the times $t_{ij}, \ j = 1, \ldots, n_i$, and here $Z_i = X_i$.

In this case, assuming Eq. (6) is equivalent to assume that there is a linear trend in time ($\beta_0 + \beta_1 t_{ij}$) on the average HIRWA, but that each patient may deviate from the average-population linear trend through patient-specific parameters $b_{i1}$ and $b_{i2}$, so that the HIRWA has patient-specific linear trend $(\beta_0 + b_{i1}) + (\beta_1 + b_{i2})t_{ij}$ (with random deviation from this value because of the error component $\epsilon_{ij}$). The assumption that $b_i = (b_{i1}, b_{i2})$ is random produces a joint marginal distribution for $Y_i$ with correlated components.

In our application, each $y_{ij}$ represents the value of a single measurement, but the model can be generalized to multiple measurements, as, for instance, in [12], where a linear mixed-effects model is fitted to longitudinal measurements of average RNFL thickness using two different techniques.

A more realistic model than (6) would include fixed-time or time-varying covariates in modeling the expectation of each measurement $Y_{ij}$.

## 4 Case Studies

Let us consider the linear mixed-effect model (1)–(2), described in Sect. 3, applied to the IGPS dataset of glaucoma patients. We first consider HIRWA as the response $Y_i$, to understand the effect of other longitudinal and fixed-time covariates on the structural progression of glaucoma. In this model, we assume that the covariates included in $X_i$ are *age, gender, diab, fh, MD, YoG*, and *IOP*, whereas the subject-specific contribution is assumed to depend on *mOPP*, which represents the average of $OPP_i$ over the $n_i$ visits each patient $i$. Ultimately, the model can be written as

$$HIRWA_{ij} = \beta_0 + \beta_1\, age_i + \beta_2 gender_i + \beta_3 diab_i + \beta_4 fh_i$$
$$+ \beta_5 MD_{ij} + \beta_6 YoG_{ij} + \beta_7 IOP_{ij} + b_i\, mOPP_i + \epsilon_{ij}, \qquad (7)$$
$$\epsilon_{ij} \stackrel{iid}{\sim} \mathcal{N}(0, \sigma^2), \quad j = 1, \ldots, n_i, \quad i = 1, \ldots, M.$$

Thus, in this model $p = 8$ and $q = 1$; note also that $\boldsymbol{\beta} = (\beta_0, \beta_1, \ldots, \beta_7)$. The prior (3)–(5) is assumed to translate vague information (i.e., with large variances), in order to let the posterior to depend more on data than on prior hyperparameters. Continuous covariates have been standardized, i.e., a linear transformation of covariates has been applied, such that sample mean and sd of each covariate are 0 and 1, respectively.

*Remark 3* We have verified that including the rest of fixed-time covariates, namely *cardio*, *hypt*, *race*, leads to a model where these variables are not significant. Variable *MAP* has not been included in the model because, by definition, it is strongly correlated to *IOP* and *OPP*.

We have computed the posterior distribution through a standard MCMC algorithm, which we implemented in JAGS [17]. Table 1 reports the posterior estimates of the parameters related to the fixed effects. As usually done in the Bayesian framework, we have assumed that a covariate in (7) is significant when the marginal posterior credible interval of the corresponding regression parameter in (7) does not contain 0. By posterior credible interval we mean a real interval such that the posterior probability that the parameter belongs to it is equal to 0.95. Figure 1 shows the corresponding graph.

The results reported in Table 1 and Fig. 1 show that the credible intervals for age at baseline, MD, indicator of diabetes, IOP, family history, and gender do not contain

**Table 1** Marginal posterior credible intervals (95%) of fixed-effects parameters in (7)

|            | 2.5%         | 50%         | 97.5%        |
|------------|--------------|-------------|--------------|
| Intercept  | 1.302749     | 1.347538    | 1.386376     |
| Age        | −0.0654611   | −0.04531992 | −0.02458655  |
| Gender     | −0.1965639   | −0.1516144  | −0.1095712   |
| Diab       | −0.15046510  | −0.10415077 | −0.04746762  |
| fh         | 0.04551027   | 0.08686653  | 0.13775995   |
| MD         | 0.05747048   | 0.07872459  | 0.0.09809502 |
| YoG        | −0.002562961 | 0.018036118 | 0.035113279  |
| IOP        | −0.04717201  | −0.03094900 | −0.01342119  |

The columns labeled as 50% correspond to the posterior median (i.e., approximately the center of the interval), whereas the columns labeled as 2.5 and 97.5% correspond to the left and right end of the interval. The intervals are visualized in Fig. 1

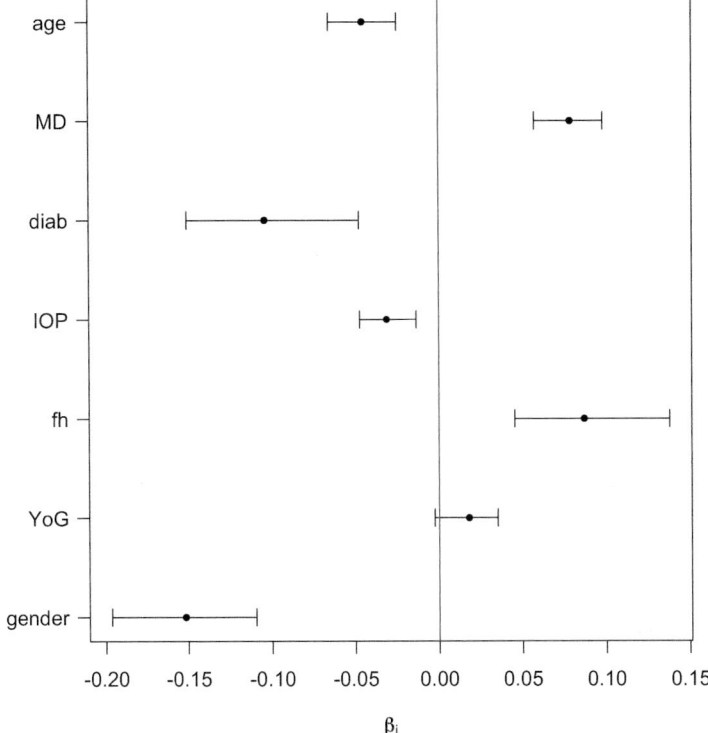

**Fig. 1** Plot of the marginal posterior credible intervals (95%) of fixed-effects parameters in (7)

0 and therefore are significant variables in the model for glaucoma progression based on the parameter HIRWA. In contrast, the value 0 is contained in the credible interval of *YoG*; however, most of the length of the interval is contained in the positive real line, so that we could consider the variable *YoG* as weakly significant to explain HIRWA.

As expected, age, indicator of diabetes, and IOP have a negative effect on the structural progression of glaucoma, and mean deviation has a positive effect. For instance, this means that larger IOP values are associated with lower HIRWA values, which correspond to worse structural conditions. Furthermore, larger MD values, corresponding to better functional conditions, are associated with larger HIRWA values, corresponding to better structural conditions. Conversely, the estimates of the statistical model pertaining to the influence of family history and gender on structural progression are quite unexpected, since we found that patients with family history show larger HIRWA values, while male patients display lower HIRWA values, adjusting for case-mix (i.e., being the other covariates fixed). Glaucoma patients at higher risk are typically considered to be of female gender [21] and with a family history of glaucoma [13]. The findings of our statistical analysis might be related to the particular dataset we have considered or to the particular type of

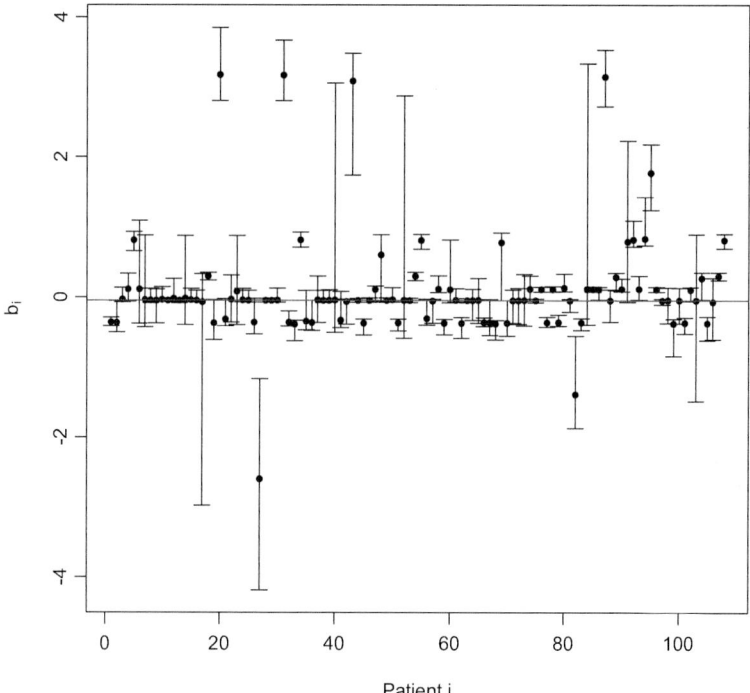

**Fig. 2** Plot of the marginal posterior credible intervals (95%) of all the random-effects parameters in (7). Each interval corresponds to a patient. The bullet represents the posterior median

statistical methods utilized for the analysis. To date, however, the roles of gender and family history in determining the glaucoma risk remain elusive. Population-based studies present inconsistent results regarding the role played by gender differences in the glaucomatous disease process, since it remains unclear whether women are at higher risk for glaucoma because of hormonal factors or because of they tend to live longer than men [25]. The contribution of family history is also difficult to assess, since many glaucoma cases remain undiagnosed [13].

Figure 2 displays posterior credible intervals (95%) of the random effects for each patient expressed in terms of the coefficients $b_i$ for $mOPP_i$. The results show noticeable variability among patients, which can be clustered in three main groups:

- *Group N:* group of 39 patients exhibiting negative random effects (i.e., the whole credible interval is negative). For the patients in this group, larger $mOPP$ values are associated with lower HIRWA values, corresponding to worse structural conditions;
- *Group P:* group of 22 patients exhibiting positive random effects (i.e., the whole credible interval is positive). For the patients in this group, larger $mOPP$ values

are associated with larger HIRWA values, corresponding to better structural conditions;
- *Group Z:* group of 47 patients corresponding to the rest of the dataset.

A closer look to the covariates in each of the groups N, Z, and P reveals that the sample average values of MD, mOPP, and HIRWA are statistically different in the two groups P and N, unlike those of IOP. In addition, the percentage of people with family history of glaucoma in the P group is much larger (59%) than those in the other two groups (36%). Clinical and population-based studies have shown inconsistent results regarding the role of mOPP in glaucoma, with both high and low mOPP values being associated with some degree of glaucoma risk [4, 22]. A clustering analysis, such as the one presented above, may help identify groups of patients exhibiting similar behaviors, thereby providing valuable information to be used as a guide for the design of new clinical and experimental studies.

As a second case study, we briefly present the results obtained by choosing the mean deviation as a response, namely as the variable describing functional progression of glaucoma. We apply the following model to the same dataset:

$$MD_{ij} = \beta_0 + \beta_1 t_{ij} + \beta_2 IOP_{ij} + \beta_3 HIRWA_{ij} + \epsilon_{ij},$$
$$\epsilon_{ij} \overset{iid}{\sim} \mathcal{N}(0, \sigma^2), \quad j = 1, \ldots, n_i, \quad i = 1, \ldots, M. \tag{8}$$

Here, $t_{ij}$ is a non-negative integer denoting the time of the $j$-th visit for patient $i$, for $j = 1, \ldots, n_i$, for any $i$. This model should be understood as a first attempt to find a simple regression model explaining the functional progression of glaucoma as a function of some clinical measurements such as IOP and HIRWA, as well as time. As such, we have not included patient-specific random effects in the model.

We assume a vague prior for the parameters $\boldsymbol{\beta}$ and $\sigma^2$ as before. Posterior credible intervals for the regression parameters $\beta_1, \beta_2, \beta_3$ are reported in Fig. 3. From this figure, it is clear that HIRWA has positive effect on MD, similar to IOP. Note that the effect of time is negative, confirming that MD decreases in time, i.e., the functional abilities of patients deteriorate as time goes by.

# 5 Discussion

Statistical models represent a major application of mathematics in medicine. Specifically, in this chapter we utilized an example applied to glaucoma research to illustrate the information that can be obtained by analyzing a dataset using a class of statistical models, namely linear mixed-effects models. Overall, we can say that statistical models, also known as *data-driven models* (see chapter "Mathematical and Physical Modeling Principles of Complex Biological Systems"), aim at inferring associations among variables, or probabilities of past and future events, as well as update those probabilities in light of new evidence or new measurements.

**Fig. 3** Plot of the marginal posterior credible intervals (95%) of fixed-effects parameters in (8)

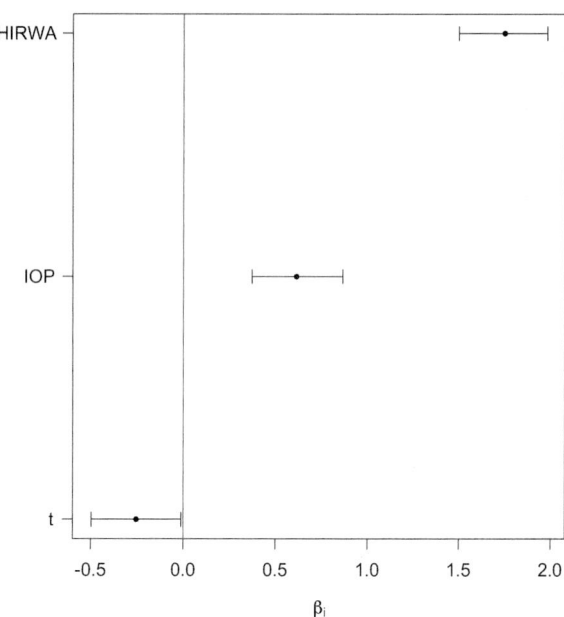

For example, the case study discussed in Sect. 4 showed that age, indicator of diabetes, and IOP significantly contribute to determine the structural progression of glaucoma, as measured by HIRWA. Mixed-effects models are key to analyze variability within each patient and among different patients, while the Bayesian approach allows us to include in the analysis patients with very few measurements, sharing information across patients with more measurements.

It is important to emphasize that, starting from the same dataset, different results could be obtained if different statistical methods are utilized to analyze it. For example, the results may depend on the statistical model we use and on the covariates included in the regression model.

The statistical models we have considered in this chapter *aim* at explaining *how*, and *not why*, the covariates in the dataset are correlated, thereby providing a perfect complement to *mechanism-driven models*, whose objective is to elucidate the mechanisms, or cause-to-effect relationships, that give rise to the observed correlations. Indeed, different models could be applied to data from glaucoma patients to have different or deeper insights on the disease, closer to the understanding of the underlining mechanism. Graphical models in statistics [2, 20] give an intuitive way of representing and visualizing the directed relationships between variables related to glaucoma, while causal inference [16] defines a framework to answer causal queries about the effects of potential intervention or about direct and indirect effects. These topics are currently a very active line of research in Applied Statistics.

# References

1. Agresti, A.: Foundations of linear and generalized linear models. Wiley, New York (2015)
2. Airoldi, E. M.: Getting started in probabilistic graphical models. PLoS Computational Biology, 3, e252 (2007)
3. Caprioli, J., Coleman, A. L.: Intraocular pressure fluctuation: a risk factor for visual field progression at low intraocular pressures in the Advanced Glaucoma Intervention Study. Ophthalmology **115**, 1123–1129 (2008).
4. Costa VP, Harris A, Anderson D, Stodtmeister R, Cremasco F, Kergoat H, Lovasik J, Stalmans I, Zeitz O, Lanzl I, Gugleta K. Ocular perfusion pressure in glaucoma. Acta ophthalmologica, **92**(4):e252–66 (2014)
5. Cowles, M. K.: Applied Bayesian statistics: with R and OpenBUGS examples. Springer, New York (2013)
6. Gelman, A., Hill, J.: Data analysis using regression and multilevel/hierarchical models. Cambridge university press, Cambridge (2006)
7. Guglielmi, A., Guidoboni, G., Harris, A.: Role of ocular perfusion pressure in glaucoma: the issue of multicollinearity in statistical regression models. Journal for Modeling in Ophthalmology **1**, 89–96 (2016)
8. Heijl, A., Leske, M. C., Bengtsson, B., Hyman, L., Bengtsson, B., Hussein, M.: Reduction of intraocular pressure and glaucoma progression: results from the Early Manifest Glaucoma Trial. Archives of ophthalmology **120**, 1268–1279 (2002)
9. Laird, N. M., Ware, J. H.: Random-effects models for longitudinal data. Biometrics **38**, 963–974 (1982)
10. Leske, M. C.: Open-angle glaucoma–an epidemiologic overview. Ophthalmic epidemiology **14**, 166–7172 (2007)
11. Leske, M. C., Heijl, A., Hyman, L., Bengtsson, B., Dong, L., Yang, Z.: Predictors of long-term progression in the early manifest glaucoma trial. Ophthalmology **114**, 1965–1972 (2007)
12. Medeiros, F. A., Leite, M. T., Zangwill, L. M., Weinreb, R. N.: Combining structural and functional measurements to improve detection of glaucoma progression using Bayesian hierarchical models. Investigative ophthalmology & visual science, **52**, 5794–5803 (2011)
13. McMonnies CW. Glaucoma history and risk factors. Journal of optometry, **10**(2), 71–8 (2017)
14. Molenberghs, G., Verbeke G.: A review on linear mixed models for longitudinal data, possibly subject to dropout. Statistical Modelling **1**, 235–269 (2001)
15. Musch, D.C., Gillespie, B.W., Lichter, P.R., Niziol, L.M., Janz, N.K.; CIGTS Study Investigators.: Visual field progression in the Collaborative Initial Glaucoma Treatment Study the impact of treatment and other baseline factors. Ophthalmology, **116**(2), 200–207 (2009)
16. Pearl, J.: Causal inference in statistics: An overview. Statistics surveys, 3, 96–146 (2009)
17. Plummer, M: JAGS: A program for analysis of Bayesian graphical models using Gibbs sampling (2003)
18. Russell, R. A., Malik, R., Chauhan, B. C., Crabb, D. P., Garway-Heath, D. F.: Improved estimates of visual field progression using Bayesian linear regression to integrate structural information in patients with ocular hypertension. Investigative ophthalmology & visual science, **53**, 2760–2769 (2012)
19. Spagnolo, V.: Modelli statistici per 1 aprogressione del glaucoma, *Master Degree Thesis*, Politecnico di Milano (2017)
20. Scutari, M., Strimmer, K.: Introduction to Graphical Modelling. In: Stumpf, Balding, Girolami (eds.) Handbook of Statistical Systems Biology, 235–254. Wiley, New York (2011)
21. Tehrani S. Gender difference in the pathophysiology and treatment of glaucoma. Current eye research, **40**(2),191–200 (2015)
22. Tham YC, Lim SH, Gupta P, Aung T, Wong TY, Cheng CY. Inter-relationship between ocular perfusion pressure, blood pressure, intraocular pressure profiles and primary open-angle glaucoma: the Singapore Epidemiology of Eye Diseases study. British Journal of Ophthalmology. 2018 Jan 13:bjophthalmol-2017.

23. Thomas, R., Walland, M., Thomas, A., Mengersen, K.: Lowering of intraocular pressure after phacoemulsification in primary open-angle and angle-closure glaucoma: a Bayesian analysis. The Asia-Pacific Journal of Ophthalmology, **5**, 79–84 (2016)
24. Tobe, L. A., Harris, A., Hussain, R. M., Eckert, G., Huck, A., Park, J., Siesky, B.: The role of retrobulbar and retinal circulation on optic nerve head and retinal nerve fibre layer structure in patients with open-angle glaucoma over an 18-month period. British Journal of Ophthalmology, **99**(5), 609–612 (2015)
25. Vajaranant TS, Nayak S, Wilensky JT, Joslin CE. Gender and glaucoma: what we know and what we need to know. Current opinion in ophthalmology, **21**(2), 91 (2010)

Printed by Printforce, the Netherlands